GPS/GNSS Antennas

For a listing of recent titles in the
Artech House GNSS Technology and Applications Series,
turn to the back of this book.

GPS/GNSS Antennas

B. Rama Rao
W. Kunysz
R. Fante
K. McDonald

ARTECH
HOUSE

BOSTON | LONDON
artechhouse.com

Library of Congress Cataloging-in-Publication Data
A catalog record for this book is available from the U.S. Library of Congress.

British Library Cataloguing in Publication Data
A catalog record for this book is available from the British Library.

ISBN-13: 978-1-59693-150-3

Cover design by Vicki Kane

10 9 8 7 6 5 4 3 2 1

Chapters 1, 2, 3, 5, and 6 of this book of which I am the author are dedicated to the memory of my grand-uncle, Sakharam R. Tombat, who took care of me when I lost my mother at age four months and then continued to provide much needed guidance, help, and encouragement throughout my life.
—B. R. R.

To Clara, the love of my life, Ronnie.
—R. F.

To my wife Shawna for continuous love, understanding and support.
—W. K.

Contents

Preface

Many varieties of GPS/GNSS antennas have been developed in recent years to increase their versatility and to make them suitable for different types of navigational applications. They vary from simple, miniaturized antennas used in handsets and smart phones for location-based services and emergency 911 applications to complex adaptive and beamforming arrays used for overcoming effects of interference and jamming. These antennas come in various sizes from handset antennas measuring no more than a few millimeters to large parabolic reflectors that have diameters as large as 110m for monitoring the signal quality of GNSS signals. The requirements for these antennas also vary greatly with their intended application: for example antennas used for surveying and other geodetic measurements impose the most stringent requirements to meet the millimeter to centimeter level of precision accuracy that is expected. In contrast, the requirements for antennas used in mobile phones are governed more by their size, cost, cosmetic appearance, and ease of integration with other complimentary systems; this may require making compromises in their requirements. A recent development that has spurred interest in new types of multiple band GNSS antennas is the entry of several new satellite navigational systems. Antennas developed up to now were designed primarily to meet the demands of GPS (U.S.) and GLONASS (Russia), which currently are the only two fully operational satellite navigation systems. However this scenario will change soon with the introduction of new systems such as Galileo (European Union) and Compass (China) and also the ongoing modernization of the two legacy systems. Modernization of GPS for example will include introduction of a new L5 civilian signal as well as a new military code. The emergence of these new or refurbished GNSS will mean introduction of new navigational codes and frequencies covering a frequency band from 1.1 to 1.6 GHz. GNSS antennas covering this expanded frequency range will soon be needed to make full use of their potential advantages for interoperability and satellite availability, especially in restricted and difficult urban environments.

The threats of interference and jamming are ever present in GNSS since the signals received from the navigational satellites are very weak. The most recent example is the recent scare, fortunately only temporary, generated by the Lightsquared terrestrial broadband network. But a more potent threat that is emerging is jamming caused by low-cost car cigarette plug jammers that are currently available. Active antennas are another important category of GNSS antennas that are being increasingly used because of their compactness and versatility. In these, a conventional passive antenna is combined with active electronic components such as bandpass filters, low-noise amplifiers, and bias tees, and directly integrated into the antenna package. This makes it difficult to separate the contributions from the passive antenna from other active electronic components integrated with it inside the package. The interaction between the antenna and the platform on which it is mounted is also important since the platform can have a strong influence on the backlobes and polarization that determines the severity of multipath.

Several recent developments have increased interest in new types of high-precision antennas for GNSS. The first is the decision in May 2000 by the U.S. Department of Defense to deactivate Selected Availability (SA) in the Global Positioning System (GPS) L1 channel. Prior to this decision the major contributor to the largest errors in high-precision code and carrier phase measurements in GPS was SA, which could have caused almost a fivefold increase in positioning error. With the SA turned off, the emphasis has now turned to reducing the other prominent error sources that degraded navigation accuracy. Besides propagation effects and multipath errors, antenna phase center and phase offset variations (PCV and PCO) and antenna group delay variations have been considered to be the next level of limiting factors preventing GNSS from achieving a breakthrough to the next higher accuracy level. One of the methods available for mitigating these errors and achieving higher precision in GNSS is by selecting an appropriate high-performance antenna with a proper calibration of its properties.

This book has attempted to cover the latest developments in the GNSS antennas mentioned above to acquaint the reader with recent fast-moving developments in this field. A large number of references to journal articles, books, and patents are listed at the end of each chapter; this will allow the reader to obtain more details on the latest antenna designs. Many of these new designs have been published recently, and some are even too new to be implemented into commercial GNSS systems. The antenna engineer can prioritize the requirements needed for a particular application and may be able to select from the many types of antennas discussed in this book the one that best matches these needs.

Chapter 1 of this book provides an introduction to the important requirements for GNSS antennas. It discusses polarization properties, including polarization efficiency and mismatch loss, the effect of polarization on reception of multipath, optimum radiation pattern, beamwidth and gain of the antenna and their effects on geometric dilution of precision, phase center of an antenna including PCV and PCO, carrier phase windup effects, and group delay variation. Antenna-induced errors in GNSS measurements and techniques for their removal are also discussed. This chapter closes with the susceptibility of a commercial receiver to interference and jamming, and potential sources of interference and jamming and their effects on the tracking loop threshold of a GPS receiver.

Chapter 2 is devoted to a discussion of fixed reception pattern antennas (FRPA) and high-gain directional antennas. FRPA antennas discussed in considerable detail include GPS microstrip patch antennas of various kinds, including the most popular and widely used of all GNSS antennas such as the quadrifilar helix antenna, crossed drooping bow-tie dipoles, and so forth. High-gain directional antennas discussed in this chapter include helical antennas with reflectors, which are also used as transmitting antenna elements in satellite antenna arrays, hemispherical helix antennas, and beamforming antenna arrays.

Chapter 3 is devoted to three important classes of GNSS antennas: multiband antennas, handset antennas, and active antennas. Various types of geodetic grade, commercial, and noncommercial multiband GNSS antennas including a large variety of spiral antennas are described. Novel multiband ground plane designs discussed in this chapter are electronic bandgap (EBG) and artificial magnetic conductor ground planes. Handset antennas discussed include planar inverted F, inverted L, coplanar balanced dipoles, miniature ceramic antennas including those that use low-temperature cofired ceramics, and fractal antennas. The influence of the receiver chassis on antenna performance, isolation techniques for mitigating EMI/EMC effects on antennas, and the effect of the human hand and body on antenna performance are some of the issues addressed in this chapter. A representative list of commercial GPS handset antennas is provided in a table. The measurement of G/T ratio, which is used as a metric for evaluating the performance of an active GPS antenna, is explained in great detail.

Chapter 4 is titled "Adaptive GPS Antennas" and it describes the design and signal processing techniques used in adaptive antenna arrays for mitigating the effects of interference and jamming. Topics discussed include the theory behind power minimization, space time adaptive processing, and polarimetric antenna arrays.

Chapter 5 is titled "Platform, Aircraft Fuselage, and Other Platform Effects of GPS Antennas." Topics in this chapter are the design of different types of ground planes such as choke rings, EBG, and tapered resistive ground planes for suppressing multipath. This chapter also describes in great detail the effects of the aircraft fuselage on the radiation pattern of a GNSS antenna. It also considers briefly the effects of an automobile on the radiation pattern of a GPS antenna.

Chapter 6 is titled "Measurement of the Characteristics of GNSS Antennas." It describes techniques used for measuring important antenna parameters such as Gain, polarization axial ratio, PCO and PCV. and group delay variation.

Chapter 7 is titled "Antennas and Site Considerations for Precise Applications." This chapter discusses typical antenna and site dependent errors that occur during very precise GNSS measurements and methods for avoiding such errors.

Acknowledgments

B. Rama Rao, the author for chapters 1, 2, 3, 5 and 6 of this book, would like to acknowledge help he has received from several people while writing this book.

The initial suggestion for writing this book came from Dr. Christopher Hegarty and Elliott Kaplan of the MITRE Corporation. The author owes a deep debt of gratitude to both of them for their help and encouragement during the several years it took to complete this book. He is particularly beholden to Dr. Hegarty, for expediting the complex and lengthy procedure for obtaining public release approval from the MITRE Corporation which at times seemed almost impossible to overcome.

Rama Rao also wishes to offer his thanks to the following people who contributed in many different ways to the successful completion of this book:

- Eddie N. Rosario, of the MITRE Corporation for providing the many drawing and illustrations appearing in these five chapters and also for his assistance in the building and testing of some of the antennas described in this book.

- Thomas M. Hopkinson, Associate Technical Director of the Electronic Systems and Technologies Tech Center of the MITRE Corporation for having provided me with an opportunity to investigate GPS antennas and for this continued interest and encouragement over the years.
- Eddy Emile of the Advanced Technology Branch of the Global Positioning Systems Directorate for providing funding to conduct the research and development of various types of GPS antennas at the MITRE Corporation over a span of several years—some of this research is included in this book
- To my many colleagues and friends at MITRE who made contributions: Janet Werth, Dr. David Lamensdorf, Dr. David N. Jones, Dr. Steven R. Best,
- Dr. Sean McKenna, Dr. Keith McDonald, Robert J. Davis, Dana Whitmer and Joe Giannetti.
- To Ms. Sheila Robertshaw for help in generating and improving many of the figures included in this book and also for typing the entire Chapter 4 for Dr. Fante and Dr. McDonald
- To Ms. Barbara A. Collins, Information Management Specialist at the MITRE Corporation for her invaluable assistance in providing me with with copies of the many journal articles and conference papers referred to in this book
- Ms. Susan Carpenito, Information Release Officer at the MITRE Corporation in Bedford, for expediting the security reviews of my five chapters in this book

Finally, but not the least, I would like to thank my wife, Dr. Jyoti Rao, for her cheerful support and patient endurance during the many years it took for me to complete the book.

CHAPTER 1

Introduction to GNSS Antenna Performance Parameters

Basrur Rama Rao

1.1 Role of an Antenna in GNSS and its Key Requirements

The antenna is the first element in the Global Navigation Satellite System (GNSS) receiver to process signals received from multiple GNSS satellites. Since it operates both as a spatial and a frequency filter, it can have a profound impact on many aspects of the navigational system, which requires that it satisfy key performance requirements. This is the topic of the first chapter of this book. One of the key requirements is that the antenna provide the minimum gain and the broad beamwidth needed by the antenna/receiver to acquire a minimum of at least four satellites; some of these acquired satellites also need to be at low elevation angles for achieving the best geometrical dilution of precision (GDOP)—a critical parameter that affects the accuracy of all GNSS measurements. The design of these fixed reception pattern antennas, the most popular and widely used in all GNSS, is discussed in Chapter 2. The bandwidth of the antenna also needs to be adequate for processing all available GNSS signals—the frequency of GNSS now extends from 1.1 to 1.6 GHz, and for many applications wideband antennas will soon be replacing the old legacy antennas that only covered either the Global Positioning System (GPS) or the Russian GNSS, GLONASS. These wideband antenna designs are discussed in Chapter 3. The antenna also needs to ensure good rejection of multipath and interference by having low backlobes, a sharp cutoff at low elevations close to the horizon and also a good polarization axial ratio. These platform-dependent antenna issues and also the antenna designs to mitigate these multipath effects are discussed in Chapters 3 and 5. The phase offsets, phase center variations, and carrier phase windup of the antenna can set limitations on the accuracy of differential GPS measurements. Hence, they need to be properly calibrated and taken into account; this will be discussed in Chapter 6. Some applications need the antenna to be small in size and also provide low cost, size, and weight—all important considerations for antennas used in commercial devices such as mobile phones, avionics systems, personal digital assistants (PDAs), and other handset devices. The design of these handset antennas is discussed in Chapter 3, as is the calibration of active antennas that are being used increasingly in all GNSS devices. The threats of interference and jamming are ever-present in GNSS since the signals received from the satellites are very weak. The most recent example is the concern raised by interference from the proposed Lightsquared terrestrial broadband network and several low-cost car cigarette plug jammers currently being sold. Adaptive antenna designs to mitigate such interference are discussed in Chapter 4. Some GNSS applications require very high gain antennas required for testing or for more detailed monitoring and troubleshooting of signals received from newly launched satellites. These high gain antennas include helical antennas and also large parabolic reflector antennas; these designs are discussed in Chapter 2, as are beamforming antenna arrays that allow multiple beams to be steered towards selected satellites for obtaining the highest carrier-to-noise ratios and simultaneously minimize multipath and interference.

The antenna is often required to satisfy multiple and often conflicting requirements mandated by the specific system for which the antenna is intended. For example, most GNSS antennas are required to have a very broadbeam with adequate gain in the upper hemisphere down to a low elevation masking angle in order to acquire the maximum possible number of visible satellites. However, the antenna is also simultaneously required to act as a spatial filter with a sharp cutoff in gain at elevations immediately below the masking angle to reduce susceptibility to multipath and interference. A sharp cutoff at low elevations may also not be desirable for certain GPS antennas located on certain platforms, such as ships, which can experience significant pitch and roll caused by heavy sea conditions, and also in antennas located on aircraft, especially during critical takeoff and landing operations when the aircraft is likely to undergo similar changes in its orientation in the pitch, roll, and yaw planes. High-profile antennas such as quadrifilar helices provide good gain at low elevation angles but may be too tall for aircraft where low profile antennas are required for reduced drag and for interference reduction.

Before discussing details of the various antenna performance parameters, we provide a brief summary of the current satellite-based navigation systems and also the fundamentals of differential GNSS systems where the antenna plays a key role; this will help in getting a better understanding of the requirements on the various performance parameters of a GNSS antenna.

1.2 Summary of Satellite Navigation Systems

Current satellite navigation systems can be grouped into three broad categories: those providing global coverage and referred in this book as Global Navigation Satellite Systems (GNSS), those providing regional coverage for general use and referred to as regional navigation satellite systems (RNSS), and aviation-specific regional systems referred to as satellite-based augmentation systems (SBAS). Only a summary of each of these three categories is given below; all three systems have been described in great detail in several recent books [1–3]. By 2013–2015, these new GNSSs are expected to provide nearly three times the number of navigation satellites and four to six times the number of individual signals compared to currently operational systems. It is expected that these extra satellite and signals will offer better positioning performance in terms of availability especially in an urban environment and also improved accuracy, continuity, integrity, and efficiency [4, 5]. It is important to emphasize these new systems since the next generation of GNSS antennas will be expected to cover all these systems so that they can be used in conjunction with the next generation of advanced multisystem receivers—the so called "system of systems" receivers [6]. These advanced receivers and multiband antennas that can support them are already entering the market to take advantage of their multiple benefits [7].

1.2.1 GNSS

In the near future there will be four major operating GNSSs. The two current legacy systems are GPS, a modernized version of which is currently being deployed by the United States, and GLONASS, operated by Russia, which is also currently being revitalized with the launching of several new satellites and the addition of a new frequency band (G3). These two systems have operated for nearly two decades now but will soon be joined by two new systems: Galileo by the European Union and COMPASS, also referred to as Beidou, by China. Beidou is currently a regional system (Beidou-1) serving mainly China, but in 2006 China announced its decision to develop its own GNSS, which is now called COMPASS (or Beidou–Phase 2). It is expected to reach full operational capability around 2020. Table 1.1 lists the important features of these four GNSS systems.

1.2.2 Regional Satellite Navigation Systems

In addition to these four global systems there are other RNSSs that provide navigational coverage to a specific region or a nation; they can work either independently or may augment another GNSS system, in which case they are required to be fully

Table 1.1 Details on Current GNSSs

GNSS	*Modernized GPS*	*GLONASS*	*Galileo*	*Compass (GNSS in 2020)*
Number of SVs	24	24	30	35
Orbits	6	3	3	?
Orbit radius (km)	26,560	25,508	29,601	MEO/GEO/IGSO
Orbit time	11 h, 57.96 min	11 h, 15.73 min	14 h, 4.75 min	?
Inclination	55°	64.8°	56°	55° (MEO)
Carrier frequency (GHz)	L_1: 1.5754 L_2: 1.2276 L_5: 1.176	G1: 1.602.000 Increment (MHz):0.5625; G2 :1.246.000 Increment (MHz): 0.4375; G3:1.204.704 Increment (MHz): 0.4230	E1: 1.575 E5a: 1.176 E5b: 1.207 E6: 1.278	B1: 1.5611 B1-2: 1.5897 B3: 1.2685 B2: 1.2071
Modulation	BPSK, BOC(1,1) BOC(10,5)	BPSK	BPSK, BOC	QPSK(2), QPSK(10), BPSK(10)+ BPSK(2)

MEO = ; IGSO = ; BPSK = binary phase shift keying; QPSK = quadrature phase shift keying; BOC = binary offset carrier.

interoperable with the selected GNSS system. The regional constellations are much smaller—generally about 5 to 7 satellites. The three major RNSSs that are currently planned are the Quasi-Zenith Satellite System (QZSS) from Japan, the Indian Regional Navigation Satellite System (IRNSS) by India, and the current Chinese regional system Beidou-1 mentioned earlier. QZSS is based on three satellites in highly inclined elliptical orbits (HEOs) with a figure-eight ground track that increases the number of satellites available at high elevation angles over Japan. It will broadcast over the GPS L_1, L_2, and L_5 frequencies and also the Galileo E6 frequency.

IRNSS is estimated to be a seven-satellite system, in a combination of geostationary (GEO) and HEO orbits designed to provide coverage only to the Indian region. It is expected to be a two-frequency system transmitting in the L5 and S bands. The third regional constellation is Beidou-1 which will eventually evolve from a regional system into the global system COMPASS.

1.2.3 Satellite-Based Augmentation Systems

It may be neither suitable nor safe to be solely reliant on GNSS systems alone for providing accurate position information needed during certain critical missions such as during flight operations in civil aviation, maritime operations such as harbor approach and entrance, or navigation through narrow waterways. The role of these space- and ground-based augmentation systems is to improve the accuracy, integrity, and availability of the basic GNSS signals during these critical missions. SBASs include a ground network to monitor the GNSS satellites and geostationary satellites to broadcast corrections and integrity data to the end user. Accuracy is enhanced through the use of wide area corrections for GNSS satellite orbits and ionospheric errors. Integrity is enhanced by the SBAS network quickly detecting satellite signal errors and sending alerts to receivers to not use the failed satellites. Availability is improved by providing additional ranging signals to each SBAS geostationary satellite. SBAS is made up of a series of reference stations, master stations, ground uplink stations, and geostationary satellites. The reference stations, which are geographically distributed, pick up the GNSS satellite data and route it to the master stations where wide area corrections are generated. These corrections are sent to the ground uplink stations that uplink them to the GEO satellites for retransmission on the GPS L1 frequency. These GEO satellites transmit these signals, which carry accuracy and integrity messages, so that they become available to users spread over a wide area. These GEO satellites act as bent-pipe transponders when fulfilling this role. The space segment of SBAS uses at least two geostationary satellites for the purpose of redundancy.

These are four principal publically operated SBASs that have been developed mainly to support en route air navigation [2, 8]. These systems are the Wide Area Augmentation System (WAAS), whose public sponsor is the U. S. Federal Aviation Administration, the European Geostationary Navigation Overlay Service (EGNOS), whose public sponsor is the European Space Agency (ESA), Multifunctional Satellite-Based Augmentation System (MSAS), whose public sponsor is Japan, and GPS and Geoaugmented Navigation (GAGAN) system supported by India. These four systems combined provide nearly worldwide SBAS service. In addition to these public SBAS there are also several commercial systems such as OMNISTAR, Star Fire, and CDGPS. A more detailed description of these SBASs has been given by

Lachapelle et al. [8] and by Hofmann-Wellenhof [2]. These SBASs are also supported by several ground-based augmentation systems, such as the local area augmentation system (LAAS) [2].

The frequency bands for many these GNSS systems are shown in Figure 1.1; their principal features are also tabulated in Table 1.1.

This will significantly increase both the number of available satellites and frequency bands available for navigational purposes, requiring new types of antennas that can operate in multiple frequency bands and be used to support new receiver architectures developed for new applications to follow from this abundance.

1.3 Polarization and Radiation Pattern of a GNSS Antenna

The GNSS receive antenna (i.e., the user antenna) as well as the transmitting satellite antenna have right-hand circular polarization (RHCP). Circular polarization is preferred since a linearly polarized signal undergoes changes in polarization while traveling through the ionosphere due to Earth's magnetic field. A second advantage is that it allows the receiving antenna to discriminate between the direct signal from the satellite and strong multipath signals arriving at the antenna from below after reflection from the surface of a good electrical conductor (e.g., sea water or a building). The polarization of the reflected multipath signal is changed to left-hand circular polarization (LHCP), thus allowing the antenna to discriminate against it, especially when it has a good axial ratio. The antenna also needs to have very low radiation below the horizon in order to better reject these multipath signals. This will be discussed in greater detail later in this section. LHCP is the cross-polarization state of a GNSS antenna; it is the unwanted polarization in an antenna that needs to be held to a minimum. The spatial distribution of the far-field radiation field of a GNSS antenna can be expressed in terms of two orthogonal polarization components of the electric field $\vec{E}_\theta$ and $\vec{E}_\phi$ along the $\hat{\theta}$ and $\hat{\phi}$ unit vector directions of a spherical coordinate system as shown in Figure 1.2, and $\hat{r}$ is the radial vector that represents the direction of propagation.

To allow optimum signal reception the polarization of the user antenna should match that of the signals received from the satellites. The polarization state of the antenna denotes the sense of rotation of the radiated electric field vector as it

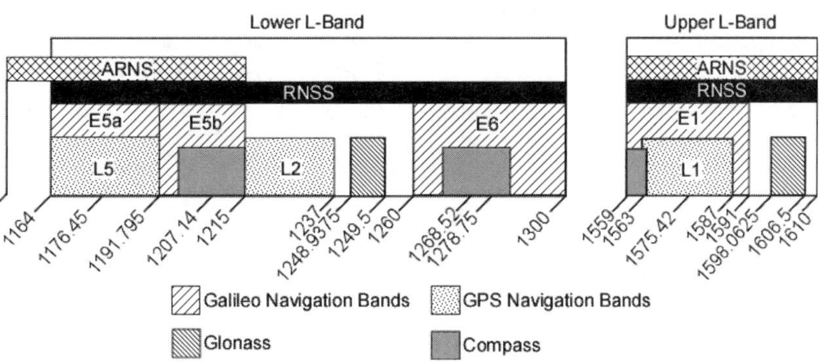

Figure 1.1 GNSS frequency bands.

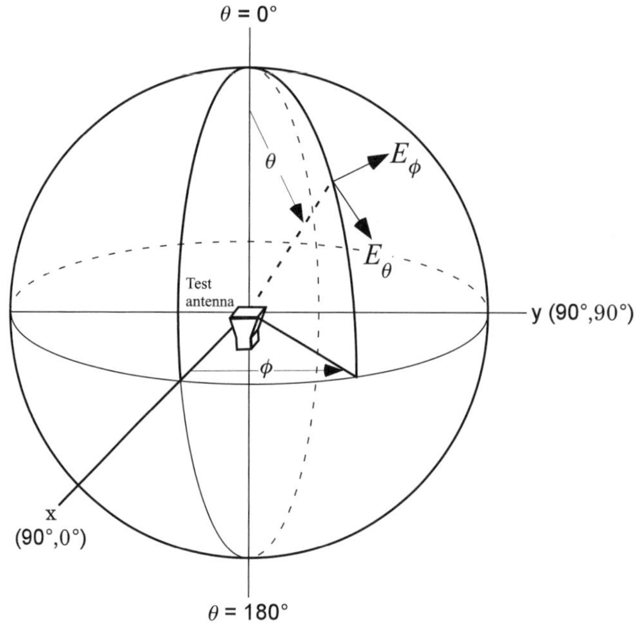

Figure 1.2 Electric field polarization in a spherical coordinate system.

rotates over time, tracing out an ellipse in the polarization plane as shown in Figure 1.3.

Following the definition for polarization in the IEEE standard 149-1979, the electric field vector for RHCP rotates counterclockwise (right-hand sense) with time for a wave approaching the observer. One easy way of determining the sense of rotation is by using the simple "thumb rule": If the thumb of the right hand is pointed in the direction of propagation of the wave, the fingers will curl in the direction of the vector rotation of the electric field. This is illustrated in Figure1.4. Hence, the electric field vector crosses the $\hat{\theta}$ axis before it crosses the $\hat{\phi}$ axis during

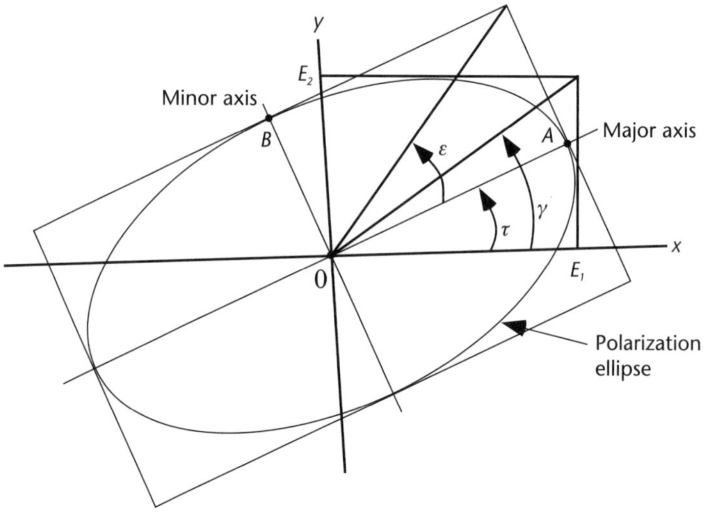

Figure 1.3 Ellipse representing the polarization state of the radiated fields of a GNSS antenna.

a cycle. Therefore for RHCP the phase of the $\hat{\phi}$ component lags against that of the $\hat{\theta}$ component of the electric field by a phase of –90°. For LHCP the electric field vector has the opposite sense of rotation and rotates clockwise with respect to time for a wave approaching the observer, as shown in Figure 1.4. Following the thumb rule again, if the thumb of the left hand points in the direction of propagation, the fingers will curl in the direction of rotation of the electric field; the phase of the $\hat{\phi}$ component leads $\hat{\theta}$ component in phase by +90° for LHCP.

The far-field radiated electric field $\vec{E}(\theta,\varphi)$ of the antenna can denoted in terms of the unit vectors for RHCPs and LHCPs, the principal and the cross-polarized states, respectively:

$$\vec{E}(\theta,\varphi) = \left[E_R(\theta,\varphi,f)\right]\left[\hat{e}_r\right] + \left[E_L(\theta,\varphi,f)\right]\left[\hat{e}_l\right] \tag{1.1}$$

In the above equation the unit vector for RHCP is

$$\hat{e}_r = \frac{e^{-j\varphi}}{\sqrt{2}}\left[\left(\hat{\theta}\right) - j\left(\hat{\varphi}\right)\right] \tag{1.2}$$

The unit vector for LHCP is

$$\hat{e}_l = \frac{e^{+j\varphi}}{\sqrt{2}}\left[\left(\hat{\theta}\right) + j\left(\hat{\varphi}\right)\right] \tag{1.3}$$

Note also that

$$\left(\hat{e}_r\right) * \left(\hat{e}_r^*\right) = \frac{1}{2}\left[\left(\hat{\theta}\right) - j\left(\hat{\varphi}\right)\right] * \left[\theta + j\hat{\varphi}\right] = 1$$

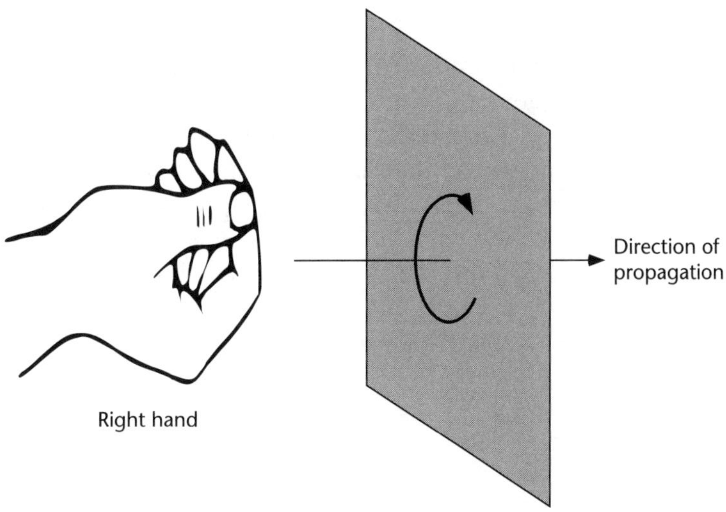

Figure 1.4 Simple right-hand rule for determining direction of rotation of electric field and direction of propagation for RHCP.

and

$$\left(\hat{e_l}\right)*\left(\hat{e_r}^*\right)=\frac{1}{2}\left[\left(\hat{\theta}\right)+i\left(\hat{\varphi}\right)\right]*\left[\left(\hat{\theta}\right)+i\left(\hat{\varphi}\right)\right]=0$$

Hence the RHCP and LHCP components of the electric field are orthonormal.

As indicated in the above equations, the phase of the RHCP signal is a function of azimuth angle ϕ and will increase by 2π radians for each 360° rotation in azimuth angle. As a result, the carrier phase observed measured in the plane of the receiving antenna depends on the direction of the line of sight between the satellite and user antennas as well as their relative orientations. Due to yawing of the GNSS satellites and also if the user antenna is mounted on a platform that is rotating, such as a projectile or a missile, the accumulated change in carrier phase is called carrier phase windup or sometimes also as carrier phase wrap-up. This can lead to position errors in GNSSs that use carrier phase measurements. It will be discussed in greater detail later in this chapter.

The polarization ellipse representing the complete polarization state of the antenna is shown in Figure 1.3. The shape of the ellipse is determined by the axial ratio; it is

$$\begin{aligned} |R| &= \frac{Major\ Axis\ Length}{Minor\ Axis\ Length} = \frac{E_{\max}}{E_{\min}} \\ R(dB) &= 20\log|R| \end{aligned} \tag{1.4}$$

The factor 20 is used in the decibel definition because $|R|$ is a ratio of field quantities and not a ratio of power quantities. For $|R|=1$ the polarization ellipse becomes a circle and represents circular polarization. The ellipse degenerates into a line when $|R| = \infty$; it then represents the special case of linear polarization. The axial ratio R is also given a sign, which is $R = +$ for right-hand polarization and $R = -$ for left-hand polarization. The orientation of the ellipse is given by its tilt angle τ. The tilt angle is the angle of the major axis of the polarization ellipse with the horizontal θ axis; it becomes an important parameter when determining the maximum and minimum polarization coupling between two circularly polarized antennas with different axial ratios. The polarization ellipse can have any shape (depending on its axial ratio) and orientation (depending on the tilt angle). The angle $\varepsilon = \cot^{-1}(R)$ is known as the ellipticity angle. The pair of angles (ε, τ) can completely describe the polarization ellipse [10]. The polarization tilt angle is particularly important when considering the polarization coupling between elliptically polarized antennas, as explained later on.

The axial ratio R of a circularly polarized antenna is defined by

$$R(\theta, \varphi)=\frac{|E_R|+|E_L|}{|E_R|-|E_L|}=\frac{|\hat{\rho}_c|+1}{|\hat{\rho}_c|-1} \tag{1.5}$$

Where the circular polarization ratio $\hat{\rho}_c$ is defined by

$$\hat{\rho}_c(\theta,\varphi) = \frac{E_R}{E_L} = \rho_c e^{i\delta_c} \tag{1.6}$$

Various terms have been used to describe the polarization properties of a GNSS antenna. The axial ratio R of a GNSS antenna at an elevation angle θ and azimuth angle ϕ can be calculated by measuring the cross-polarization ratio (CPR) XPD (θ, ϕ), defined as

$$XPD(\theta,\ \varphi) = \left[\frac{E_R(\theta,\varphi)}{E_L(\theta,\varphi)}\right] \tag{1.7}$$

The axial ratio R can be calculated from XPD by the following equation:

$$\rho_c = 10^{\frac{XPD(dB)}{20}} \tag{1.8}$$

Another parameter is the CPR, which is expressed as

$$CPR = \frac{|E_L|^2}{|E_R|^2} \tag{1.9}$$

$$CPR(dB) = 20\log\frac{|R| - 1}{|R| + 1}$$

$$CPR = \frac{|E_L|^2}{|E_R|^2} = \left[\frac{|R| - 1}{|R| + 1}\right]^2 = \frac{1}{XPD} \tag{1.10}$$

The parameters R, XPD, CPR, and ρ_c are all strong functions of the elevation angle; they generally do not vary as much with azimuth angle if the antenna is well designed. The axial ratio R is a metric of several critical performance requirements expected from the antenna:

- It indicates the polarization efficiency with which the receiver antenna is able to receive RHCP signals transmitted by the GNSS satellites.
- The ability of the antenna to reject a single-bounce multipath signal that is LHCP caused by reflections from conducting surfaces such as a nearby building , cars, or seawater. This is generally the strongest multipath signal received by the antenna, hence its suppression greatly benefits GNSS performance.
- The axial ratio of receiving antennas can vary widely with elevation and azimuth angles depending on its design and the feed network used to generate RHCP.

The current specifications for the axial ratio for a GPS avionics antenna is that is should not exceed 3.0 dB for all operating frequencies as measured at the boresight of the antenna.

Generally the axial ratio of receiving antennas most commonly used in GNSS, such as microstrip antennas, degrade with elevation angle. All microstrip antennas need a ground plane. The horizontal component of the electric field becomes zero at the surface of the ground plane, making the antenna linearly polarized; its polarization is vertical to the ground plane. The axial ratio R becomes very large (with an axial ratio approaching close to ∞ for a very large, metallic ground plane). Figure 1.5(a) shows the RHCP and LHCP components in the radiation pattern of a typical GPS microstrip antenna that was measured at 1.5754 GHz on a metallic ground plane that was 51" in diameter with rolled edges to reduce edge diffraction effects. The axial ratio R and XPD values for this antenna are shown in Figure 1.5(b). Due to this degradation in axial ratio signals received by such antennas from low elevation, satellites can suffer a polarization loss of up to 3 dB as compared to signals received from higher elevation satellites where the axial ratio is relatively much better (i.e., smaller). The axial ratio of microstrip antennas at lower elevation can be greatly improved by mounting the antenna on a choke ring ground plane; this decreases the cross-polarized LHCP radiation, resulting in improved axial ratios.

Figure 1.6 shows the measured pattern of a NovAtel GPS antenna measured on a choke ring ground plane; notice that the LHCP component is lower than for the corresponding microstrip patch antenna shown in Figure 1.5(a). The axial ratio of certain low-cost GPS patch antennas that use ceramic substrate antennas can also vary with the azimuth angle since they use single-probe designs to reduce cost [9]. The axial ratio of such antennas is a strong function of both the elevation angle as well as azimuth angle [9].

Some of the commercially available GNSS antennas, for example the NovAtel 704X with a pinwheel type of ground plane, have an extremely good (i.e., low) axial ratio requirements for their antennas, at least in the GPS L_1 band. The axial

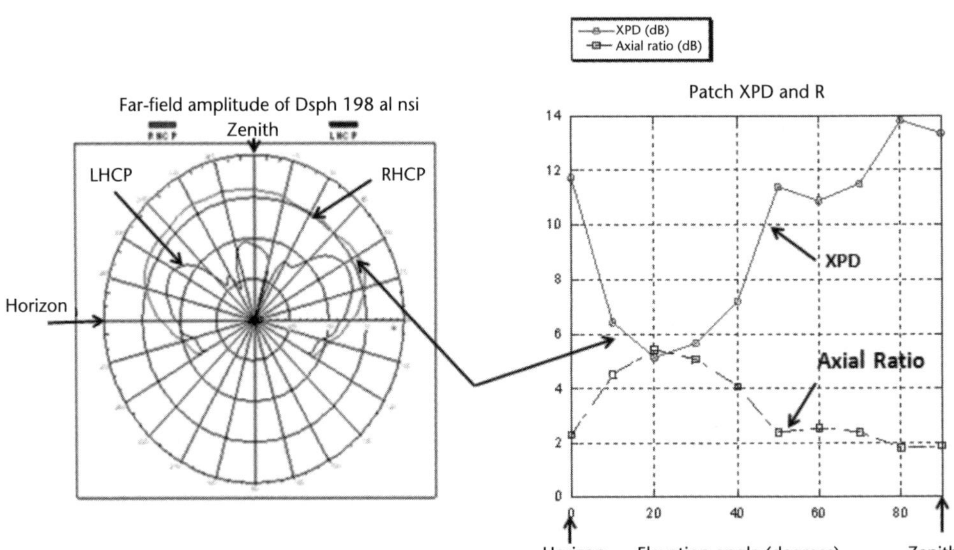

Figure 1.5 Measured radiation pattern of a GPS microstrip patch antenna at 1.5754 Ghz.

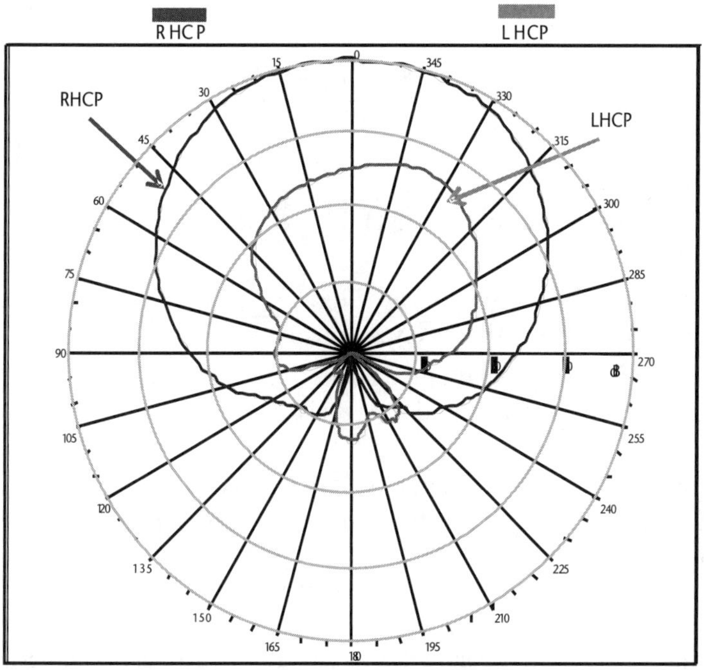

Figure 1.6 Measured radiation pattern of a NovAtel choke ring antenna (model #503) showing RHCP and LHCP components.

ratio requirements for the NovAtel 704X antenna in the GPS L1 and L2 bands are given in Table 1.2.

1.3.1 Polarization Efficiency and Polarization Mismatch Loss

Optimum reception of the satellite signals is achieved when the polarization of the receiver antenna matches that of the satellite antenna; it is assumed that both antennas are perfectly RHCP under ideal conditions. The receive antenna "absorbs" no power if it is orthogonally polarized to the satellite antenna (i.e., if it is LHCP). However, in reality neither the satellite nor the receiver antenna are perfectly RHCP but are elliptically polarized instead with finite axial ratios. Hence the power received is a function of the axial ratios of the satellite and receiver antennas. This

Table 1.2 Axial Ratio Requirements for NovAtel 704X Antenna

GPS Band	*Elevation Angle*	*Axial Ratio in dB*
L_1	45°	1.0
L_1	15°	1.4
L_1	5°	1.6
L_2	45°	2.0
L_2	15°	3.8
L_2	5°	5.0

results in a polarization mismatch loss defined as η_p, the polarization efficiency [10].

$$p = \frac{1}{2} + \frac{4R_S R_A + (R_S^2 - 1)\cdot(R_A^2 - 1)\cdot\cos 2\Delta\tau}{2\cdot(R_S^2 + 1)\cdot(R_A^2 + 1)} \tag{1.11}$$

The terms appearing in the (1.11) are defined below:

R_s = Axial ratio of satellite antenna

R_A = Axial ratio of the receiver antenna

$\Delta\tau = \tau_S - \tau_A$ Relative tilt angle between the major axes of the polarization ellipses of the satellite antenna and the receiver antenna.

The polarization mismatch loss associated with polarization efficiency is often expressed in decibels and is formed from 10 log of the power ratio.

$$L_P = -10\log p \text{ [dB]} \tag{1.12}$$

The polarization mismatch loss can also be calculated precisely using the above equation or determined approximately from a convenient nomograph referred to as the Ludwig chart after its originator A. C. Ludwig [10, Figure 6.3]. The relative tilt angle in the above equation is important while estimating the maximum and minimum loss in polarization efficiency. The loss is minimum when the relative tilt angles between the major axes of the polarization ellipses of the satellite and receiving antennas is zero ($\Delta\tau = 0$) (i.e., when the two axes are aligned); it is maximum when the major axes are orthogonal to one another (i.e., $\Delta\tau = 90°$). Some special cases of (1.11) that were mentioned earlier can now be derived from the above equation to estimate polarization loss resulting from different axial ratios and polarization of the satellite and receiving antennas.

1.3.1.1 Case 1: Polarization Coupling between Satellite and User Antenna, Both of Which are RHCP

Consider the ideal situation in which both satellite and receiving antennas are matched in polarization and perfectly aligned (i.e., both are RHCP) and their major polarization axes are aligned to be parallel. $R_S = R_A = 1$; also $\Delta\tau = 0$.

Hence

$$p = \frac{1}{2} + \frac{4R_A^2 + (R_A^2 - 1)^2}{2(R_A^2 + 1)^2} = \frac{1}{2} + \frac{1}{2} = 1$$

and for this ideal situation, the polarization mismatch loss suffered by the receiving antenna while receiving the satellite signal is $L_p = 0$ dB.

1.3.1.2 Case 2:2(a) Coupling between RHCP Transmit Antenna and an LHCP Receive Antenna, or 2(b) between a User Antenna that is RHCP and the Multipath Signal that is LHCP

Polarization of the satellite antenna is RHCP but polarization of the receiving antenna is LHCP; that is, their polarizations are orthogonal. A different and more relevant scenario can also be used to the above mentioned conditions. The signal from the satellite is reflected off a perfectly conducting surface and arrives at the receiving antenna as multipath with the same axial ratio but is LHCP. The same equations apply if the R_S and R_A designations are interchanged for this second case 2(b).

Hence for Case 2(a) $R_S = -R_A$. In addition, if the major axes of their polarization ellipses are also orthogonal to one another, $\Delta\tau = 90°$. Then

$$p = \frac{1}{2} + \frac{-4R_A^2 - \left(R_A^2 - 1\right)^2}{2\left(R_A^2 + 1\right)^2} = \frac{1}{2} - \frac{1}{2} = 0$$

Hence the polarization mismatch loss is $L_p = \infty$ for this case and no signal is received from the satellite by the receiving antenna. Similarly if the receiving antenna is perfectly RHCP it will be able to reject a multipath signal that is perfectly LHCP. This would be the case if the satellite signal were perfectly RHCP—which it is not.

1.3.1.3 Case 3: User Antenna is Linearly Polarized and is Receiving RHCP Satellite Signals

In certain cases, such as in handsets and other low-cost commercial GPS, the receiving antenna is a linearly polarized dipole or monopole and is used to receive RHCP signals from the satellite.

$$R_S = 1(RHCP);\ R_A = \infty(\text{Linear Polarization});\ \Delta\tau = \text{arbitrary}$$

$$p = \frac{1}{2} + \frac{4\frac{R_S}{|R_A|} + \left(1 - \frac{1}{R_A^2}\right)\left(R_S^2 - 1\right)\cos 2\Delta\tau}{2\left(1 + \frac{1}{R_A^2}\right)\left(R_S^2 + 1\right)} = \frac{1}{2}\left[1 + \frac{\left(R_S^2 - 1\right)\cos(2\Delta\tau)}{R_S^2 + 1}\right]$$

Hence if the signal from the satellite were to be perfectly RHCP such that $R_s = 1$, then $p = \frac{1}{2}$.

Hence, when a linearly polarized antenna such as a dipole or a monopole is used to receive RHCP signals from the satellite, one can expect a polarization mismatch loss of 3 dB or half the power being lost due to mismatch in polarizations. We can expect such a 3 dB loss even in microstrip patch antennas most commonly used in GPS when they receive signals transmitted from satellites at low elevation angles close to the horizon; these patch antennas are almost linearly (vertically) polarized at these lower elevations since the horizontal component of the radiated field becomes zero due to the presence of a metallic ground plane. The same situation also occurs in GNSS antenna mounted on the fuselage of aircraft where the metallic fuselage "shorts" out the horizontal polarization. The satellite antenna is, of course, not perfectly RHCP but instead is elliptically polarized with a finite axial ratio, generally around 1.8 dB. The actual polarization loss can be calculated from the above equation by substituting the appropriate value for $R_{s.}$

As explained in Chapter 3, many handset antennas use a linearly polarized monopole or dipole type antennas for reducing their size and cost and for cosmetic appeal. These handset antennas are most often used in urban settings where the multipath environment is very dense and where the polarization state of GNSS signals received by the handset antenna is less clearly defined due to the presence of multiple reflecting surfaces contributing to multipath. It has been shown by Pathak et al. [11] that in such a multipath environment a linearly polarized antenna can perform nearly as well as a purely RHCP antenna.

1.3.1.4 Case 4: Polarization Coupling between a Linearly Polarized Lightsquared Transmit Antenna and an Elliptically Polarized GNSS User Antenna

Another interesting case that can be considered is one that has become particularly relevant recently: this is interference between a signal transmitted by a Lightsquared antenna [12] that is assumed to be linearly polarized and a GNSS user antenna that is RHCP—or more correctly elliptically polarized or linearly polarized—such as a GNSS antenna used in a handset. As mentioned previously many user antennas become linearly polarized, with vertical polarization, at very low elevation angles. As explained in Section 3.9.3, many handset antennas use a linearly polarized monopole or dipole type antennas for reducing their size and cost and for cosmetic appeal. These handset antennas are most often used in urban settings where the multipath environment is very dense and where the polarization state of GNSS signals received by the handset antenna is less clearly defined due to the presence of multiple reflecting surfaces contributing to multipath. It has been shown by Pathak et al. [11] that in such a multipath environment a linearly polarized antenna can perform nearly as well as a purely RHCP antenna.

For Case 4: $R_s = \infty$ (linear polarization) and R_A the polarization of the user antenna can either be elliptical or linear (handset antenna). The polarization coupling for this case then is:

$$p = \frac{1}{2}\left[1 + \frac{\left(R_A^2 - 1\right)\cos 2\Delta\tau}{\left(R_A^2 + 1\right)}\right]$$

If the user antenna is also linearly polarized, as in a handset user antenna, then

$$p = \frac{1}{2}[1 + \cos 2\Delta\tau]$$

Hence when $\Delta\tau = 0$ (i.e., when the polarization of the Lightsquared and user antennas are aligned to be parallel to one another the polarization coupling is 1), all the energy transmitted by the Lightsquared antenna will be absorbed by the user antenna with no polarization mismatch loss, resulting in maximum interference. If one the other hand if $\Delta\tau = 90°$ (i.e., the axes of the two antennas are orthogonal to each other), then $p = 0$; hence the polarization mismatch loss is $L_p = \infty$ and there will be no energy exchange between the two antennas.

If the user antenna is elliptically polarized then the actual coupling between the two antennas can be calculated using the above equation and will depend on the alignment of their polarization axes.

1.3.2 Effect of Axial Ratio on the Reception of Satellite Signals by the Receiving Antenna

Both the satellite and the receiver antennas are in reality elliptically polarized with finite axial ratios, leading to some loss in reception. This problem is also of concern when calculating C/N_0 in satellite communications [13]. The axial ratios of the GPS satellites is required to be no worse than 1.8 dB over an angular range of 14.3° from boresight; the corresponding axial ratio requirement for the Block II/IIA satellites is 1.2dB [14]. In the GPS L_2 band the requirement for axial ratio for GPS satellites is no worse than 2.2 dB and for the Block II/ IIA satellites no worse than 3.2 dB [14]. In the GPS L_5 band the axial ratio is no worse than 2.4 dB in an angular range of ± 13.8° from nadir. Satellite antennas are only required to maintain a good axial ratio over a fairly narrow beamwidth that is required by the earth coverage beam; this is generally around ±14°. The receiver antenna, on the other hand, has a more difficult requirement since it has to maintain a good axial ratio over a much wider beamwidth that can extend from zenith down to an elevation angle of 10° to 5° in elevation; this is difficult to achieve.

Figure 1.7 shows the maximum and minimum polarization loss suffered while receiving a satellite signal with an axial ratio of 1.8 dB by a receiving antenna whose axial ratio varies from 0.5 to 10 dB. In these calculations the axial ratio of the satellite antenna is 1.8 dB, which corresponds to the axial ratio of GPS satellites in the GPS L_1 band [14]. The values shown in this graph were calculated from (1.11) and (1.12). The minimum polarization loss of 0.0008 dB can occur when the relative tilt angle $\Delta\tau = 0°$ (i.e., when their major axes of polarization are aligned parallel) and the axial ratios of the receiving antenna is the same as the satellite signal (i.e., 1.8 dB).

1.3.3 Effect of Axial Ratio of a GNSS Antenna on Reception of Different Types of Multipath Signals

Figure 1.8 shows the ability of the receiving antenna through polarization discrimination to reject single-bounce multipath signals generated by reflection of the satellite signals from a perfectly conducting smooth surface. These are the strongest multipath signals received by the antenna and can occur as a result of reflections from a building or from seawater for example. The polarization of such multipath signals is LHCP if the incidence angle is less than the Brewster's angle; their axial ratio is the same as the satellite signal (i.e. 1.8 dB).

Figure 1.8 shows the maximum and minimum rejection of these multipath signals as the axial ratio of the receiving antenna is varied from 0.5 to 10.0 dB.

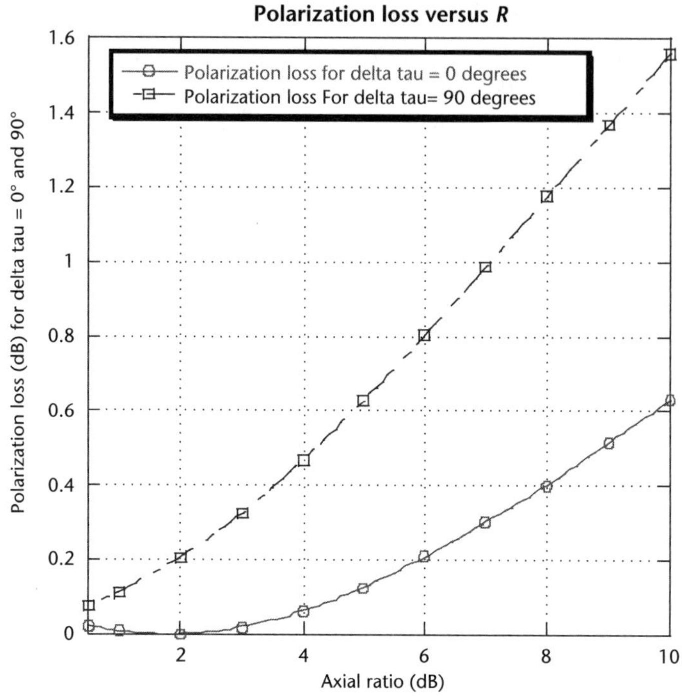

Figure 1.7 Polarization loss for signals received from the GPS satellite in the L1 band (axial ratio = 1.8 dB) by a GNSS receiving antenna with an axial ratio that varies from 0 to 10 dB.

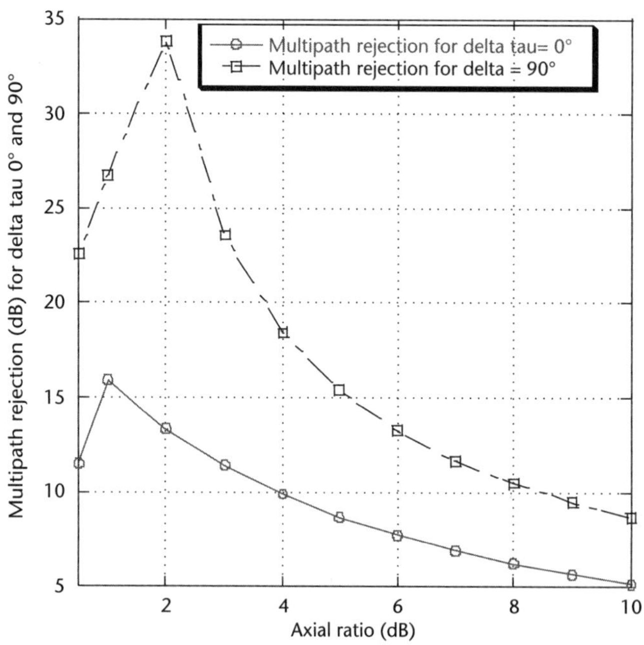

Figure 1.8 Rejection of single-bounce multipath signals from the satellite (axial ratio = 1.8 dB) received by a GNSS user antenna with axial ratios that vary from 0 to 10 dB.

The maximum and minimum rejection of multipath for a specific axial ratio occurs when $\Delta\tau = 90°$ and $\Delta\tau = 0°$, respectively. For example, when the axial ratio of the receiving antenna is 2.0 dB the maximum rejection of multipath by the antenna can be as large as -33.8 dB for $\Delta\tau = 90°$; however, the rejection level drops to −13.3 dB when $\Delta\tau = 0°$. We also notice that the best rejection of multipath occurs when the axial ratio of the receiver antenna is the same as that of the satellite signal (i.e., 1.8 dB)—but with opposite polarization.

If the multipath signal is generated by reflection from a rough surface, the polarization characteristic of the reflected signal can be random and this can result in rejection that is much less than the results shown in Figure 1.8; the rejection in such cases can be around 3 dB.

1.4 Directivity and Gain of a GNSS Antenna

The gain and the beamwidth are perhaps the two key performance parameter of a GNSS antenna. As will be described later, the antenna needs to meet certain minimum requirements on the gain so as to provide an adequate C/N_0 for a commercial GNSS receiver to acquire and track the GNSS satellites—both under normal quiescent conditions but also in the presence of interference or jamming (defined as the effective C/N_0 ratio in this case) if necessary. The beamwidth of the antenna in the elevation plane needs to be wide enough and the gain should also be greater than the threshold minimum gain above the specified masking elevation angle for all azimuth angles so as to meet these requirements This will provide the best geometrical dilution of precision (GDOP), which is a metric on the accuracy of pseudorange measurements to be described later in this section. Since the gain varies inversely as the beamwidth, this requirement sets an upper bound to the optimum permissible gain in a GNSS antenna.

Two definitions are used to describe the gain of the antenna and need to be distinguished from each other. The first is directivity, a measure of the concentration of RHCP radiation in the direction of the maximum of the antenna beam denoted by the angle (θ_0, ϕ_0). It is purely a power pattern shape parameter and describes the distribution of radiated energy by the antenna in space. It is defined as the ratio of the maximum radiation intensity for RHCP in the direction of the maximum normalized by the average radiation intensity for both polarizations RHCP and LHCP. It is defined by the equation

$$\text{Directivity for RHCP radiation (dBic)} = D(\theta_0,\varphi_0)10\log\left[\frac{P_{RHCP}(\theta_0,\varphi_0)}{P_{AVG}}\right] \tag{1.13}$$

The "c" in dBic denotes circular polarization radiation.

The numerator is the maximum RHCP radiation intensity defined by

$$P_{RHCP}(\theta_0,\varphi_0) = \left|E_{RHCP}(\theta_0,\varphi_0)\right|^2$$

The denominator is the average radiation intensity defined by

$$P_{AVG} = \frac{1}{4\pi}\int_{\varphi=0}^{2\pi}\int_{\theta=0}^{\pi}\left[\left|E_{RHCP}\right|^2 + \left|E_{LHCP}\right|^2\right]\sin\theta d\theta d\varphi$$

The integration for calculating the average radiation intensity is performed in the elevation plane over $0 \le \theta \le \pi$ and in the azimuth plane over $0 \le \varphi \le 2\pi$. The directivity as defined above does not include any of the losses in the antenna, such as the input impedance mismatch loss between the antenna and its coaxial feed cable, the losses in the polarization feed network necessary for generating RHCP in the antenna, polarization mismatch loss between the user's receive antenna and the satellite transmit antenna, and other relatively smaller dielectric and ohmic losses in the antenna. The realized peak gain of the antenna is defined by

$$G(\theta_0, \varphi_0) = \eta_{ant} D(\theta_0, \varphi_0) \tag{1.14}$$

where η_{ant} is the total antenna efficiency.

$$\eta_{ant} = \eta_i \cdot \eta_f \cdot \eta_P \cdot \eta_r \cdot \eta_d \tag{1.15}$$

where

η_i = Impedance mismatch loss between the feed cable and the antenna

η_f = Loss in the feed network for generating RHCP in the antenna

η_p = Polarization efficiency is caused by a polarization mismatch between the user's receiving antenna and the satellite transmitting antenna, both of which can be elliptically polarized with finite axial ratios

η_r = Ohmic (or resistive) loss in the antenna

η_d = Dielectric and surface wave losses in the antenna

The realized gain is the term that should be used when calculating either the C/ N_0 or the link margin when evaluating the system performance.

1.4.1 RHCP Gain of a GNSS Antenna for an Optimum Radiation Pattern Defined by Its Low Elevation Masking Angle

The optimum radiation pattern for a RHCP GNSS antenna is restricted by the minimum elevation angle masking requirement in the upper hemispheric. The antenna is expected to radiate only within this upper sector of the hemisphere and have no radiation below this masking angle since such radiation would make the antenna susceptible to deleterious effects such as multipath, interference, and jamming. In his section we derive the gain of a GNSS antenna that produces such an optimum antenna pattern.

The directivity or the directive gain of the antenna increases as we constrain the angular region over which the antenna can radiate. Thus the directivity D of the antenna is given by

$$D = \frac{Area\ of\ Sphere}{Area\ of\ the\ Radiation\ Pattern\ Of\ the\ Antenna} \tag{1.16}$$

Consider an ideal (but fictitious) GNSS antenna with a perfect antenna efficiency of 1 (no losses) and also with a perfect RHCP isotropic radiation pattern distributed uniformly over an entire sphere. We will use the reciprocity relationship between transmit and receive antennas in this illustrative example. Then the power density P_D at any given point would be the input power P_{in} to the antenna divided by the surface area of an imaginary sphere of radius R enclosing the antenna. This is illustrated in Figure 1.9. Thus the power density for this case is $P_d = \frac{P_{in}}{4\pi R^2}$. Since P_{in} is uniformly distributed over the entire sphere the gain of this isotropic antenna is $G = 0$ dBic. Now supposing an infinitely large metal ground plane is placed beneath the antenna that prevents the antenna from radiating into the lower hemisphere. Then antenna is now constrained to radiate only over the upper hemisphere, and the spatial area over which the antenna is now allowed to radiate is half as much as the previous case but the power input to the antenna is the same, so we now have $P_d = \frac{P_{in}}{2\pi R^2}$; hence the gain would now be double that in the previous example and would be 3 dBic. Now consider the case relevant to a GNSS antenna where we place further constraints so that it radiates only over a portion of the upper hemisphere down to an elevation angle of θ_m, which is the low elevation masking angle. This is the optimum radiation pattern for GNSS antenna; any radiation below this minimum masking angle is undesirable for reasons explained earlier. Similarly, a narrower antenna beam that cuts off above the desired masking angle is also undesirable since it prevents acquisition of low elevation satellites needed for a good PDOP. The optimum radiation region has now been restricted to a spherical sector in the upper hemisphere whose surface area has been reduced to $2\pi R^2[1 - \sin(\theta_m)]$. Since the surface area of radiation has been decreased this implies that RHCP gain of the antenna compared to the previous hemispherical case has now increased to $3dB - 10\log[1 - \sin(\theta_m)]$. Since we are considering an ideal, purely

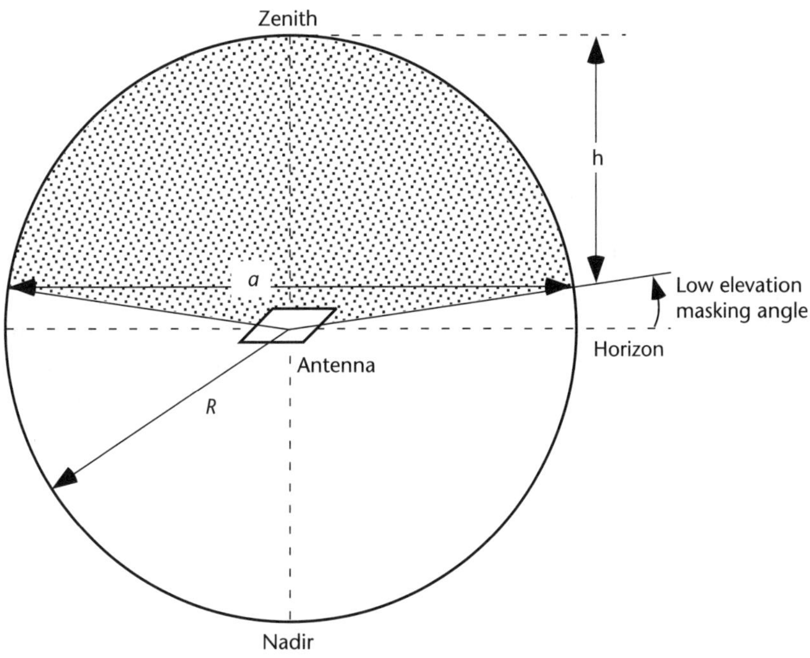

Figure 1.9 Optimum directivity of a GNSS antenna.

polarized RHCP antenna with no losses, this represents the optimum gain of an ideal GNSS antenna that provides coverage over the entire upper hemisphere down to the low elevation masking angle defined by θ_m but no radiation below this angle. If the masking angle is $\theta_m = 5°$ the optimum directivity of the antenna would be 3.396 dBic. For a 24-satellite GPS constellation, a 5° elevation masking angle for the antenna always allows at least five satellites in view and at least seven satellites more than 80% of the time for a user located at all altitudes [15]. Any increase in directivity beyond this maximum limit implies that the radiation beam of the antenna has now become narrower and will not be able to satisfy the minimum 5° masking angle. For example, for a slightly higher masking angle of $\theta_m = 10°$, the gain of the antenna could increase to 3.826 dBic. However for a 10° masking angle, an observer located in the vicinity of latitude 35° to 55° has only four satellites in view only a small fraction of less than 0.5% of the time. A failure of one of these satellites can create an outage.

The results of this analysis indicate that there is a linkage between the maximum gain and the ideal beamwidth of a GNSS antenna, which places an upper bound on gain if the antenna has to comply with low elevation masking requirements of 5°: it is 3.396 dBic. Any increase in antenna gain above this maximum limit will result in a narrower antenna beam, resulting in degradation of gain at an elevation angle of 5 degrees, which is the required minimum masking angle.

Real GNSS antennas do not have such ideal radiation patterns nor are they perfectly polarized. They also suffer from various losses. They mostly also radiate in the "broadside" direction (90° in elevation), achieving peak gain at zenith but then the gain drops off as the elevation angle is decreased. In most commonly used GNSS antennas, such as microstrip patch or spiral antennas that need ground planes, the polarization is almost linear (vertical) rather than RHCP at lower elevations closer to the horizon, resulting in an additional decrease in RHCP gain. The gain of these antennas may decrease by as much as than 10 dB as the elevation angle decreases from zenith down to the horizon. The only antenna that continues to produce reasonable RHCP radiation even at low elevation angles is a quadrifilar helix antenna, but this has a very poor cutoff in radiation at lower elevations, making the antenna vulnerable to multipath, interference, and jamming—all three of which are prevalent at low elevations.

The current minimum elevation masking angle $\theta_{mask} = 5°$ (or 85° from vertical) and the corresponding minimum gain for RHCP at this masking angle in the GPS frequency bands was –4.5 dBic [16, 17]. This specification is only for airborne antennas. The gain specification has since been relaxed to –5.5 dBic at 5° in DO-301. This specification applies to avionics antennas set by RTCA [17] and is also the specification for the commercial aeronautical antenna (model number S67-1575-16) built by Sensor Systems and meets the ARINC 743 A characteristics for size and so forth. The corresponding minimum antenna gain specification for other GPS systems has also been discussed by Van Dierendonck [16]. The elevation masking angles in these requirements are specified at a higher elevation angle, at $\theta = 10°$ (or 80° from vertical) and the gain at this higher masking angle is specified to be –2.5 dBic. This specification in its full detail is as follows: "The GPS antenna shall provide a minimum gain of –2.5 dBic to a RHCP signal over a 160° solid cone angle of coverage (above 10° elevation angle) for signals in both the L_1 and L_2 bandwidths." The specification also goes on to state that the combined effect of the environment will not cause the gain to be less than –3.3 dBic [16]. Although the gain at low elevation angles in the vicinity of the masking angle is of concern since it is a more difficult requirement to meet, there are also specifications on the antenna gain at higher elevation angles as well as requirements for maximum gain at various elevation angles.

These antenna gain versus elevation angle requirements for airborne antennas have been specified by the RTCA for the GPS L_1 band [17] and also by Europe (EUROCAE) for the Galileo E5 band [18] and are listed in Tables 1.3 and 1.4. The elevation angle of 5° is highlighted in Table 1.3 since it is the low elevation masking angle that yields the best dilution of precision (DOP) and availability of satellites. Later in this chapter we will calculate the minimum required carrier-to-noise ratios (C/N_0) at this masking angle that is needed to acquire and track the satellites depending on the components used in the front end of a GNSS receiver.

There are also requirements for maximum gain for the L_1 –E5 band. These are given in Table 1.4.

1.4.2 Relationship between Beamwidth, Aperture Size, and Directive Gain of a GNSS Antenna

The energy radiated within the antenna beam varies with angle off the boresight direction and losses also occur from radiation lost in the backlobes and sidelobes. We can obtain a simple relationship between the directive gain of an antenna and its beamwidth in the two orthogonal planes by considering an idealized model of antenna pattern. Assuming that the antenna pattern is uniform, the directive gain, as was derived in Section 1.4.1, is equal to the area of the isotropic sphere ($4\pi r^2$) divided by the cross-sectional area of the sector representing the radiated beam.

Table 1.3 Minimum Gain Versus Elevation Angle for the L_1 - E5 Band

Elevation Angle (Degrees)	*Minimum Gain (dBic)*
>15	–2
10	–3
5	–4.5
0	–7.5

From: [18].

Table 1.4 Maximum Gain Versus Elevation Angle for the L_1–E5 Band

Elevation Angle (Degrees)	*Maximum Gain (dBic)*
5	5.0
0 (horizon)	–2
Below –30	–10

From: [18].

$$D = \frac{Area\ of\ Sphere}{Area\ of\ Antenna\ Pattern} = \frac{4\pi}{\alpha_{\varphi AZ}\alpha_{\theta EL}}$$

The directive gain of the antenna is inversely proportional to product of the two orthogonal beamwidths $\alpha_{\varphi AZ}$ and $\alpha_{\theta EL}$ are the beamwidths in the azimuth and elevation planes respectively expressed in radians.

If the antenna beam is circular in shape (good symmetry in the azimuth plane) as is ideal for GNSS applications, then

$$D(\theta, \varphi) = \frac{\text{constant}}{\alpha^2}$$

The beamwidth α is also proportional to the wavelength and inversely proportional to the aperture size of the antenna. If the GNSS antenna were circular in shape with an uniform aperture illumination with no tapering of amplitude, then

$$\alpha = constant\left(\frac{\lambda}{d}\right) \text{ and directivity } D(\theta,\phi) = constant\left(\frac{d^2}{\lambda^2}\right)$$

where λ = wavelength and d = diameter of the circular aperture of the antenna.

As previously explained, GNSS antennas are required to have very wide beam covering almost the entire upper hemisphere from zenith down to a very low making angle of only 5° to 10°. To meet such a requirement, the above equations indicate that the aperture size of a GNSS antenna needs to be small compared to the wavelength. Increasing the gain will only result in the beamwidth becoming very narrow, resulting in high loss of gain at lower elevations causing poor GDOP as described later.

The beamwidth of an antenna is defined in most textbooks as the half-power or 3 dB beamwidth. This is the angle between two directions in the radiation pattern of an antenna in which the directive gain of the principal radiation lobe of the antenna is one-half its maximum value (one-half the directivity). However, this conventional definition of half-power antenna beamwidth, which is important when considering point-to-point communications such as terrestrial and satellite communications, is of little significance in evaluating performance of a GNSS navigational antenna. Very wide beams are necessary in GNSS to acquire the maximum number of satellites, especially those at very low elevation angles. Their gain within this sector has to comply with minimum gain requirements that are critical for satellite acquisition and for achieving precision. There are two minimum gain conditions to consider. The first applies to the case of quiescent conditions in the absence of interference or jamming, and the second is the minimum gain needed to acquire the satellite in the presence of interference or jamming. The two minimum gain requirements are derived a little later in Sections 1.15 and 1.16. The ability of the antenna to acquire and track low elevation satellites is dependent on its gain, which generally decreases with elevation angle, as indicated earlier. The maximum directivity—and hence gain for an antenna with no other losses—that provides coverage over the entire upper hemisphere down to an elevation angle of 5° (or 85°

from zenith) is 3.396 dBic at zenith, as indicated earlier. The current requirement for gain for a GNSS antenna at 5° is –4.5 dBic. Hence the rate of decrease in gain with elevation is 0.0928 dBic for every 1° of change in the elevation.

A new definition is needed to describe the beamwidth and the associated gain that is more appropriate and meaningful for a GNSS antenna; the standard definition of half-power or 3-dB beamwidth so commonly used for antennas used in point-to-point communications or in radar cannot be applied for a broadbeam antenna needed for GNSS. In this book it will be called the minimum gain antenna beamwidth, defined as the beamwidth that extends from zenith down to the minimum elevation masking angle θ_m and provides within this coverage sector the minimum gain necessary for the antenna to acquire and track at least four satellites capable of providing the best GDOP. Figure 1.10 shows the minimum gain beamwidth required by a GNSS antenna.

1.4.3 Effect of Minimum Gain Beamwidth and Antenna Pattern Contour on GDOP

The minimum gain beamwidth of a GNSS antenna and the rate of variation in gain in the elevation plane determine the number and the elevation angle of satellites that are acquired by the antenna/receiver. Most GNSS antennas, such as patch antennas, spiral antennas, and crossed dipoles radiate in direction that is broadside to the antenna aperture and therefore their peak gain and best polarization properties occur at zenith. The gain drops off rapidly as the elevation decreases towards the horizon and the polarization also becomes linear (vertical polarization) when a planar metal ground plane is used. This decrease in gain can be as large as 10 dB or even greater.

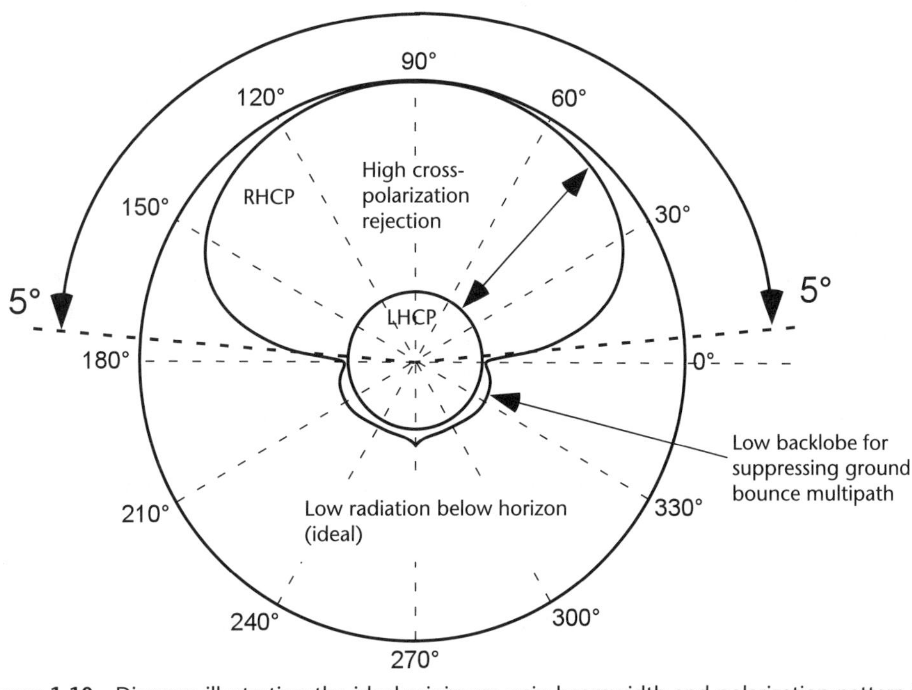

Figure 1.10 Diagram illustrating the ideal minimum gain beamwidth and polarization pattern for a GNSS antenna.

This is particularly true for high gain antennas with narrower beams. If the gain of the antenna at these low elevation angles decreases below the minimum required gain for the antenna to acquire signals from the satellite with the associated receiver then there is a risk that the antenna will only acquire satellites at higher elevation angles and reject satellites at lower elevation angles within its view. This can have a large impact on a factor, described as dilution of precision (DOP). DOP is a numerical value that is an expression of the confidence of the position solution obtained from the geometry of the satellites acquired by the antenna and receiver. Very good descriptions of DOP and its impact on position error have been provided by Rizos and Brzezinska [19], Misra and Enge [20], and by Seeber [21]. There are six different types of DOPs:

1. GDOP expresses the uncertainty of all parameters such as latitude, longitude, height, and the clock offset;
2. PDOP is the uncertainty of the 3-D parameters (latitude, longitude, and height), which are important in airborne receivers
3. HTDOP is the uncertainty of 2-D and time parameters (i.e., latitude, longitude, and time);
4. HDOP is the uncertainty in 2-D parameters (latitude and longitude), which are important for maritime receivers;
5. VDOP is the uncertainty in the height parameter, which is important for aircraft attempting precision landings;
6. TDOP is the uncertainty in the clock offset, which is important for military personnel, scientists, and engineers attempting to synchronize their clocks using GNSS.

The contributions from the various error sources on the accuracy of the GNSS measurements is always of concern to the user. The combined effect of all the errors projected on to the line connecting the user and the satellite is called the user equivalent range error (UERE) or just as the user range error (URE) [19, 21]. The standard deviation of the overall solution is expressed as

$$\sigma_S = URE \cdot GDOP = \sqrt{\left[\sigma_e^2 + \sigma_n^2 + \sigma_U^2 + \sigma_{clk}^2\right]} \tag{1.17}$$

where σ_e, σ_n, σ_U are the standard deviations of the estimated position along the east, north, and up directions and σ_{clk} is the standard deviation of receiver clock solution.

The lower the value of GDOP, the greater is the confidence in the solution. The effect of GDOP on positioning can be great since it multiplies the standard deviations of all other errors. Therefore URE is 5m and the PDOP is 3m, then σ^* for the total position error is 15m. Van Diggelen [22] has indicated that the same type of relationship also applies to the Doppler navigation solution where the Doppler navigation PDOP is multiplied by the standard deviation for Doppler σ_D

Of the various DOP, GDOP is perhaps the most important. It is assumed that the measured errors are independent with zero mean and that all measurements have the same rms value σ_r. GDOP is calculated by the following formula [23]

$$\text{cov}(position) = \sigma_r^2 \cdot \left[G^T G\right]^{-1} \quad (1.18)$$

In the above equation G is the direction cosine matrix and is expressed in terms of the direction cosines of the satellite location relative to the user. When the location of the satellite is defined in terms of an east, north, up coordinate system shown in Figure 1.11 its azimuth Az is measured 360° clockwise from true north, and its elevation E is measured up from the local horizontal.

Hence the G matrix becomes

$$\begin{bmatrix} (\cos(E_1)*\sin(Az_1)) & (\cos(E_1)*\cos(Az_1)) & \sin(E_1) & 1 \\ (\cos(E_2)*\sin(Az_2)) & (\cos(E_2)*\cos(Az_2)) & \sin(E_2) & 1 \\ (\cos(E_3)*\sin(Az_3)) & (\cos(E_3)*\cos(Az_3)) & \sin(E_3) & 1 \\ (\cos(E_4)*\sin(Az_4)) & (\cos(E_4)*\cos(Az_4)) & \sin(E_4) & 1 \end{bmatrix} \quad (1.19)$$

Hence cov(Position) =

$$\sigma_R^2 \begin{bmatrix} (East\,DOP)^2 & & Covariance\ Terms & \\ & (North\ DOP)^2 & & \\ & & (Vertical\ DOP)^2 & \\ Covariance\,Terms & & & (Time\,DOP)^2 \end{bmatrix}$$

The scalar GDOP is defined to be the square root of the trace of the GDOP matrix.

$$\begin{aligned} &Horizontal\ HDOP = \sqrt{(North\ DOP)^2\,(East\ DOP)^2} \\ &Postion\ PDOP = \sqrt{(HDOP)^2 + (VDOP)^2} \end{aligned} \quad (1.20)$$

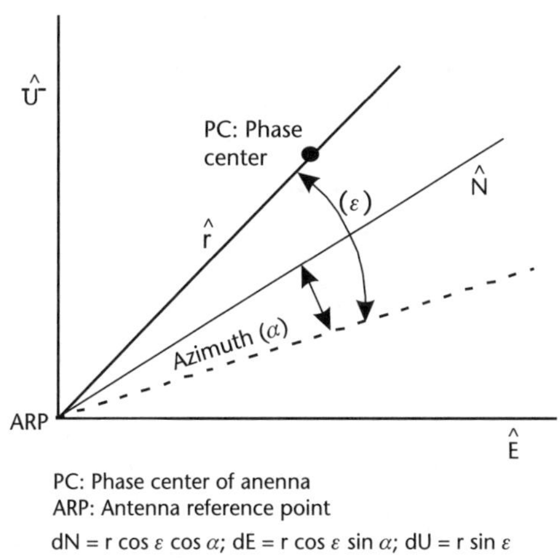

Figure 1.11 North, east, up (NEU) coordinate system.

The position accuracy along a given axis is maximized when the angle between that axis and the line of sight towards the satellite signal is minimized. Therefore the accuracy of GNSS position in the horizontal direction, or HDOP is optimized when signals from low elevation satellites are acquired by the antenna and the line-of-sight vectors are evenly distributed in azimuth. Similarly, the position accuracy along the vertical (up) axis or VDOP is optimized when the antenna is able to acquire signals from the higher elevation satellites. Analysis conducted by Spilker [15] indicates that for a GPS constellation of 24 satellites that the number of satellites varies between 6 and 10 for an elevation masking angle of 5°. Parkinson [23] has shown that the best GDOP for 4 satellites is obtained when one satellite is overhead (i.e., at an elevation angle of 90° and azimuth angle of 0°) and the other three satellites are equally spaced at the minimum elevation of 5° and at azimuth angles of 0°, 120°, and 240°. When these values of azimuth and elevation are inserted into G matrix for the four satellites and the square roots for the appropriate diagonal terms of the GDOP matrix are taken we obtain the following values for the various DOP parameters:"

HDOP (Horizontal DOP) = 1.16; VDOP (Vertical DOP) = 1.26; PDOP (Position DOP) = 1.72

TDOP (Time DOP) = 0.64; GDOP = 1.83.

Parkinson has indicated that a PDOP of 1.72 is considered good for GNSS measurements and is in fact the optimum for four satellites with a minimum elevation angle of 5°. When the masking angle is dropped to 0°—which means essentially coverage of the entire upper hemisphere by the antenna—Spilker [15] has shown that GDOP = 1.73, PDOP = 1.63, HDOP= 1.15, VDOP=1.15 and TDOP= 0.57. These are the best DOP values that can be measured. In reality this is difficult to achieve because constraints on antenna performance and from terrain blockage, multipath, and atmospheric loses. The satellite geometry is considered good if the antenna is able to acquire satellites that can provide PDOP values of less than 3 and HDOP values of less than 2.

When the gain of the antenna to low elevation angles drops below the minimum gain at elevation angles above the masking angle, it is unable to acquire sufficient number of low elevation satellites to have a PDOP that is less than the maximum desired value of 3. This happens when the antenna gain is too high and the beamwidth is narrow. This situation is illustrated in Figure 1.12.

In the inner figure, the antenna beam is very narrow since this antenna has a high peak gain which decreases rapidly with elevation angle; the gain is lower than needed to acquire at low elevation satellites. The outer figure shows a more optimum situation where the antenna beam is broader—with a smaller peak gain at zenith but with sufficient gain to acquire the needed number of low-elevation satellites to provide a low PDOP.

To illustrate consider a situation when the four satellites acquired by the antenna have the following azimuth and elevation angles shown in Table 1.5 [24, 25].

None of the satellites acquired by the antenna/receiver shown in this example have elevation angles lower than 22°, although they seem fairly evenly distributed in azimuth. Inserting the values these value of azimuth and elevation angles for the satellites in (1.18) and (1.19) we obtain a PDOP = 3.44; this is above the desired limit.

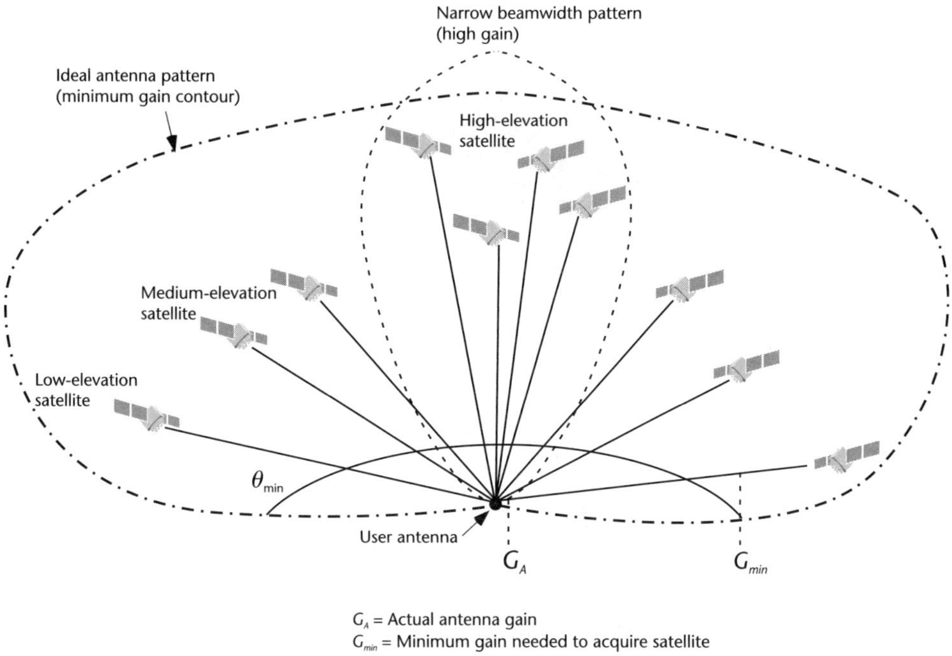

Figure 1.12 Antenna gain and beamwidth effects on PDOP.

Table 1.5 Data on Satellites for Calculating PDOP

Satellite Number	*Elevation (Degrees)*	*Azimuth (Degrees)*
1	37.6	130.9
2	65.1	326.6
3	22.8	289.5
4	40.6	39.8

Hence (1.18) and (1.19) are useful for evaluating the DOP obtained from GNSS antennas with less than optimum antenna patterns. One technique is to select the three satellites at the lowest elevation angle from all that are acquired by the antenna (i.e., antenna gain needs to be equal to or above the minimum gain limits satellites at this lowest elevation) at any instant of time; a fourth satellite at the highest elevation angle is then selected. The azimuth and elevation angles for these four selected satellites are then used to calculate the G matrix and then evaluate their DOPs from the covariance matrix to check on their respective DOP values.

A method of geometrically determining the position dilution of precision has been suggested by some authors [24, 25] which provides a better insight than the purely mathematical determination from the equations given above. This method of geometrical determination is shown in Figure.1.13.

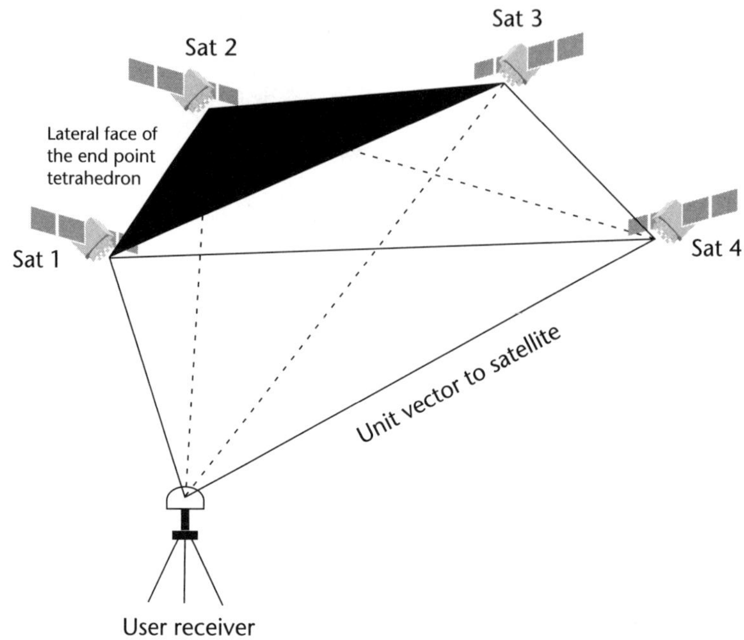

Figure 1.13 Determination of DOP from geometry of satellites acquired by a GNSS antenna.

The four unit vectors in this diagram point from the user antenna towards the respective satellites acquired by the antenna/ receiver. The ends of these unit vectors are then connected resulting in 6 line segments that form a tetrahedron. PDOP is then the RSS (square root of the sum of squares) of the four faces of the tetrahedron divided by three times the volume of the tetrahedron.

$$PDOP = \frac{\sqrt{\sum_{1}^{4} b_i^2}}{3V} \tag{1.21}$$

where V is the volume of the tetrahedron formed by connecting the end points of the four unit vectors

$$b_i = \text{area of the } i^{th} \text{ lateral face of the end point tetrahedron}$$

The volume of the tetrahedron is equal to 1/3 base multiplied by its altitude. Thus PDOP decreases as we maximize the volume of the tetrahedron; this is possible by increasing either the base of the tetrahedron or by increasing its altitude. The base can be increased by selected three satellites evenly spread out in azimuth at approximately 120° intervals and the altitude can be increase by selecting one the four satellites that is closer to the zenith. This is the optimum satellite geometry that provides the best PDOP [23].

1.5 Phase Center of a GNSS Antenna

The phase center of a receiver GNSS antenna is defined as a point where the signal transmitted from the satellite is collected; hence, it is also becomes the reference point at which all measurements derived from GNSS signal measurements are referred to [26, 27]. In an ideal antenna it is at the center of a spherical surface with equal phase surrounding the antenna as shown in Figure 14(a); in reality the phase contour is not a uniform sphere but has a distorted shape with the phase that can change with frequency (e.g., differences in the phase centers between the L_1, L_2 and L_5 bands), azimuth and elevation angles as shown in Figure 1.14 (a). However since the phase center of the antenna cannot be identified physically through a tape measure, a geometrical point on the antenna known as the antenna reference point or ARP is selected. The IGS has defined the ARP as the intersection of the vertical axis of symmetry and the bottom of the antenna.

One should also note that PCV in both user antenna and the satellite antennas can contribute to these PCV errors as shown in Figure 1.14(b). The phase of the carrier as measured at the receiver involves three components: $A_j + B_j + N_i$.

Where, A_j = fractional initial phase at the receiver antenna

B_j = fractional initial phase at the satellite antenna

N_i = ambiguous (or unknown) integer number of cycles.

It can be resolved through the use of ambiguity resolution (AR) techniques [1, pp. 202- 237] or can be removed through double differencing methods if there are no cycle slips. A cycle slip occurs when signal tracking to a satellite is interrupted for whatever reason; when this occurs the ambiguities involving the satellite before the break are not equal to the ambiguities after the break. Note that A_j and B_j are measured in the receiver during carrier phase measurements. PCO and PCV of both the user and the satellite antenna will be discussed in greater detail in Chapter 6 of this book. Single and double differencing techniques will be explained later in this chapter. Single differencing helps to remove antenna PCO and PCV errors when either the same or accurately calibrated antennas are used at either end of a baseline during these differencing methods. Ambiguity errors are removed by double differencing techniques, in the absence of cycle slips.

An antenna model which is particularly convenient for calibration of the absolute phase center of a GNSS antenna has been proposed by Gorres et al. [25] and is shown in Figure 1.15.

The total antenna phase center correction for phase pseudorange measurements as depicted in this model consists of two parts as shown in this figure: 1) Δ_{PCO}, which is the projection of the vector $\vec{a}$ between the mean electrical phase center of the antenna and the antenna reference point (ARP) and the unit vector $\hat{r}_0$, the direction of arrival of the satellite signal at the receiver ; and (2) $\Delta_{PCV}(\theta, \phi, f)$ which is phase variation of the antenna relative to the mean electromagnetic phase center as a function of the elevation θ and azimuth angle ϕ of the direction of arrival of the satellite signal and its frequency f. Note that in this model there are two separate sets of coordinate systems—the first whose origin is at the ARP and a second system, in spherical coordinates, whose origin is located at the mean phase center of the antenna. This antenna model for the total phase correction from both PCO and is represented by the following equations:

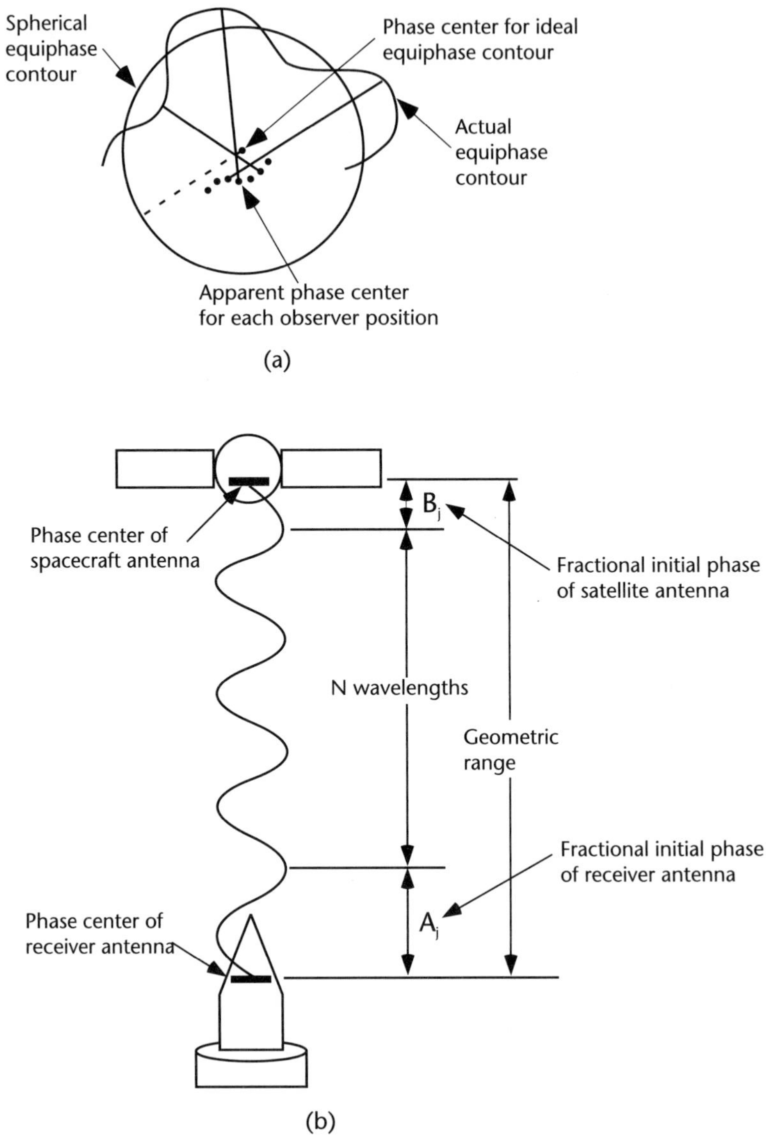

Figure 1.14 Phase variation in a GNSS antenna.

$$\Delta r(\theta, \varphi) = \Delta_{PCO} + \Delta_{PCV} = \vec{a} \cdot \vec{r}_0 + d\tau(\theta, \varphi, f) \cdot \left(\frac{\lambda}{360^\circ}\right) \tag{1.22}$$

where the PCO is defined as

$$\vec{a} = \hat{x}(a_x) + \hat{y}(a_y) + \hat{z}(a_z) \tag{1.23}$$

In the above equation $\Delta r(\theta, \phi)$ = total correction for phase pseudorange in the direction of the satellite; θ, ϕ are the elevation angle and azimuth angle of the direction of arrival of the satellite signal; λ is the wavelength of the satellite signal and $\vec{r}_0$ is the unit vector between the satellite and the mean electromagnetic phase center of the antenna, $\vec{a}$ is the vector direction of the PCO between the ARP and the

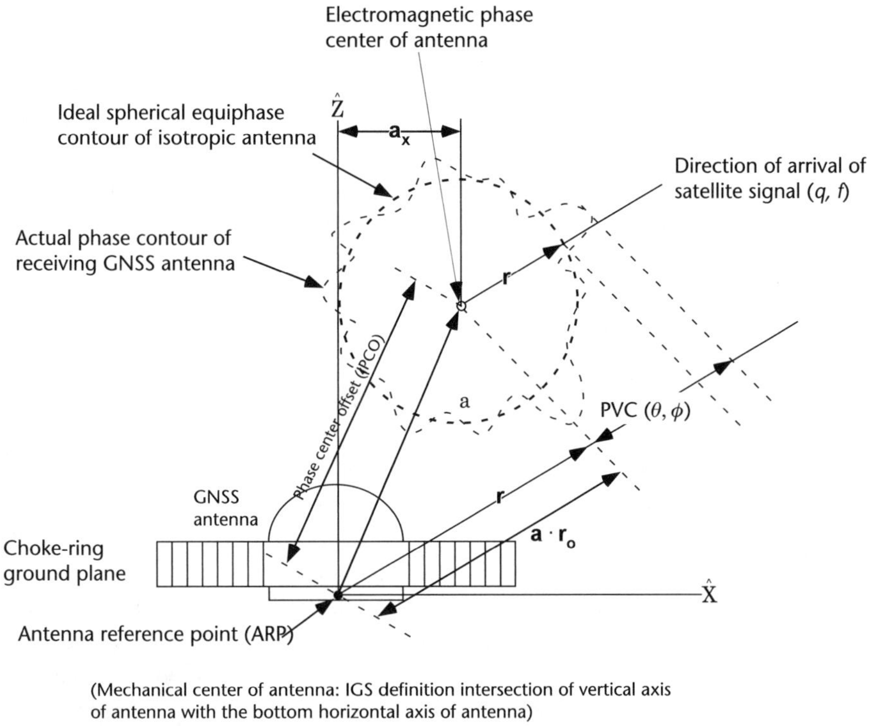

Figure 1.15 Diagram of an antenna model proposed by Gorres, et al. [25] for determining phase center offset (PCO) and phase center variations (PCV).

mean electromagnetic phase center and $d\tau(\theta, \phi)$ is the phase center variation (PCV) in degrees which is a function of the elevation and azimuth angles of the satellite signal. Note that although a simple Cartesian x, y ,z coordinate system has been used in this model, it is more customary to use other local level systems(LLS) such as the East, North, Up (ENU)) coordinate system for antenna calibration that was shown in Figure 1.11.

GNSS users who use carrier-phase observables for very high precision position measurements for applications such as surveying, geodesy, and mapping/ GIS in applications will experience errors if the phase center variations (PCV) of the antenna is not accounted for properly. These applications use differential GPS (CDGPS) or real-time kinematic (RTK) techniques. In these applications the user measures the phase of the GPS carrier phase relative to the carrier phase at a reference site thus achieving precision in range measurements that are a few percent of the carrier wavelength—typically less than 1 centimeter or even a few millimeters. As explained earlier ephemeris, ionospheric and tropospheric errors in these differential GNSS measurements can be eliminated by single and double differencing techniques; satellite clock errors also cancel out by differencing between receivers and satellite clock errors can be removed by differencing. These differencing techniques will be explained in greater detail later in this chapter.

Errors caused by antenna PCO and PCV can be eliminated or minimized in RTK measurements by using antennas and receivers of the same make and model at both ends of a baseline and also by orienting both antennas towards the same direction—generally the magnetic north. The vertical height of a specific point on the exterior of the antennas above the geodetic markers needs to be measured so that appropriate corrections can be made. The reference and rover receivers in these difference measurements need to be either of the same type of antenna /receiver or accurate antenna calibration methods, also described in Chapter 6 need to be used to remove these errors through software such as RINEX [1, p. 48]. Seeber has indicated that by neglecting PCV corrections, which can only be determined through accurate antenna calibration, one can introduce position errors of up to 3 cm. and height errors of up to 5 cm [28]. The fundamental premise upon which differencing is based is that errors that can be eliminated by the operation of differencing must be similar and systematic—that is, observations that need to be differenced only errors having the same sign and magnitude will cancel. If such errors are not identical, differencing the observations will leave some residual component of that error source—or the difference between the errors. Antenna PCV and PCO change with elevation, azimuth, and frequency and can differ from antenna to antenna. Alemu [29] has conducted an analysis on the effects of mixing different types of antennas and receivers at opposite ends of a baseline in precise geodetic measurements. However if a mixture of antennas of different make or model are used in a baseline or in a network, then a data processing software such as RINEX (Receiver Independent Exchange) will need to know the relative positions of the phase centers of these antennas so that appropriate correction obtained through calibration can be applied to compensate for these errors. The calibration procedures are discussed in greater detail in Chapter 6.

1.5.1 Carrier Phase Windup

The GPS carrier wave transmitted by GPS satellites is right-hand circularly polarized (RHCP); the GPS receiver antenna is also RHCP to allow optimum reception. The receiver processes the carrier phase and the pseudorandom code from the transmitted signal and derives the range between the transmitter and the receiver. The electric field vector associated with this transmitted electromagnetic wave can be visualized as a rotating vector that rotates 360° for one spatial wavelength or every temporal cycle. The carrier phase measurement conducted by the receiver is actually the measurement of the difference between the phase of incoming electromagnetic wave and the phase of a either an L_1 or a L_2 reference signal generated in the receiver. Therefore any relative rotational motion between the transmitter and receiver antennas will have an influence the carrier phase measurement and result in a spurious variation of the measured range even though the relative range distance between the antennas has not changed. This effect is known as carrier phase windup [30, 31, 32]. A GNSS antenna mounted on a rapidly rotating missile [31, 33], as shown in Figure 1.16, can experience such phase windup effects. A full circle of relative rotation between the satellite and receiver antennas can generate a range error of one wavelength or an error of 19.029 cm at L_1 and 24.21 cm at L_2.

A simple test to demonstrate carrier phase wind up effects in GPS has been conducted by Muellerschoen of JPL and has been described by Leick [30]. Two RHCP antennas—one for transmitting and the other for receiving were mounted opposite each other 5 meters apart. Both antennas were connected to the same transmitter and receiver so that phase and change in relative range (i.e., height in this case) could be determined once per second for half an hour. One of the antennas was

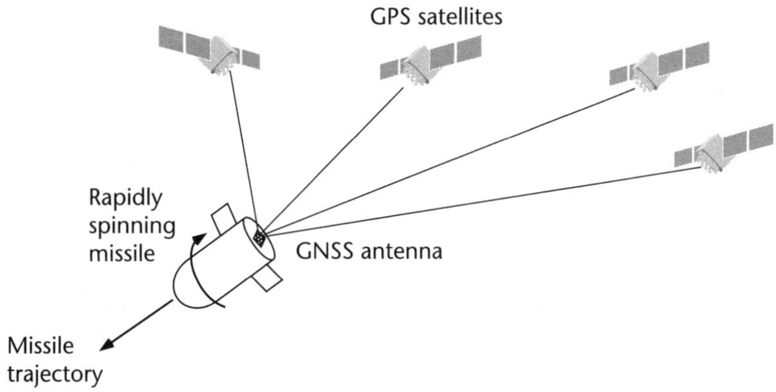

Figure 1.16 Carrier phase wind-up experienced by a GNSS antenna on a rapidly spinning missile.

rotated 360° in azimuth four times clockwise with 1 minute between rotations, as shown in the left portion of Figure 1.17.

It was then rotated four times counterclockwise again with 1 minute between rotations as shown in the right side. Measurements were made in both L_1 and L_2. Results shown in this figure indicate that each complete rotation in azimuth causes a change in relative height of one wavelength. Since the phase is measured in the plane of the receiving antenna, the phase of the received carrier signal is also dependent on the direction of the line of sight to the satellite antenna, in addition to the relative rotation between the antennas. Carrier phase windup can be a problem for GNSS antennas mounted either at the base or on the circumference of a rapidly rotating platform, such as artillery shells or rockets that use spinning rotation for attitude stabilization. GPS carrier phase wind up problems have been pointed out by Tetewsky and Mullen [31] and also by Garcia/- Fernandez, Markgraf and Motenbruck [33] who have used the windup contribution to compute the spin rate

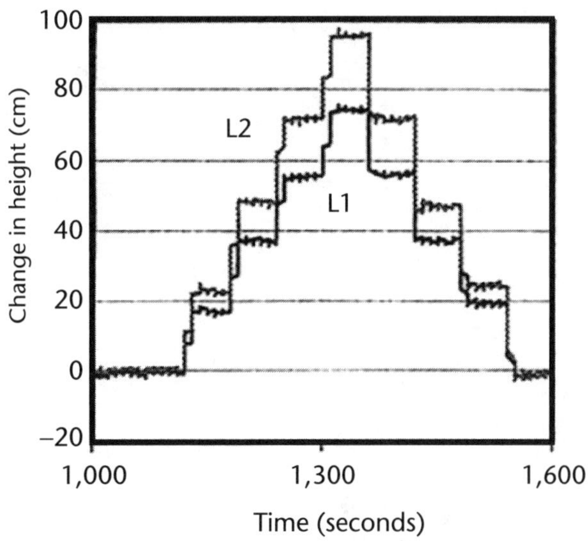

Figure 1.17 Results from carrier phase wind-up experiments conducted by R. Muellerschoen (courtesy of R. Muellerschoen and NASA, JPL/Caltech, [30]).

of these rapidly rotating platforms by using an appropriate combination of GPS observables. The accumulated wind up in these platforms with high rotation rates can actually be larger than even from variations due to the ionosphere or from receiver noise.

Wu et al. [34] has derived a model for estimating the carrier phase windup by representing the RHCP polarized fields of the transmitting and receiving antennas in terms of "effective dipole vectors." The RHCP transmit and receiving antennas are represented as "crossed dipoles." The analysis by Wu takes into account the direction of propagation relative to the orientation of unit vectors representing the crossed dipoles as shown in Figure 1.18. The windup correction is calculated by taking the vector dot product between these effective dipole vectors.

The receiver RHCP antenna is represented as a crossed dipole with $\hat{x}$ and $\hat{y}$ denoting the orthogonal unit vectors along the elements of the crossed dipoles. The signal from $\hat{y}$ is delayed n phase by 90° relative to the $\hat{x}$ dipole. $\hat{k}$ is the unit vector pointing from the satellite transmit antenna to the receiving antenna. The satellite antenna can be represented similarly by the unit vectors $\hat{x}'$ and $\hat{y}'$. The signal from the $\hat{y}'$ arm of the dipole lags that of the signal from the $\hat{x}'$ arm of the dipole by 90°. The "effective dipole vector" $\hat{d}$ representing the receiving antenna is

$$\hat{d} = \hat{x} - \hat{k}\left(\hat{k} \cdot \hat{x}\right) + \hat{k} \times \hat{y}$$

Similarly the effective dipole vector representing the transmit antenna is

$$\hat{d}' = \hat{x}' - \hat{k}\left(\hat{k} \cdot \hat{x}\right) - \hat{k} \times \hat{y}'$$

The carrier phase windup correction is determined by the angle between these two effective dipole vectors and its past history

$$\Delta\Phi = 2N\pi + \Delta\varphi$$

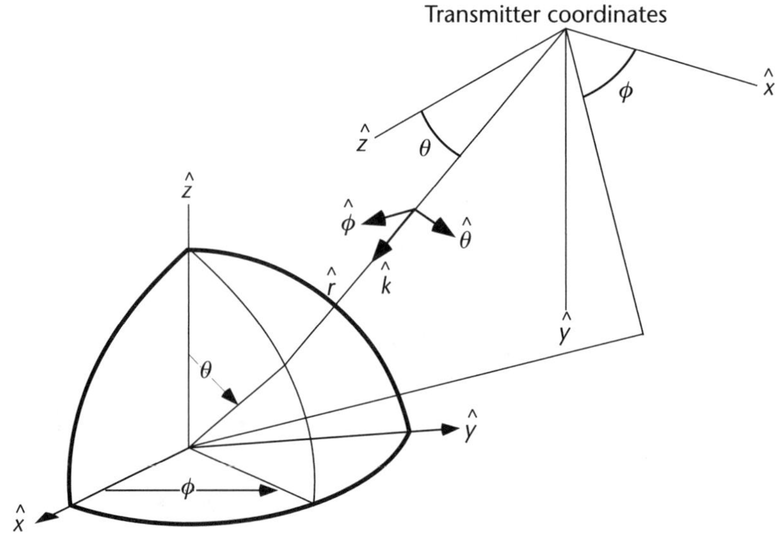

Figure 1.18 Diagram for illustrating carrier wind-up analysis conducted by Wu [34].

Where $\Delta\phi$ is the fractional part of a cycle given by

$$\Delta\varphi = sign(\zeta)\cos^{-1}\left[\frac{\hat{d}'\cdot\hat{d}}{\|\hat{d}'\|\cdot\|\hat{d}\|}\right] \tag{1.24}$$

And

$$\zeta = \hat{k}\cdot\left(\hat{d}'\times\hat{d}\right) \tag{1.25}$$

and N is an integer.

Kim, Serrano and Langley [35] have conducted a study on the effects of carrier phase windup on RTK-based vehicle navigation systems and observed that it can cause degradation in system performance under certain conditions. Kim et al. [36] have also developed single and double differencing techniques for removing or at least reducing these errors in RTK. Bisnath [32] has indicated that in a majority relative positioning applications where differencing techniques are used, a bulk of these phase windup effects may cancel out since effects observed by both receiving antennas (i.e., the rover and the reference antennas) are very similar when the base line distances are small. However for very long baselines, in the vicinity of thousands of kilometers, the transmitter and receiver geometry can no longer be considered to be similar and phase windup errors can increase to as large as several centimeters.

1.6 Group Delay Variation in GNSS Antennas

In (1.1) the total far field radiation of a GNSS antenna was expressed in terms of the principal, right-hand circular polarization (RHCP) component and the cross-polarized, left-hand circular polarization (LHCP) component.

The group delay of the GNSS antenna is defined in terms of the phase behavior of the RHCP component, which can be expressed in its complex form as:

$$\vec{E}_R = [\hat{e}_R]|E_R(\theta,\varphi,f)|\angle\Phi(\theta,\varphi,f) \tag{1.26}$$

where $\Phi(\theta, \phi, f)$ expressed in degrees represents the angular part of the phase pattern. The group delay is a measure of the time delay experienced by a narrowband signal packet and is defined as [37]

$$\delta\tau(\theta,\varphi,f) = \frac{1}{360}\frac{\partial\Phi(\theta,\varphi,f)}{\partial f}\text{ in seconds} \tag{1.27}$$

The group delay can also be expressed in terms of distance (meters) by multiplying it by the velocity of light "c". If the phase of the antenna does not change with frequency its group delay is zero. But in reality the group delay does change with frequency and also with elevation and azimuth that determine the aspect angle of the signal received by the antenna from the satellite.

Active GNSS antennas can introduce large group delays because of the frequency response of active and passive components used in conjunction with the passive antenna. These components can include the low noise amplifier (LNA), pre-select and post-select bandpass filters which may be needed in multiband GNSS antenna to separate out the different frequency bands and also to limit mutual interference and intermodulation between the bands, and the diplexers. Even passive microstrip patch antennas can introduce group delays because they are very high Q circuits in narrow bandwidths and also because of the feed probes and polarization networks used for generating RHCP [38].

The group delay introduced by an antenna can also affect and distort the spectral quality of certain binary off-set carrier (BOC) Modulations used in current GNSS systems [39]. The dispersive effects of group delay, particularly the quadratic component of the variation in frequency, can raise the sidelobes of the correlation function but does not appear to effects the peak value of the main lobe [39].

There are two group delay parameters that are used in the evaluation of the performance of a GNSS antenna: the first deals with spectral variation of group delay (i.e., with frequency) within the operating bandwidth of a specific GNSS band; the second deals with variation of group delay with the aspect angle (i.e., elevation and azimuth angles) of the arriving GNSS signal.

The specified limits for both of these and also the methods used for their measurement are discussed in Chapter 6.

1.6.1 Requirements for Group Delay Variation with Frequency

The boresight differential group delay (BDGD) is defined individually for each of the GNSS frequency bands that are listed in Chapter 3.

$$\Delta T_B = Max\left\|\delta\tau(\theta_B,\varphi_b,f) - \delta\tau(\theta_B,\varphi_B,f)\right\| \tag{1.28}$$

where θ_B, f_B denote the direction of boresight; f_i and f_J are any two frequencies within the individual bandwidth of the GNSS bands on either side of the center frequency of the selected GNSS band. For example if the operating bandwidth of the GPS L_1 band and the center frequency is denoted by f_C, then f_i and f_J are any two frequencies such that:

$$f_{C-10.23\,MHz} \le f_i, f_J \le f_{C+10.23\,MHz}$$

For a GNSS antenna: $\theta_B = 0$; it can be averaged over several azimuth angles of ϕ at this fixed elevation angle if significant variation with an azimuth rotation is noticed.

The BDGD represents the combined effects of both the passive antenna as well as the built-in preamplifier and also includes other embedded filtering functions. The BDGD in any of the GNSS bands is expected to be less than 25 nanoseconds and applies equally to both passive as well as active antennas. For example the new MOPS issued by RTCA requires that the group delay difference between any two frequencies within the operating bandwidth of ±10.23 MHz to be less than 25 nanoseconds [40, 41]. This new requirement is required to provide measurement consistency between satellite signals that may have different amplitude and phase characteristics. In the case of a simple passive antenna without any embedded additional filtering BDGD is expected to be less than 2 nanoseconds. The 25 nanoseconds specification is meant to be small relative to variations expected in the rest of the RF chain within a GNSS receiver where the maximum tolerable differential group delay variation for airborne equipment for example can be as large as 150 to 600 nanoseconds [40, 41]

1.6.2 Group Delay Variation with Aspect Angle

The differential group delay (DGA) versus angle specifies a requirement on the maximum variation in group delay as a function of the azimuth and elevation angle of the satellite signal arriving at the receiving GNSS antenna; it is specified as:

$$\Delta T_A = \max \left\| \delta\tau(\theta, \varphi_A, f_C) - \overline{\delta\tau(\theta = 85°, \varphi_A, f_C)} \right\| \tag{1.29}$$

Where f_C = Center frequency of the GNSS band.

ϕ_A = Azimuth angle of the arriving satellite signal at the receiving antenna in the antenna reference frame.

$\theta = 85° = \textit{Elevation Angle} = 5°$ is the elevation angle of the arriving satellite signal at the antenna in the antenna reference frame. $\overline{\delta\tau(\theta = 85°, \varphi_A, f_C)}$ is the group delay averaged over all azimuth angle of the pattern at $\theta = 85° = \textit{Elevation Angle} = 5°$.

The requirements are that the DGA calculated for each GNSS band should not exceed the limit

$$\Delta T_A \leq 3 \text{ nanoseconds}$$

The requirement for maximum group delay set by the RTCA for the maximum DGA [40, 41] is:

$$\Delta T_A = 2.5 - 0.04625(\textit{Elevation} - 5°) \text{in nanoseconds for } 5° \leq \textit{Elevation} \leq 45°$$

and

$$\Delta T_A = 0.65 \text{ nanoseconds for Elevation} \geq 45°$$

Active GNSS antennas can introduce large group delays because of the frequency response of active components and also some passive components used in conjunction with the passive antenna. These components include the low noise amplifier, the pre-select and post-select bandpass filters used to limit the overall noise and the diplexer in dual band antennas. Even passive antennas such as microstrip patch antennas can contribute to phase delay because of the types of feeds used to generate right-hand circular polarization including hybrid networks. The measured variation of group delay versus frequency for two commercial avionics quality antennas over 1565 to 1585 MHz (discussed in Chapter 6) indicates that the group delay is well within the specified 25 nanoseconds in this portion of the band although it increases significantly outside this range of frequency [40, 41].

1.7 Propagation and Multipath Errors in GNSS Measurements

The measurement of pseudorange and carrier phase—the two most important observables obtained from GNSS measurements—are affected by several errors such as propagation errors experienced by the signal as it travels from the satellite to the receiver antenna. The largest contribution to such errors comes during propagation through the troposphere and the ionosphere as illustrated in Figure 1.19.

The ionosphere is dispersive and will delay the code measurements (i.e., pseudorange) while advancing the signal phase by the same amount (i.e., the carrier range obtained from phase measurements). The GPS signal in addition can experience multipath, also shown in Figure 1.19, where both the direct signal and one or more reflected signals arrive at the antenna causing errors in phase and amplitude. There are additional errors due to variety of other sources such as ephemeris errors, and errors contributed by the clocks in both the satellite and in the receiver. All of these errors have been well described by Leick [30] and

Parkinson and Spilker [15, Chapters 12–14] Many of these errors, if not all, can be either removed or at least reduced by differencing techniques used in DGPS as explained by Abidin [42] and Leick [30]. Multipath errors present more of a problem as it is caused by the environment that is unique to the reference and user antenna in DGPS; hence it can affect each antenna differently. Hence every effort should be made in picking a location for each of these two antennas that avoids multipath if that is possible or an effort to reducing multipath errors through suitable receiver software may need to be used. Since the strongest multipath signals generally arrive in directions close to the horizon or below it, the antenna needs to be designed to reduce backlobes. Antennas using choke ring, EBG or resistivity tapered ground planes help in the reduction of such backlobes; they are good candidates for use as Reference antennas which are stationary. These antennas with good MPR (multi-path ratio) will be discussed in Chapter 3 of this book. But they are too large, heavy, and expensive for use in most rover-type receivers.

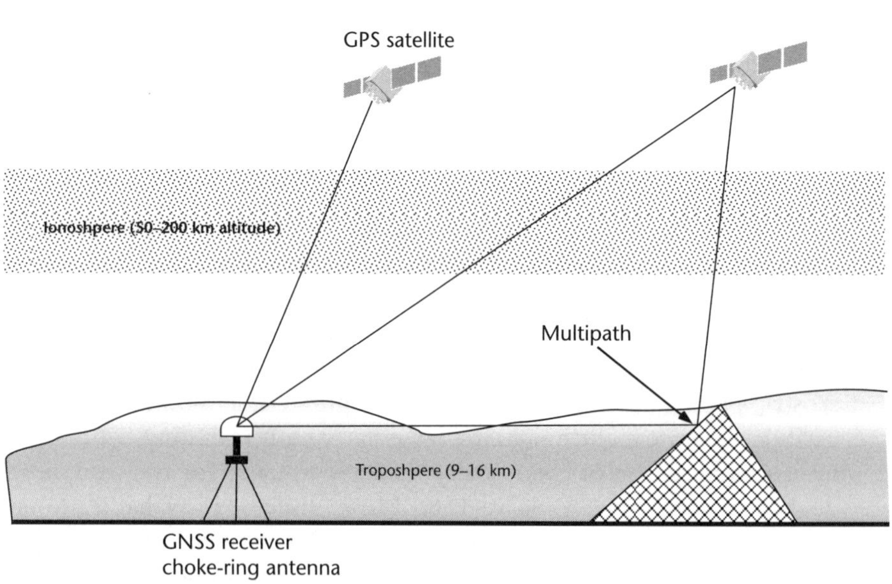

Figure 1.19 Propagation errors, multipath errors, and satellite and receiver clock errors.

1.8 Antenna-Induced Errors in GNSS Measurements

In addition to all of these are errors contributed by the antenna—primarily errors introduced by the group delay in the antenna which affects the pseudorange measurements and phase errors which affect carrier phase measurements [30, 27]. The latter become important when precise positioning is needed such as in surveying, geodesy, mapping, aircraft landing systems, precise positioning service (PPS) or real time kinematic (RTK). If the group delay varies with the observation angle the delays will be different for the signals received from different satellites causing other positional errors.

1.8.1 Group Delay Error

$\delta\tau_g$ is the time delay expressed in seconds that is introduced by the group delay in the antenna. It is given by

$$\delta\tau_g(\theta, \varphi, f) = \frac{1}{360}\frac{\partial\Phi(\theta, \varphi, f)}{\partial f} \tag{1.30}$$

where f is the frequency and Φ is the phase. It can be translated into distance (range) error "A_G" by multiplying it by the velocity of light c.

$A_G = c\delta\tau_g$ (meters). Since there are several other errors that enter into pseudorange and carrier phase measurements which are discussed next, it is best to differentiate the errors induced by the antenna by the letter "A" denoting an antenna with a prefix of either "G" or "P" to denote whether the error is caused by a group delay or a phase delay. A_G (meters) is an antenna induced error that needs to be added to the pseudorange observable when making range measurements using the code delay measurements. It is a function of the frequency as well as elevation and azimuth angles.

1.8.2 Phase Center Error

Similarly, as explained earlier, the phase variation in the antenna caused by both the PCO (phase center offset) and the PCV (phase center variation) should also be translated into distance and added as an error in range that is induced by the antenna in the carrier phase observable.

$$A_P(\omega, \theta, \varphi) = \Delta_{PCO} + \Delta_{PCV} = \vec{a} \cdot \vec{r}_0 + \frac{\lambda}{360}\delta\tau(\theta, \varphi, f) \text{ (meters)} \tag{1.31}$$

to the carrier phase observable when range is measured using carrier phase measurements.

1.8.3 Carrier Phase Windup Error

When the platform on which the GNSS antenna is rotating rapidly, such as for example an antenna mounted on a spinning missile, as shown in Figure 1.15, there is an additional error caused by carrier phase windup, described in an earlier section, that needs to be accounted for in carrier phase evaluations. This is known as the carrier phase wind-up error or sometimes as the carrier phase wrap-up error. The GNSS antenna which is circularly polarized has a phase pattern that is a function of the azimuth angle. Such an increase in phase can occur when the vehicle on which the antenna is mounted is rotating; this was discussed in greater detail earlier in this chapter. The phase change can also due to the rotation of the satellite around a stationary antenna due to the satellite orbit. The measured phase will increase by exactly 360 degrees for one rotation in the azimuth plane. This phase change needs to be accounted for to prevent errors in measurements that involve the carrier phase. It can be added as an error A_w in the observed phase measurement and is given by:

$$A_w(w,\theta,\phi) = \frac{\lambda \Phi_w(w)}{360} \text{ meters} \tag{1.32}$$

where Φ_w is the phase change in degrees caused by the carrier wrap due to the antenna rotation [30-36]. Carrier phase windup does not affect pseudorange measurements using code delay; it only appears as an error in the carrier phase observables.

While discussing the removal of these antenna induced errors through differencing techniques, it is best to separate the carrier wrap error caused by antenna rotations from the other two errors caused by group and phase delay since carrier wrap needs a different method for its determination and elimination.

1.9 Differencing Techniques for Removal of Bias Errors in GNSS Measurements

As explained in these previous sections the principal sources of errors in GNSS measurements are:

- Errors introduced by propagation effects, primarily tropospheric and ionospheric errors
- Multipath errors
- Antenna induced errors
- Satellite and receiver clock errors

The most effective method for either eliminating or reducing these errors is by differencing observations made to several satellites from two or more receivers located in a relative positioning mode at the same time. Differencing of such GNSS data can be carried out between different receivers, between different satellites or between different epochs as illustrated in Figure 1.20. It should be noted that differencing between receivers or between satellites are made at the same observation epoch. Single, double, and even triple differencing techniques have been employed to remove various bias errors. These have been well described by Abidin [42] and by Leick [30]. In

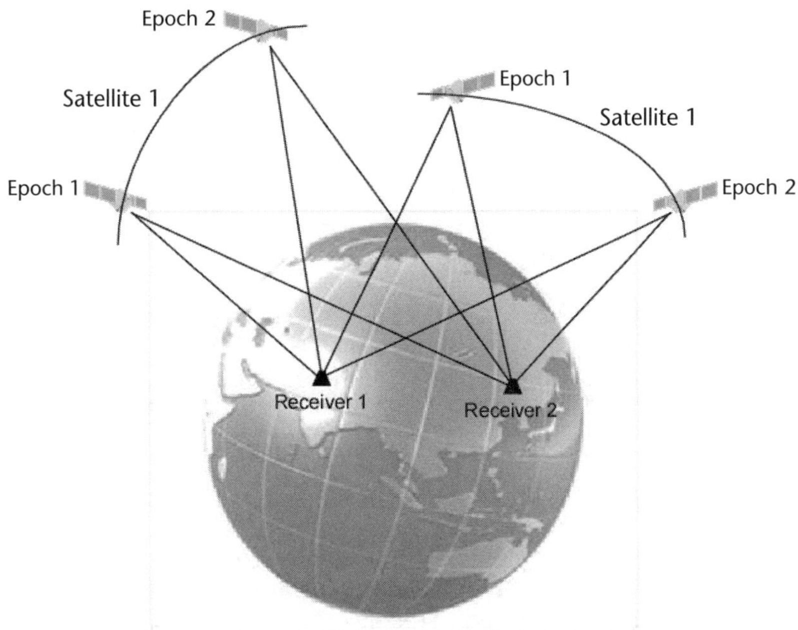

Figure 1.20 Various GNSS differencing models—between receivers, between satellites, and between epochs.

this book we shall concentrate primarily on the removal of antenna-induced errors at GPS frequencies using single differencing between two receivers using either data collected in the GPS L_1 and L_2 bands; This is illustrated in Figure 1.20.

Double differencing techniques have also been used to estimate and eliminate carrier phase windup errors by Kim et al. [35, 36].

1.10 Differencing of GPS Data for Removal of Antenna-Induced Carrier Phase Errors

The pseudorange and carrier phase measurements obtained from one-way (OW) measurements can be differenced using appropriate software to create new data combinations that are either free from many errors or at least have the errors vastly reduced if they cannot be completely eliminated completely [42,30]. Generally the penalty to be paid by this differencing are increased noise levels which can increase by a factor of $\sqrt{2}$.

The two receivers whose signals are being differenced are stationed at some distance at the opposite ends of a baseline as shown earlier in Figure 1.21.

Antenna induced carrier phase errors can be most easily removed by just using "single difference" techniques on the data gathered from two similar antenna / receivers combinations stationed at opposite ends of a baseline.

The following equations represent the pseudorange P_I^J and carrier phase $L_I^J = \lambda_Q \cdot \varphi_I^J$ when both of these GNSS data observables are expressed in units of meters [42, 30, 33]. The superscript J in this equation identifies the satellite whose signals are being processed by the receiver in these measurements and the subscript "I" identifies the receiver. Subscript "q" identifies the carrier frequency used in

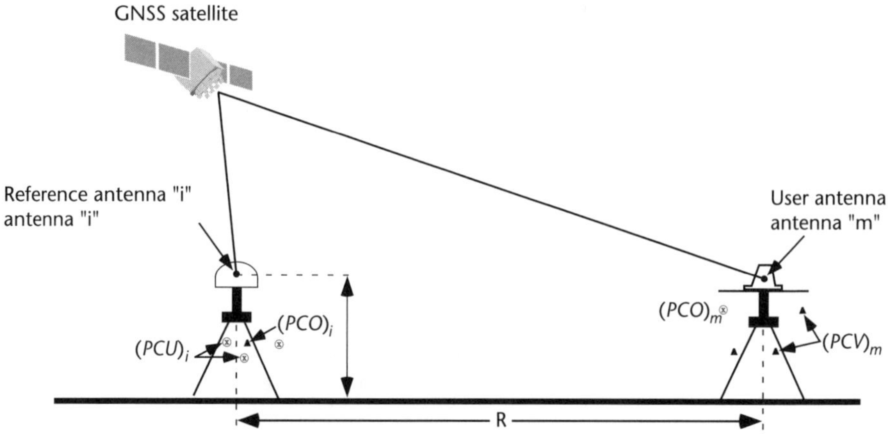

Figure 1.21 Single differencing techniques between two antennas "i" and "m".

the measurements and can be either the GPS L_1 or the GPS L_2 band. Sometimes a combination of both L_1 and L_2 is used for special applications such as estimation of carrier wind up explained in the next section. For the purpose of this discussion we shall only consider the L_1 band.

$$P_I^J = \rho_I^J + d_{tropI}^J + D_{IonqI} + c.\left(dt_I - dt^J\right) + mP_I^J + \varepsilon P_I^J + A_{GI}^J \tag{1.33}$$

$$L_I^J = \lambda_q \cdot \Phi_I^J = \rho_I^J + d_{tropI}^J - d_{IonqI}^J + c\left(dt_I - dt^J\right) + \lambda_q N_{qI}^J + mC_I^J + \varepsilon C_I^J + A_{PqI}^J \tag{1.34}$$

The terms appearing in the above equation are defined as follows:
Pseudorange at frequency f_q in meters = P_I^J
Carrier phase at frequency f_q in meters = $L_I^J = \lambda_q \cdot \Phi_I^J$
Carrier phase at frequency f_q in cycles = Φ_I^J
Carrier wavelength at frequency f_q in meters = λ_q
Geometric range between receiver and satellite = ρ_I^J
Range bias effect caused by tropospheric delay = d_{tropI}^J
Range bias effect caused by ionospheric delay = $d_{Ion\,q\,I}$
Clock offsets for the receiver and satellite clocks respectively = dt_I and dt^J
Velocity of electromagnetic radiation in free-space = c
Multipath error in the P_I^J and L_I^J observables respectively = mp_I^J and mC_I^J
Cycle ambiguities of the L_1 signal (integer cycles) = N_{ql}^J
Noise of the P_I^J and L_I^J observables = εp_I^J and εC_I^J

Let the reference receiver and the user receiver located at opposite ends of the baseline be denoted as receiver "i" and "m" respectively. The mathematical equation for the single differencing data between these two receivers for the same satellite can be expressed as:

$$\begin{aligned} &\lambda_q\left(\Phi_i^J - \Phi_m^J\right) = \\ &\left(\rho_i^J - \rho_m^J\right) + \left(d_{tropi}^J - d_{tropm}^J\right) - \left(d_{ionqi}^J - d_{ionqj}^J\right) + c\left(dt_i - dt_m\right) + \\ &\lambda\left(N_i^J - N_m^J\right) + \left(mc_i^J - mc_m^J\right) + \left(\varepsilon C_m^J - \varepsilon C_i^J\right) + \left(A_{pqi}^J - A_{pqm}^J\right) \end{aligned} \tag{1.35}$$

The last term on the right-hand side of this equation is the difference between the antenna induced carrier phase errors of the reference receiver "i" and the user receiver "m" at a carrier frequency f_q . If both of these receivers use the same antenna and receiver then the antenna induced phase error would cancel out. For this condition to apply $(PCO)_i = (PCO)_m$ and also $(PCV_i) = (PCV)_m$ (i.e., their phase center offsets and phase center variations need to be identical. If the antenna / receivers are a "mixed pair" with different PCO and PCV then each antenna/ receiver needs to be calibrated accurately as described later in Chapter 6 of this book. A large number of geodetic antennas have been calibrated by NGS and information on these calibrations is available from http://www.ngs.noaa.gov/ANTCAL/. This data can be obtained from RINEX files [43]. However, it is only available for GPS frequencies. Antenna calibration at other GNSS frequency bands has only been done recently and will also be discussed in Chapter 6; such data is still very limited. The errors resulting from using mixed pair of antenna/receivers has been discussed by Alemu [29]. Seeber has indicated [28] that by neglecting PCV corrections one can introduce position errors of up to 3 cm and height errors of up to 5 cm. To reduce or eliminate these errors in such cases an accurate calibration of both the phase center offset (PCO) and phase center variation (PCV) of both GNSS antenna/ receiver is essential in high precision applications so that positions determined from GNSS measurements can be related to physical antenna mounts. Since the phase center of the antenna changes with the azimuth angle and zenith angle (which is 90° minus the elevation angle) of the arriving satellite signal and also with its frequency, its calibration becomes especially important when different antennas and receivers are used in RTK differential measurements and in continuously operating permanent reference station networks [2]. Smaller errors also occur when two antennas that are used at opposite ends of a baseline are oriented at other than at their previously calibrated directions. Antenna calibration is also important when estimating the effects of the troposphere on GNSS measurements since both the PCV and delay from the troposphere depend on the elevation angle. PCO and PCV must be used together as a set since this will allow different sets to lead back to the same ARP (antenna reference point). Generally the largest off set occurs in height with variations as large as 10 centimeters in the vertical plane and only about 3 centimeters in the horizontal plane; in the horizontal plane the variation in elevation can be larger than in azimuth.

1.11 Differential GPS and RTK Measurements

Differential GPS (DGPS) was developed mainly to meet the high-precision positioning and distance measuring requirements of the surveying and geodesy uses a reference receiver (RR) which is located at a known location (a "monumented point") that has been previously surveyed, and one or more either stationary user receivers (UR) or a mobile receiver—the real time kinematic (RTK)—as illustrated in Figure 1.22

The reference receiver at the known surveyed location continuously tracks all visible satellites to measure systematic GPS errors.

The errors that are calculated from such spatial correlation will then communicated to the various UR antennas which can then make the necessary corrections to

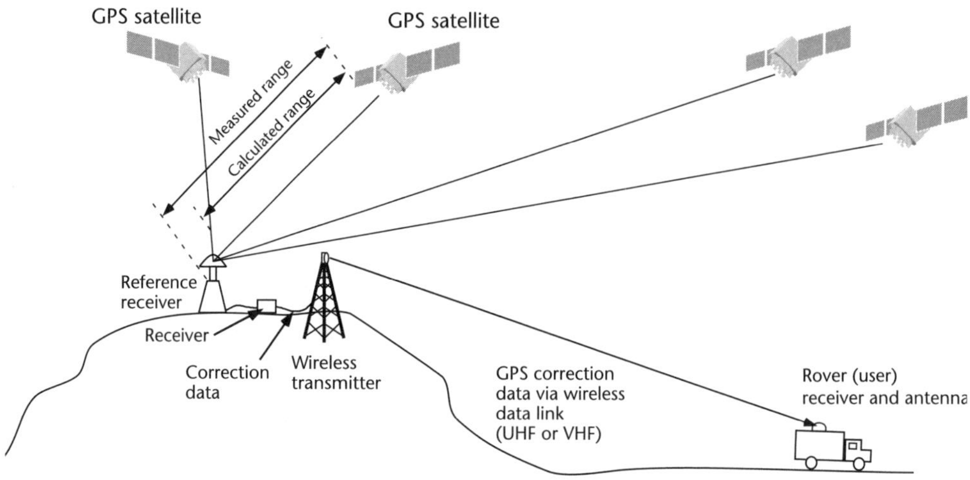

Figure 1.22 Real time kinematic (RTK) differential GPS technique.

remove these errors. This correction data is transmitted to the users via a wireless data link. The separation between the RR and UR antennas should not be too long to ensure that the elevation angles to the satellites and other atmospheric propagation errors are similar so that the errors cancel. Position precision of typically in the centimeter level can be achieved. These common errors include signal path delays through the atmosphere, the satellite clock and ephemeris errors.

RTK is an attractive alternative that was developed to promote more rapid surveying; the RR receiver transmits carrier-phase data to the user receivers permitting high accuracy, real-time positioning even when the user is moving. The software within the user receiver is able to resolve ambiguities in the shortest time possible, using on the fly-ambiguity resolution (OTF-AR) algorithms and the resulting carrier-range data is then used to derive centimeter level positioning accuracy. This technique is highly beneficial in field surveying mining, and agricultural machinery. The lengths of the baseline used in RTK are generally 5–10 kilometers; these baselines need to be small since the degree of cancellation of atmospheric errors depends on the length of the baseline. A large increase in the length of the baseline can de-correlate these errors, although these lengths have been extended to 20–30 kilometers more recently.

1.12 Techniques for Removal and Estimation of Carrier Phase Windup Errors Caused by a Rotating Antenna

It has been indicated previously [33] that carrier phase windup error is independent of the elevation angle of the satellite for pure rotation about the antenna boresight axis; it is usually absorbed by the clock error in the navigation algorithms. Therefore the impact of carrier phase windup is neglected for applications that do not require accuracies to the centimeter level. The cumulative nature of these effects can become important while making precise GNSS measurements using antennas mounted on rotating platforms, such as rapidly spinning rockets or missiles, which rely on the spinning rotation for attitude stabilization [33,31]; the magnitude of these carrier windup errors become important in such cases when used to predict their impact point of the rocket. Garcia–Fernandez et al. [33] have indicated that for antennas mounted on receiving platforms with high rotation rates, the accumulated wind-up error can actually be higher than errors introduced by the receiver noise or even ionospheric variations and need to be accounted for in precision measurement. Phase wrap-up effects can also become important in RTK for antennas mounted on vehicles prone to rotation or turning. Kim et al. [36] have investigated the effects of carrier wrap in RTK navigational systems. Carrier phase windup may also be important for systems which rely on GNSS navigations systems for attitude determination.

Several authors have investigated the use of single or double differenced measurements to estimate and remove the errors introduced by carrier phase windup in rotating platforms.

Kim et al. [35, 36] have considered the use of single and double differencing techniques for both estimation and removal of these errors. The first step in their recommended process is the high pass filtering of the single difference carrier phase measurements to remove or reduce constant error components such as integer ambiguity and also low frequency components such as tropospheric, ionospheric errors, and the multipath and phase center variation effects in the equations described previously in the previous section. The second step is generating geometry free combinations of L_1 and L_2 as such frequency combinations can further reduce residual low frequency error components from the tropospheric delay and receiver clock bias left over after high pass filtering. Kim et al. [35, 36] have shown that the carrier phase windup error can be obtained by the integration of this geometry free carrier phase component. A full description of this method and the results achieved are beyond the scope of this book and details can be found in their original paper.

Garcia-Fernadez et al. [33] have developed a method of using the results of the GPS L_1 and L_2 code and carrier phase variables to estimate the spin or rotation angle and also the spin-rate of a rapidly rotating rocket. Since the carrier phase windup affects the code and carrier phase variables differently and causes different range contributions depending on the wavelength, its value can be estimated by forming code-carrier divergence or by using L_1/L_2 combinations if dual frequency tracking data is available.

1.13 The Susceptibility of a Commercial GPS C/A Code Receiver to Interference and Jamming

GPS signals are very weak due to their long travel distances from satellites 20,000 kilometers above the earth resulting in a space-path loss of 185 dB. They are spread spectrum signals that are more than 25 dB below the noise floor of the receiver. Tables 1.6 and 1.7 provides the minimum received signal power from satellites in the modernized GPS and the Galileo systems. The extremely low power of these GPS signals makes an unaided C/A code commercial GPS receiver extremely vulnerable to interference and jamming from a large number of sources including both unintentional RF interference (RFI) from telecommunications, handsets, and wireless

Table 1.6 Minimum Received Signal Power in Modernized GPS System

Signal Description	*Center Frequency (MHz)*	*Minimum Received Signal Power (dBW)*	*Bandwidth (MHz)*
L_1 C/A - Civilian	1575.42	−158.5	2
L_1 P(Y) - Military	1575.42	−161.5	20
L_2 P(Y) - Military	1227.40	−164.5	20
L2C - Civilian	1227.40	−160.0	2
L_5 - Civilian	1176.45	−154.9	20
L_1 M Code - Military	1575.42	−158.0	24
L_2 M Code - Military	1227.40	−158.0	24

Table 1.7 Minimum Received Signal Power in Galileo

Signal Description	*Frequency Band (MHz)*	*Minimum Received Signal Power (dBW)*
E5	1164-1215	−155
E6	1260-1300	−155
E2 – E1	1559-1592	−155

data transmission systems as well as hostile jamming. Some of these are listed in Table 1.8. these sources are generally ground based and hence much closer to the antenna with less $1/r^2$ path loss. Their power can often be several tens of decibels higher than the satellite signals and can easily overwhelm the GNSS signals. It will be demonstrated later in this section that a relatively low power of interference or jamming signal, greater than 1.5 picowatts, received by an FRPA antenna is sufficient to disable a GPS C/A code receiver. FRPA antennas used with GNSS receivers have very broad antenna patterns to allow reception of signals from low-elevation satellites; this feature also has drawbacks since it also allows reception of signals from jammers, interference, and multipath, which generally are located closer to the horizon.

The possibility of either inadvertent interference or hostile jamming are of particular concern in commercial GPS receivers that are used in safety-of-life applications such as precision approach and landing operations in avionics. The frequency proximity of the L_5 band to a large number of potential sources of RFI in the ARNS band such as DME, TACAN, and on-board as well as ground based VHF communications systems can be particularly troublesome in avionics [44]. In these systems the principal source of interference to GNSS antennas is from signals transmitted by antennas mounted on both the top and bottom surfaces of the aircraft. Signals transmitted by these antennas can “creep around” the surface of the aircraft through surface diffraction and arrive at a GNSS antenna mounted on the top surface of the aircraft. These effects are discussed in greater detail in Chapter 5 of this book.

Table 1.8 Sources of Interference & Jamming to GNSS

Source	*Frequency (MHz)*	*Effects on GNSS*
Radio Navigation on Earth	1215–1240	In-Band; GPS L_2
Exploration Satellites & Radar		Galileo E5b (1215 MHz)
Aeronautical Radio Navigation Service; DME, TACAN, Link 16	960–1164 1164–1215	In Band; GPS L_5 & Galileo E5A at 1176 MHz
Harmonics of VHF Comm. for ATC (Air Traffic Control)	118-137.5 MHz; 760 channels; 25 KHz bandwidth	12th. and 13th harmonics at GPS L_1 at 1575 MHz
UHF TV	782–788 MHz; 525 MHz	2nd and 3rd Harmonics in GPS L_1
Amateur Radio	220–225 MHz	7th Harmonic in GPS L_1
Jamming	Broadband, Partial Band, Swept Band, Narrowand	Receiver Tracking Loop Loses Lock From High Power Signal
Spoofing	All GNSS Bands	Deception by False Signals
WB - Impulse Radar	L Band > 25% Bandwidth	In-Band Interference From High Power Pulses
"LightSquared" Downlink Signals*	Two 10 MHz Wide Downlink_ Signal Channels: 1526-1536 MHz; and 1545.2–1555.2 MHz Uplink_Channel: 1626.5 – 1660.5 MHz	Out of band emissions from signals in the upper channel of the downlink band; blocking interference, Intermodulation
Civil "Car" Jammers	Frequency Swept Car Jammers, GPS L1 and Galileo E5	Effective at a range of 1 km. to 8 km. depending on power of jammer

*Permission to operate denied by the FCC in 2012.

1.13.1 LightSquared Transmissions

The latest and the most potent source of interference to GPS is from LightSquared [12] whose frequency plans indicate a downlink frequency range from 1525 – 1559 MHz and an uplink frequency range from 1626.5 to 1660.5 MHz. Their latest frequency plan indicates a 5 MHz wide frequency band between 1552.7 and 1528.8 MHz and a phase 2 10 MHz wide band between 1550.2 MHz and 1531MHz. This second phase 2 band is the closest to the GPS L_1 band and has raised concerns about potential interference from out-of-band emissions and intermodulation products [12].

1.13.2 Civil GPS Jammers

On September 30, 2011 the Federal Communications Commission (FCC) issued citation and order DA 11-1661 against 20 manufacturers for illegal marketing signal jamming devices that operate in the GPS, cellular and W-FI bands and pose significant risks to public safety [45]. Mitch et. al. [46] have conducted a study on the signal properties and effective jamming range of 18 of these devices. These test results show that almost all are swept tone jammers (or "chirp jammers")—all of which operated in the GPS L_1 and only six in the L_2 band and none in the L_5 band.

Tests were conducted to test the effective range of 4 jammers—the weaker group disrupted tracking at an effective range of 300 meters and acquisition at about 1 kilometer. The most powerful jammer disrupted tracking at an effective range of 6.1 kilometers and acquisition at an effective range of about 8.7 kilometers. Tests have also been conducted by Bauernfeind et al. [47] at the University FAF in Munich [46] on these widely available car jammers. Open field tests measured with an experimental software receiver enabled a detailed analysis from these looming threats to GPS. The test results showed that the majority of these low-cost in-car jammers transmit a chirp signal with a bandwidth between 9.4 to 44.9 MHz in the E1/ L1 bands. Some were sine wave oscillators with a 3dB bandwidth of around 0.92 KHz. While both types of jammers can be considered as narrowband devices the chirp jammers are more effective in jamming the signal within the GNSS receivers. Various static and dynamic tests were also conducted in the Galileo Test Range (GATE) in Berchtesgaden, Germany where the impact of the jammer signals on the reception of both GPS and Galileo RF signals could be evaluated in a realistic manner. These measurements show that currently available in-car jammers can degrade receive performance in a radius of about 1 kilometer around the interference source and disable position determination within a radius of 200 meters.

1.14 Architecture of a GNSS Receiver

We next need to evaluate the interference and jamming effects of these signals on the performance of a commercial GPS receiver that uses a non-adaptive, fixed reception pattern antenna (FRPA) type antenna discussed in subsequent chapters 2 and 4 of this book To conduct this analysis a basic understanding of the receiver architecture is necessary. Almost all of the GNSS receivers being used today are "digital"—this means that the signal processing is performed with sampled digital signals. The technology being used in GNSS receivers for such signal processing has changed dramatically during the last few years with highly sophisticated chip sets being built through application-specific integrated circuit (ASIC) technology. Good descriptions of new technologies such as field programmable gate arrays (FPGA) and digital signal processors (DSP) for reducing cost and improving performance and also for making it applicable for processing multiple GNSS signals have been described by Hein et al. [48] and also by Sand [49].

A detailed description of the design of a GNSS receiver is outside the scope of this book but has been well described in several recent books and papers [16, 50, 51]. Figure 1.23 shows a greatly simplified functional block diagram of a GNSS receiver connected to a GNSS antenna: RF front end, digital signal processor which calculates the code range, the carrier phase and Doppler frequencies as well as generates a navigational data stream, and finally the navigation processor which calculates the position, velocity and time. In the RF front end the satellite signals received by the antenna are first amplified by a low noise amplifier (LNA), then sent through a bandpass filter to remove out-of-band interference and noise. The analog signal is then down-converted to an intermediate frequency (IF) signal using a frequency mixer. In this down conversion only the frequency of the carrier wave is decreased but the PRN codes and Doppler information in the signal is preserved,

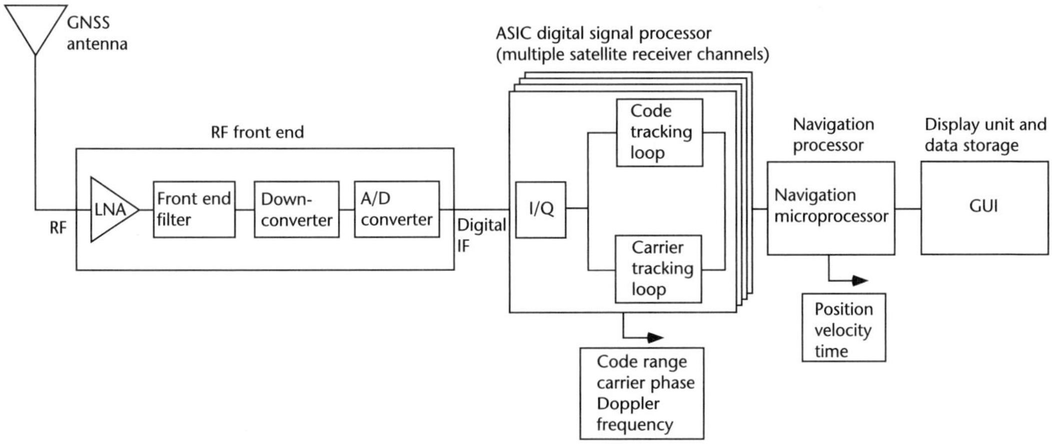

Figure 1.23 Functional block diagram of a GNSS receiver.

In the next step, the IF signal is sampled to yield a digital signal by an A/D converter—which discretizes the incoming analog signals in time (sampling) and in magnitude (quantization). The next functional block is the digital signal processor, which is normally an ASIC (application-specific integrated circuit) chip, whose function is channel multiplexing, correlation, acquisition and tracking. To demodulate the received signal to extract the navigation data, the receiver needs to first determine the satellites that are visible and then determine the carrier frequency and code phase of the signal; this is processed in two stages: acquisition and tracking. Acquisition consists of "finding" the signal transmitted from a certain satellite. During acquisition, frequency and code phase are determined coarsely using a search technique. The "tracking" process is used to maintain lock to both the spreading code and the carrier, with the output being the phase transitions of the navigation data. The final third stage of the receiver is the navigation processor which conducts three main tasks: 1) Decoding the navigation message and to compute the satellite positions; 2) Use the code, phase and Doppler lock measurements to compute position, velocity and time information; and 3) provide aiding information to the tracking loops and to the filters.

1.14.1 Software Defined Radio (SDR)

An alternative receiver processor called the software defined radio (SDR) has recently come in for increased attention, where band pass sampling, signal correlation and data processing is integrated into a software controlled unit called the software receiver [52, 53, 54]. A block diagram of the SDR is shown in Figure 1.24.

SDR incorporates digitization closer to the front end of the receiver antenna and accomplishes all digital processing using a programmable microprocessor or general purpose CPU instead of an ASIC chip as in the previously discussed conventional version of the receiver. This completely separates analog signal conditioning hardware from digital signal processing software, allowing wider bandwidth provided by the analog signal conditioners with increased flexibility in processing through appropriate software the sampling of multiple frequency bands of interest. An SDR allows new and optimized signal processing algorithms to be more

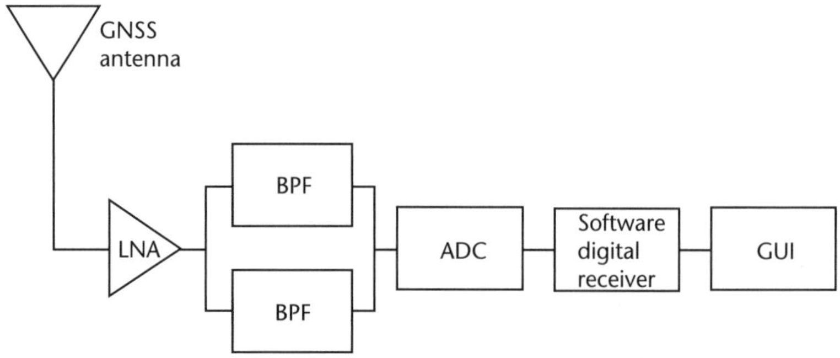

Figure 1.24 Functional block diagram of a software defined radio (SDR) receiver.

easily developed and deployed into the receiver so as to provide better performance at increasingly higher frequencies and for wider bandwidths. The expectation is that future improvements in processor technology will provide large improvements in processing power allowing these improvements to occur as the receivers are adapted to handle signals from multiple GNSS—the so called "System of System" receiver of the future [2, 4, 6]. However, in this book we shall only consider the receiver which uses as ASIC type of DSP unit.

1.15 Effects of Interference and Jamming on Tracking Loop Thresholds of a GPS Receiver

The pseudorange is measured by the tracking loop of a GPS receiver by correlating the pseudorandom noise (PRN) code of the SV (satellite vehicle) that is being received with an internally generated replica of the code generated by the receiver. The receiver must shift the phase of the replica code until it achieves maximum correlation with the phase of the incoming SV PRN code, whereas minimum correlation occurs when the phase of the replica code is offset by more than one chip on either side of the SV code that is being received. The offset for the maximum correlation is a measure of the propagation time and hence of the distance; this is the manner in which the GPS receiver acquires and tracks the SV signal in the code-phase dimension. The GPS receiver must also detect the SV in the carrier phase dimension by replicating the carrier frequency plus Doppler by obtaining carrier phase lock with the SV signal. Hence, the acquisition and tracking of a GPS signal by a receiver is a two-dimensional code-plus-carrier-signal replication process.

Due to the movement of both the satellite and the receiver, the pseudorange will change with time requiring the receiver to continually adjust the offset of the internally generated code to obtain maximum correlation with the received signal. This function is performed by two tracking loops in the receiver—the carrier tracking loop and the code tracking loop. The function of the code tracking loop is to synchronize the received coded signal with an internally generated signal. The function of the carrier tracking loop is to adjust the frequency of the signal used for down-converting, which is necessary because the received signal can be Doppler-shifted due to the movement of the receiver. If the frequency of the received signal is changed, the discriminator in the carrier loop will sense the change and try to adjust the down-conversion frequency to the correct value. For a slow-moving or stationary receiver this frequency change is caused by the movements of the satellites. During periods of large dynamic stress, the frequency may need large adjustments. A carrier tracking loop can be implemented with either a phase locked loop (PLL) or a frequency locked loop (FLL). An FLL is less sensitive to jamming and dynamic stress than a PLL; but a PLL provides more accurate Doppler measurements. The tracking loops need a certain signal strength compared to the interfering signals and noise in the receiver to be able to track the satellite signal. The quality of the satellite signal is determined by the C/N_0. It is computed as the ratio of the recovered power, C in watts (W), from the desired GNSS signal-to-noise-density N_0 in (W/Hz). The C/N_0 ratio is reduced in the presence of either interference or jamming to an equivalent ratio $(C/N_0)_{eq}$ and will be derived later in this chapter. The lowest (C/N_0) that the receiver can track is the tracking loop threshold. GNSS measurement errors and the tracking threshold level are closely related. When the $(C/N_0)_{eq}$ drops below the tracking threshold the receiver loses lock and is unable to function because the measurement errors exceed a certain limit [3, pp. 247-281, 56, pp. 219-236]. The first part of the receiver to lose track when the $(C/N_0)_{eq}$ drops below the threshold value is normally the carrier tracking loop because of increased phase errors in the PLL. The dominant sources of phase error in the PLL of the GPS receiver are the phase jitter and the dynamic stress errors. The PLL is nonlinear and will lose lock for phase errors larger than 45° [56]. Considering this as a 3-σ limit, the 1-σ rule-of-thumb threshold for the PLL tracking loop is therefore

$$\sigma_{PLL} = \sqrt{\sigma_{tPLL}^2 + \sigma_v^2 + \sigma_A^2} + \sigma_e/3 \le 15° \tag{1.36}$$

Where: σ_{tPLL} = standard deviation of thermal noise in degrees
σ_v = standard deviation of vibration induced oscillator jitter in degrees
σ_A = Allan variance-induced oscillation error in degrees
σ_e = dynamic stress error in the PLL tracking loop

Often the PLL thermal noise with a standard deviation of σ_{tPLL} is often treated as the only source of PLL jitter since the other terms are either transient or negligible. The thermal noise jitter for a PLL is computed as follows [56]:

$$\sigma_{tPLL} = \frac{360}{2\pi}\sqrt{\frac{B_n}{C/N_0}\left[1+\frac{1}{2T\frac{C}{N_0}}\right]} \tag{1.37}$$

Where B_n = carrier loop noise bandwidth (Hz)
$\frac{C}{N_0}$ = carrier-to-noise-power ratio $= 10^{\frac{C}{N_0}/10}$ for C/N_0 expressed in dB/Hz
T = Predetection integration time (sec).

In theory, better anti-jam performance can be achieved by narrowing the bandwidth of the carrier tracking loop. Narrowing the tracking loop bandwidth, however, results in a sluggish response time. If the vehicle is undergoing high acceleration, a tracking loop with a narrow bandwidth cannot keep pace with vehicle dynamics; widening the tracking loop bandwidth on the other hand would improve the responsiveness to high acceleration but would increase the jitter. Hence a bandwidth that is a compromise between threes two extremes is needed.

The 3σ carrier phase tracking jitter in degrees versus the carrier to noise ratio (dB/Hz) for a typical Costa's loop applied to a GPS L1 signal has been calculated [56] and is shown in Figure 1.25.

The curve shown in this figure was calculated for a loop bandwidth $B_n = 20$ Hz. and a predetection integration time $T = 20$ milliseconds. A first order magnitude of the functioning limitation of a carrier tracking loop can be obtained by setting the 3σ jitter to less than 45° (or 1σ jitter of less than 15° as indicated in equation (1.36); this figure leads to a working threshold level of 25 dB/Hz from Figure 1.25. Approximately the same value for the 3σ error can also be deduced from the $1\ \sigma$ error shown by Kaplan and Hegarty in Figure 5.22 on page 186 [3].

The nominal rule-of-thumb for the threshold value for C/N_0 used most frequently for evaluating effects of RF interference on a GPS receiver is 28 dB/Hz [3} also [56, Table 6.2]. This is based on the 1-σ error of 15° for a third-order loop taking into account all the tracking errors described earlier in equation 63 and as shown in Figure 5.24, page 191, Kaplan and Hegarty [3].

The values used in these calculations were: bandwidth B_n = 18Hz, a predetection integration time $T = 20$ milliseconds, and maximum line of sight (LOS) jerk dynamic stress = 10g/s = 98 m/s^3 plus contributions from the other two error sources, the vibration, and Allan variance-induced oscillation errors, mentioned in (1.36).

It is to be noted that equation for the standard deviation of the carrier thermal noise error is independent of factors relating to the C/A code or P(Y) codes in the GPS system and is also independent of the carrier frequency. It is dependent on three factors: the carrier-to-power-noise ratio (C/N_0), the noise bandwidth B_n, and the predetection integration time T. Decreasing the noise bandwidth reduces the standard deviation; this is the reason, for example, for trying to narrow the loop bandwidth by using external aiding. However, lowering the bandwidth can increase the Allan variance-induced oscillation error σ_A as indicated in Figure 5.23 of

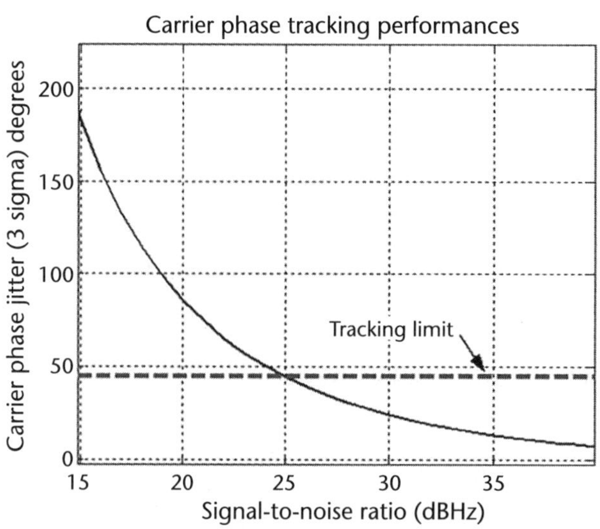

Figure 1.25 Carrier tracking loop threshold level in a GPS receiver.

Kaplan and Hegarty [3]; hence for very-low bandwidths σ_A can become the dominating error. Part of equation (1.36) involving the predetection time T, is called the squaring loss. Increasing the pre-detection integration time reduces the squaring loss and also decreases the standard deviation. The (C/N_0) ratio plays a key role in determining both the standard deviation of the phase litter in the PLL and the overall performance of the receiver. The gain of a GNSS antenna has an important role in maintaining the (C/N_0) ratio above the critical tracking threshold level of the carrier tracking loop to allow the receiver to function properly. The (C/N_0) in dB/Hz for the quiescent conditions in the absence of jamming as well as the effective $(C/N_0)_{eff}$ in dB/Hz in the presence of jamming are derived below. These C/N_0 values are then used to illustrate the vulnerability of a commercial GPS receiver to jamming while receiving C/A code signals in GPS L_1 band when using a FRPA type antenna.

1.16 Carrier to Noise Ratio of an Antenna and Receiver Front End at "Masking Elevation" of 5° in the GPS and Galileo Frequency Bands for Quiescent Conditions (Absence of Interference or Jamming)

The minimum antenna gain required to enable the receiver to acquire and track GNSS satellites can also be determined by calculating the minimum carrier to noise density or the (C/N_0) ratio. The C/N_0 can be related to the more familiar term SNR (signal to noise ratio) which is usually expressed in dB and it can be defined as signal power divided by the noise power in a 1 Hz bandwidth. The C/N_0 is function of several factors: the components used in the front end of the receiver such as the LNA, filters, loss in the cables and most importantly includes the antenna gain, it is also a function of the processing techniques used in the receiver including integration times, loop bandwidths and various other receiver design trade-offs. Due to these different contributions the C/N_0 derived in the section below should be considered only as a conservative (i.e. safe) minimum estimate of the ability of the receiver to acquire and track the satellite. It provides a useful index for calculating the minimum antenna gain requirements in the absence of a better estimation method.

The ratio of the received carrier power level to the noise power level in a 1 Hz. Bandwidth is called the carrier-to-noise ratio (C/N_0); this is a normalized measure of the signal-to-noise ratio and is an important parameter that describes the precision of the pseudorange and carrier phase measurements in GNSS. The un-jammed carrier-to-noise-power ratio in a 1-Hz bandwidth (dB/Hz) under quiescent conditions is represented by $\left[\frac{C_S(\theta_S,\varphi_S)}{N_0}\right]$ and can be derived from the equation below [3, 56]:

$$\left[\frac{C_S(\theta_S,\phi_S)}{N_0}\right]_{dB} = \left(C_{RS}(\theta_S,\phi_S)\right)_{dB} + \left(G_S(\theta_S,\phi_S)\right)_{dB} - L_{dB} - 10\log_{10}\left[k\left(T_{eff}\right)\right]_{dB} \qquad (1.38)$$

$(C_{RS}(\theta_S, \phi_S))_{dB}$ = Signal power (in dBW) received from a GPS satellite located at elevation θ_S and azimuth ϕ_s. It is –158.5 dBW (minimum) for the GPS L_1 C/A code.

$(G_S(\theta_S, \phi_S))_{dB}$ = Gain of the antenna at 1575 MHz towards a GPS satellite.

L_{dB} = receiver implementation loss including the analog-to-digital (A/D) converter loss; it is approximately –1.9 dB [22].

k = Boltzmann's constant = 1.38×10^{-23} (J/K)

T_{eff} = effective temperature of the entire front end of the receiver; this can be calculated using the Friis equation; this procedure is described in greater detail below.

Figure 1.26 shows a schematic diagram of the antenna and the basic front end of the receiver frontend. It consists of both passive and active components. Each component in the active section is represented by either its gain "G" and by its noise figure "F". If the component is a passive and lossy two port system such as a coaxial cable, bandpass filter or a bias -tee, it is represented as having a gain G= 1/L with L>1; L is the insertion loss and its equivalent noise temperature is represented by $T_e = (L - 1)\, T_0$ or by a noise figure F = L=1/G. In other words the noise figure is simply the inverse of the gain. T_0 is the ambient temperature, and is normally assumed to be 290 degrees K. If the component in this section is active, such as an LNA, it will be represented by a gain G_L and by its noise figure F_L; the noise figure of an LNA is generally provided as part of its component specifications. The effective input noise temperature T_{eff} and the net gain G_R of the active section that is connected to the passive antenna can be calculated using Frii's formula [22, 57].

In this formulation, the contribution to the effective temperature T_R from each component in the active section of the antenna is calculated as follows: the ambient temperature T_0 is multiplied by the noise figure of that component minus 1 and then divided by the gains of the earlier components in the active section. The LNA which has a large gain amplifies both the signal and the noise inputs to component in the chain. The LNA being an active device also needs a bias T, which precedes the LNA and provides it with direct current (DC) power from an external power supply. In some cases this DC power is provided directly from the co-located GNSS receiver through a coaxial cable connecting the receiver to the antenna.

The effective temperature T_{eff} of the entire front end, calculated using Friis formula [Reference 22 page 133–145, Reference 20, pages 408–411, Reference 57, pages 82–88] is given below:

$$T_{eff} = T_A + (F_1 - 1)T_0 + \frac{(F_2 - 1)T_0}{G_1} + \frac{(F_L - 1)T_0}{G_1 G_2} + \frac{(F_3 - 1)T_0}{G_1 G_2 G_L} + \frac{(F_4 - 1)T_0}{G_1 G_2 G_L G_3} \tag{1.39}$$

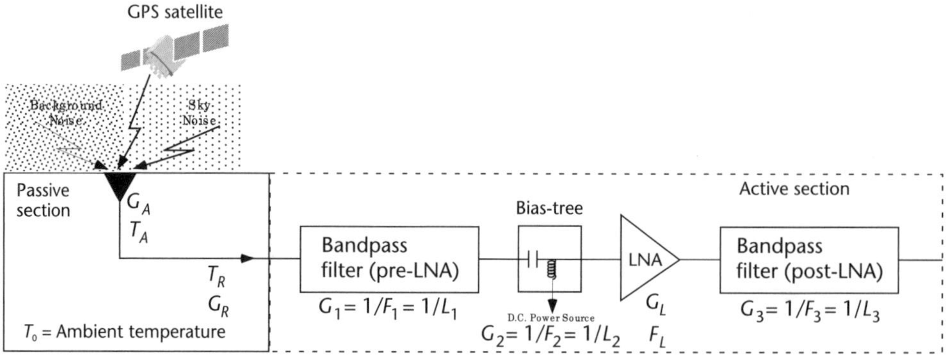

Figure 1.26 Receiver model used for calculating the effective temperature of a GNSS receiver.

Note that the large gain of the LNA which occurs in the denominator of the last two terms on the right hand side in the above equation helps to minimize the effects of noise sources that come after it. Hence it is advisable to locate the LNA as close as possible to the output terminal of the antenna. The effective antenna temperature T_A is the temperature of a resistor that would produce the same thermal noise power of its environment that is coupled into the antenna. The radiative noise sources that contribute to it can be grouped into three categories: sky noise, man-made noise and the background noise due to due to terrestrial objects. For GPS antennas used outdoors the value generally used is $T_A = 130°K$.

The C/ N_0 ratio of a GPS antenna that is connected to a receiver front-end has been calculated below using two different GNSS receiver models. The first model is for a stand alone GPS receiver, the Maxim 2742 [58]. This receiver uses a very low noise figure LNA – the MAX 2641 SiGe LNA with a noise figure of just 1.3 dB. Current LNA's used in GNSS receivers have noise figures that vary from a low of 1 dB and up to 3 dB for the less expensive types. The noise figure of the LNA has a large effect on T_{eff}; hence for the second case we will consider a LNA with noise figure of 3 dB and a gain of 15.9 dB. For the first case will consider a complete stand-alone GPS receiver, the MAXIM 2742. Table 1.9 provides a convenient data sheet for calculating the effective temperature T_{eff} In the calculations we shall assume that both the loss in the short coaxial cable connecting the antenna to the front end and the bias T have zero losses. An effort is always made to keep such losses in the circuit that precedes the LNA to the lowest level possible to prevent an increase in T_{eff}.

Substituting these values into equation we obtain $T_R = 137.28$ and the effective temperature

$T_{eff} = 267.23$ degrees K.; hence

Thermal noise power density (dBw – Hz) = 10 log($k * T_{eff}$) = – 204.33 dBw-Hz.

Where k = Boltzmann's constant = 1.38×10^{-23} J/K. The implementation loss can be assumed to be –1.9 dB [22]. The minimum power received for the L_1 C/A code at the user terminal unity gain RHCP antenna is –158.1 (dBW).

If we assume the minimum gain of the antenna to be–4.5 dBic—the current specification for a GPS airborne antenna specified by RTCA DO-228 we obtain for

Table 1.9 Data Sheet for Receiver Front End for Calculating T_{eff}

Component in Receiver Front End	*Model*	*Property of Component*	*Term Used in Equation and its Value*
Cable from Antenna to Front End	N/A	Low Loss 0 dB	$F_1 = 1/G_1 = 1.0$
Bias –T for LNA	N/A	Low Loss 0 dB	$F_2 = 1/G_2 = 1.0$
LNA	MAXIM Model 2641; Si-GE Low Noise LNA	Noise Figure 1.3 dB	$F_L = 1.3489$
LNA	Ibid	Gain 15.7 dB	$G_L = 37.15$
SAW Filter	N/A	Filter Loss = 3 dB	$F_3 = 1.9952$ $G_3 = 1/F_3 = 0.5017$
Down Converter	MAXIM Model 2742 150 dB gain	Noise Figure 4.5 dB	$F_4 = 2.818$

the C/A code. Hence the carrier to noise ratio $\frac{C_S}{N_0}$ = 39.43 dB-Hz in an un-jammed environment. The carrier to noise ratios in the other GPS bands can be calculated similarly assuming that that the minimum gain requirement at an elevation angle of 5° is specified to be the same in all these bands. The results are shown in Table 1.10. Similar results can also be calculated for the frequency bands of Galileo and are shown in Table 1.11.

According to Hoffman–Wellenhoff [Reference 1, page 86] a carrier to noise power ratios below 34 dB-Hz characterizes a weak signal although a performance analysis conducted previously by others [3, 56] indicate a carrier to noise ratio of 28 dB-Hz at the input to the correlators of a DSP in the receiver as adequate to acquire the satellite signals. Some manufacturers consider even 36.5 dB to be a weak signal [59]. For receivers with low dynamics such as antennas that are stationary or are moving very slowly (pedestrians) a C/N_0 of 28-dB-Hz that was previously calculated appears to be acceptable. For antennas mounted on platforms with high dynamics that are not aided by external sensors, such as IMUs, C/N_0 threshold levels of between 30–35 dB-Hz appears acceptable. The general range of values depends on the receiver used and on the platform dynamics etc.—hence the accepted range for C/N_0 is between 28–35 dB-Hz [60].

There is also no general accepted value for the minimum threshold level of C/N_0 needed for acquisition since the sensitivity is a function of the number of correlators and the integration times that are used. This subject has been investigated in great detail by Van Diggelen [22]. High sensitivity GNSS receivers developed recently for indoor GPS and "Assisted GPS" use massive parallel correlation and achieve processing gains as much as 30 dB higher than standard GPS receivers

Table 1.10 Carrier to Noise Ratios for GPS Bands at the Minimum Elevation Angle of 5° [Noise Figure of LNA = 1.3 dB; Gain of LNA = 15.7 dB]

GPS Band	*Center Frequency (MHz)*	*Bandwidth (MHz)*	*Specified Minimum Signal Power (dBW)*	*Carrier to Noise Ratio (dB- Hz) at 5° Elevation*
L_1 C/A	1575.42	2	−158.5	39.43
L_1 P(Y)	1575.42	20	−161.5	36.43
L_2 P(Y)	1227.40	20	−164.5	33.43
L2C	1227.4	2	−160.0	37.93
L5	1176.45	20	−154.9	43.03

Table1.11 Carrier to Noise Ratios for Galileo Bands at the Minimum Elevation Angle of 5° [Noise Figure of LNA = 1.3 dB; Gain of LNA = 15.7 dB]

Galileo Band	*Frequency Range (MHz)*	*Carrier to Noise Ratio (dB-Hz) at 5° Elevation*
E5	1164–1215	42.93
E6	1260–1300	42.93
E2 – E1	1559–1592	42.93

[22]. This allows the receiver front end to achieve signals that are much weaker than previously considered possible. The values given in the tables above are for conventional receivers using LNAs with noise figures between 1 to 3 dB which use standard correlation processors and integration times and provide the required TTFF (time to first fix).

These tables above indicate a very narrow margin for C/N_0 in some GPS bands, if the conventional threshold level of with the results in the L_2 P (Y) code even being slightly below the weak signal level. The results shown in this table is for an LNA with a super low noise figure of only 1.3 dB. When a less expensive LNA with a Noise figure of 3 dB is used instead, the carrier to noise ratios degrade by 2.31 dB. Thus for the L1 C/A code , the C/N_0 will decrease to 37.12 dB and for the L2 P(Y) code will only be 31.12 dB.

The corresponding C/N_0 values for Galileo look much better because of the minimum received signal powers are higher with a good margin to compensate for a LNA with poorer noise figures.

1.17 Effective Carrier-to-Noise Ratio in the Presence of Interference or Jamming

The relationship between the effective carrier-to-noise ratio $\left[\frac{C_S(\theta_S,\varphi_S)}{N_0}\right]_{eff}$ in the presence of a jammer (or an interference source) with a jammer/signal ratio (J/S) can be determined by the following equation [3]:

$$\left[\frac{J(\theta_J,\varphi_J)}{S(\theta_S,\varphi_S)}\right]_{dB} = G_S(\theta_S,\varphi_S) - G_J(\theta_J,\varphi_J) + 10\log_{10}\left[QR_C(A-B)\right]$$

Where

$$A = 10^{-\frac{\left[\frac{C_S(\theta_S,\varphi_S)}{N_0}\right]_{eff}}{10}} \text{ and } B = 10^{-\frac{\left[\frac{C_S(\theta_S,\phi_S)}{N_0}\right]}{10}} \tag{1.40}$$

$\left[\frac{C_S(\theta_S,\varphi_S)}{N_0}\right]_{eff}$ = Effective carrier-to-noise ratio in dB/Hz when jamming or interference is present.

$J(\theta_J, \phi_J)$ = Strength of the jammer signal arriving at the antenna in direction θ_J, ϕ_J

$S(\theta_J, \phi_J)$ = Strength of satellite signal arriving at the antenna in direction θ_S, ϕ_S

R_c = spreading code rate of the GPS code generator in chips per second

Q = jamming resistance quality factor (dimensionless) that can be determined for various types of jammers and signal modulators. The derivation of formulas for R_C and Q are beyond the scope of this book; they are explained in greater detail in Section 6.2.2.5 on pages 256–278 of Kaplan and Hegarty [3].

1.17.1 Effects of Jamming or Interference on a GPS Commercial C/A Code Receiver

The example described in this section calculates the amount of jamming power needed in the C/A code of the GPS L_1 band to disable a commercial GPS receiver which uses a simple FRPA antenna with no adaptive capabilities. The upper hemispheric gain pattern of a typical FRPA antenna was shown earlier in Figure 1.5. The RHCP antenna gain towards a satellite located at $\theta_S = 30°$ (or at an elevation angle = 60°) and azimuth angle $\phi_S = 0°$ is $G_S = 1.5$ dBic. The gain of this antenna when directed towards a jammer located at elevation $\theta_J = 75°$ (or at an elevation of 15°) and azimuth angle $\phi_J = 0°$ is $G_J = 0°$ is -3 dBic. For the C/A code in the GPS L_1 band

$R_C = 1.023$ Mchips/sec and Q = 2.22 for a broadband white-noise jammer as indicated in Table 6.3 of [3]. The tracking loop threshold level of the PLL in an unaided receiver is $\left[\frac{C_S(\theta_S,\varphi_S)}{N_0}\right]_{eff} = 28$ dB-Hz. and $\left[\frac{C_S(\theta_S,\varphi_S)}{N_0}\right]$ = is 41.9 dB-Hz.

Hence (J/S) = 1.5 +3.0 +10 $\log_{10}$ [2.22 x 1.023x 10^6 ($10^{-2.8}$ - $10^{-4.19}$)] = 39.9 dB. This means that a jammer of strength J/S = 39.9 dB lowers the C/N_0 from its unjammed value of 41.9 dB/Hz to an effective value of 28 dB/Hz. The jammer power needed to disable to GPS receiver can now be calculated

$$\left[\frac{J}{S}\right]_{dB} = J_{dBW} - S_{dBW} \tag{1.41}$$

J_{dBW} = incident jammer power at the input terminals of the receive antenna (dBW)

S_{dBW} = incident satellite signal power at the input terminals of the receive antenna (dBW)

$$J = 10^{\frac{\left(\frac{J}{S}\right)_{dB} + S_{dBW}}{10}}$$

$S_{dBW} = -158.5$ dBW—the minimum specified signal power for the C/A code received from the GPS satellite. Hence

$$J = 10^{\frac{39.9+(-158.5)}{10}} = 1.38 \times 10^{-12} \text{ watts}$$

The results of this calculation indicate that less than 1.5 picowatts of either RF interference or jammer power incident on the antenna is sufficient to disable a commercial un-aided GPS receiver.

The transmit power of the jammer (or RF interference source) and its distance relative to the GNSS receiver needed to disable the receiver can be calculated from the equation for the effective isotropic radiated power (EIRP).

$$(EIRP)_{dB} = (J_r)_{dB} - (G_J)_{dB} + (L_P)_{dB} + (L_f)_{dB} \tag{1.42}$$

Where $(EIRP)_{dB} = (J_t)_{dB} + (G_t)_{dB}$ = EIRP of the jammer

Where $(J_t)_{dB}$ = jammer transmitter power into the jammer antenna (dBW) = $10\log_{10} J_t$ = (with J_t expressed in watts)

$(G_t)_{dB}$ = Gain of the jammer transmit antenna (dBic)
$(J_r)_{dB}$ = Incident (received) jammer power (dBW)
$(L_p)_{dB} = 20\log_{10}\left[\frac{4\pi d}{\lambda_j}\right]$ free space propagation loss (dB
$(G_t)_{dB}$ = Receiver antenna gain towards the jammer (dBic)
D = range to jammer (meters)
λ_j = Wavelength of the jammer frequency (meters)
$(L_f)_{dB}$ = Jammer power loss due to front-end filtering in receiver (dB)

For these sample calculations we will assume a jammer operating in the GPS L_1 band with a transmit power of 2 watts and a jammer antenna gain of 3 dB (EIRP = 4 watts). The jammer will be assumed to be a band-limited white-noise jammer null-to-null jammer with a J/S ratio = 49.7 dB. The corresponding $(J_r)_{dB} = -111.8$ dBW and $(L_p)_{dB} = 114.8$ dB [3, 56]. Since the jammer is in the receiving frequency band of the receiver it will be assumed that $(L_f)_{dB} = 0$ dB. The range "d" from the jammer to the receiver that will produce a threshold power level at the receiver to disable it can next be calculated:

$$d = \frac{\lambda_j 10^{\frac{(L_p)_{dB}}{20}}}{4000\pi} = 8.3 \text{ kilometers} = 4.5 \text{ nmi}$$

The above calculations show that a commercial GPS receiver with a FRPA antenna with a gain of –3 dBic can be disabled by a jammer with a transmitter power of just 2 watts and an antenna gain of 3 dB. (EIRP = 4.0 watts) which is located at a distance of 4.5 nautical miles (8.3 kms) from the victim receiver. Further details of these calculations are given in Kaplan and Hegarty [3, 56].

The receiver can be made more resistant to jamming and interference by reducing the gain of the receiving antenna to increase $G_S(\theta_S, \phi_S) - G_J(q_J, \phi_J)$. For example, a reduction of 20 dB in G_J will increase the jammer power that is required to disable the receiver to 138 picowatts, a significant improvement in resistance to jamming and interference.

References

[1] Hoffman-Wellenhof, B., H. Lichtenegger, E. Wasle; " GNSS—Global Navigation Satellite Systems; GPS, GLONASS, Galileo and More"; © 2008, Springer Verlag Wien, , Chapter 9 to 11, pp. 309–430.

[2] C. Rizos; " GPS, GNSS and the Future," Manual of Geospatial Science and Technology, Editor: J. D. Bossler; Second Edition; CRC Press; Boca Raton, 2010, pp. 259–284.

[3] Kaplan, E. D. and Hegarty C. J.; "Understanding GPS"; Second Edition, 2006, ARTECH HOUSE Inc, Norwood, MA., Chapter 6, pp. 243–279.

[4] Hein, G, W., J. A. Avila-Rodriguez, S. Wallner, B. Eisfeller, P. Thomas, P. Hartl; "Envisioning a Future GNSS System of Systems", Inside GNSS, Part 1, January/February 2007, pp. 58–67, Part 2; March/April 2007, pp. 64–72.

[5] Feng, Y., C. Rizos, " Impact of Multiple Frequency GNSS Signals on Future Regional GNSS Services"; International Global Navigation Satellite Systems Society, IGSS Symposium 2007, Sydney, Australia, December 2007.

[6] Hein, G, W., J. A. Avila-Rodriguez, S.Wallner, B. Eisfeller, P. Thomas, P. Hartl; "Envisioning a Future GNSS System of Systems," Inside GNSS, Part 2, March/April 2007, pp. 64–73, March/April 2007.

[7] "2010 Receiver Survey" GPS World, Vol. 21, Number 1, January 2010, pp. 35–57; and "2010 Antenna Survey" GPS World, Vol. 21, Number 2, February 2010, pp. 36–49

[8] Lachapelle, G., P. Heroux, S. Ryan, "Serving the GNSS/GPS User," Manual of Geospatial Science and Technology, Editor: J. D. Bossler; Second Edition; CRC Press; Boca Raton, 2010; Chapter 14, pp. 235–257.

[9] Wirola, L., I. Kontola, I. Syrjarrine, "The Effect of the Antenna Phase Center Response," *Proceedings IEEE/ION PLANS Conference,* May 2008, pp. 606–615.

[10] Stutzman, W. L., "Polarization in Electromagnetic Systems," 1993 Artech House, Inc., Norwood, MA. 02062.

[11] Pathak, V., S. Thornwall, M. Kriz, S. Rowson, G. Poilasne, L. Desclos, "Mobile Handset Performance Comparison of a Linearly Polarized GPS Internal Antenna With a Circularly Polarized GPS Internal Antenna," *IEEE Antenna and Propagation Society Symposium,* Vol. 3, 22-27 June 2003, pp. 666–669.

[12] Boulton, P., R. Borsato, B. Butler, K. Judge, "GPS Interference Testing, Lab, Live and Light-squared," *Inside GNSS,* July/August 2011, pp. 32–45.

[13] Morgan, W. L. , G. D. Gordon, "Communication Satellite Antenna Handbook," 1989, John Wiley & Sons, p. 250.

[14] Betz, J., "Link Budgets," Paper Presented to International Working Group A, Torino, Italy, 19 October 2010.

[15] Spilker, J. J., "Satellite Constellation and Geometric metric Dilution Of Precision," Chapter 5, " Global Positioning Systems: Theory and Applications," Vol. 1, Editors: B. W. Parkinson, J. J. Spilker, 1996, American Institute of Aeronautics Inc., pp. 177–207.

[16] Van Dierendonck, A. J., "GPS Receivers," Chapter 8, "Global Positioning Systems: Theory and Applications," Vol. 1; Editors: B. W. Parkinson and J.J. Spilker, 1996, AIAA Inc., pp. 340–341.

[17] RTCA, "Minimum Operational Performance Standards For Global Navigation Satellite Systems (GNSS) Airborne Antenna Equipment," RTCA /DO-228, October 20, 1995: Replaced by D0-301 for active antennas.

[18] Phillippakis, M., M. Patel, D. Moore, D. Kemp, "Experimental Results for an Airborne Multi-standard GNSS Antenna," EUROCAE WG62, Toulouse, France, 18–19 June 2008.

[19] Rizos, C., D. A. Grejner-Brzezinska, "GPS Positioning Models For Single Point and Baseline Solutions," Chapter 9, "Manual of Geospatial Science and Technology," Editors: J. Bossler, Second Edition, 2010, CRC Press, Boca Raton, FL., pp. 136–149.

[20] Misra, P, P. Enge, " Global Positioning Systems: Signals, Measurements and Performance," (Second Edition), 2006, P. Misra and P. Enge, Ganga–Jamuna Press, Lincoln, Massachusetts, pp. 208-212.

[21] Seeber, G., "Satellite Geodesy," Second Edition, 2003, Walter de Gruyter GmBH&Co KG, Berlin, pp. 300-304.

[22] Van Diggelen, F., "A-GPS: Assisted GPS, GNSS, and SBAS," 2009, Frank van Diggelen, Artech House Ltd., Boston/London.

[23] Parkinson, B.W., "GPS Error Analysis," Chapter 11, "Global Positioning Systems: Theory and Applications," Vol. 1, Editors: B. W. Parkinson, J. J. Spilker, 1996, American Institute of Aeronautics Inc., pp. 177–207

[24] Phillips, A.H.; "Geometrical Determination of PDOP"; Navigation, Vol. 41, No.4, Winter 1994-95, pp. 329–337.

[25] Massat, P., K. Rudnick, "Geometric Formulas For Dilution Of Precision Calculations," Navigation, Vol. 37, No.4, Winter 1990–1991, pp. 379–391.

[26] Gorres, B., J. Campbell, M. Becker, M. Siemes, "Absolute Calibration of GPS antennas: Laboratory Results and Comparison with Field and Robt Techniques," GPS Solutions, Vol. 10, No. 2, 2006, pp. 136–145.

[27] Kunysz, W, "Antenna Phase Center Effects and Measurements in GNSS Ranging Applications," 2010 14th International Symposium on Antenna Technology and Applied Electromagnetics Conference [AMREM].

[28] Seeber, G., "Real–Time Satellite Positioning on the Centimeter Level in the 21st Century Using Permanent Reference Stations," Paper given during the Nordic Geodetic Summer School, Fevik, Norway, 30.08.2000.

[29] Alemu, G. T., "Assessments on the Effects Of Mixing Different Types of GPS Antennas and Receivers," Master Of Science Thesis in Geodesy No. 3106, TRIGAT-GIT–EX 08-009, Royal Institute of Technology (KTH), Stockholm, Sweden, September 2008.

[30] Leick, A., "GPS Satellite Surveying," (Third Edition), Copyright 2004, John Wiley & Sons, Hoboken, New Jersey, 2004, pp. 231–233 and 170–177.

[31] Tetewsky, A. K., F. E. Mullen, "Carrier Phase Wrap-up Induced by Rotating GPS Antennas," GPS World, Vol. 8 (2), pp. 51–57, 1997.

[32] Bisnath S., "What is Carrier Phase Wind-Up? What is its Effect on GNSS Performance," *Inside GNSS,* July- August 2007, pp. 32–35.

[33] Garcá-Fernandez, M., M. Markgraf, O. Montenbruck, "Spin Rate Estimation of Sounding Rockets Using GPS Windup," *GPS Solutions,* Vol.12 (3), pp. 155–161, 2008.

[34] Wu, J. T., S. C. Wu, G. A. Hajj, W. I. Bertiger, and S. M. Lichten, "Effects of Antenna Orientation on GPS carrier phase," Manuscripta Geodaetica, Vol. 18 (2), pp. 91–98, 1993.

[35] Kim, D., L. Serrano, and R. Langley, "Phase Wind-up Analysis: Assessing Real-Time Kinematic Performance," *GPS World,* Vol. 17 (9), 2006, pp. 58–64.

[36] Kim, D., L. Serrano, R. B. Langley, "Compensation of the Effects of the Phase Wind-up for Improving Performance Of a GPS RTK-based Vehicle Navigation System," *Proceedings 2005 ION GNSS Conference,* pp. 346–354.

[37] Van Graas, F., C. Bartone, T. Arthur, "GPS Antenna Phase and Group Delay Corrections," *Proceedings ION NTM 2004,* San Diego, CA., pp. 399–408.

[38] Dong, W., J. T. Williams, D. R. Jackson, L. R. Basilio, "Phase and Group Delays for Circularly Polarized GPS Microstrip Antennas," *Proceedings ION 63rd Annual Meeting,* Cambridge, Massachusetts, April 23–25, 2007, pp. 545–554.

[39] Betz, J., "Effect Of Linear Time-Invariant Distortions on RNSS Code Tracking Accuracy," *Proceedings 2002 ION GNSS Conference,* pp. 1636–1647.

[40] Van Dierendonck, A. J., R. J. Erlandson; "RTCA Airborne GPS Antenna Testing and Analysis for a New Antenna Minimum Operational Performance Standards (MOPS)," Proceedings of ION National Technical Meeting, January 2007.

[41] Murphy, T., P. Geren, T. Pankaskie, "GPS Antenna Group Delay Variation Induced Errors in a GNSS Based Precision Approach and Landing Systems," *Proceedings 2007 ION GNSS Conference,* Fort Worth, Texas, September 2007, pp. 2974-2989.

[42] Abidin, H. Z., "Fundamentals of GPS Signals and Data," Chapter 8, Manual of Geospatial Science and Technology," Editor: J. Bossler, CRC Press, Second Edition, 2010

[43] Rizos, C., D. Smith, S. Hilla, J. Evjert, W. (Bill) Henning, "Carrying Out a GPS Surveying/ Mapping Task," Chapter 13, *Manual of Geospatial Science and Technology,"* Editor: J. Bossler, CRC Press, Second Edition, ©2010.

[44] RTCA Inc., SC-159, "Assessment of Radio Frequency Interference Relevant to GNSS L5/ E5A Frequency Band," RTCA /DO 292, July 29, 2004.

[45] Federal Communications Commission Omnibus Citation and Order, DA 11-1661, September 30, 2011.

[46] Mitch, R. H., R. C. Dougherty, M. L. Psiaki, S. P. Powell, B. W. O'Hanlon, "Signal Characteristics of Civil GPS Jammers," *Proceedings 2011 ION GNSS Conference,* Portland, Oregon, September 19–23, 2011, pp. 1907–1919.

[47] Bauernfeind, T. Kraus, D. Dotterb rock, B. Eisfeller, "Car Jammers: Interference Analysis," GPS World, October 2011, pp. 28–35.

[48] Hein, G. W., T. Pany, S. Wallner, J-H Won, "Platforms for a Future GNSS Receiver," *Inside GNSS,* March 2006, pp. 56–62.

[49] S. Sand, "Challenges in Multi-System Multi-Frequency GNSS Receiver Design—Introduction," GRAMMAR (Galileo Ready Advanced Mass Market Receiver), 13 June 2010, pp. 1–22.

[50] Van Dierendonck, A. J., Van Dierendonck A. J., "GPS Receivers," Chapter 8, *Global Positioning Systems: Theory and Applications,* Vol. 1, Editors: B. W. Parkinson and J.J. Spilker, 1996, AIAA Inc.

[51] Langley, R. B., "The GPS Receiver: An Introduction," *GPS World,* January 1991, pp. 50–53.

[52] Van Dierendonck, A. J., "Understanding GPS Receiver Terminology: A Tutorial," *GPS World,* January 1991, pp. 34–44.

[53] Akos, D. M.; "Real-Time GPS Software Radio Receiver," *Proceedings Institute of Navigation National Technical Meeting (ION NTM) 2001,* Long Beach, CA., January 22-24, 2001, pp. 809–816.

[54] Tsui, J. B. Y., "Fundamentals of Global Positioning System Receivers: A Software Approach," 2nd Edition, New York: Wiley, 2005.

[55] Borre, K., D. Akos, N. Bertelsen, and S.H. Jensen, "A Software-Defined GPS and Galileo Receiver: Signal Processing Approach," Boston, Basel, Berlin, Birkhauser, 2006.

[56] Kaplan, E. D. (Editor), "Understanding GPS, Principles And Applications," 1996, Artech House Ltd., Norwood, MA.

[57] Ha Tri, T., "Digital Satellite Communications," 1986, MacMillan Publishing Co., New York, pp. 82–88.

[58] MAXIM-IC Application Note 3477, "Complete Stand-Alone GPS Receiver with MAX 2742 Integrated CMOS Front-End GPS Receiver," www.maxim-ic.com; /an/3447.

[59] Santere, R., "GPS SNR Observations," Memorandum Geodetic Research Laboratory, Date: 21/07/99; page #3.

[60] Communications to the author by Dr. A. Cerruti of the MITRE Corporation.

CHAPTER 2

FRPAs and High-Gain Directional Antennas

Basrur Rama Rao

2.1 Categories of GNSS Antennas

Many types of GNSS antennas have been developed in recent years to make them suitable for different applications. Since their designs and performance requirements vary depending on their application, they have been grouped into six different categories in this book:

1. FRPA;
2. High-gain directional antennas;
3. GPS adaptive antennas;
4. Multiband antennas;
5. Handset antennas;
6. Active antennas.

The design and performance of each type of antenna will be described here and throughout the rest of the book.

This chapter discusses the design of the first two varieties: FRPAs and high-gain directional antennas. Figure 2.1 shows a representative sample of these antennas. FRPAs are the most popular and widely used of all GNSS antennas. There are many different types of these antennas; hence they will need two chapters—this chapter and Chapter 3—to fully cover all the important types. This chapter will discuss the following FRPA designs: microstrip patch antennas, which are the most ubiquitous of all GNSS antennas, and quadrifilar helix antennas (QHAs), which are also popular for handheld receivers and crossed "drooping" bow-type dipoles, which provide relatively wider bandwidths and good circular polarization. Spiral antennas such as the Archimedean spiral antenna will be discussed in Chapter 3 under multiband antennas since they are ultrawideband and can cover the entire GNSS band from 1.1 to 1.6 GHz. The conical spiral antenna is also a high-gain di-

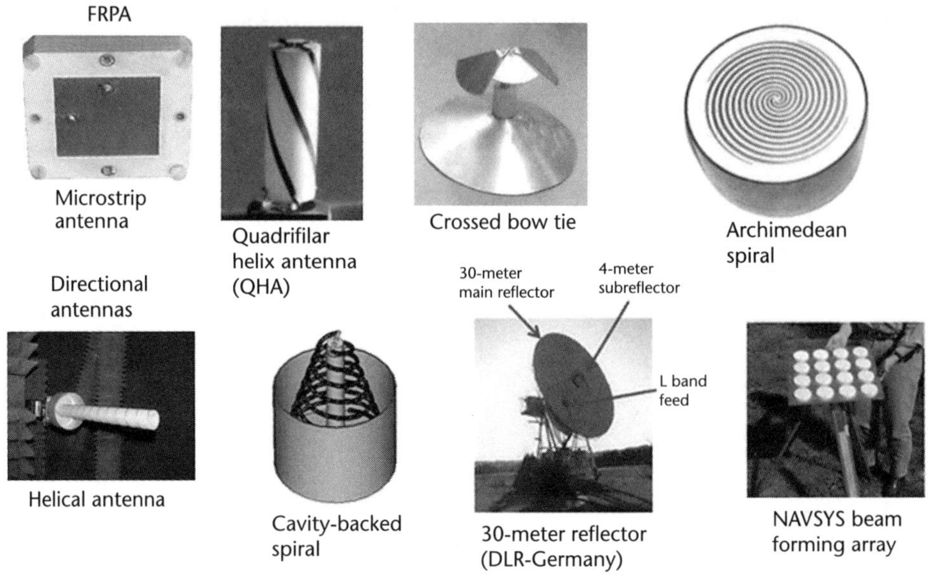

Figure 2.1 Popular types of fixed radiation pattern antennas and directional antennas used in GNSSs.

rectional antenna with very wideband characteristics; it is also discussed in Chapter 3 under multiband antennas.

Directional antennas have radiation characteristics that are distinctly different from FRPAs. FRPAs are receiving antennas that have a broad antenna pattern for acquiring four or more satellites needed for a PVT solution. Directional antennas can be used as either receiving or transmitting antennas. When operated in the receiving mode they have high gain but narrowbeam patterns for selecting only signals of interest (SOI)—the GNSS satellite signals—while rejecting signals not of interest(SNOI) such as multipath signals. When used as transmitting antennas they generate a high-gain, narrowbeam pattern that can be pointed towards either a specific target or region of operation. Three types of GNSS directional antennas are described in this chapter: (1) helical antennas, (2) reflector antennas, and (3) beamforming antenna arrays. Helical antennas are used as elements in transmitting antenna arrays in GNSS satellites for generating an Earth-coverage beam [1–3] and for other special GNSS applications such as transmit antennas for pseudolites [4] and also during laboratory testing. Several large reflector antennas, which range in diameter from 1.8 to 110m, have also been used as receiving antennas for monitoring GNSS signals transmitted by recently launched satellites (such as GIOVE-A&B), and the Compass M1 [5–8]. Beamforming antenna arrays are another type of directional receiving antenna capable of steering four or more beams towards selected satellites, thereby greatly increasing the signal-to-noise ratio while simultaneously reducing unwanted signals such multipath [9–12].

2.1.1 FRPA

An FRPA antenna has a nearly omnidirectional pattern in the upper hemisphere and is designed for acquiring almost all, but at least a minimum of four, satellites visible to the antenna above a certain masking angle. A representative elevation plane pattern of a conventional FRPA antenna of the first category is shown in Figure 2.2. It is expected to meet a number of desired performance requirements with the objective of being able to provide high precision in GPS measurement. These requirements are described in several recent publications [13, 14] and can change depending on the intended application. The antenna is expected to be RHCP so as to efficiently receive signals from GNSS satellites. The antenna is also expected to provide a better than the minimum required gain over much of the upper hemisphere covering 360° in azimuth and from zenith down to a masking angle of generally between 5° or 10° in elevation; this assures that the receiver has high satellite availability and is able to acquire at least four or more satellites within its view. Good PDOPs, well below the required maximum limit of six, can be achieved if the antenna is capable of receiving signals from low-elevation satellites that are also widely separated in azimuth. For airborne GNSS antennas the gain requirements at low-elevation angles are particularly daunting [20] since they require –3 to –4.5dBic of RHCP gain at 10° and 5° in elevation, respectively. Achieving good gain and good circular polarization (CP) axial ratio at such low-elevation angles is generally a challenge since in most types of GNSS antennas that are located on metal ground planes (or the aircraft fuselage for avionics antennas) the gain drops off sharply from its peak value at zenith as the elevation decreases. The metallic ground plane nulls out the horizontally polarized component at its surface (i.e., the horizon); hence, the polarization of the antenna becomes linear instead of RHCP and is oriented vertical to the ground plane representing a further 3-dB loss in gain. These antennas should also preferably be able to discriminate against multipath by having good front-to-back ratios with reduced gain at low-elevation angles at and below the horizon. Since multipath reflections change the polarization state from RHCP to LHCP, these antennas are required to have good axial ratios with low-LHCP cross-polarization level at low-elevation angles where multipath effects are most prominent. The desired axial ratio expected from airborne GNSS antennas should be no

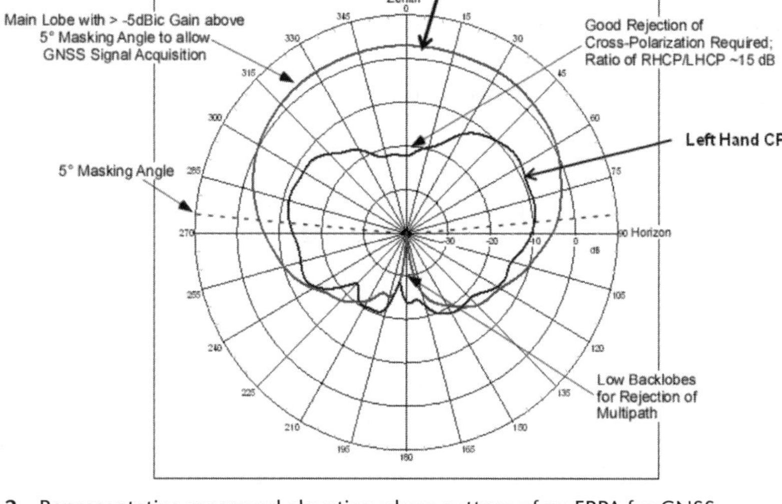

Figure 2.2 Representative measured elevation plane pattern of an FRPA for GNSS.

greater than 3 dB for all operating frequencies at elevation angles greater than 10° nor exceed 6 dB for all operating frequencies for elevations between 5° and 10° [13, 14]. The antenna therefore needs to have a cross-polarization ratio of 15.3 dB to be able to meet the 3 dB in the axial ratio specification; the cross-polarization ratio is defined by the ratio of the power density in the cross-polarized LHCP component of the incoming signal to that of the principal RHCP component. High-precision geodetic quality antennas when used for carrier phase tracking are also required to have a stable phase center that varies only minimally with elevation. This same requirement is also needed for GPS-based attitude determination systems. The antenna is also required to limit the group delay variation with frequency over the operating bandwidth of the antennas. Group delay becomes particularly important in maintain the fidelity of BOC codes that are currently being used for M code in modernized GPS as well as for several of the frequency bands in Galileo [15].

2.2 Microstrip Antennas

Microstrip antennas, commonly called patch antennas, are the most popular type of GNSS antennas used in a large variety of civilian and military systems. Their very low-profile, compact size, ability to conform their shape to that of the host surface, ease of obtaining RHCP, and low cost of manufacture gives them unique advantages that are difficult to match for GNSS applications with any other antenna design. They are universally used in avionics since their low profile and compactness easily allows them to meet the ARINC 743 size requirements, which is a voluntary aircraft industry specification; this requirement restricts the lateral sizes of the antenna to be no more than 4.7" × 3" and their height to no more than 0.73". The ARINC 743 cross section is shown in Figure 2.3. Another form specification is called "teardrop," which is even smaller than ARINC 743. These same qualities also makes them best suited for use in all types of adaptive antenna arrays used in GPS military navigation systems for combating jamming and interference and for beamforming. Patch antennas are also popular for handset antennas since miniaturized antennas can be built using high-dielectric constant ceramic substrates; this allows them to be easily integrated into a variety of popular handheld navigation devices such as cell phones and PDAs. Shorted annular ring microstrip antennas have been proposed for multipath limitation; these are smaller in size, less complex to manufacture, and lower cost than the more expensive choke ring antennas. Microstrip antennas will therefore be discussed here in much greater detail than the other GNSS antennas considered in this book owing to their importance and popularity.

The basic microstrip antenna consists of a metallic conductor of a specific shape that is etched on the top surface of dielectric substrate that is physically bigger than the metallic patch. Both the metallic patch and the substrates are placed over an even larger metallic ground plane that can sometimes be many times the GNSS wavelength. The copper cladding of the bottom surface of the dielectric substrate becomes the central part of the ground plane. Single-band microstrip antennas use a single dielectric substrate but dual- and triple-band patch antennas can use two or more substrates.

Figure 2.4(a) shows a picture of a basic RHCP microstrip patch antenna; the conducting ground plane underneath the patch antenna is not shown in this figure.

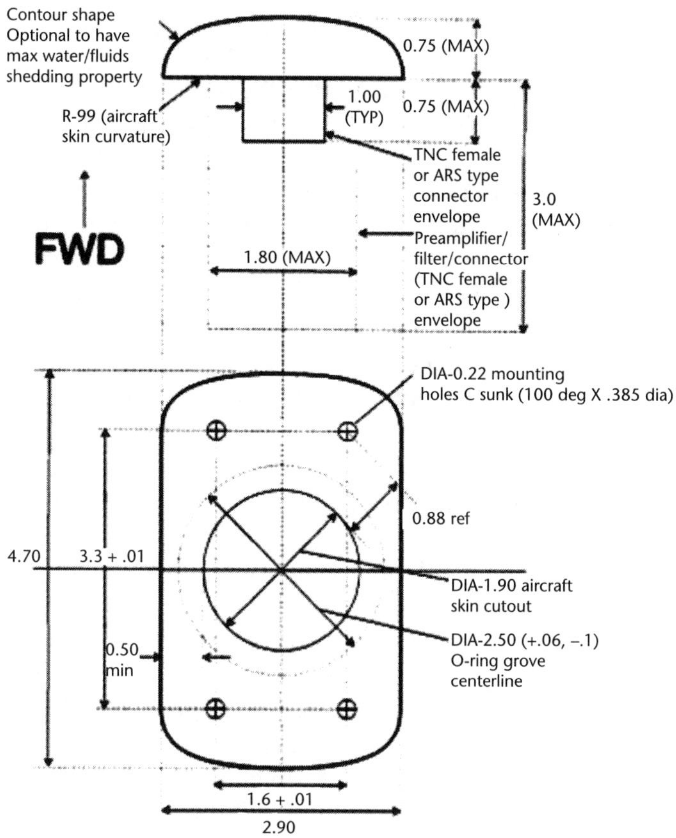

Figure 2.3 ARINC 743 dimensions for an avionics GPS antenna.

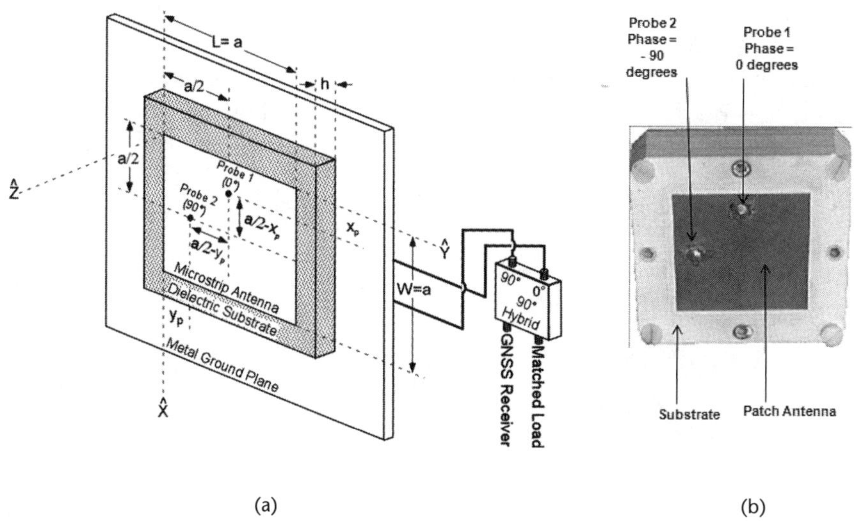

Figure 2.4 RHCP square-shaped single-band GNSS microstrip antenna.

RHCP is achieved by using two coaxial probes that are clearly visible in this picture. A schematic sketch of this antenna is shown in Figure 2.4(b). The metallic patch and the ground plane are assumed to be good electrical conductors and form the top and bottom surfaces of a high Q resonant RF cavity that is tuned to resonance at the desired GNSS frequency. The four edges of the patch act as perfect magnetic conductors and form the sides of the cavity. The energy stored inside the resonant cavity leaks out from the edges of the metallic patch. The resonance frequency of the patch antenna for a desired cavity mode is dependent on the size and shape of the conducting patch, the dielectric constant ε_r of the dielectric substrate, and the thickness of the substrate layer. The directional properties of radiation pattern and its polarization are determined by the electromagnetic fields of the cavity modes generated within the RF cavity and the method of feeding the patch antenna for receiving the satellite signals. To be suitable for GNSS applications the patch antenna needs to have all principal characteristics of a typical FRPA antenna noted earlier.

The three most commonly used shapes for the metallic patches to achieve maximum circular symmetry in azimuth are a square, a circle, or an annular ring, as shown in Figures 2.5(a), (b), and (c), respectively.

A majority of GPS microstrip antennas currently used in low-cost, commercial GPS receivers operate only over a single-frequency band—the GPS L_1 band with a center frequency of 1.5754 GHz—and use a single layer of the dielectric substrate. Some of these antennas have a 2-MHz bandwidth only just sufficient to receive the C/A code. These single-band antennas are the simplest to design and are the least expensive. Dual-band microstrip antennas that operate at either the L_1 and L_2 bands or the L_1 and L_5 bands and also triple-band antennas that operate in all three frequency bands—L_1, L_2, and L_5—will be needed soon to meet the demands of the modernized GPS. Multiband patch antennas designed to meet these requirements are discussed later in this chapter and in Chapter 3. These multiband patch antennas consist of a combination of two or more patch antennas with each patch antenna resonating at a different frequency, generally either stacked on top of each other or parasitically coupled to one another. These antennas may require a bandwidth of 20 MHz in each band or 24 MHz if reception of a GPS M code signals is needed. All the microstrip antennas shown in Figure 2.5(a), (b), and (c) are dual-band " stacked patch" antennas, whose design will be explained in greater detail later; they operate in the GPS L_1 and L_2 bands and use two dielectric substrate layers—one for each patch antenna.

Square-shaped patch antenna (two feed probes)	Circular-shaped patch antenna (four feed probes)	Annular ring patch antenna (four feed probes)
(a)	(b)	(c)

Figure 2.5 (a) Square, (b) circular, and (c) annular ring GPS microstrip antennas.

2.2.1 Selection of the Dielectric Substrate for Microstrip Antennas

The selection of a suitable dielectric substrate is a critical first step in the design of microstrip patch antennas and involves a trade-off between different application-driven requirements such as size and height, bandwidth, and gain coverage in the upper hemisphere. Small-size antenna elements with a low profile are required in avionics and for miniaturized adaptive antenna arrays used in military airborne navigation systems; very compact antennas that can be unobtrusively integrated into the receiver are used in handsets. A substrate with a higher dielectric constant reduces the size of the patch, but at the cost of decreasing both the bandwidth and gain of the antenna with increased radiation near the horizon from surface waves. A variety of substrates with dielectric constants ranging from as low as 1.07 to as high as 88 are available for building GNSS antennas depending on the application. More popular substrate materials are listed in Table 2.1 along with their dielectric constant and manufacturer. Three of these substrates with different dielectric constants will be selected later to illustrate the effects of the dielectric properties of the substrate on the performance of the GNSS patch antenna. Many of these substrates have electrodeposited copper cladding with a thickness of ½ mil to 1 mil on their upper and lower surfaces; this corresponds to a copper foil thickness of 0.0007 inches to 0.0014 inches, respectively. The desired shape and size of the patch can be obtained by either photo etching or milling the top copper cladding and the bottom copper cladding serves as the ground plane. High dielectric constant ceramic substrates with a single-probe feed are particularly popular for compact, narrowband GNSS antennas used in handsets. Since these antennas have very narrow bandwidths, the temperature stability of the dielectric constant is very important in ceramic substrates for preventing detuning if a large variation in the ambient temperature were to occur, as for example in avionics systems. Lower dielectric constant substrates such as foam or foam derivatives are also often used in conjunction with other higher dielectric constant substrates for improving bandwidth and are also used as substrates for multiband antennas and in antenna feeds. More

Table 2.1 Dielectric Substrates for GNSS Microstrip Antennas

Name of Substrate	*Dielectric Constant*	*Loss Tangent*	*Manufacturer*
Rohacell Foam	1.07	0.001	Rohm
Arlon Foam-Clad 100	1.15–1.35	0.002–0.004	Arlon; www.arlom-med.com
Duroid 5870	2.35	0.005	Rogers Corp.;www.rogerscorporation.com
RO3003	3.00	0.0013	Rogers Corp.
TMM4	4.50	0.0017	Rogers Corp.
TMM6	6.0	0.0018	Rogers Corp.
TMM10	9.2	0.0017	Rogers Corp.
RO3010	10.2	0.0023	Rogers Corp.
TMM13i	12.78	0.002	Rogers Corp.
SM200 (ceramic)	20	0.001	Kyocera North America; americas.kyocera.com
SB 350 (ceramic)	35	0.001	Kyocera North America
D88 (ceramic)	88 ± 2	Temperature coefficient = 0 ± 5 ppm^0 C^2	Morgan Electro Ceramics, U.K.; www.morgan-electroceramics.com

recently textile materials, including synthetic fabrics also called e-textiles, have been considered as flexible substrates for building body-worn, circularly polarized GPS antennas [17].

2.2.2 Effects of Surface Waves on Microstrip GNSS Antennas

All microstrip antennas need a grounded dielectric substrate as indicated in Figure 2.4; they therefore generate surface waves [24], which have a significant impact on performance when used in GNSS. As implied by its name these are electromagnetic modes of propagation that are trapped in the grounded dielectric substrate and travel via successive reflections between the dielectric-air boundary and the metallic ground plane underneath the substrate, as shown in Figure 2.6 Their field amplitudes decrease slowly with distance r, is the propagation distance. Surface waves propagate until they reach the truncated edge of the dielectric substrate or the edge of a finite (substrate-covered) ground plane where they are either diffracted or reflected. The diffracted signals then interact with the primary space wave radiation from the patch antenna and can cause numerous undesirable effects on GNSS performance. These effects include increased multipath from antenna back lobes, higher cross-polarization levels, phase variations, and ripples in the antenna pattern that are noticeable even at higher-elevation angles close to zenith. Surface waves have a deleterious impact on the performance of a GNSS adaptive antenna array due to increased mutual coupling between adjacent elements in the array; these effects are illustrated in Figure 2.6 and are also discussed in greater detail in Chapter 5. In beamforming arrays they can cause a change in the antenna pattern response of each array element requiring careful calibration of the amplitude and phase of each array element as a function of azimuth and elevation angle to compensate for their effects on beamforming. They can also produce enhanced interaction with the human operator in handset antennas.

Surface waves in microstrip antennas can be divided into two types: transverse magnetic (TM) mode and the transverse electric mode (TE) mode [18]. In TM surface waves, the magnetic field is parallel to the surface, whereas the electric field forms loops that extend vertically out of the surface. In TE surface waves the

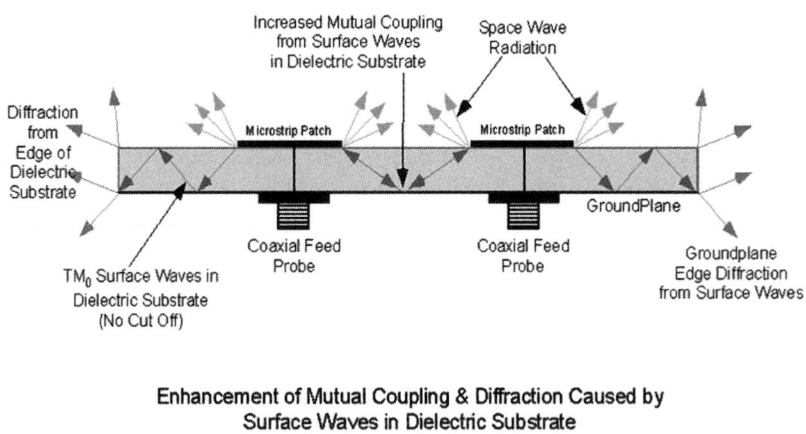

Figure 2.6 Diffraction effects and enhanced mutual coupling from surface waves in microstrip antennas.

electric field is parallel to the surface and the magnetic field forms vertical loops out of the surface. Cutoff frequencies for the different surface wave modes are given by $f_C = \frac{nc}{4h\sqrt{\varepsilon_r - 1}}$, where h is the thickness of the substrate, c is the velocity of light = 3×10^{10} cms per second, and ε_r = permittivity of the substrate; $n = 1, 3, 5$ for the TE_n modes and $n = 0, 2, 4$ for the TM_n modes. The lowest order TM_0 mode has no cutoff frequency, can be excited at any frequency, and has the greatest effect on GNSS antennas. It can only be mitigated by using a specially designed shorted annular ring microstrip antenna [19, 20]; this will be discussed later in this chapter under GPS multipath limiting microstrip antennas. Propagation of the lowest order TE_1 mode can however be avoided by keeping the thickness of the dielectric substrate less than h_c where $h_C = \frac{0.3c}{2\pi f_U \sqrt{\varepsilon_r}}$ and f_U is the maximum operating frequency. When $f_U = 1.5754$ GHz, $h_C \le \frac{0.3576}{\sqrt{\varepsilon_r}}$ inches. Surface wave energy propagated by the TE_1 mode can be decreased by keeping the thickness of the substrate below h_c. However, this can reduce the bandwidth and also the radiation efficiency. Alternatively, surface waves can be reduced by selecting a substrate with a lower dielectric constant but this can increase the size of the patch antenna, which is not desirable for many GNSS applications.

Despite its many disadvantages, not all of the surface wave effects created by the grounded dielectric substrate are necessarily detrimental for GNSS use. In an isolated antenna, surface waves can provide some unexpected benefits for GNSS applications such as increasing the gain of the antenna closer to the horizon, thereby promoting the acquisition of low-elevation satellites for improving PDOP. This is generally found difficult to achieve with other low-profile antenna designs where, due to their limited vertical height, they mainly produce broadside radiation directed towards zenith and away from low elevations.

The impact of surface waves on antenna performance is determined by the surface wave efficiency defined by $\eta_{sur} = 1 - \frac{P_{SUR}}{P_r}$, where P_{SUR} is power that is trapped within the surface wave and Pr is the total radiated power from the antenna. Surface wave efficiency will play a prominent role in the performance parameters of GNSS antennas as discussed later.

2.2.3 Design of Dual-Probe-Fed RHCP Single-Band Microstrip GNSS Antenna

Several types of feeding techniques have been devised for microstrip antennas with the goal of generating RHCP radiation for their use in GNSS systems. Some of these are quite complex as discussed later and can have a profound impact on its performance. We will first consider the design and performance of one of the simplest of the many methods for generating RHCP; namely, two direct-contact coaxial probes feeding a square-shaped RHCP microstrip antenna for operation over a single GNSS frequency band. We will also conduct a parametric study of this antenna to illustrate how various parameters affect its performance when used in GNSS. A schematic diagram of this antenna is shown in Figure 2.4(a). The antenna dimensions of the patch antenna are L along the x axis, W along the y axis, and h is the thickness of the dielectric substrate. The patch antenna is placed on top of a large square metallic ground plane with dimension = b. A square patch is needed for generating RHCP, so $W = L = a$. However, although the dimensions of this square patch along the y and x axes are the same, it is nevertheless useful to distinguish between these two sides of the patch by using letters W and L to aid in the discussion of TM and TE modal electric fields and their currents. An RHCP-radiated field is obtained by placing the two probes at orthogonal positions within the patch, as shown in Figure 2.4. Each probe is just an extension of the center conductor of the coaxial feed line and is soldered to the conducting patch antenna with the outer conductor of the coaxial line soldered to the ground plane. The coordinates of Probe 1 is $X = X_P$, $Y = +a/2$; and of Probe 2 is $x = a/2$, $y = y_p$. x_p is the distance of the center of Probe 1 from the upper horizontal edge of the patch antenna and Y_P is the distance of the center of Probe 2 from the left vertical edge of the patch. The locations of these two probes are precisely selected so that the input resistance of each probe is close to 50 ohms and the input reactance is close to zero, ensuring a good impedance match and low return loss to the feeding coaxial cables connected to the GNSS receiver. The two probes are connected to a 90° hybrid such as a branch-line coupler or a coaxial hybrid to generate RHCP signals. Probe 1 has 0° phase, whereas Probe 2 has –90° phase. The design of the branch-line coupler used with a microstrip antenna will be discussed later in this chapter. The electric fields inside the cavity are directed along the z axis and are independent of the z coordinate but vary along the x and y axes depending on the electromagnetic modes generated within the resonant microwave cavity of the patch antenna. These can be described in terms of the TM_{mn} resonant cavity modes identified by their specific double index (m,n). The integer mode index m of the TM_{mn} mode is related to the half-cycle variations of the electric field under the square patch over the width W along the x axis. The mode index n is related to the number of half-cycle electric field variations over the length L parallel to the y axis. L is approximately one-half wavelength in the dielectric substrate and the length of the patch $a \sim \lambda_0/2\sqrt{\varepsilon_r}$, where λ_0 is the wavelength at the resonant frequency of the antenna and ε_r is the relative dielectric constant of the substrate. For the TM_{mn} cavity mode of the rectangular patch, the z-directed electric field has the form

$$E_z(x,y) = A_{mn}\cos\left(\frac{m\pi x}{W_{eff}}\right)\cos\left(\frac{n\pi y}{L_{eff}}\right) \tag{2.1}$$

where A_{mn} is the modal amplitude of the cavity mode (m, n). The walls of the cavity are slightly larger electrically than they are physically due to the fringing electric field at the edges of the patch. To compensate for this the boundaries of the metallic patch are extended outwards and the new dimensions become W_{eff} and L_{eff} as indicated in (2.1). For GNSS applications, the lowest fundamental order modes, the TM_{10} or the TM_{01}, are selected to obtain an azimuthally symmetric RHCP pattern. To generate RHCP required for receiving the GNSS signals, the phase of Probe 2 is 0° and the phase of Probe 1 is –90°. Probe 1 excites the fundamental TM_{10} mode and is polarized linearly along the $+X$ axis. The electric field for radiation Probe 2, which is resonant in the TM_{01} mode, is polarized along the $+Y$ axis. The direction of the magnetic current flows and the radiating fields from each of these two probes is illustrated and explained in greater detail in Figure 5.1. The radiation fields of this patch antenna from these two pairs of slots parallel to the X and Y axes, respectively, generate orthogonally polarized electric field, which when combined in phase quadrature such as a branch-line coupler generate the required RHCP radiated signals for GNSS applications.

2.2.3.1 Computer Codes for Designing Microstrip Antennas

Despite its structural simplicity, an accurate design of a microstrip antenna that is needed for meeting various GNSS requirements requires the use of computational electromagnetic codes. A discussion of these codes is beyond the scope of this book, but they are well described in handbooks [21, 22] and also in two excellent textbooks devoted exclusively to microstrip antennas [18, 23]. A large number of computer codes are now available to the antenna engineer for designing microstrip antennas; some of the more frequently used codes are listed in Table 2.2. A comparison of the accuracy of some of these codes for microstrip antenna design has been conducted by Pozar et al. [24]. Due to the long computer run times that may be needed, these codes should be considered as suitable for only verifying and optimizing the final design; the starting initial design is often based on simplified computer-aided design (CAD) formulas or on engineering intuition of the antenna designer. Several hardware iterations may also be needed for achieving an optimum design that meets most of the antenna requirements since the computer models may not account for all the nuances in the actual design. Some specific examples of GNSS dual-band microstrip patch antennas designed using the High Fidelity System Simulation (HFSS) code will be described later in this section; the performance of these antennas have been verified through measurements and agree fairly well with design predictions.

2.2.4 A Parametric Study of a Single-Band RHCP Square-Shaped Microstrip Antenna

We will first consider the procedure for designing a basic GNSS microstrip antenna: an RHCP, single-band GNSS microstrip antenna fed with a single substrate layer and two feed probes as shown in Figures 2.4(a) and (b). Designing this simple antenna will help the reader to obtain a better understanding of how the various parameters can first be selected for the initial design needed for meeting the many

Table 2.2 Electromagnetic Codes Used for the Design and Analysis of Microstrip Antennas

Name of Software Code	*Analysis Technique*	*Software Company*
Ensemble (Designer)	Method of moments	Ansoft
HFSS	Finite element	Ansoft
IE3D	Method of moments	Zeland
Microwave Studio Suite & Microstrips	FDTD TLM	CST CST/Flomerics
FEKO	Full-wave method of moments	EM Software & Systems–S.A. (Pty) Ltd.

FDTD = finite difference time domain; TLM = transmission line modeling.

important GNSS requirements such as size, location of feed probes in the antenna, bandwidth, radiation efficiency, gain, beam width, polarization axial ratio, and stability of the antenna phase center.

Simple analytical formulas [21] and numerical calculations obtained from a CAD design tool by Sainati [25] will be used to provide a preliminary insight into selecting the parameters. The CAD design tool by Sainati is based on curve fitting of rigorous full-wave solutions and provides good initial estimates that compare well with several actual verification measurements. Three temperature-stable, low-loss dielectric substrates obtained from Table 2.1 will be considered in this study to show their effects on antenna performance; all three are made by the Rogers Corporation and have dielectric constants of 2.35 (Duroid 5870), 6.0 (TMM6) and 9.2 (TMM10). Although slightly less accurate, this approach will provide the reader with good physical insight into the role played by the various design variables in determining the performance of the microstrip antenna. It also serves as an initial starting point in the antenna design whose accuracy can be refined by using more advanced electromagnetic codes mentioned in Table 2.2 or through iterative measurements on prototype antenna models for arriving at a final optimized design. Later in this section we will consider dual-band stacked microstrip antennas designed using the more accurate HFSS code and provide examples of the measured performance of these antennas.

2.2.4.1 Resonance Frequency and Size of a Single-Layer Square-Shaped GNSS Microstrip Antenna

The resonance for the TM_{10} (or the TM_{01}) mode is given by

$$f_0 = \frac{c}{2a_{eff}\sqrt{\varepsilon_r}} \tag{2.2}$$

where c is the speed of light and ε_r is the relative permittivity of the dielectric substrate. To account for the fringing of the electric fields of the cavity at the edges of the patch, an effective length a_e for the dimensions of the square patch while calculating the resonance frequency is chosen as

$$a_{eff} = a + 2\Delta a \tag{2.3}$$

The Hammerstad formula for the effective length extension [26] and the effective dielectric constant caused by the fringing fields proposed by Schneider [27] can be used for resonance frequency calculation

$$\Delta a/h = 0.412\left[\frac{\left(\varepsilon_{eff} + 0.3\right)\left(\frac{a}{h} + 0.264\right)}{\left(\varepsilon_{eff} - 0.258\right)\left(\frac{a}{h} + 0.8\right)}\right] \tag{2.4}$$

where

$$\varepsilon_{eff} = \frac{\varepsilon_r + 1}{2} + \frac{\varepsilon_r - 1}{2}\left[1 + 10\frac{h}{a}\right]^{-\frac{1}{2}} \tag{2.5}$$

Figure 2.7 shows the variation in the size of a square-shaped patch antenna as a function of the thickness of the three selected dielectric substrates with permittivity $\varepsilon_r = 2.35$ (Rogers Duroid 5870), $\varepsilon_r = 6.0$ (Rogers TMM6), and $\varepsilon_r = 9.2$ (Rogers TMM10). Notice that the length of the sides of this square patch is reduced by nearly 50% as the dielectric constant is increased from 2.35 to 9.2. The reduction in size caused by increasing the dielectric constant of the substrate increases the beamwidth but reduces both the gain and bandwidth, as discussed later.

2.2.4.2 Locations of Feed Probes in an RHCP Patch Antenna

The location of the two feed probes to generate RHCP is the next critical step in the design of a GNSS patch antenna. As explained earlier, the two feed probes need to be located orthogonally at the correct distances from the edges of vertical and horizontal sides of the patch to obtain a good impedance match to the GNSS receiver. The two probes are connected to a quadrature hybrid, such as branch-line coupler, so as to have equal amplitudes but with a phase of 90 degrees relative to each other to generate RHCP radiation from the patch antenna.

The resonant radiation resistance R_r of a square or rectangular shape microstrip fed at an edge of the patch is given by [21, pp. 277–279]:

$$R_r = \frac{V_0^2}{2P_r} = \varepsilon_r \frac{Z_0^2}{120 I_2} \tag{2.6}$$

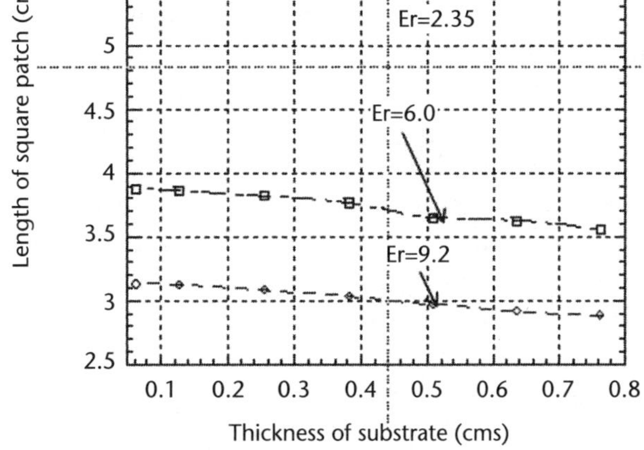

Figure 2.7 Variation in size of a square RHCP GPS microstrip antenna versus thickness and dielectric constant of the substrate; resonance frequency = 1.5754 GHz.

V_0 is the voltage across the representative edge slot in the radiating microstrip antenna, P_r is power radiated by the patch, and Z_0 is the characteristic impedance of the microstrip line of which the patch is a segment. The power P_r radiated by the antenna can be obtained by integrating the real part of the Poynting vector over the hemisphere above the disk.

$$P_r = \frac{1}{2\eta_0} \int_0^{2\pi} \int_0^{\frac{\pi}{2}} \left(|E_\theta|^2 + |E_\varphi|^2 \right) r^2 \sin\theta \, d\theta \, d\varphi \tag{2.7}$$

where E_θ and E_φ are the radiated electric field components along the $\hat{\theta}$ and $\hat{\varphi}$ vector directions of the far field. I_2 is a complicated function of several parameters such as ε_{re} the real part of the complex dielectric constant of the dielectric substrate, h is the thickness of the dielectric substrate, and $W = a$ is the width of the patch antenna [21, pp. 279].

As shown in Figure 2.4(b), the patch antenna is fed with the feed Probe 1 located at a distance on the X axis and at Y = a/2 from the top, horizontal radiating edge of the patch; its input resistance is obtained as

$$R_{in} = R_r \cos^2\left(\frac{\pi x_P}{L}\right) \tag{2.8}$$

A good impedance match to the receiver is obtained when input resistance of the Probe R_{in} = 50 ohms. Similarly the location of the other orthogonal feed Probe 2 is located at a distance y_P along the Y axis and at $X = a/2$ from the vertical, left edge of the patch. y_P can be calculated from (2.8) by substituting y_P for x_P and L for W, noting that for a square patch since $L = W = a$. In Figure 2.8 the offset distance of the probe from the edge of the patch, which is X_P for Probe 1 or Y_p for Probe 2, are plotted for three different substrates with dielectric constants e_r = 2.35, e_r = 6.0, and e_r = 9.2 as a function of the thickness of the dielectric substrate. Not surprisingly, the offset distance of the probe from the edge decreases as the dielectric constant of the substrate is increased since the size of the patch antenna also decreases. For a specific substrate the probe offset distance increases as the thickness is increased to about 0.25 cms, but then plateaus out with little change as the thickness is increased further.

2.2.4.3 Bandwidth of a GNSS Microstrip Antenna

The bandwidth of the antenna is the frequency range over which a selected performance parameter of the antenna has satisfactory operation. Depending on the specific performance parameter selected, three different definitions of bandwidth are possible: the impedance bandwidth (or return loss bandwidth), gain bandwidth, and axial ratio bandwidth. The impedance or the return loss bandwidth denotes a frequency range over which 89% of the signal power received by the patch antenna is transferred to the GNSS receiver, representing a return loss of –9.5 dB (11% reflected power). The gain bandwidth is defined in terms of the frequency range over which the antenna provides a gain better than the minimum threshold gain needed

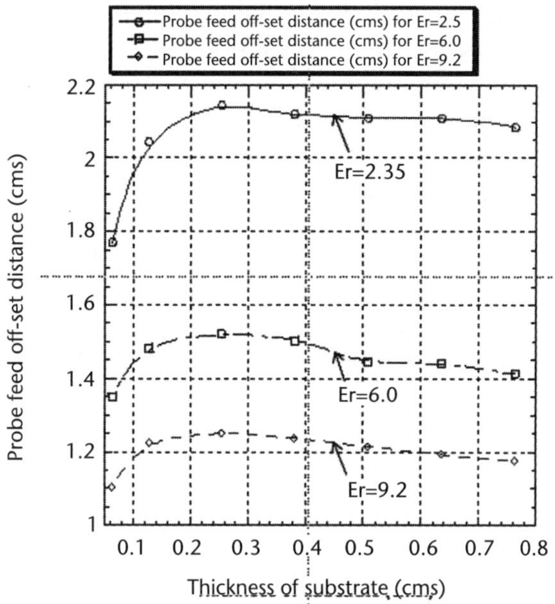

Figure 2.8 Offset distance X_P or Y_P of probe feed from edge of RHCP GPS microstrip antenna versus substrate thickness and dielectric constant. Resonance frequency = 1.5754 GHz.

to acquire the GNSS satellites within a viewing region covered by the entire upper hemisphere down to the minimum specified low-elevation masking angle. The axial ratio bandwidth is defined by the frequency range in which the antenna is able to meet the maximum cross-polarization axial ratio level. Generally the impedance bandwidth of patch antennas is much narrower than either its gain or axial ratio bandwidth. The return loss bandwidth, which can often be smaller than the gain bandwidth by nearly a factor of 10 [28, pp. 17–18] is therefore a more difficult requirement to meet. The impedance bandwidth is also relatively easier to measure using a vector network analyzer whereas the measurement of gain or axial ratio needs some form of antenna range that is more expensive and less readily available. Hence the bandwidth of GNSS antennas is often defined only in terms of impedance bandwidth, although in terms of actual GNSS performance the gain bandwidth of the antenna is probably a far more meaningful parameter for determining satellite acquisition and also the signal quality (carrier-to-power noise ratio [C/N_0] based on the sensitivity of the receiver. The impact of the impedance bandwidth on the phase characteristics also need to be considered carefully while considering group delay effects, especially in wideband signal BOC waveforms such as the M code in modernized GPS and similar waveforms in Galileo. A good polarization axial ratio is an important factor in GNSS since it reduces multipath. This is because the signal generated by the first reflection off a multipath source is LHCP and this is reduced by an antenna with a good axial ratio.

The impedance bandwidth (BW) is determined by the maximum voltage standing-wave ratio (VSWR) defined by the symbol S.

$$BW = \frac{S-1}{Q_T\sqrt{S}} \tag{2.9}$$

In the above equation, Q_T is the total quality factor of the patch antenna [21] and S is defined in terms of the input reflection coefficient Γ as

$$S = \frac{1 + |\Gamma|}{1 - |\Gamma|} \tag{2.10}$$

The reflection Γ coefficient is a measure of the reflected signal at the antenna feed point before it is attached to the quadrature hybrid or branch-line coupler used for generating RHCP. It is defined in terms of the input impedance Z_{in} of the antenna and the characteristic impedance Z_0 of the transmission line feeding the antenna as given below:

$$\Gamma = \frac{Z_{in} - Z_0}{Z_{in} + Z_0} \tag{2.11}$$

The impedance bandwidth BW is normally specified as the frequency range over which S is less than 2. This corresponds to a power loss of 0.454 dB in the power delivered from the antenna to the receiver or to a return loss of –9.5 dB or to 11% of reflected power. The bandwidth of a microstrip antenna can be increased by reducing the Q factor as indicated in (2.6). The Q factor is proportional to the dielectric constant of the substrate and inversely proportional to its thickness. Hence the bandwidth can be increased by selecting a thicker substrate with a low-dielectric constant among the list of substrates listed in Table 2.1. However, this would increase the size of the microstrip patch antenna, which may not be desirable or allowable in certain commercial and military applications.

A good closed-form approximation for the impedance bandwidth of a patch antenna is [21]

$$BW = \frac{16}{3\sqrt{2}} \frac{p}{e_r} \left(\frac{1}{\varepsilon_r}\right) \left(\frac{h}{\lambda_0}\right) \left(\frac{W}{L}\right) q \tag{2.12}$$

where

$$p = 1 - \frac{0.16605}{20} (k_0 W)^2 + \frac{0.02283}{560} (k_0 W)^4 - 0.0019142 (k_0 L)^2 \tag{2.13}$$

$$q = 1 - \left(\frac{1}{\varepsilon_r}\right) + \frac{2}{5\varepsilon_r^2} \tag{2.14}$$

e_r = radiation efficiency of the antenna. Expressions for e_r are derived in detail in the next section.

Figure 2.11 shows the change in bandwidth as a function of the thickness of the dielectric substrate for three different values of the dielectric constant of the substrate: $\varepsilon_r = 2.35$, $\varepsilon_r = 6.0$, and $\varepsilon_r = 9.2$. Notice that for a fixed thickness of the substrate, the bandwidth of the patch antenna decreases as the dielectric constant is increased; also for a fixed dielectric constant of the substrate the bandwidth increases as the substrate increases in thickness from 0.025" to 0.3".

While measuring the return loss of a GNSS antenna to determine its bandwidth using a network analyzer, it is important to note many commercial GNSS antennas already have a quadrature hybrid, such as a branch-line coupler, built directly across the antenna terminals and concealed inside the antenna package to obtain RHCP. Hence it is impossible to separate out the effects of this hybrid coupler on the antenna impedance. This prevents a measurement of the actual return loss of the antenna since the signal reflected from the input port of the antenna is diverted to the fourth port of the hybrid coupler, which has a 50-ohm matched termination. The hybrid coupler has a much broader bandwidth than the antenna; the signal reflected back from the antenna into the input terminal, where the return loss is being measured, would be negligible, resulting in a very flat response over a large frequency band. This would then lead to the erroneous conclusion that the antenna bandwidth is much broader than its true bandwidth. In cases where a hybrid coupler is built into the antenna module, a measurement of the antenna gain versus frequency is a more accurate measure of the true bandwidth of a GNSS antenna than the conventional definition using reflection loss at the input port.

In the case of an active GNSS antenna with a low-noise LNA built into the antenna package, the definition of antenna bandwidth becomes even more ambiguous since it is impossible to separate the frequency behavior of the antenna from the LNA. In such cases the only criteria for defining antenna performance becomes the gain/noise temperature or the G/T ratio as explained in Chapters 1and 4.

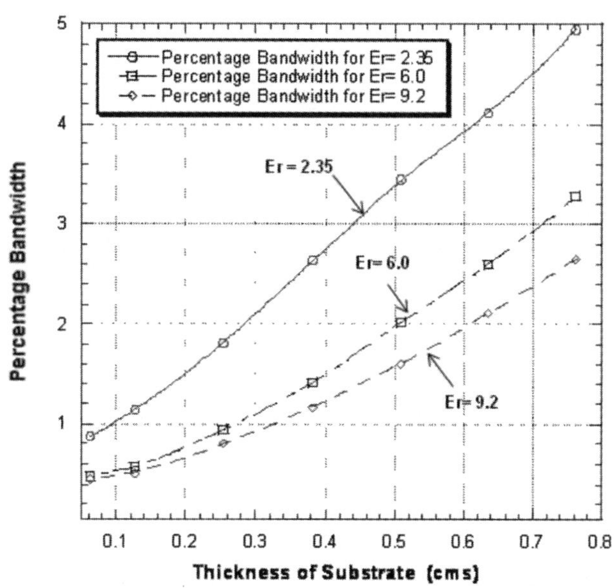

Figure 2.9 Variation in percentage bandwidth versus thickness and dielectric constant of the substrate. Resonance frequency = 1.5754 GHz.

2.2.4.4 Radiation Efficiency of an Antenna

The radiation efficiency e_r is defined by the ratio of radiated power P_r to the input power $P_{i.}$ The input power P_i is distributed between the radiated power P_r, P_{sur} is the surface wave power propagated in the grounded dielectric substrate, Pc is the power dissipated in the metallization used for the patch antenna, and P_d is the power dissipated in the dielectric substrate. P_c is proportional to the square root of the conductivity of the patch metallization and P_d is proportional to the loss tangent of the dielectric substrate. For low-loss dielectric substrates with copper cladding, the dielectric loss P_d and conductor loss P_c are both very small so the radiation efficiency can be simplified and expressed as

$$e_r = \frac{P_r}{P_i} = \frac{P_r}{\left(P_r + P_c + P_d + P_{sur}\right)} \approx \frac{P_r}{\left(P_r + P_{sur}\right)} \tag{2.15}$$

Closed form expressions for P_r and P_{sur} has been derived [21, pp. 285].

$$P_r = 40k_0^2\left(k_0 h\right)^2\left[1 - \frac{1}{\varepsilon_r} + \frac{2}{5\varepsilon_r^2}\right] \tag{2.16}$$

$$P_{sur} = 30\pi k_0^2 \frac{\varepsilon_r\left(x_0^2 - 1\right)}{\varepsilon_r\left[\frac{1}{\sqrt{x_0^2 - 1}} + \frac{\sqrt{x_0^2 - 1}}{\varepsilon_r - x_0^2}\right] + k_0 h\left[1 + \frac{\varepsilon_r^2\left(x_0^2 - 1\right)}{\varepsilon_r - x_0^2}\right]} \tag{2.17}$$

where

$$x_0 = \frac{\beta}{k_0} \approx 1 + \frac{1}{2}\left(\frac{\varepsilon_r - 1}{\varepsilon_r}k_0 h\right)^2$$

x_0 is the normalized phase constant of the TM_0 of the surface mode which has no cutoff frequency. The suppression of the surface wave mode will be discussed in greater detail in Section 2.2.8.1.

Figure 2.10 shows the variation in radiation efficiency of the square-patch antenna versus the dielectric constant and thickness of the dielectric substrate. Notice that the radiation efficiency is poor for thin substrates, but it improves rapidly as the substrate thickness increases; it then levels out for low-dielectric substrates such as for $er = 2.35$ since these lower dielectric substrates do not generate much surface waves except for the fundamental TM_0 mode. However, as the thickness increases for substrates with higher dielectric constants, as shown for $er = 6.0$ and $er = 9.2$ in Figure 2.10, the increase in surface wave radiation P_{sur} decreases the radiation efficiency as indicated in (2.15) when the thickness in increased beyond a certain level. This increase in surface wave radiation is not necessarily detrimental for GNSS

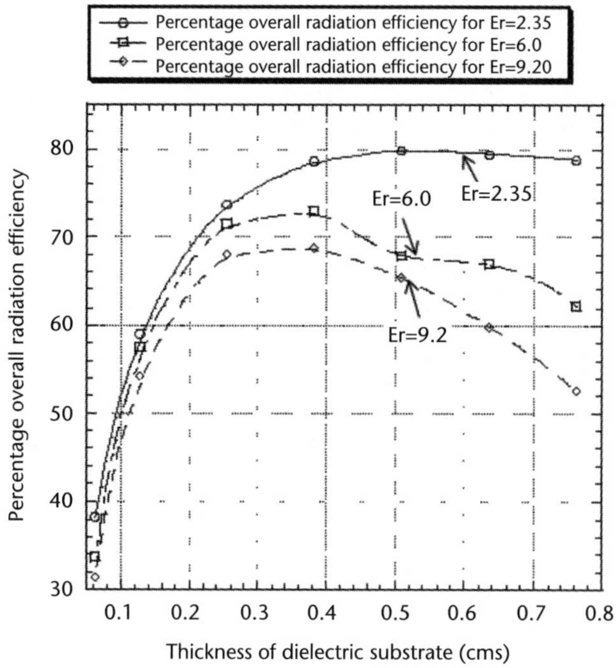

Figure 2.10 Percentage overall radiation efficiency versus substrate thickness and dielectric constant. Resonance frequency = 1.5754 GHz.

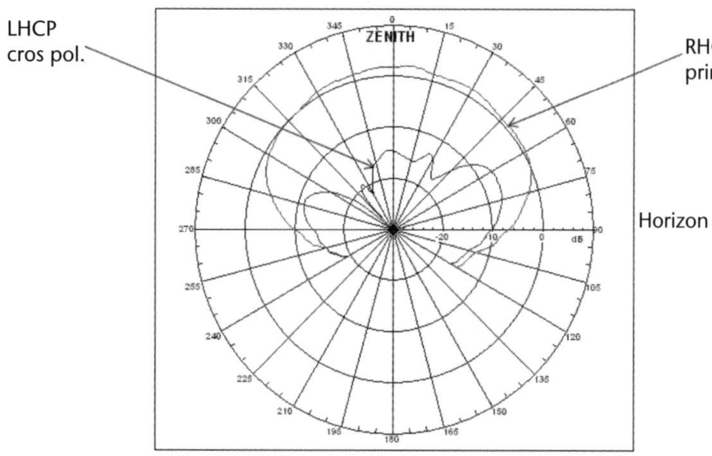

Figure 2.11 Measured radiation pattern of a typical GPS microstrip FRPA antenna. Resonance frequency = 1.5754 GHz.

applications since it boosts the gain at lower-elevation angles (makes the antenna beamwidth broader and lowers directivity), allowing the antenna to acquire low-elevation satellites and thereby improving PDOP.

2.2.4.5 Radiation Pattern and Axial Ratio of an RHCP Square Microstrip Antenna

The radiation pattern of an RHCP square-shaped microstrip antenna of length L and width W is equal to a is placed on a very large (infinite) ground plane can be calculated by modeling the antenna as a combination of two orthogonal pairs of slot antennas with each pair of slots excited separately by a directly coupled probe, as shown in Figures 2.4(a) and (b). The first probe, Probe 1, excites the TM_{10} mode in the two parallel slots of length $L = a$ parallel to the Y axis and are spaced at a distance $W = a$ along the X axis. The radiation pattern of the two parallel slots is linearly polarized with the electric field directed along the X axis parallel to the spacing distance W. The second probe, Probe 2, excites the TM_{01} mode in the other pair of parallel slots of length $W = a$ oriented parallel to the X axis and spaced at a distance $L = a$ apart along the Y axis. The radiation pattern of this second parallel slot pair is also linearly polarized but with the electric field directed along the Y axis, orthogonal to the fields produced by the first pair of slots. RHCP is obtained by making the amplitude of the two orthogonal linearly polarized fields generated by the two slot pairs to be equal in amplitude but with a relative phase difference of 90°, with Probe 1 with 0° phase, and Probe 2 with –90° phase. If the voltage across either radiating slot is taken as V_0, the radiation fields for the pair of slots can be obtained by multiplying the radiation pattern of a single slot with an array factor to represent the parallel pair. For Probe 1 exciting the TM10 mode, the component E_θ and E_φ of the far-field radiation is given by:

$$(E_0)_{TM_{10}} = -jk_0 V_0 L \frac{e^{-jk_0^r}}{4\pi r} \cos\varphi F_1 F_2 \tag{2.18}$$

$$(E_0)_{TM_{10}} = -jk_0 V_0 L \frac{e^{-jk_0^r}}{4\pi r} \cos\theta \sin\varphi F_1 F_2 \tag{2.19}$$

where

$$F_1 = \sin c\left\{\frac{k_0 h \sin\theta \cos\varphi}{2}\right\} \sin c\left\{\frac{k_0 L \sin\theta \sin\varphi}{2}\right\} \tag{2.20}$$

and

$$F_2 = 2\cos\left\{\frac{k_0 W \sin\theta \cos\varphi}{2}\right\} \tag{2.21}$$

Similarly, for Probe 2 exciting the TM_{01} mode, the component E_θ and E_φ of the far-field radiation is given by

$$\left(E_0\right)_{TM_{01}} = jk_0V_0W\frac{e^{-jk_-^r}}{4\pi r}\sin\varphi F_1^I F_2^I \tag{2.22}$$

$$\left(E_\varphi\right)_{TM_{01}} = jk_0V_0W\frac{e^{-jk_-^r}}{4\pi r}\cos\theta\cos\varphi F_1^I F_2^I \tag{2.23}$$

where

$$F_1^I = \sin c\left\{\frac{k_0 h\sin\theta\sin\varphi}{2}\right\}\sin c\left\{\frac{k_0 W\sin\theta\cos\varphi}{2}\right\} \tag{2.24}$$

and

$$F_2^1 = 2\cos\left\{\frac{k_0 L\sin\theta\sin\varphi}{2}\right\} \tag{2.25}$$

Since we are considering a square-shaped patch antenna, $W = L = a$ in the above equations.

Signals for the TM_{10} and TM_{01} modes generated in the patch antenna are parallel to the $\hat{x}$ and $\hat{y}$ axes, respectively. RHCP is generated when these two orthogonal signals are combined with equal amplitudes but in phase quadrature such that their ratio $\frac{E_X}{E_Y} = j$; this is achieved through a branch-line coupler.

By examining the equations above the following features are noticed:

1. At zenith $\theta = 0$ and, $\cos\theta = 0$, $\vec{E}_{RHCP} = [C]\left\{\hat{\theta}[\sin\varphi + j\cos\varphi] + \hat{\varphi}[\cos\varphi - j\sin\varphi]\right\}$ where $\vec{E}_{RHCP} = [C]\hat{e}_r$ is the RHCP electric far field from the patch antenna expressed in spherical coordinates with the wave traveling in the +r direction and. $C = k_0V_0a\frac{e^{-jk_0^r}}{4\pi r}$ $\hat{e}_r = \hat{\theta}[\sin\varphi + j\cos\varphi] + \hat{\varphi}[\cos\varphi - j\sin\varphi]$ is defined as the RHCP complex unit vector [29, p. 59]; hence at zenith the patch antenna is purely RHCP and the CP axial ratio is 1. Similarly, for a wave traveling in the +r direction we can also define (using this same type of derivation) an LHCP complex unit vector $\hat{e}_l = \hat{\theta}[\sin\varphi + j\cos\varphi] + \hat{\varphi}[\cos\varphi - j\sin\varphi]$. The LHCP electric far field from the patch antenna expressed in spherical coordinates can now be represented as $\vec{E}_{LHCP} = [C]\hat{e}_l$. Two parameters are often used to define purity of CP. The most commonly used parameter is the axial ratio "R" for CP defined as $R = \frac{\vec{E}_{RHCP} + \vec{E}_{LHCP}}{\vec{E}_{RHCP} - \vec{E}_{LHCP}}$; a second parameter used less often is the polarization ratio for CP $\rho_C = \frac{\vec{E}_{RHCP}}{\vec{E}_{LHCP}}$. Figure 2.13 shows the measured RHCP (i.e., principal polarization) and LHCP (i.e., cross-polarization) radiation patterns of a microstrip patch antenna measured in the upper hemisphere down to

an elevation of −30° below the horizon at a frequency of 1.5754 GHz. A 51" diameter rolled edge ground plane was used in this pattern measurement; this ground plane is shown in Figure 2.12(d). The edges of the ground plane are rolled inwards below the top surface of the ground plane to prevent the signals diffracted from the edges from affecting the radiation pattern of the antenna in the upper hemisphere.

2. We notice that for a microstrip patch antenna located on an infinitely large ground plane, the horizontally polarized $\hat{\phi}$ component $(E_\varphi)_{TM_{10}}$ and $(E_\varphi)_{TM_{01}}$ of the far-field radiation for the TM_{10} and the TM_{01} modes are both zero at the horizon since $\cos\theta = 0$. Hence the far-field radiation from the patch antenna at the horizon is linearly polarized with just the vertically polarized E_θ component. This can also be concluded by noticing that the horizontal E_ϕ component being tangential to the large conducting ground plane needs to be zero. The axial ratio for CP of the antenna is ∞ at the horizon. Hence the purity of CP and axial ratio of a patch antenna on an electrically large ground plane is elevation angle-dependent with good RHCP obtained only near zenith but with progressive degradation in axial ratio as the horizon is approached. The polarization becomes vertically (i.e., linearly) polarized at the horizon with no perceptible difference between RHCP and LHCP.

2.2.4.6 Half-Power Beamwidth Gain and Directivity of an RHCP Microstrip Antenna

The half-power beamwidth (HPBW) of a receiving antenna is defined as the angular width between directions where the power of the received radiation is decreased by 3 dB. A more appropriate performance measure for a GNSS antenna is its minimum

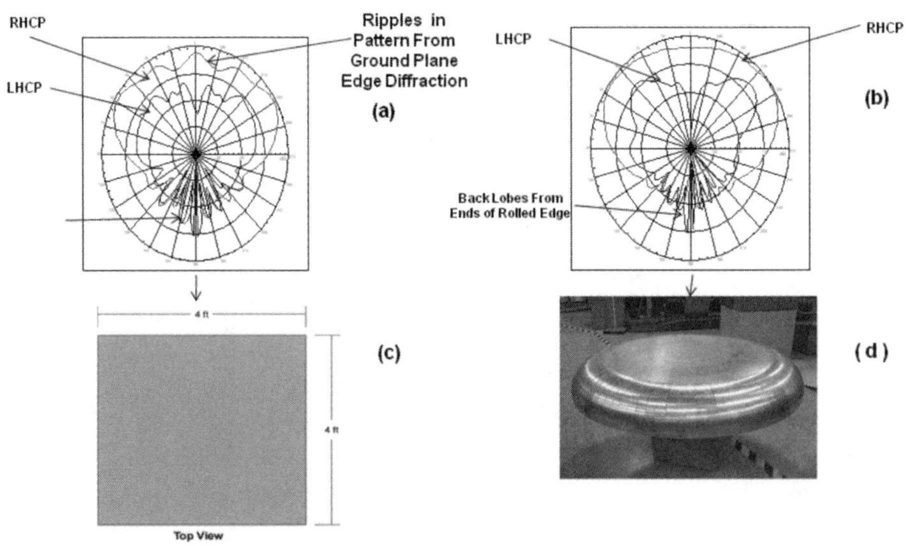

Figure 2.12 Measured radiation pattern of a GNSS patch antenna on two different types of finite size ground planes. (a) Measured pattern of 4-foot square-shaped ground plane, (b) measured pattern on rounded edge ground plane, (c) diagram of 4-foot square ground plane, and (d) actual rounded edge ground plane. Resonance frequency = 1.5754 GHz.

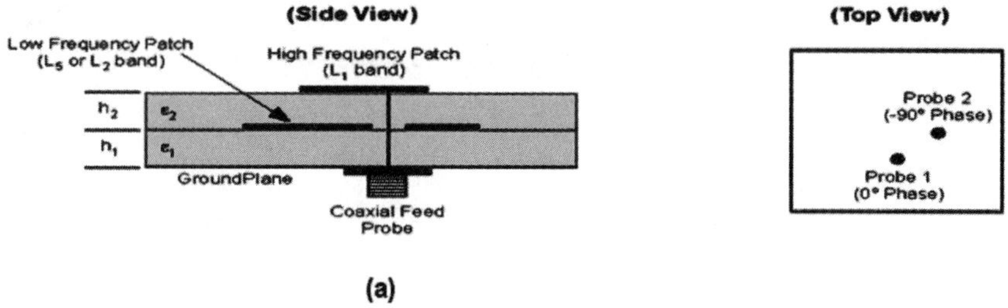

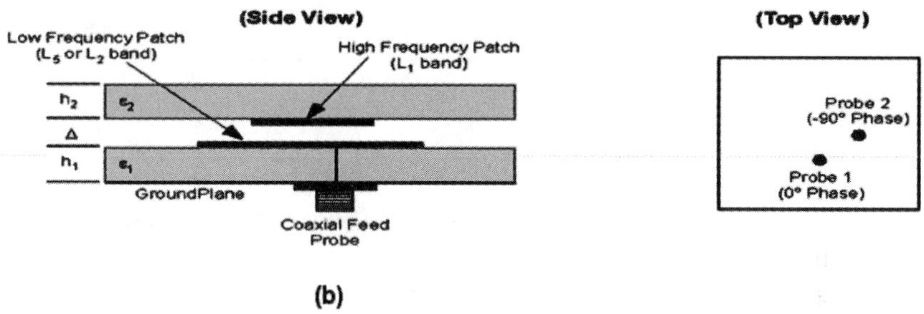

Figure 2.13 GNSS dual-band RHCP microstrip antennas with direct contact feed probe. (a) Feed connected to top patch, and (b) feed connected to bottom patch; inverted configuration.

gain beam width or the angular region in the upper hemisphere above the minimum masking elevation angle where the antenna is able to meet the minimum gain needed to acquire GNSS satellites when used with a specific receiver (see Figure 2.2). The minimum gain requirement that the antenna is required to meet to acquire the satellite was discussed in Chapter 1. The HPBW θ_E in the E plane where $\phi = 0$ and the HPBW θ_H in the H plane where $\phi = 90°$ of a single-layer patch antenna can be determined from its dimensions [21, p. 276–277].

$$\theta_H = 2\ \arcsin\left\{\frac{1}{2 + k_0 L}\right\}^{\frac{1}{2}} \tag{2.26}$$

$$\theta_E = 2\arcsin\left\{\frac{7.03}{3k_0^2\, W^2 + k_0^2 h^2}\right\}^{\frac{1}{2}} \tag{2.27}$$

Since we are considering a square-shaped patch antenna $W = L = a$, where a is the dimension of the patch and h is the thickness of the dielectric substrate. The beamwidth of the antenna can be increased by decreasing the size of the patch antenna by reducing a; this can be accomplished by selecting a substrate with a high dielectric constant. This helps the antenna to meet the minimum gain beam width required to acquire GNSS satellites at even at lower-elevation angles. A reduced size GNSS antenna is also a requirement for many commercial and military applications, especially for avionics. The smaller size, however, needs to be balanced against the resulting reduction in bandwidth and decrease in antenna gain and directivity. The resulting reduction in bandwidth can be recovered by either increasing the thickness of the dielectric substrate or by using stacking patch antennas tuned to different frequencies on top of each other as discussed later in this chapter.

The directivity of the antenna is a measure of its directional properties compared to that of an isotropic antenna and is defined as the ratio of the maximum power density in the main beam direction to the average radiated power density. The directivity D of the patch is expressed as

$$D = \frac{\left(\frac{r^2}{2\eta_0}\left\{\left|E_\theta\right|^2 + \left|E_\varphi\right|^2\right\}_{\theta=0}\right)}{\frac{P_r}{4\pi}} \tag{2.28}$$

where P_r is the radiated power, $\eta_0 = 120\,\pi$, and the radiated fields E_θ and E_ϕ were defined earlier in (2.22) through (2.26). A simple approximate expression for the directivity D of the patch antenna is given by

$$D = \frac{4\left(k_0 a\right)^2}{\pi\eta_0 G_r} \tag{2.29}$$

where G_r is the radiation conductance of the antenna and $G_r = \frac{1}{R_r}$, where R_r is the radiation resistance defined earlier. The gain G of the antenna is defined as $G = e_r D$ where e_r is the radiation efficiency of the antenna. Gain is always less than the directivity because the efficiency e_r is in the range $0\langle e_r \rangle 1$.

2.2.4.7 Finite Ground Plane Effects on the Antenna Pattern of Microstrip GNSS Antennas

The interaction between a microstrip antenna and a finite-size ground plane on which it is mounted is a complex problem with serious consequences on GNSS performance. This topic is discussed in greater detail in Chapter 5, including methods used for both the analysis and the mitigation of such effects. The equations given above for the radiation patterns should be considered as valid only for an infinitely large ground plane and dielectric substrate. A finite ground plane has significant effects on the radiation pattern of a GNSS antenna from diffraction caused from the edges of the ground plane. These diffracted signals cause ripples to occur in the antenna pattern at higher-elevation angles close to zenith, affects the RHCP gain at lower elevations by changing the axial ratio, and also creates backlobes that can degrade the front-to-back ratio of the pattern and make the antenna susceptible to multipath and interference. Approximate expressions to account for these ground plane effects have been provided [21, pp. 293–296].

Figure 2.12 shows the radiation pattern of a microstrip patch antenna measured on two different types of ground planes, which illustrates how the size and shape of the ground plane influences the antenna pattern. Figure 2.12(a) shows measured radiation pattern of the microstrip antenna at a frequency of 1.5754 GHz when it is placed at the center of a 4-foot square planar ground plane. The cross section of this 4-foot square ground plane is shown in Figure 2.12(c).The large ripples seen in the main beam pattern of the antenna shown in Figure 2.12(a) are caused by straight edge diffraction from the ends of this planar ground plane. Figure 2.12(b) shows measurements made on the same antenna when it is placed at the center of a 51" diameter rolled-edge ground plane whose edges have been rolled underneath the ground plane to reduce the impact of diffraction effects on the antenna pattern in the upper hemisphere. A picture of the rolled-edge ground plane is shown in Figure 2.12(d). The main beam is broader and smoother since the diffraction effects have been reduced due to rolling the edges. Notice the neither of these two ground planes is able to suppress the antenna backlobes; this would require the use of more sophisticated ground planes designs such as choke ring ground planes, electronic bandgap ground planes, and resistivity tapered ground planes designed specifically to mitigate the diffraction effects from the edges of the ground plane. The design of these special ground planes are discussed in greater detail in Chapter 5.

2.2.5 GNSS Dual-Band Stacked Microstrip Patch Antennas

Antennas needed for modernized GPS, Galileo, and GLONASS may need to operate in two or even three separate frequency bands for providing the highest measurement accuracy. Modernized GPS applications, for example, will require the antenna to have a minimum operating bandwidth of 20 MHz for civilian use and 24 MHz for military applications (for GPS M code) in each frequency band; the antenna also needs to be RHCP to allow optimum reception of the satellite signals. The instantaneous bandwidth needed to cover the entire frequency range of modernized GPS, Galileo, and GLONASS extending from 1166 MHz (the lower end of the L_5 band) to 1607 MHz (the upper end of the GLONASS G1 band) would be 31.8%, covering a frequency span of 441 MHz. It would be difficult to meet this wide instantaneous bandwidth with a single microstrip patch antenna, which being a resonant device with a high Q generally has a bandwidth of just a few percent, as shown in Figure 2.9. Since these antennas also need to be small in size for most GNSS applications, substrates with high dielectric constants are frequently used in their construction, which further restricts their bandwidth. However each specific GNSS band that the patch antenna is needed to cover is no more than 24-to 30 MHz wide; one method of circumventing the large instantaneous bandwidth problem is by integrating two or more narrowband microstrip antennas with each antenna covering only the required bandwidth of 4 MHz with a performance similar to that of a notched passband filter. This technique also provides the added advantage of avoiding out-of-band interference with each antenna acting as its own filter. However an integrated feeding technique is needed to connect the two or more patch antennas operating in different frequency bands to a common GNSS receiver that is often used to process signals in the various receiving bands. One popular technique that has been used to obtain dual-band performance is by "stacking" multilayer resonant patch antennas vertically, one on top of the other with the upper patch resonating at the higher frequency and the lower patch at a lower-frequency, as shown in Figure 2.13. The stacked patch antennas are all fed with a common feed and polarization network that is connected to the GNSS receiver. These stacked patch designs have been described in several textbooks [33, 22]. Stacking also prevents increasing the lateral dimensions of the patch antenna, a key requirement in many GNSS applications such as avionics and military systems.

Two feed techniques have been used to feed multilayer patch antennas with the same set of two orthogonally placed probes. The first method is called the top-feed technique and is shown in Figure 2.13(a). The inner conductor of each feed probe passes through the bottom patch without making electrical contact; a small circle of the lower patch is removed for this purpose. The inner conductor proceeds through the dielectric substrate of the top patch and is then soldered to the top patch. The location of the two probes within the patch is optimized to provide a fairly good impedance match at both frequency bands. The larger bottom patch acts as a ground plane for the smaller top patch; the upper patch when resonant at the higher frequency has negligible reactance effect on the bottom patch effect and vice versa. Since there is strong electromagnetic coupling between the top and bottom patches, their resonant sizes can be determined accurately only through the use of advanced electromagnetic codes listed in Table 2.2. The HFSS code (described in Section 2.3.5.1) has been used for designing a dual-band patch antenna.

A second direct feed method that is less commonly used is the bottom-feed technique shown in Figure 2.13(b). Here the feed probe is connected only to the lower-band patch antenna at the bottom of the stack and the higher-frequency patch at the top is parasitically coupled to the bottom fed patch. The bandwidth of these bottom probe-feed stacked patch antennas can also be increased further by using the hi-lo stacked patch design obtained by selecting the appropriate substrate materials for the top and bottom patch antennas [28, pp. 56–67]. In this design the lower patch antenna element is mounted on a high dielectric constant substrate and the upper patch is mounted on a substrate of foam with $\varepsilon_r \sim 1.07$. This design would cause the size of the patch antenna to increase. However, this technique does not appear to have been tried for building GNSS antennas. Even larger bandwidths can be obtained by using aperture coupled, dual-band stacked patches [30]. Triple-band antennas have been developed by stacking up to three patch antennas and by using aperture-coupling techniques; these will be discussed under multiband GNSS antennas in Chapter 3.

2.2.5.1 Design of a Dual-Band Microstrip Antenna through HFSS Computer Simulations

As mentioned earlier there are no convenient CAD tools currently available to simplify the design of a dual-band stacked patch antenna with common feed probes for both bands; this is due to the complex coupling between the top and bottom patches. This can only be done accurately by using one the advanced computer design codes listed in Table 2.2.

In this section we will describe the design of a dual-band stacked patch antenna obtained from simulations using the HFSS computer code. A picture of the antenna that was designed and built is shown at the left in Figure 2.14(a); the corresponding HFSS simulation model of this antenna is shown at the right in Figure 2.14(b). The dielectric substrate used for this antenna was Rogers TMM4 with a dielectric constant of 4.5 and loss tangent of 0.002. The feed design used was the top-feeding technique shown earlier if Figure 2.13(a). Two direct-contact top-feed probes are

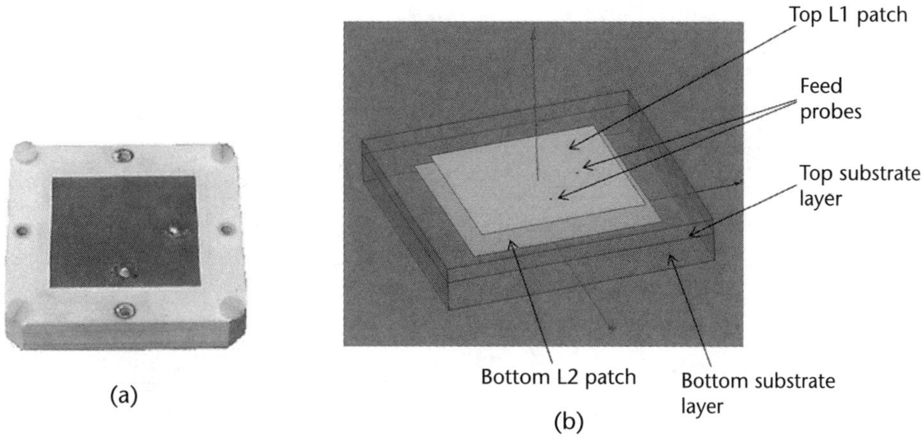

Figure 2.14 (a) Dual-band L_1/L_2 stacked microstrip GNSS antenna, and (b) antenna and HFSS simulation model.

used to feed both the L1 patch antenna at the top and the also the L2 patch antenna at the bottom of this stacked structure. The top L_1 patch is 1.68" square and the bottom L_2 patch is 1.94" square. The top and bottom substrate layers are 0.150" and 0.3" in thickness, respectively. The size of the truncated dielectric substrate is 2.75". The return loss at the two frequency bands, the Smith chart showing dual resonances as calculated by the HFSS program, are shown in Figures 2.15(a) and (b), respectively. The measured return loss for this antenna is shown in Figure 2.16 and agrees well with the HFSS simulations shown in Figure 2.15(a); the measured RHCP and LHCP antenna patterns and gain at 1.5754 and 1.2276 GHz, and the center band frequencies of the GPS L_1 and L_2 bands are shown in Figure 2.17.

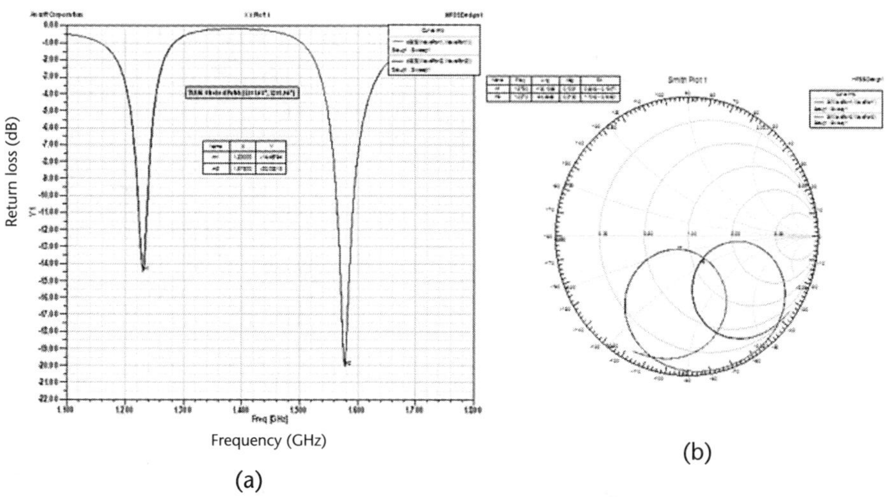

Figure 2.15 Results of HFSS simulations of a dual-band GPS L_1 and L_2 microstrip antenna: (a) return loss, and (b) Smith chart resonances.

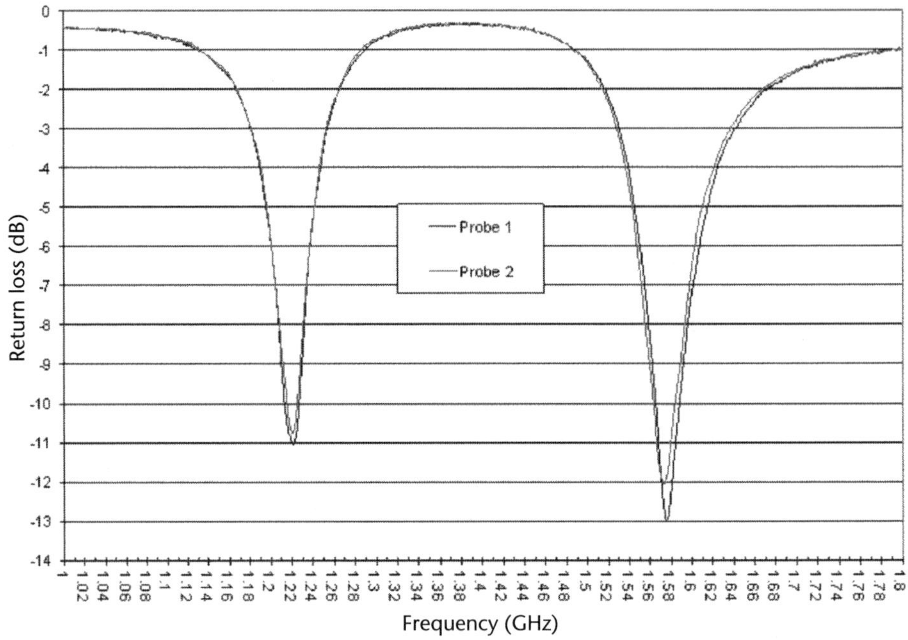

Figure 2.16 Measured return loss of the GPS dual-band L_1/L_2 stacked microstrip antenna; substrate: Rogers TMM4.

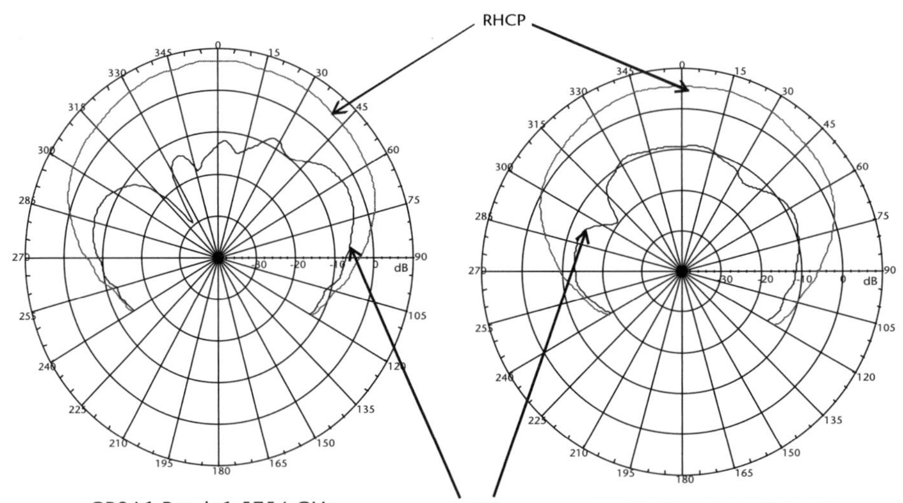

Figure 2.17 Measured radiation pattern of a dual-band stacked microstrip antenna; GPS L_1/L_2 bands.

2.2.5.2 Coplanar Dual-Band GPS Patch Antennas

A second method for designing a compact, dual-band RHCP microstrip antenna for the L_5/L_1 bands of GPS is shown in Figure 2.18. This design uses a thin circular ring antenna resonant in the L_5 band (the lower-frequency band) that is parasitically coupled to a concentrically located circular patch antenna resonant in the

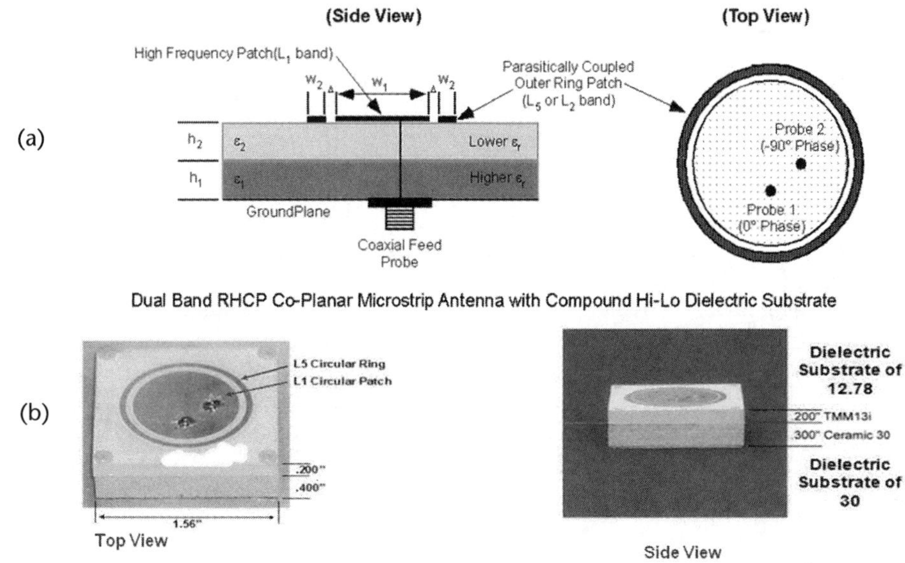

Figure 2.18 Coplanar dual-band RHCP microstrip patch antenna for the GPS L_1/L_5 bands. (From [31] ©2009 ION.)

higher-frequency band L_1 [31]. Schematic sketches of the top and side view cross section of this dual-band antenna is shown in Figure 2.18. Pictures of the top and side views of this antenna are shown in Figure 2.18(b). This compact antenna meets the ARINC 743 height and width requirement needed for avionics; the ARINC cross-sectional requirement was shown earlier in Figure 2.3. The centrally located circular patch in this design is fed by two or more direct-contact feed probes for generating RHCP in both patch antennas. In this design antennas operating in both frequency bands are coplanar, which allows them to share a common dielectric substrate including a compound substrate consisting of two or more different substrate materials with different thicknesses and dielectric constants.

The new design will allow increased flexibility in controlling the size of the antenna elements and provide wide antenna patterns with good gain coverage at the lower-elevation angles in both the frequency bands. Good gain at lower elevations is particularly important in the L_5 band to better withstand potential interference from other in-band ARNS transmissions. Both antennas are RHCP although only the central element has feed probes connected to it with the outer element being parasitic. The antenna dimension without the radome is 1.56" square with a height of 0.6". It uses a compound substrate: the top substrate layer is 0.2" thick and has a dielectric constant of 12.78 (Rogers TMM 13i) and the bottom substrate is 0.4"thick and is made from a ceramic-type material with a dielectric constant of 30 manufactured by the Emerson & Cummings Corporation. Figure 2.19 shows the measured return loss of this antenna. Figure 2.20 shows the measured RHCP and LHCP antenna patterns at 1.5754 GHz and 1.176 GHz—the center-band frequencies in the GPS L_1 and L_5 bands.

Another coplanar design for a circularly polarized single-layer dual-band microstrip antenna has also been proposed by Fan and Rahmat-Samii [32]. This uses a square patch with stubs attached on opposite ends as well as four slots parallel to the edges of the main square-patch antenna with switches installed in each slot

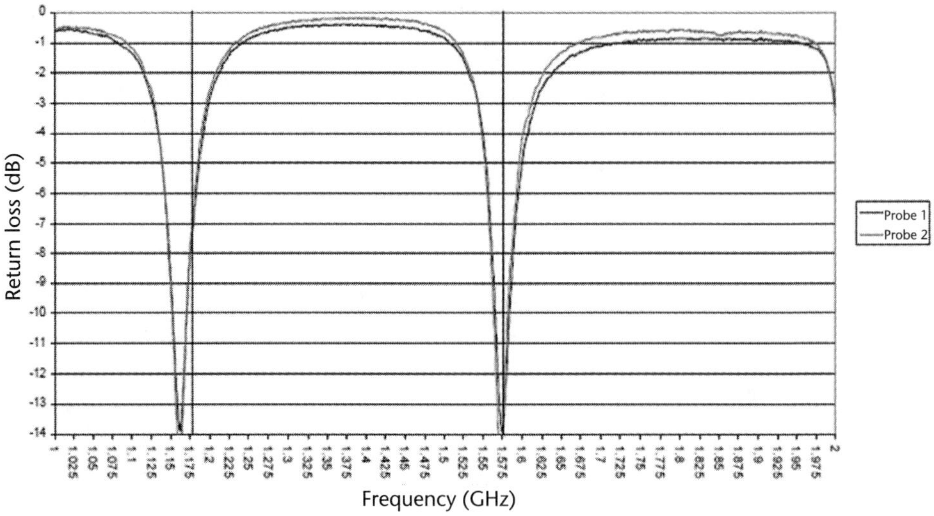

Figure 2.19 Measured return loss of coplanar dual-band microstrip antenna for the GPS L_1/L_5 bands. (From [31] ©2009 ION.)

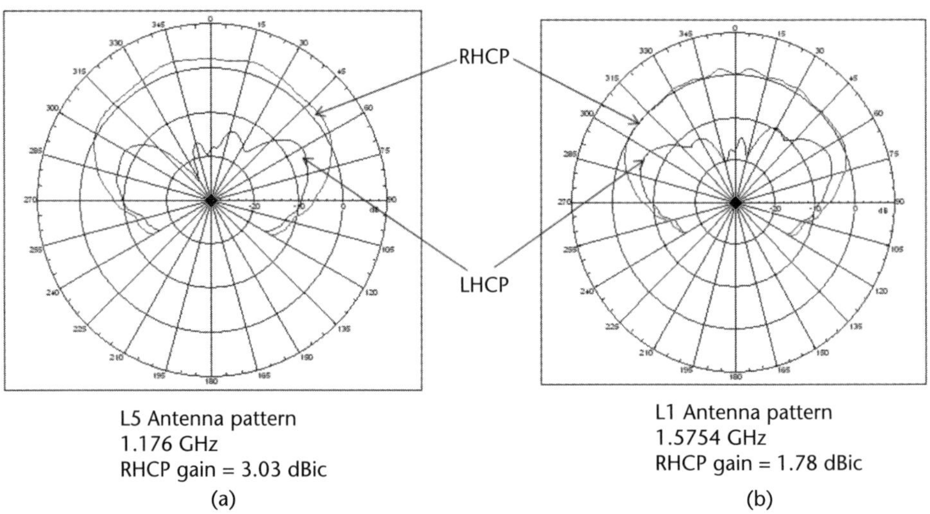

Figure 2.20 Measured radiation pattern of a coplanar GPS dual-band (a) L_5 and (b) L_1 microstrip antenna.

to control their lengths. The design is complex, and also the axial ratio appears to be poor. This design does not allow the antenna to operate simultaneously in both bands since the antenna needs to be switched from one band to the other through the PIN diodes.

2.2.6 Feed Techniques for Generating RHCP in GNSS Microstrip Antennas

The polarizing feed network used for generating RHCP in a GNSS patch antenna is another important feature in its design. It has an impact on many GNSS performance parameters such as CP axial ratio bandwidth, the PCV relative to look angle and frequency, and RHCP antenna gain and pattern symmetry. Different feeding techniques have been developed varying from a simple single feed for low-cost, narrowband applications to broadband multilayer aperture coupled networks covering all three frequency bands of modernized GPS. They can be grouped into four broad categories: direct-contact probes, aperture-coupling, edge-coupling, and proximity-coupling techniques; these are illustrated in Figures 2.21, 2.22, and 2.23(a) and (b), respectively. Each feed technique offers some advantage in terms of performance, complexity, and cost that is best suited for a specific application. The probe and aperture coupling techniques are two that are most popular. In the

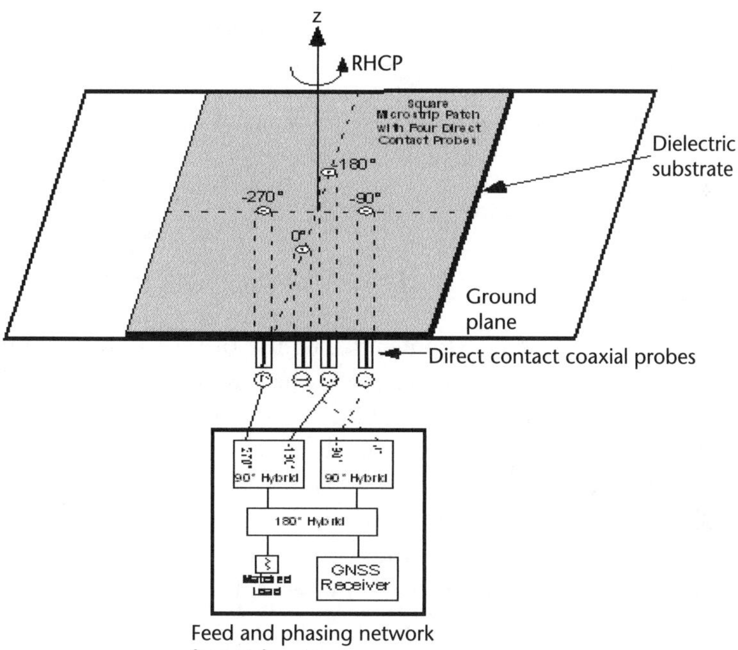

Figure 2.21 Four direct-contact feed probes for generating low-axial ratio RHCP in GNSS microstrip antennas.

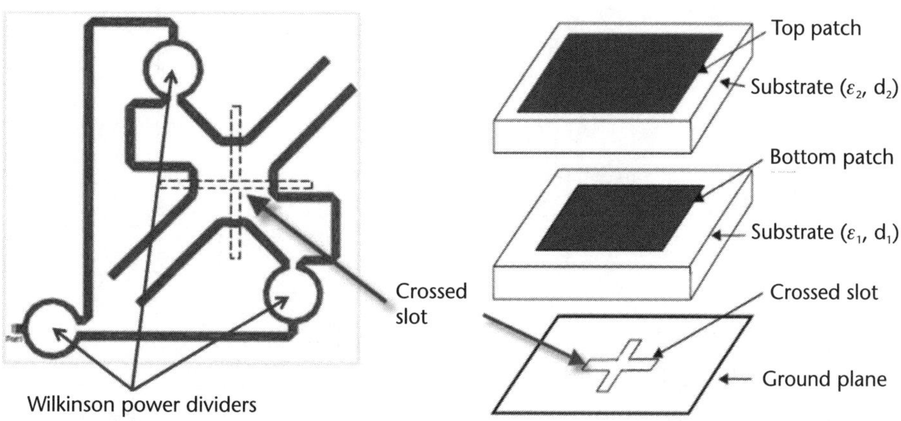

Figure 2.22 Aperture-coupled dual-band microstrip antennas. (From [33] ©2008 IEEE.)

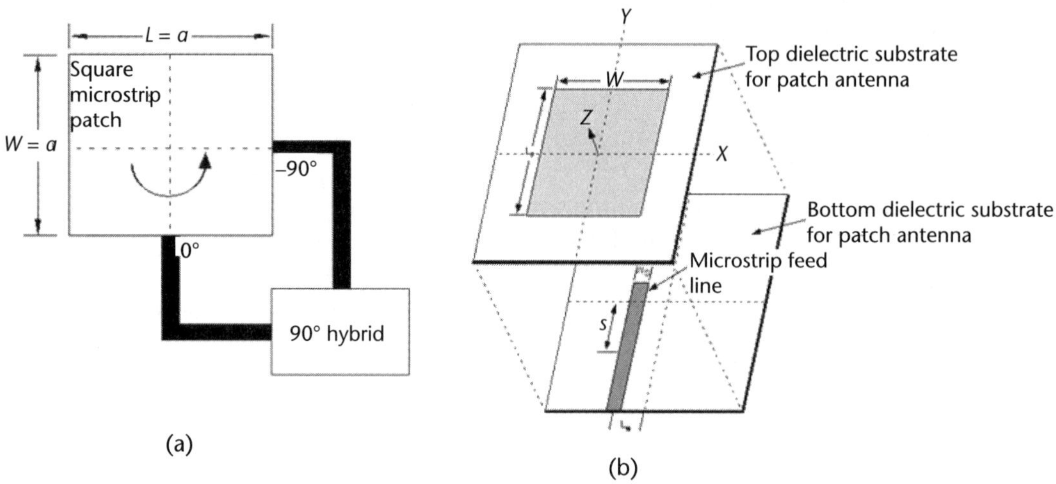

Figure 2.23 (a) Edge- and (b) proximity-coupling techniques for GNSS microstrip antennas.

direct-probe feeding method the RF signals received by the patch antenna from the GNSS satellites are fed to the receiver using one or more coaxial probes in direct contact with one or more patch antennas. Direct-contact probe-feed techniques are popular because of their simplicity. The use of one, two, or four probes are most common in many GNSS applications. They can generate RHCP in square, circular, and annular ring patches. Figure 2.5(a through c) shows pictures of RHCP GPS microstrip antennas for these three different geometries fed either by two or four direct-contact probes.

Two- and four-feed probes, shown in Figures 2.24 and 2.21, respectively, are more popular for general use since they are less complex and provide adequate performance for many applications. The two-probe technique was discussed in detail earlier in this chapter. The PCV variation can be reduced further by using four-feed probes [34]. This feed design is used in high-quality GNSS antennas and provides

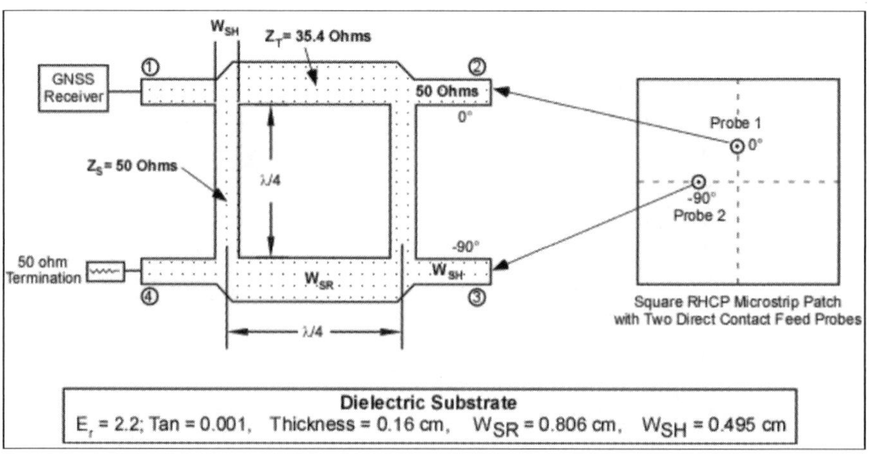

Figure 2.24 Microstrip branch-line hybrid coupler for generating RHCP in a two-probe feed microstrip antenna.

the best compromise between simplicity and good pattern symmetry and low PCV; it will be discussed in detail later in this chapter.

Up to eight probes (called an N-point feed) has been used by Trimble recently to develop high-quality geodetic antennas with very good phase center stability such as in their Zephyr antenna [35, 36]. Using more probes helps to suppress higher-order modes and reduces cross-polarization levels by preserving modal purity. This is achieved by better adherence to the optimum phase relationship between the probes required for exciting the fundamental mode. It produces greater symmetry in the azimuth pattern with a more stable phase center and lower CP axial ratio. The disadvantage of a multipoint feed is the reduction in gain due to losses in the feed network and increased cost and complexity.

In the aperture coupling technique shown in Figure 2.22, the microstrip feed line that is connected to the GNSS receiver is on a separate dielectric substrate from the patch antenna; it is in close physical proximity but not in direct physical contact with the patch. The transfer of RF power between the patch antenna and the microstrip line is through electromagnetic coupling from intervening resonant slots cut into the ground plane. The advantage provided by these noncontact techniques is the vast improvement in bandwidth over what direct-contact probe feeds provide. Crossed slots for aperture coupling can produce very wide bandwidths that can cover multiple GNSS bands by using stacked patch antennas excited by crossed resonant slots. Dual-band [37] and triple-band [33] GNSS antennas using these aperture feeding techniques have been built and will be discussed in Chapter 3 under multiband antennas.

Edge coupling is another type of a direct-contact feed where two microstrip transmission lines are connected to the patch antenna at its edges, as shown in Figure 2.23(a) [23, pp. 85]. The other ends of the microstrip lines are connected to a coplanar broadband hybrid on the same substrate to obtain a 90° phase difference between the two ports. The advantage is that the feed network can be colocated with the patch antenna on the same printed circuit board, obviating the need for soldering as in the direct feed probes and reducing the manufacturing cost. The disadvantage is the large lateral size needed to support the feed around the radiating patch. Proximity coupling, shown in Figure 2.23(b), is similar to aperture coupling; the microstrip feed line that is connected to the GNSS receiver is on a separate dielectric substrate from the patch antenna but is in close physical proximity to but not in direct contact with the patch.

We will now consider the design of some of these feed techniques in greater detail.

2.2.6.1 RHCP Feed Design Using Two Direct-Contact Probes

The design of a square-shaped microstrip patch antenna with two direct-contact feed probes was covered in a previous section. This type of feed technique provides several advantages:

1. The quadrature hybrid to which these probes are connected has a very wide bandwidth, sometimes covering a full octave. Hence RHCP with a low axial ratio can be obtained over a wide range of frequencies with a limitation on bandwidth imposed only by the impedance variation with frequency of

the feed probe, which is at a fixed location inside the patch and therefore provides an optimum match only at one frequency.

2. The feed network is isolated from the patch antenna by the ground plane, resulting in a good front-to-back ratio especially when the ground plane is designed to reduce edge diffraction effects as explained in Chapter 5. This decreases the susceptibility of the GNSS antenna to multipath.
3. The coaxial probes are normal to the surface of the patch and do not increase the lateral dimensions of the antenna; this is a huge advantage when multiple patch antennas need to be located close to each another within the small aperture of an adaptive array needed for antijam applications. The available space for adaptive antenna arrays in tactical military aircraft is often smaller than a wavelength in diameter at GPS frequencies for miniaturized arrays. These space restrictions only allow antenna elements with only very small lateral dimensions including the feed networks to be used in such small antenna arrays.
4. The quadrature hybrid, a schematic sketch of which is shown in Figure 2.24, can be used for extracting both RHCP and LHCP signals from the patch antenna using Ports 1 and 4, respectively.
5. Direct-contact probe feed techniques also avoid misalignment of multiple dielectric ayers encountered in aperture coupled patches.

A key component in the feed network for generating RHCP is the quadrature hybrid a schematic cross section of which is shown in Figure 2.24 and whose scattering matrix is given by:

$$[S] = \frac{1}{\sqrt{2}}\begin{bmatrix} 0 & j & 1 & 0 \\ j & 0 & 0 & 1 \\ 1 & 0 & 0 & j \\ 0 & 1 & j & 0 \end{bmatrix} \tag{2.30}$$

For generating RHCP, Port 1 is connected to the GNSS receiver and Port 4 is terminated in a matched load. Alternatively, a cross-polarized LHCP output signal can be obtained from Port 4 when it is not terminated in a matched load; this is useful when obtaining both RHCP and LHCP signals from the antenna element for use in dual polarized adaptive antenna arrays to be discussed in Chapter 4. Ports 2 and 3 are connected to the two orthogonally placed probes in the patch antenna as shown in the Figure 2.24. The signal is evenly divided in amplitude but in phase quadrature at Ports 2 and 3, with a phase of 0° from Port 2 and a phase of –90° from Port 3. For obtaining a 3-dB split between these two ports with the reference impedance Z_0 generally set to 50 ohms, the shunt branches of the hybrid $Z_S = Z_0$ and the through branches $Z_T = \frac{Z_0}{\sqrt{2}} = 35.4$ ohms. The lengths of the branches are all a quarter wavelength at the center design frequency. Port 4, which is the isolated port if not used for dual polarization application, is generally terminated in a matched load to absorb any imbalance between Ports 2 and 3. Any signals reflected as a result of an impedance mismatch between the patch antenna and Ports 2 or 3

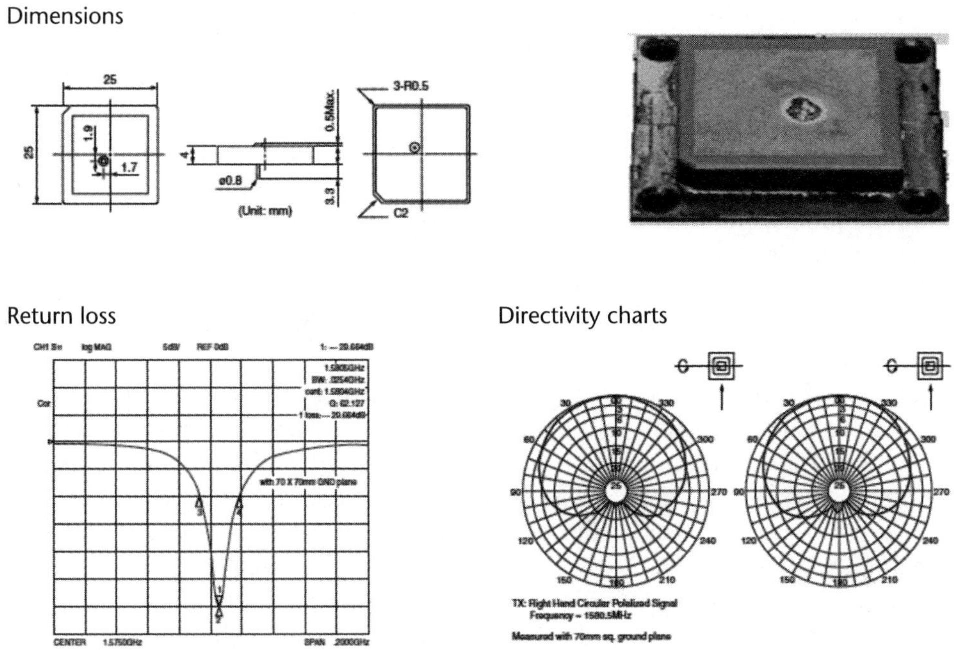

Figure 2.25 Toko/America's GPS ceramic microstrip antenna with a single-feed probe. GPS L_1 band; C/A code. (*From:* [39]. Printed with permission from Toko/America.)

will be returned to Port 4 and absorbed by the match termination. When a quadrature hybrid is connected to the patch antenna and integrated within the antenna package, a feature most common in commercial GNSS antennas, the reflected signals from the antenna ports are not reflected back into the input Port 1. Therefore a relative measurement of the forward and reflected powers is not possible, and the bandwidth of these antennas cannot be estimated by measuring its VSWR or return loss. Some of these antennas can have very poor gain and bandwidth but still show deceptively low return loss as a function of frequency indicating good bandwidth.

2.2.6.2 RHCP Feed Design Using Four Direct-Contact Probes

One method of increasing the bandwidth of a patch antenna is by increasing its thickness as described in the previous sections. However, due to the unbalanced and asymmetrical nature of the dual probe feed arrangement for generating RHCP, this can result in the generation of unwanted higher-order modes inside the patch antenna. The mode closest to the desired TM_{01} and TM_{10} modes is the TM_{21} mode, which causes coupling between the dual feed probes in the patch antenna and affects the amplitude and phase balance between the probes even when the antenna is connected to the quadrature hybrid. This imbalance increases the cross-polarized LHCP signal and distorts the radiation pattern. These effects can be reduced by replacing the dual feeds by a balanced four-probe feed system as shown in Figure 2.21. In this feed arrangement, two additional probes have been added to the two original ones and feeding all four terminals with equal amplitude but 90° out of phase from each other. A new feed circuit consisting of two 90° hybrids that are combined with a 180° hybrid. The four probes have equal amplitudes but phases of 0°, −90°, −180°, and −270° varying in a counterclockwise direction to generate RHCP. These two orthogonally placed pairs of probes have opposite phases but equal amplitudes and result in a cancellation of the unwanted TM_{21} modal fields. Four coaxial probes provide good symmetry in the azimuth plane of both the phase and the radiation pattern of the antenna. This type of feed also provides the low phase and group delays [34]. The disadvantage of the four-probe feed method besides the increased complexity is a lower antenna gain than obtained with the dual probe feed network because of the losses contributed by the additional hybrid components needed in the feed network.

Difficulties can arise in impedance matching when a thick dielectric substrate is used for increasing the bandwidth and gain. The extension of the center conductor of a directly coupled feed probe through the dielectric substrate introduces a series of inductive reactance, which is proportional to the substrate thickness. This increased inductance can limit the impedance bandwidth of the patch antenna. Modified types of coaxial probes for feeding patch antennas have been designed recently to compensate for the increase in inductance by using capacitive coupling between the probe and the patch antenna, thereby avoiding direct contact between the two. These techniques are more similar to proximity coupling techniques although the probes are embedded in the substrate beneath the patch antenna. Various methods have also been developed to tune out the probe inductance to increase the bandwidth [30, 31]. They include etching a small hole in the patch at the top around the probe, or attaching a small conducting strip to the top of the probe and then coupling it to the patch either from above or from below, or by using a proximity-coupled L-shaped probe. In an L-shaped probe the horizontal part of the probe runs underneath the patch and provides capacitive coupling to it [40]. Two orthogonal L-shaped probes have been used recently in a broadband patch antenna covering all three of modernized GPS frequency bands [41].

2.2.6.3 Single Direct-Contact Feed Probe Design for RHCP GNSS Patch Antennas

A single direct feed probe placed along the diagonal axis of the patch antenna at a suitable distance from the center is often used in low-cost and compact antennas produced for the mass market that operate only over a single GNSS band. It is especially popular with high-permittivity ceramic substrates [38, 39]. Figure 2.27 shows a popular, single-probe fed GPS antenna (part number DAKC1575MS74T) manufactured by Toko America Inc. Use of a single direct-contact probe for generating CP removes the need for an external phasing network and greatly reduces cost, complexity, and size. However, there are two penalties paid for this reduction in complexity:

1. It generates the largest variation in the phase center with elevation and azimuth angle, resulting in an increase in the phase center error ellipsoid [34, 38].
2. It has a very narrow axial ratio CP bandwidth. The PCV variation results from the generation of higher-order modes in the patch antenna due to the asymmetric excitation; this increases the cross-polarized levels contributing to the pattern asymmetry [34, 38].

The narrow RHCP axial ratio bandwidth makes this design suitable for use only over a single GNSS frequency band, generally over only the C/A code in the GPS L_1 band. The single-probe patch antenna is able to generate RHCP by exciting mutually orthogonal TM_{10} and the TM_{01} modes with the same amplitude but with a phase difference of 90°. This is done by several different methods illustrated in Figures 2.26 (a through d). A more detailed description of these various single-feed probe designs for generating RHCP can be found in two recent books [30, pp. 319–339] and [23, pp. 37–49]. Two single-probe designs for a GPS patch are discussed below; the resonance frequency of the antenna is 1.5754 GHz.

One of the techniques used is by altering the shape of the perfectly square patch to "nearly square" by changing the aspect ratio of width/length as shown in Figure 2.26(a). When the patch antenna is perfectly square, it will excite the orthogonal TM_{01} and the TM_{10} modes with identical amplitudes and in phase; the phase centers of these modes also coincide and the far field will have a linear polarization directed diagonally. However, a slight departure from a perfectly square shape causes the two spatially orthogonal modes, the TM_{10} and the TM_{01} modes,

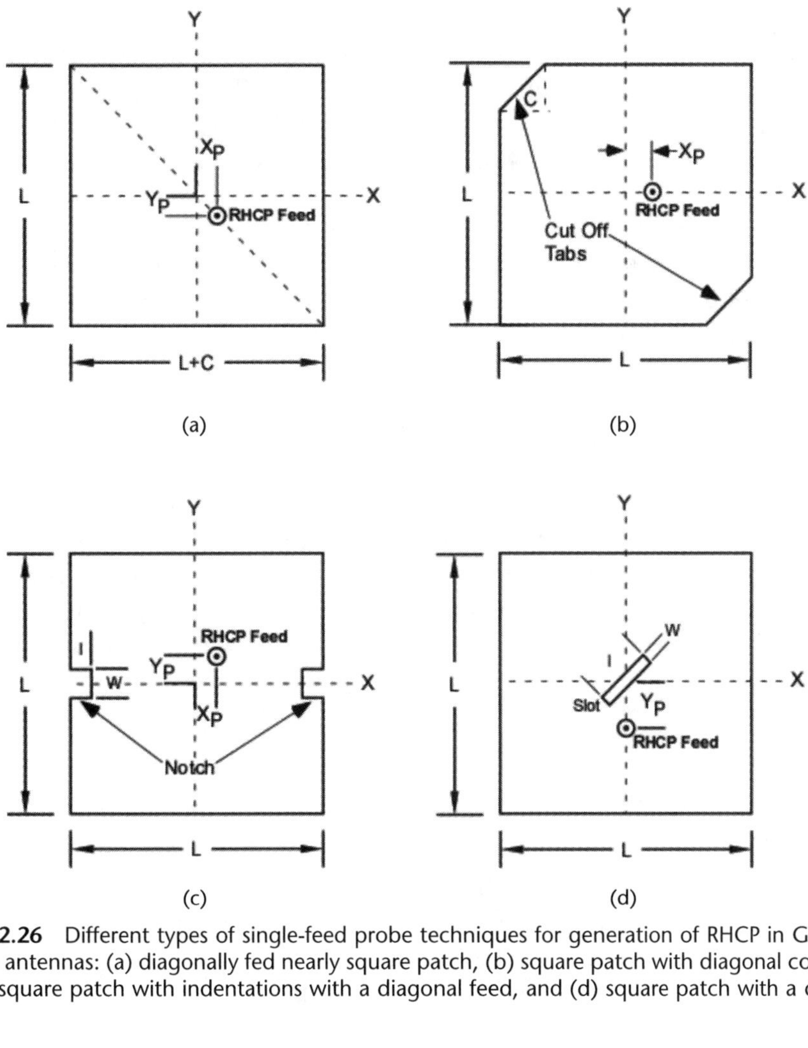

Figure 2.26 Different types of single-feed probe techniques for generation of RHCP in GNSS microstrip antennas: (a) diagonally fed nearly square patch, (b) square patch with diagonal coners cut off, (c) square patch with indentations with a diagonal feed, and (d) square patch with a diagonal slot.

to have different resonant frequencies that are close to each other. The resonance frequency of the TM_{01} mode f_2 is higher than the desired signal frequency whereas the resonant frequency of the TM_{10} mode f_1 is lower than the desired signal frequency f_0. The feed location of the probe is next located at a point where the antenna is excited at the desired frequency f_0 in between the resonance frequencies f_1 and f_2 so that the two modes have equal amplitudes but a phase difference of +45° or −45° with respect to the feed point to meet the conditions needed for obtaining RHCP. However, these conditions of amplitude and phasing can only be met exactly over a narrowband where the frequency ratio of the two orthogonal modes is $1.01\left\langle \frac{f_2}{f_1} \right\rangle 1.10$.The axial ratio bandwidth is small generally around 1%. To illustrate this method for generating RHCP, consider a nearly square microstrip patch on a 0.254-cms thick dielectric substrate with a dielectric constant $\varepsilon_r = 6.0$ and loss tangent $\tan\delta = 0.0018$. If the patch was square, the dimensions of each side would be equal to 3.8245 cms to obtain resonance at a frequency of 1.5754 GHz, and RHCP could be generated by using two feed probes located at orthogonal positions in the patch as illustrated in the previous section. To obtain RHCP with a single probe at the same design frequency of 1.5754 GHz, the shape of the patch antenna can be made nearly square with a width along the x axis equal to $L+c = 3.8757$ cms and length along the Y axis equal to $L = 3.8245$ cms or an aspect ratio of 1.0138 corresponding to a width extension $c = 0.0512$ cms. The feed probe is located along the diagonal at $X_P = 0.3039$ cms and $Y_P = -0.2783$ cms. The best RHCP is obtained at a frequency of 1.5634 GHZ, slightly below the desired frequency.

A second popular technique for generating RHCP with a single feed, shown in Figure 2.26(b), is to change the shape of the patch by truncating the two diagonal opposite corners of the square-patch antenna by a length c. This again creates a pair of diagonal modes that can be adjusted to have identical magnitudes and a 90° phase difference between the modes. By placing the feed probe at an appropriate distance from the center along the central x axis it is possible to excite each of these diagonal modes to have equal amplitude so as to provide the best RHCP as well as a good impedance match. To illustrate this method consider the same dielectric substrate used in the previous example of the nearly square patch. The dimension of the patch is 3.8245 cms square except that two opposite corners are truncated by distance $c = 0.3129$ cms. The feed is located at $X_P = 0.3650$ cms and $Y_P = 0.0$. The best RHCP is obtained at 1.5634 GHz, again slightly below the desired frequency of 1.5754 GHz. Two other methods for obtaining RHCP with a single feed probe are shown in Figures 2.26(c) and (d). These methods change the geometry of the square patch by inserting indentations or notches or by cutting a diagonal slot across the patch.

2.2.6.4 Aperture Coupled Feed Design for RHCP GNSS Patch Antenna

In this feed technique, shown in Figure 2.22, the radiating patch and the microstrip feed line are separated by a common ground plane that they share. Coupling between the patch and the feed line is made through a crossed slot aperture in the ground plane that is located directly below the center of the patch. This type of feed technique is sometimes also called aperture stacked patches (ASPs); for GNSS applications dual-polarized ASPs are of interest [33, 37]. This feed configuration has the ability to isolate the patch radiator from the feed line by virtue of the common ground plane that they share. This minimizes the effects of the spurious radiation from feed on the antenna pattern of the patch and also allows the design of the feed and the antenna to be optimized relatively independently of each other. The resonant frequency is primarily determined by the size of the square-patch antenna but the resonant slot length of the aperture exciting the patch may also have a secondary effect on the resonance frequency. The resonant frequency of the patch may show a slight decrease as the length of the aperture increases. A decrease in the length of the aperture may also decrease the level of coupling between the microstrip feed line and the patch antenna. The dielectric constant of the antenna substrates are also a factor in determining antenna performance. Increasing the permittivity of the feed substrate has minimal impact on the resonant frequency, but the coupling between the feed line and the patch antenna is increased as a result of the aperture now being electrically larger than its physical dimensions. If the thickness of the feed substrate is increased the level of coupling to the patch antenna is decreased, but the resonant frequency is not affected.

An aperture feeding technique used for generating RHCP is a crossed-slot aperture fed by microstrip lines from a parallel feed configuration consisting of three Wilkinson power dividers shown in Figure 2.22. Two orthogonal linear modes are excited by the crossed slot aperture that are of equal amplitude but with a 90° phase shift for achieving CP in the patch antenna. The 90° phase shift between the two slots is produced by extending one arm of the Wilkinson power divider by a quarter wavelength. The two Wilkinson power dividers provide diametrically opposite ends of the crossed slots with a 180° phase difference by being connected to a third Wilkinson power divider. This combination provides equal amplitudes but with a phase progression of 0, 90°, 180°, and 270° across the four ends of the pair of slots to generate RHCP. This type of feeding technique has a very broad impedance and axial ratio bandwidths and allows the stacking of two square-patch antennas above the crossed-slot aperture that are resonant at different frequencies to obtain multiband performance.

Aperture-coupled feeding techniques also present two major disadvantages for GNSS applications. The first is the backward radiation into the lower hemisphere by the coupling aperture cut into the ground plane. The backward radiation degrades the front-to-back ratio (FBR) of GNSS antennas; this makes it more susceptible to multipath effects. This is apparent from the measured patterns of the two GNSS antennas that have been built using this type of feed [33, 37]. The measured backlobes in a GPS dual-band aperture coupled patch antenna is shown in Figure 2.27 [37].

Techniques for reducing backward radiation in both polarizations have been proposed by Waterhouse [18, pp. 139–145]. One of the methods suggested is to use a printed cross-back patch below the resonant crossed slots used for exciting the patch antennas. This microstrip patch element placed behind the microstrip-fed aperture acts as a reflector to substantially reduce backward radiation. By using this method Waterhouse has been able to reduce FBR typically from between 10 to 14 dB to more than 20 dB. However, additions to the feed circuit greatly add to the size, cost, and complexity. The second disadvantage of aperture coupling is the increase in lateral dimensions of the antenna resulting from the feed structure

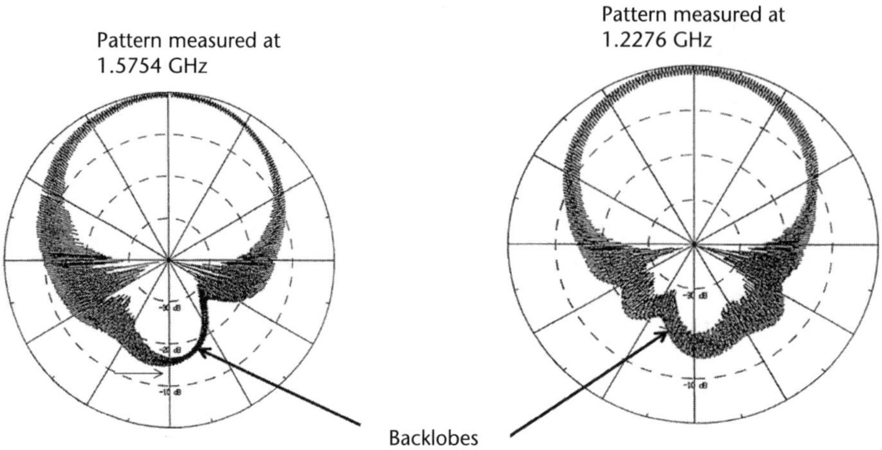

Figure 2.27 Large backlobes in the measured antenna patterns of dual-band aperture coupled microstrip antennas. (From [37]. ©1997 IEEE.)

especially for the parallel feed configuration which requires three Wilkinson power dividers.

2.2.6.5 Edge-Coupled Feed Probe Design for RHCP Patch Antenna

In this type of feeding scheme the two sides of a square-patch antenna are each connected to separate feed microstrip feed lines, which are in turn connected to the output ports of a branch-line coupler as shown in Figure 2.23(a). This ensures that they will have equal amplitudes but with phases that differ by 90° for generating RHCP from one of the two output ports of the coupler. The other port of the coupler is terminated in a matched load. Impedance matching networks can also be inserted between the antenna and the feed lines connecting the antenna to the branch-line coupler to improve the impedance bandwidth. This type of feed network provides several advantages. The patch antenna, the polarizing feed, and impedance-matching network are coplanar and on the same substrate. This makes it simpler and less expensive to build and allows for an easier integration of other components such as bandpass filters or an LNA directly into the antenna circuit board; it is especially conducive for multilayered circuit boards often used in receivers. The disadvantage is that the antenna has a larger lateral dimension, making it unsuitable for many applications such as an adaptive antenna array, where a compact size antenna is essential because of the restriction on lateral dimensions. Another disadvantage is spurious radiation from the feed lines especially if there is a mismatch in impedance at the junction between the feed line and the patch antenna.

2.2.6.6 Proximity Coupled Feed Design for RHCP GPS Patch Antenna

A single proximity coupled feed for a RHCP patch antenna that operates at 1.575 GHz, the GPS L_1 center band frequency, has been designed by Iwasaki and others [42] and is shown in Figure 2.23(b). This feed concept does not need an external hybrid; the advantage it provides is similar to that of single direct feed contact probe described earlier. This scheme also provides a choice of two dielectric substrates—one for the patch and the other for the feed line so that the performance can be optimized. The rectangular-shaped radiating patch antenna is proximity-coupled to the feed microstrip transmission line that is offset from the center of the patch; there is no electrical contact between the feed line and the patch antenna. The four design parameters available with this design for obtaining both a good impedance match as well as axial ratio is the aspect ratio of the patch (L_p/W_P), where L_P and W_p are the length and width, respectively, of the rectangular patch; the offset length L_0 from the center of the rectangular patch to the center line of the microstrip line and S; the distance from the edge of the patch antenna; and the edge of the microstrip line. By adjusting these parameters it is possible to excite two orthogonal modes of equal amplitude and with a phase difference of 90° at the selected design frequency. To obtain RHCP requires $L_P < W_P$ and $-W_P/2 < L_0 < 0$. The antenna model to demonstrate this concept was built on a copper-clad substrate with a dielectric constant of 2.6; unfortunately neither the thickness t of the patch antenna substrate nor the thickness h of the microstrip line substrate are provided in [42]. $L_P = 56$ mm and $W_P = 58$ mm for an aspect ratio (L_p/W_P) = 0.966; the width W_S of the microstrip feed line was 4 mm and $L_0 = 14.4$ mm and $S = 13.8$ mm. A 0.3-dB axial ratio at boresight and a 2-dB axial ratio over 60° angular range around boresight were measured. The bandwidth of the 2-dB axial ratio was 0.55% and the bandwidth for a VSWR< 2 was 3.5% at a center frequency of 1.575 GHz. The ground plane below the microstrip feed line prevents back radiation and yields a good FBR with this feed design. Spurious radiation from the feed line is also avoided. The resonant frequency of the antenna can be tuned by changing the dimensions of the patch antenna and also be fine-tuned further by changing the location of the open-circuit termination at the end of the microstrip feed line. If the patch substrate is made thinner a slight improvement in gain is observed and a thicker substrate produces the opposite effects. The operating frequency is increased when a thinner feed substrate is used. The disadvantage of this feed method is that accurate alignment is needed between the patch and the feed line and the axial ratio bandwidth is relatively narrow that is typical of a single-feed design.

2.2.7 Circular RHCP Microstrip Antenna

A picture of a dual-band circular microstrip patch antenna for GPS applications was shown earlier in Figure 2.5(b). Figure 2.28 shows the schematic diagram of a single-layer circular microstrip antenna fed by two feed probes; a cylindrical coordinate system (ρ, ϕ, z) represents the antenna geometry. The microstrip antenna consists of circular metal disc of radius a located at the center of a dielectric substrate of thickness h, a dielectric constant ε_r, and loss tangent tan δ. The dielectric substrate is backed by a conducting ground plane. As in the square-shaped patch antenna, RHCP in the circular patch can be generated by using two direct-contact coaxial probes located at $P_2(\rho_0, \phi)$ and at $P_2(\rho_0, \phi + \alpha)$ The two probes are placed along orthogonal radii so $\alpha = 90°$ and the radial distance ρ_0 of the two probes from the center of the patch antenna are adjusted so that their input impedance is 50 ohms and that they provide a good match to the receiver. The coaxial feeds are connected to a quadrature hybrid, which provides equal amplitudes but phases that differ by 90° to the two probes. The electric field E_Z of the circular microstrip antenna generated by a probe is given in the cylindrical coordinate system (ρ, ϕ, z) by:

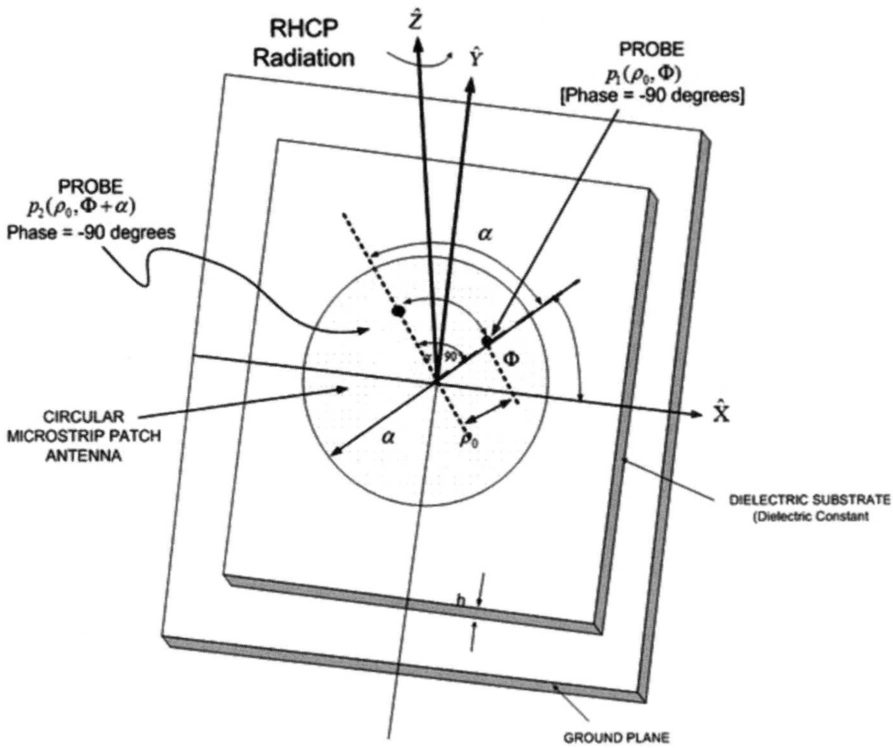

Figure 2.28 Circular microstrip antenna.

$$E_z = E_0\, J_n\left(k_0 \sqrt{\varepsilon_r}\, \rho\right)\, \cos(n\, \phi) \tag{2.31}$$

where J_n is the Bessel function of the first kind of order n, $k = k_0\sqrt{\varepsilon_r}$ is the propagation constant in the dielectric substrate with relative dielectric constant ε_r; $k_0 = \dfrac{2\pi}{\lambda_0}$ and $\lambda_0 = \dfrac{c}{f}$ where f is the frequency in cycles per sec, and c is the velocity of light in free space. E_0 is the electric field at the edge of the patch across the gap.

2.2.7.1 Resonance Frequency and Radius of a Single-Layer Circular Microstrip Antenna

For each modal configuration of the circular patch antenna, a radius can be found that results in resonance when the derivative of the Bessel function of order n is zero [23, pp. 73–85; 21, pp. 317–356].

$$J_n'\left(k_0\sqrt{\varepsilon_r}\, a\right) = 0 \tag{2.32}$$

If χ_{mn} is the mth zero of $J_n'(k_0\sqrt{\varepsilon_r}\, a)$, the derivative of the Bessel function of order n. The resonance in the antenna occurs when $k_0\sqrt{\varepsilon_r}\, a = \chi_{mn}$, $n = 0, 1, 2, \ldots$, and $m = 1, 2, 3, \ldots$.. In the TM_{mn} mode of the circular microstrip antenna the index n represents the angular mode related to f and the index m represents the radial mode related to the radial coordinate. The mode number n corresponds to the number of sign reversals in π radians in ϕ. Value of χ_{nm} for the first few TM_{nm} modes, each of which has its own unique radiation pattern, is given in Table 2.3.

The lowest-order mode in the circular patch antenna is the TM_{11} mode, with $n = 1$ and $m = 1$ is important for GNSS. This is a bipolar mode with the electric field concentrated at each end of the antenna, has the smallest radius or the lowest resonance frequency for a circular patch antenna, and is analogous to the lowest-order mode TM_{10} in the square-patch antenna. This is the primary mode of interest of the circular patch for GNSS; $\chi_{11} = 1.841$ for this mode.

The resonant frequency f_{nm} for the TM_{nm} mode of a circular microstrip antenna is

$$f_{nm} = \frac{\chi_{nm}\, c}{2\pi\, a_{eff}\sqrt{\varepsilon_r}} \tag{2.33}$$

where a_{eff} is the effective radius of the patch antenna and is given

$$a_{eff} = a\left[1 + \left[\frac{2h}{\pi a \varepsilon_r}\right]\left[\ln\left\{\frac{\pi a}{2h}\right\} + 1.7726\right]\right]^{\frac{1}{2}} \tag{2.34}$$

where a is the physical radius of the circular patch antenna.

The above two equations may now be combined to produce for the radius a for the TM_{11} mode

$$a = \frac{1.8411c}{2\pi\sqrt{\varepsilon_r}}\left[1 + \left[\frac{2h}{\pi a \varepsilon_r}\right]\left[\ln\left\{\frac{\pi a}{2 f_{11} h}\right\} + 1.7726\right]\right]^{-\frac{1}{2}} \tag{2.35}$$

In the above equation f_{11} is the resonant frequency of the TM_{11} mode of the circular patch. The above equation is of the form $a = f(a)$ and can be solved by using an iterative process to determine the radius of the patch antenna needed to obtain resonance for the TM_{11} mode at a resonance frequency f_{11}. The initial approximation for the radius can be selected to be $a_0 = \frac{1.8411c}{2\pi f_{11}\sqrt{\varepsilon_r}}$. The initial value is placed on the right-hand side of (2.35) to produce the first iteration value of the radius designated as a_1. In the second iteration a_1 is placed on the right-hand side of the equation to obtain second iterative value designated a_2. In the third iteration a_2 is now substituted on the right-hand side to obtain a fourth value. This process is continued until it converges to a stable solution; generally 5 to 6 iterations are sufficient to produce the final result for the radius.

Table 2.3 TM_{nm} Modes and Their Unique Radiation Pattern

Mode	TM_{11}	TM_{21}	TM_{31}	TM_{41}	TM_{51}
χ_{nm}	1.841	3.054	4.201	5.317	6.415

2.2.7.2 Radiation Pattern, Feed Probe Location, and Gain of a Circular Microstrip Patch GNSS Antenna

The radiation pattern of a circular patch antenna for the TM_{11} mode can be obtained from the cavity model and is given by [21]:

$$E_{\theta} = -jV \frac{ak_0}{2} \frac{e^{-jk_0^r}}{r} J_1^I(k_0 a \sin\theta)\cos(\varphi) \tag{2.36}$$

$$E_{\varphi} = jV \frac{ak_0}{2} \frac{e^{-jk_0^r}}{r} \frac{J_1(k_0 a \sin\theta)}{k_0 a \sin\theta} \cos(\theta)\sin(\varphi) \tag{2.37}$$

An RHCP-radiated field can be obtained by combining the outputs of the two feed probes located on the antenna at $P_1(\rho_f, \phi_f)$ and $P_2(\rho_f, \phi_f + \alpha)$ with equal amplitudes but with their phases differing by 90° when the offset angle between the two feed probes $\alpha = 90°$. The θ and ϕ components of the total field E_T can now be expressed as:

$$E_{T\theta} = E_{1\theta}(\varphi, \theta) - jE_{2\theta}(\varphi + 90°, \theta) \tag{2.38}$$

$$E_{T\varphi} = E_{1\varphi}(\varphi, \theta) - jE_{2\varphi}(\varphi + 90°, \theta) \tag{2.39}$$

In the above equations the subscripts 1 and 2 indicate the field contributions E_1 and E_2 made by the feed probes 1 and 2, respectively, to the total field E_T along $\hat{\theta}$ and $\hat{\phi}$. Although only two coaxial feed probes are considered in this discussion, some GNSS antenna designs use up to four probes to improve the circularly symmetry of the RHCP antenna pattern and to reduce phase center variations as shown in Figure 2.5(b). The locations of the additional two feed probes are selected to be diametrically across from the original two feed probes. However, using an increased number of probes will increase the complexity of the feed network and will reduce antenna gain because of the resulting increase in losses. The radial distance ρ_0 of the two probes on the patch antenna must be selected to provide an input impedance of 50 ohms to the input of the GPS receiver. The input resistance R_{in} of the feed probe can be calculated from the equations below:

$$R_i n = R_r \frac{J_1^2(k_{11}\rho_0)}{J_1^2(k_{11}a)} \tag{2.40}$$

where $k_{11}a = 1.84118$

$$R_r = \frac{1}{G_r} = \frac{\lambda_0^2 \eta_0}{\pi^3 a^2} \left[\frac{1}{\frac{4}{3} - \frac{8(k_0 a)^2}{15} + \frac{11(k_0 a)^4}{105}} \right] \tag{2.41}$$

where:

λ_0 = wavelength

$\eta_0 = 120\ \pi$ = free space impedance

a = radius of the patch antenna for the TM_{11} mode

R_r = radiation resistance of the antenna

G_r = the resonant radiation conductance of the circular patch antenna fed from the edge.

D is the directivity of the antenna and the gain G of the antenna are given by the following equations:

$$D = \frac{(k_0 a)^2}{120 G_r} \tag{2.42}$$

$$G_r = e_r D \tag{2.43}$$

The radiation efficiency e_r is the ratio of the radiated power to the input power and is determined by the conductor losses in the patch antenna, the dielectric losses in the dielectric substrate, and the surface wave losses. The value of e_r is within the range $0<e_r>1$. Hence, the gain is always less than the directivity.

Figure 2.29 shows the measured radiation pattern at 1.5754 and 1.2276 GHz. of a stacked, dual-band, circularly shaped patch antenna designed for operation in the GPS L_1 and L_2 bands using the HFSS code. The RHCP is obtained by exciting both patch antennas with four coaxial probes. The substrate used for this circular patch antenna is TMM13i with a dielectric constant of 12.78 and a loss tangent of 0.0027. The substrate thickness for the top and bottom patches was each 0.2"and the diameter of the dielectric substrate was truncated. The radius of the top L_1 patch was 1.138" inches and the radius of the bottom L_2 patch was 1.285".

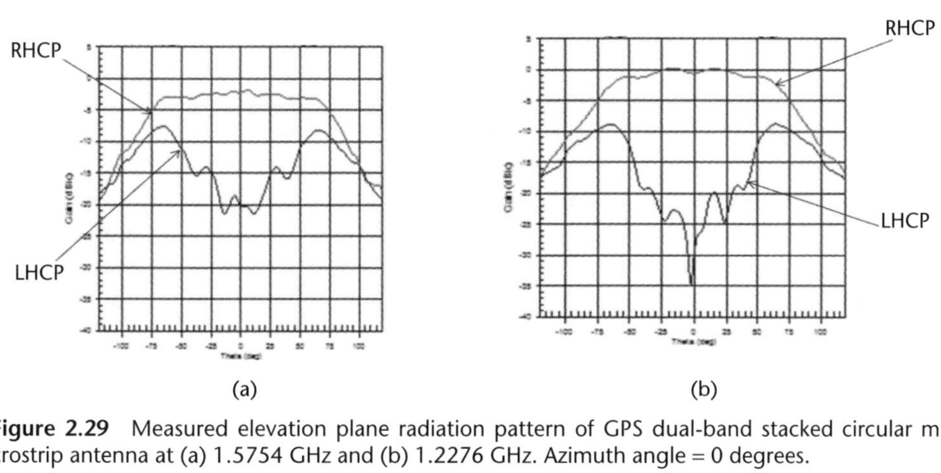

Figure 2.29 Measured elevation plane radiation pattern of GPS dual-band stacked circular microstrip antenna at (a) 1.5754 GHz and (b) 1.2276 GHz. Azimuth angle = 0 degrees.

2.2.7.3 Bandwidth of a Circular Microstrip Patch Antenna

The percentage bandwidth BW of a circular patch antenna for an input VSWR of S:1 is given by:

$$BW = \frac{100\,(S-1)}{Q_T\sqrt{S}}\% \tag{2.44}$$

where Q_T is the total quality factor [21, pp. 349–350]. For a circular patch antenna operating in the dominant TM_{11} mode and with a patch antenna made from copper

$$Q_T = \left[\frac{2.09\times10^{-4}}{\left(\frac{h}{\sqrt{f}}\right)} + tan\delta + hf\left(ak_0\right)^2\left\{\frac{I_1}{45.64}\right\}\right]^{-1} \tag{2.45}$$

where h = height of the dielectric substrate in cms, f is the frequency in GHz, is the loss tangent of the dielectric substrate, and

$$I_1 = \int_0^{\pi}\left[\{A\}^2 + \cos^2(\theta)\{B\}^2\right]\sin\theta d\theta \tag{2.46}$$

where [21, 23]

$$A = J_2(k_0\,a\sin\theta) - J_0(k_0\,a\sin\theta) \text{ and } B = J_2\,(k_0\,a\sin\theta) + J_0(k_0\,a\sin\theta)$$

The above equations show that the bandwidth of a circular patch antenna for a selected resonance frequency can be increased by choosing a thicker dielectric substrate and also by lowering the dielectric constant of the substrate material. However, lowering the dielectric constant of the substrate will increase the diameter of the circular patch antenna.

2.2.8 Annular Ring RHCP GNSS Microstrip Antenna

An annular ring microstrip antenna is formed by removing a concentric circle of the conducting surface at the center of a circular microstrip patch antenna [22, pp. 169–179]. Modified versions of this antenna have been used for several different GNSS applications including multipath suppression. A picture of a typical annular ring antenna is shown in Figure 2.30(b). A cross-sectional sketch of the annular ring patch antenna in Figure 2.30(a) shows an outer radius of b and an inner radius a The resonant frequency f_{nm} of the annular ring patch antenna for the TM_{nm} mode can be determined from

$$f_{nm} = \frac{(k_{nm})c}{2\pi\sqrt{\epsilon_r}} \tag{2.47}$$

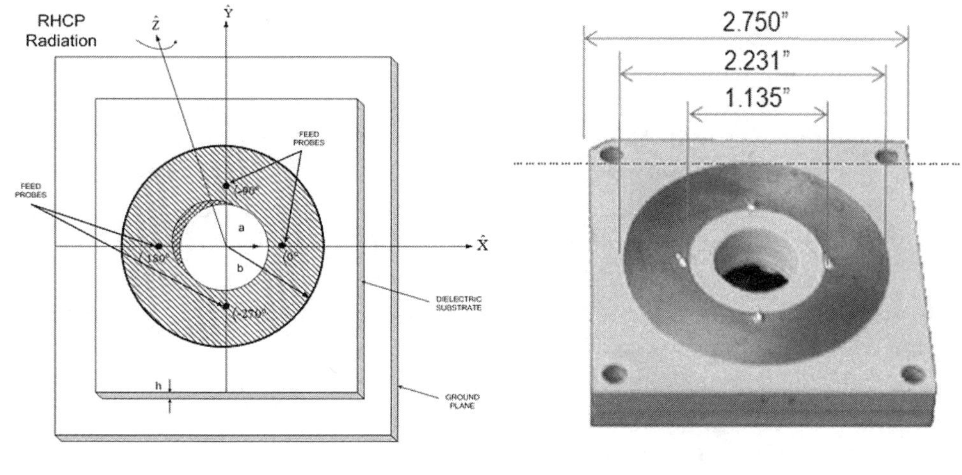

(a) (b)

Figure 2.30 Annular ring microstrip antenna.

where k_{nm} are the roots of the characteristic equation of the form

$$J'_n(CX_n m)Y'_{(n)}(X_n m) - J'_{(n)}(X_n m)Y'_{(n)}(CX_n m) = 0 \tag{2.48}$$

In the above equation a and $C = \frac{b}{a}$ and $J_n(x)$ and $Y_n(x)$ are Bessel functions of the first and second kind of order n, respectively, and the prime denotes the derivatives with respect to x. For the case $C = 2$ and for the TM_{11} mode with $n = 1$ and $m = 1$, the lowest order mode of the annular ring patch antenna, the inner and outer radii, can be determined from $x_{nm} = 0.6773$. An annular ring patch antenna excited in the TM_{11} mode is shown in Figure 2.30(b). RHCP is obtained by exciting the antenna shown in this picture by four symmetrically located direct-contact probes excited with equal amplitudes and 0°, –90°, –180°, and –270° in phase, respectively, by using suitable hybrid couplers. The four probes need to be located at appropriate positions inside the patch antenna so as to provide an input impedance of 50 ohms. The impedance is at its highest value near the outer edge of the annular patch and lowest at its inner edge. The correct location of the probe is determined by using one of the computer design codes mentioned earlier or through analytical formulas [21, pp. 178]. Dual-band GNSS antennas can also be built by stacking a higher-frequency patch on top of a lower-frequency patch and exciting both patches with a common set of direct-contact probes.

Additional modifications to the radiation pattern can also be achieved by shorting either the inner and outer surfaces of the annular patch antenna and adjusting the inner and outer radii a and b so as to achieve resonance at the desired GPS frequencies. Dual-band annular ring patch antennas can also be obtained by stacking. One such useful modification for limiting GPS multipath effects is by using a reduced surface wave antenna described below.

2.2.8.1 Shorted Annular Ring Microstrip Antenna for GPS Multipath Limitation

As discussed in Section 2.2.4.4, the TM_0 surface wave mode in a microstrip antenna has no cutoff frequency. Lateral waves are also launched at the surface of the ground plane [25]. Hence they can generate antenna backlobes that can increase multipath effects when both the surface wave and lateral wave that are launched by the microstrip antenna are diffracted from either the truncated edges of the dielectric substrate of the antenna or the ground plane on which the antenna is mounted (shown in Figure 2.6). Multipath arises when the signals transmitted by the GPS satellites arrive at the receiving antenna using two or more different paths—one from the main beam and the other from extraneous reflected signals picked up from the antenna backlobes—resulting in amplitude and phase differences caused by their interference. These reflected signals often impinge on the antenna at low-elevation angles. The range distance measurement from the satellite to the user is corrupted by such multipath interference and the error depends on the time delay between the direct and reflected signals and the strength of the reflected signal relative to the direct signal. Multipath effects and their impact on GNSS measurement accuracy are described in greater detail in Chapter 5 and 7.

Various types of multipath limiting antennas have been designed for GNSS applications. Unlike different types of choke ring designs that rely on complex corrugated ground planes to suppress surface waves, the shorted annular ring (SAR) antenna accomplishes the same result of reducing the horizon and backlobe radiation by using an antenna design that is simpler, lighter, and less expensive to build [19, 20]. In this design the inner circumferential edge of the annular ring patch antenna is shorted to the ground plane and both the inner and outer radii of the antenna are adjusted to suppress the TM_0 surface wave as well as the lateral wave propagation. A schematic diagram of the annular microstrip ring patch antenna with its inner circumferential edge shorted to the ground plane is shown in Figure 2.31. The TM_0 surface wave and the lateral wave are suppressed when the outer radius b of the annular ring patch is adjusted to satisfy the condition:

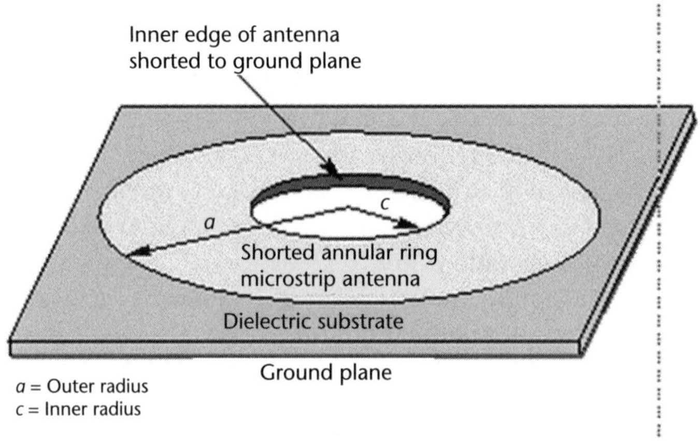

Figure 2.31 SAR microstrip antenna for reducing multipath.

$$k_0 b = k_{TM0} b = 1.8412 \tag{2.49}$$

where $k_0 = \frac{2\pi}{\lambda_0}$ and k_{TM0}wave propagation constant of the TM_0 surface wave mode in the dielectric substrate of the patch antenna, and λ_0 is the wavelength in free space. If the dielectric substrate is electrically thin (i.e., if $h << \lambda_0$), then $k_{TM0} \approx k_0$ and the condition specified in (2.49) is satisfied. Hence the outer radius b of the annular ring microstrip antenna that is necessary for achieving maximum suppression of both surface wave and lateral wave propagation is given by:

$$b = \frac{1.8412}{k_0} = 0.29\,\lambda_0 \tag{2.50}$$

The above equation indicates that for a thin substrate the outer radius of the annular ring antenna necessary for suppressing both surface wave and lateral wave radiation is independent of the dielectric constant and is 0.29 times the wavelength. This means that the outer radius needs to be 2.191" at 1.57654 GHz and 2.819"at 1.2276 GHz, the center frequencies of the GPS L_1 and L_2 band, respectively. To make the annular ring antenna resonate at the desired GPS frequency a short-circuited inner boundary needs to be incorporated into the design. In some designs the short circuit to the ground plane is obtained by a concentric array of closely spaced vias at the inner surface of the annular ring; in other designs a concentric circumferential metal ring is soldered to the top and bottom inner surface of the annular ring patch antenna. To obtain resonance the radius a of the short-circuited inner boundary should satisfy the transverse resonance equation given by:

$$\frac{[J_1(k_1 a)]}{[Y_1(k_1 a)]} = \frac{[J_1'(k_1 b)]}{[Y_1'(k_1 b)]} \tag{2.51}$$

where $k_1 = k_0\sqrt{\varepsilon_r}$ and ε_r is the dielectric constant of the substrate.

Two types of dual-band SAR antennas for multipath mitigation have also been proposed for the GPS L_1 and L_2 bands. The first design uses stacked patch antennas with elliptically shaped patches for achieving dual-band performance; both top and bottom patches are excited by a common set of direct-contact probes [43]. In the second design resonance in the GPS L_2 band is obtained by using a second inverted shorted annular ring (ISAR) microstrip antenna where the outer radius is shorted to the ground plane instead of the inner radius [43]. This "shorting" is achieved by using a concentric array of closely spaced vias similar to that used for the inner annular ring. The ISAR antenna is made to resonate in the L_2 band by adjusting the outer and inner radii of the outer ring, a_I and b_I, respectively, to satisfy the transverse resonance equation [43]:

$$\frac{[J_1(k_1 a_1)]}{[Y_1(k_1 a_1)]} = \frac{[J_1'(k_1 b_1)]}{[Y_1'(k_1 b_1)]} \tag{2.52}$$

Figure 2.32(a) shows a schematic diagram of the second design concept by combining a SAR antenna operating in the L_1 band with an ISAR antenna operating in the L_2 band. Dimensional and other design details are discussed in a paper by Basilio et al. [43]. Figure 2.32(b) shows the measured radiation pattern of a dual-band multipath limiting antenna. The first dual-band antenna design is more compact and is also compatible for use with a dual-band GPS receiver since it uses a common set of feed probes for both frequency bands. The second design is much wider in circumference and is also less compatible for use with a dual-band receiver unless used with a diplexer.

2.2.9 Mutual Coupling Effects in GNSS Microstrip Antenna

Mutual coupling is always present between elements in a microstrip antenna array and is a function of the edge separation between adjacent elements, the dielectric constant the thickness of the substrate [23, pp. 163–174]. In microstrip antenna elements mutual coupling can occur from both space wave as well as surface wave radiation, especially through the TM_0 surface wave, which has no cutoff even with thin substrates. Mutual coupling effects are generally negligible when the spacing between elements is a half-wavelength or greater. As the spacing between the elements decreases, the mutual coupling increases rapidly and can have a significant negative impact on almost all of the important parameters needed for optimum GNSS performance such impedance mismatch to the feed network, degradation in RHCP gain and axial ratio, pattern asymmetry with elevation and azimuth angle, and increased phase and group delay. Mutual coupling effects can be significant in small and finite size array antennas placed on aircraft where the antenna elements are packed much closer than a half-wavelength apart in the interests of conserving space available for deploying antennas [44]. Previous studies have shown that mutual coupling between antenna elements can affect the channel mismatch in GPS antenna arrays and can affect the interference cancellation ratio (ICR), and signal-to-interference-plus-noise ratio (SINR). Mutual coupling effects in microstrip

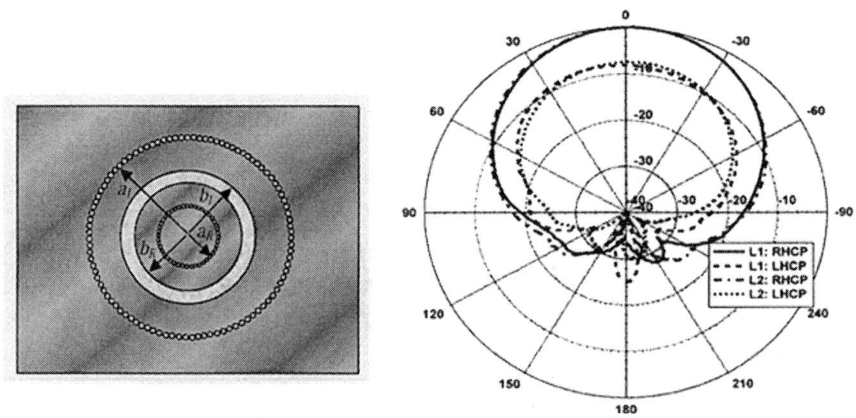

(a)

(b)

Figure 2.32 GPS dual-band multipath limiting antenna using annular ring microstrip antennas: (a) schematic diagram, and (b) measured radiation pattern. (From [43]. ©2007 IEEE.)

antennas vary differently in the E and H planes and affect the axial ratios in RHCP elements used in GNSS. For close interelement spacing, mutual coupling in the H plane are larger than in the E plane; the H-plane coupling and can be as large as -10 dB when the edge spacing decreases to 0.15 wavelengths. The differences in the measured mutual coupling between two linearly polarized square-shaped microstrip antennas in the E and H plane as a function of their relative edge spacing distance have been measured by Jedlicka [45] and are shown in Figure 2.33.

This imbalance in mutual coupling in the E and H planes can affect the amplitude and phase balance between the fields generated by the two orthogonally located feed probes in a patch antenna that generate the RHCP [44]. In an isolated antenna element, the two probes have equal amplitude with a phase difference of 90°. The E- and H-plane asymmetry in mutual coupling alters this phase and amplitude balance in an antenna element placed inside an array and increases the cross-polarization level depending on the severity of the coupling between adjacent elements. Mutual coupling between the two seperate feed probes also excites higher-order modes, especially the TM_{12} mode in a square-patch antenna that can cause the radiation pattern to degrade [46]. This a particularly true on a patch antenna fed by two feed probes because of the inherent asymmetry in excitation. The problem can be remedied to some extent by using symmetrically placed four-feed probes as mentioned in the next section. In a closed packed array of microstrip antenna elements the dielectric substrate below each patch antenna is often truncated to minimize direct surface wave coupling between adjacent elements.

Ngai and Blejer [47] have calculated the mutual coupling for a number of different element configurations used in GPS microstrip antenna arrays. Figure 2.34 shows the mutual coupling between two identical square-shape RHCP antennas at two different relative orientations for substrates with three different dielectric constants. Note that mutual coupling for the diagonal configuration is greater than for the horizontal or equivalent vertical configuration. The diagonal configuration

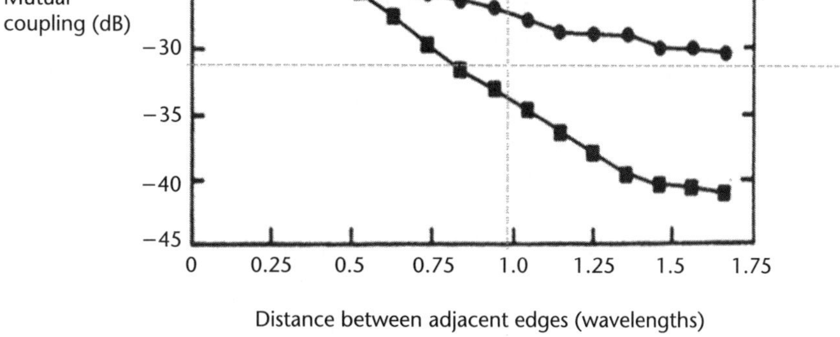

Figure 2.33 Measured mutual coupling between two microstrip antennas in the E and H planes. (From [45]. ©1981 IEEE.)

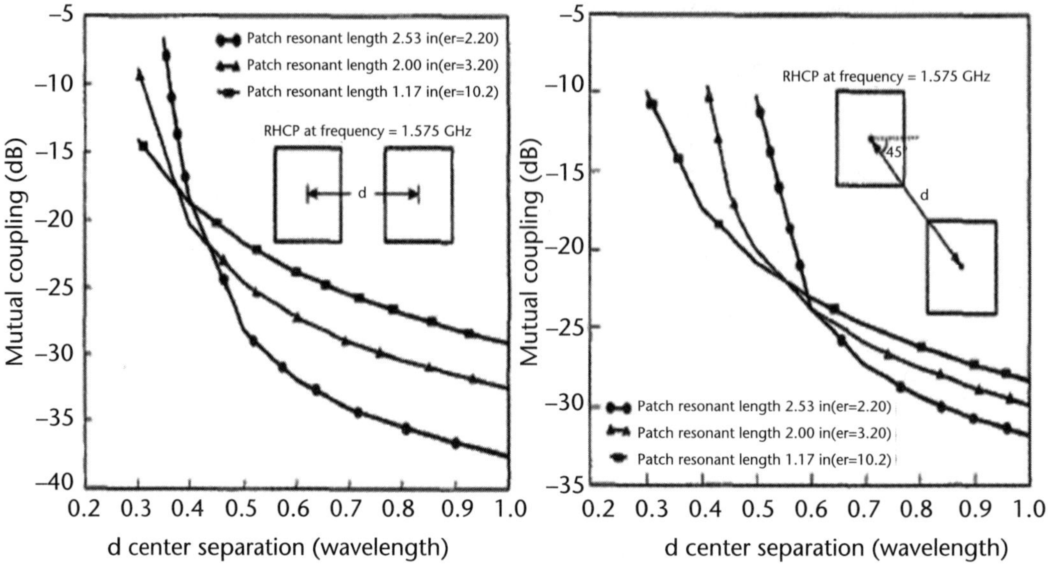

Figure 2.34 Mutual coupling between GPS microstrip antennas at 1.5754 GHz. (©2001 IEEE.)

is the geometry that is prevalent in the mutual coupling that exists between the central element (generally the key reference element of an adaptive array) and the outer auxiliary elements of a square array. Ngai and Blejer [47] have also done extensive calculations of mutual coupling effects in arrays of microstrip antennas containing from 3 to 9 antenna elements enclosed in a 7" × 7" square aperture. To keep the element size small the dielectric constant of the substrate used in these simulations was 10.2. Calculations have been made for both the GPS L_1 and L_2 frequency bands. The results for mutual coupling between elements in GPS antenna arrays measured by Nagai and Blejer are shown in Figure 2.35.

2.2.10 Phase and Group Delays in GNSS Microstrip Antennas

Phase and group delays introduced by microstrip antennas are a potential source of error in high-precision GNSS; there have been several investigations to estimate their magnitude [34, 38, 48, 49]. These errors can vary with both the elevation and the azimuth angle and thus can be different for each satellite signal that is received, leading to positional errors. Group delay errors tend to be much larger than phase delay errors in microstrip antennas due to the high Q factor of the microwave cavity formed by the dielectric substrate sandwiched between the top metallic patch and the ground plane. This restricts the antenna bandwidth and is particularly true for substrates with high dielectric constants such as ceramics. Carrier phase-based methods such as precise point positioning (PPP) and real-time kinematic (RTK) are particularly susceptible to antenna-induced errors; they can be compensated by calibrating the PCO and PCV of the antenna.

Dong et al. [34] have investigated the phase and group delay as a function of elevation and azimuth angles introduced by GPS microstrip antennas with four different types of direct feed probe configurations. The simulation was made in the

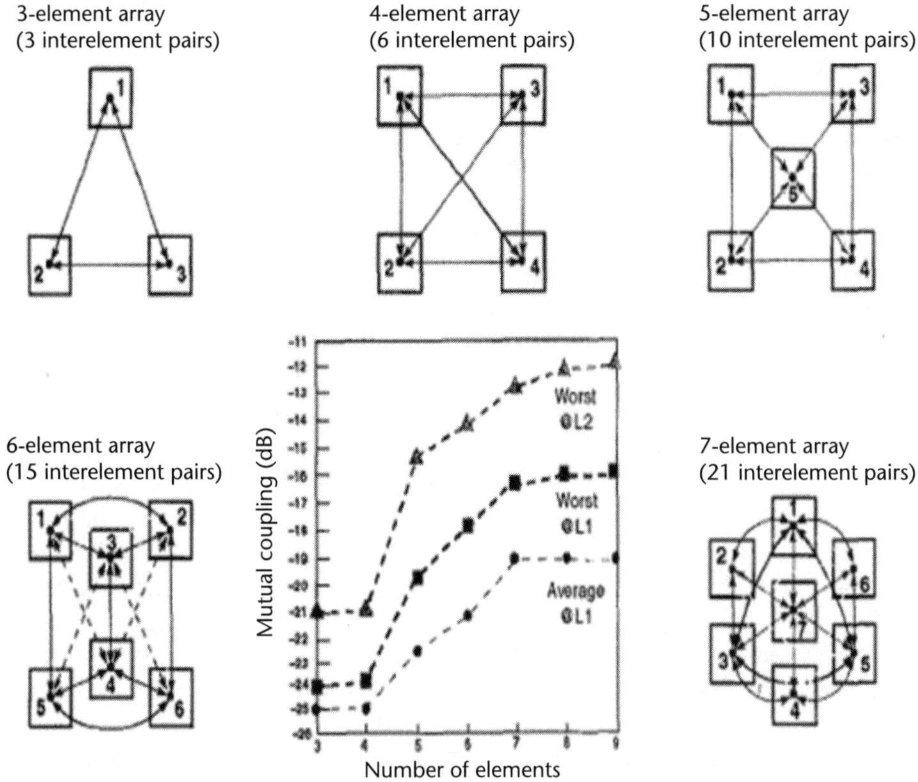

Figure 2.35 Mutual cupling in GPS antenna arrays with 3 to 7 microstrip antenna elements. (©2001 IEEE.)

GPS L_1 frequency band. The four patch antenna designs considered in this study were:

1. A square-shaped patch antenna with a pair of orthogonal probes excited with a relative phase of 0° and –90° (discussed in Section 3.6.1).
2. A nearly square patch antenna with a single diagonal probe (discussed in Section 3.6.3).
3. A square-shaped patch antenna consisting of two pairs of symmetric orthogonal probes for a total of four probes; these are excited with relative phase of 0°, –90°, –180°, and –270°, respectively (discussed in Section 3.6.2).
4. A nearly square patch antenna with two symmetric diagonal probes; these are excited with relative phase of 0° and 180°. This type of feeding scheme was not discussed earlier but is shown in Figure 1(d) in [34].

The coaxial probes were SMA connectors with a probe radius of 0.635 mm. The phase and group delays were analyzed using the full wave electromagnetic simulator HFSS by Ansoft. A cavity model was also used in the analysis but provided less accurate predictions since it is unable to fully take into account effects from higher-order modes. The dielectric constant of the substrate used in most of the cases shown in the paper was 2.94, although in one case simulations were also made for orthogonal feed patch antennas with substrates with dielectric constants of 1.0, 2.94, and 10.2. The results of this study show that the group delay for the diagonal feed patches are much higher than for the orthogonal feed patches. For both the orthogonal- and diagonal-feed configurations the symmetric version with four probes had smaller group delays than the corresponding asymmetric two-feed orthogonal or the one-feed diagonal configuration. This is because the symmetric configuration is able to suppress the higher-order modes from being excited in the patch antenna. The symmetric, orthogonal probe feed antenna with four coaxial probes, discussed in Section 2.3.6.2, showed the best symmetry for group delay variation in the azimuth plane; it does this by suppressing the excitation of the higher-order modes other than the dominant (1,0) and (0,1) modes. In contrast, the diagonal-feed patch antennas showed the greatest variation in both elevation and azimuth angles with in some cases group delays of the order of 4m; large asymmetry in the azimuth plane is also noticed with the diagonal feeds.

The primary contributor to group delay is the narrow bandwidth of the patch antennas. The group delay increases with the dielectric constant of the substrate due to the narrower bandwidth. The phase delay error is not large for all feed configurations and is of the order of millimeters for a typical patch antenna. The primary contributor to the phase delay is probe reactance of the coaxial feed probe. Murphy et al. [48] have also investigated group delay from microstrip patch antennas via modeling using the FEKO, Ver. 5.3 electromagnetic simulation code. The patch was excited by two-feed probes orthogonally located relative to each other within the patch; the patch conductor was 0.0287m square on a finite-size dielectric substrate that was 0.0445m square and 0.00254m thick. The substrate was Roger's 6010 LM with a dielectric constant of 10.2 and a loss tangent of 0.0023. The ground plane was 2-feet square with rolled edges to reduce edge diffraction effects. The FEKO model of this patch antenna was used to compute the phase and amplitude phase response of the antenna pattern in the GPS L_1 band. The variation in the group delay as a function of elevation and azimuth in 1° increments above the horizon was calculated. These computations were repeated for multiple frequencies about the center frequency $f_c = 1575.42$ MHz in the GPS L_1 band. A completed frequency domain transfer function was obtained for each angle of arrival. The differential group delay at f_c was obtained based on the first difference of phase versus frequency as predicted by the model. The variation of the differential group delay was calculated as a function of the azimuth and elevation angles of the satellite. The maximum group delay variation is less than 1.5 ns and appears to be in general agreement with the results shown by Van Dierendonck et al. [49] based on measurements made on two commercial GPS microstrip antennas used in aircraft navigation.

Wirola et al. [38] have investigated the antenna phase response and their effects on ambiguity resolution for three antennas in the GPS L_1 band. Two of these antennas were custom-made microstrip patch antennas with a single diagonal-feed probe used in Bluetooth- (BT) assisted GPS (AGPS) receivers, referred to as BAG. These are 25 × 25 mm square-patch antennas with one corner truncated. The substrate is a ceramic with a dielectric constant $\varepsilon_r = 20$ and has a $Q = 5000$. The performance of these two patch antennas was compared against that of a third antenna: Trimble Bullet^{TM} III. The full 3-D complex RHCP far-field response as a function of elevation and azimuth for all three antennas were measured using the Satimo SG128 3-D measurement system. The S_{11} parameter of the antennas is also provided in the paper. The results show a clear contrast in response between the inexpensive BAG ceramic patch antennas and the Trimble Bullet antenna. Details on the design of the Trimble Bullet antenna are, unfortunately, not provided in the paper.

The Bullet antenna shows a highly symmetric amplitude and phase response and has less than a 4° variation in phase in any given direction, which corresponds to an error of less than 1.6 mm. The standard deviation of the phase response from a best fit sphere is only 0.7°. In contrast the two BAG antennas that were tested showed maximum variations of 70° and 49° in their phase response with standard deviations around the best fit sphere of 8° and 6°, respectively. Their amplitude responses showed similar asymmetry with the observation angles. The greatest variation was observed at the lower-elevation angles. When the elevation masking angle was raised to 30° or 45°, the maximum phase response variation decreases to 32° and 19°, respectively. In addition the response for the two antennas showed differing phase responses so that a single PCV table for this type of antenna is not feasible. The large phase errors observed with the two BAG antennas did not appear to significantly affect ambiguity resolution according to Wirola [38]. These results seem to support observation by Dong et al. [34]. Diagonally fed, single-probe patch antennas create large phase response errors that can affect precision GPS measurements due to the excitation of higher-order modes. More uniform phase and amplitude response as a function of observation angle can be obtained by replacing these with patch antennas fed with four feed probes symmetrically arranged around the center of the square patch. However, the use of four probes will reduce the overall gain of the antenna due to the higher feed losses from the more complex feed network that is needed.

2.2.11 Advantages and Disadvantages of Microstrip Antennas for GNSS Applications

The popularity of microstrip patch antennas for GNSS stems from the numerous advantages discussed in the previous sections and is summarized below.

Advantages

- They can be made to have a very low profile and hence are especially suited for avionics since they can easily meet the ARINC 743 requirement for GPS avionics antennas. This requirement stipulates that the maximum antenna height be no greater than 0.73" so as not to increase aerodynamic drag. Figure 2.3 shows the ARINC 743A antenna footprint for such airborne antennas.
- By using substrates with high dielectric constants listed in Table 2.1, the size of the antenna can be made very small, which makes microstrip antennas ideally suited for densely packed antenna arrays. Very compact ceramic loaded antennas are also common in GPS handsets since they can be unobtrusively mounted inside the receiver casing; their design is described in Chapter 3.
- High dielectric constant substrates enhance the end-fire surface wave radiation from the patch antenna. This improves directivity of single element GNSS antennas at very low-elevation angles near the horizon. The antenna is

therefore able to provide better gain at the lower-elevation angles to acquire GNSS satellites that are critical for improving PDOP. Surface wave radiation from microstrip antennas can also be suppressed by using SAR patch antennas. Although this greatly reduces the gain at low elevation angles and affects satellite availability, it is nevertheless useful as a technique for limiting multipath for certain applications.

- Dual-band antennas covering the GPS L_1 and L_2 bands or the L_1 and L_5 bands needed for correcting propagation errors caused by the ionosphere can be made by either stacking patch antennas on top of one another or by using parasitic coupling of adjacent patches. Recently, triple-band microstrip antennas covering the modernized GPS (L_1, L_2, and L_5) or the Galileo system (E5a, E5b, E6) have been built using these design principles (discussed in Chapter 3). By providing narrowband spot coverage at only the selected frequency bands of operation the microstrip antenna also serves as a passband filter to reduce out-of-band interference.
- RHCP can be easily achieved by using a variety of different feed structures such as coaxial probes, aperture coupled slots, proximity coupled microstrip lines, or edge coupled microstrip lines directly attached to the patch antennas. RHCP with a good axial ratio over only a narrow bandwidth is possible by using a single feed probe coupled to a patch antenna whose external shape is changed by either truncating the corner or by inserting indentations or tabs on the side. This type of antenna feed does not need an external quadrature hybrid and is therefore popular for compact, simple, and low-cost ceramic antennas used in some handheld GPS receivers.
- Conformal mounting on surfaces of sleek aerodynamic shapes such as cones and cylinders is possible by using thin flexible substrates making microstrip antennas popular for small missiles or other projectiles used in many military applications. Special flexible fabrics have recently been used for placing GPS antennas directly on garments.
- Antennas can be very light in weight and can easily be integrated into microwave integrated circuits used in GNSS receivers. Photolithography combined with etching techniques lowers fabrication costs for high-volume production.

Disadvantages

Microstrip antennas also suffer from several disadvantages that provide a challenge to the antenna designer when optimizing their performance for GNSS systems.

- They have a very narrow bandwidth requiring special design techniques to achieve operation at two or more frequency GNSS bands such as stacking one patch on top of another, or coupling to another parasitic patch tuned to a different frequency, or by etching slots into the metallic patch.
- Dielectric substrates with high dielectric constants while reducing the size of the antenna also decrease the bandwidth and increase surface wave radiation at low-elevation angles near the horizon. While this provides some advantages such as improved PDOP due to increased beam width, it also

causes increased diffraction from the edges of the ground plane on which the antenna is mounted. This results in backlobes that increase susceptibility to multipath and interference and also ripples in the antenna gain and phase at higher elevations due to interference with the direct radiation from the patch antenna. These effects are discussed in Chapter 5.

2.3 Quadrifilar Helix Antenna

The quadrifilar helix antenna (QHA), also known as a "volute," is popular for many GNSS applications and is especially suitable for use in handheld receivers and mobile terminals. It is an array of four helically shaped resonant antenna elements wrapped around a cylinder at the desired pitch angle and driven in phase quadrature to produce a broad, RHCP beam. The QHA can be fed from its terminals at either the top end or the bottom end with the other end either shorted or left open and can therefore generate both forward and backward wave radiation patterns with different beamwidths and shapes, adding to its versatility for various GNSS applications. Its size can also be controlled by using dielectric substrates and printed versions can be made to reduce cost. Two types of QHAs are used in GNSS: self-phasing QHAs shown in Figure 2.36(a) and externally phased QHAs shown in Figure 2.36(b).

2.3.1 Self-Phasing QHA

The half-turn self-phasing QHA shown in Figure 2.36(a) consists of four helical elements connected to a split sheath coaxial balun at the top and with the bottom ends electrically shorted to the outer sheath of the coaxial line [50–53]. The QHA

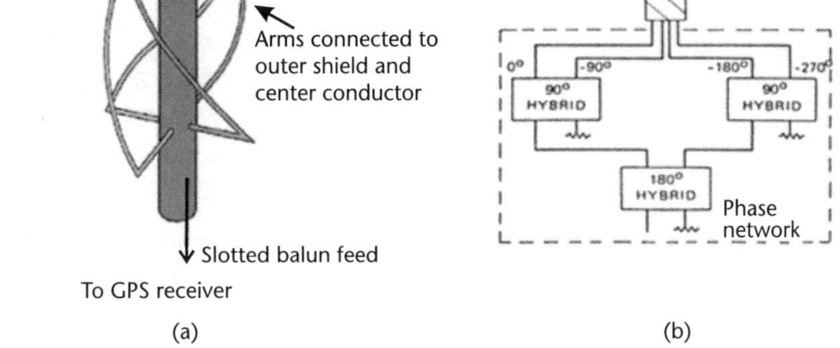

Figure 2.36 Diagram of (a) self-phasing QHA and (b) externally phased QHA. (Figure 2.36(a) Courtesy of S. Best.)

in this design is fed from the top and operates in the backfire mode since the main radiation is directed from the bottom shorted end of the antenna towards the feed section at the top. Quadrature phasing required for generating RHCP is obtained by adjusting the lengths of four elements (described in greater detail later). Figures 2.37(a) and (b) show the side and top view, respectively, of this QHA without the radome cover normally placed around the QHA. The measured principal RHCP and the cross-polarized LHCP patterns of this antenna are shown in Figure 2.37(c). They indicate a very broad antenna beam providing an almost hemispheric coverage with good visibility to all available satellites. One of the drawbacks of QHAs fed with conventional coaxial line balun is considerable backlobe radiation resulting in poor front-to-back ratios, which makes this antenna susceptible to multipath. However more recent QHA designs use infinite microstrip balun feeds that allow significant reduction in backlobe radiation and are able to achieve front-to-back ratios of 20 dB. The measured return loss of this self-phased QHA is shown in Figure 2.37(d). This antenna is resonant at a frequency of 1.5756 GHz and indicates a narrow bandwidth typical of a highly resonant, single-band antenna structure.

2.3.1.1 Design Principles of a Self-Phasing Quadrifilar Helix Antenna

The self-phasing QHA consists of four helical arms, each with an $n/4$ turn where $n = 1, 2, 3, 4$, and so forth, which are equally spaced circumferentially and then wrapped with a constant pitch around a small-diameter cylindrical dielectric core to form a helix. To make the antenna resonate, the four arms must be equal to $m\lambda/4$,

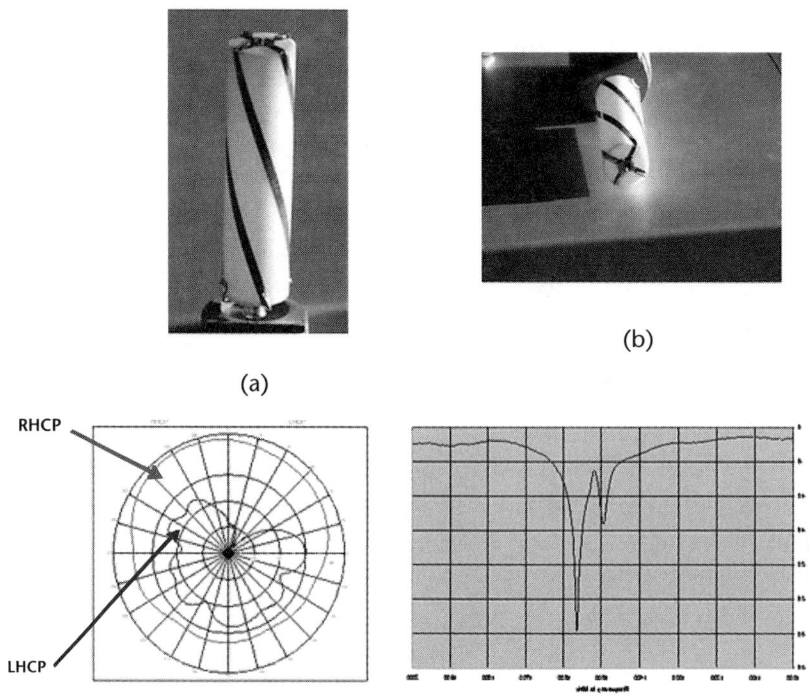

Figure 2.37 GPS half-turn self-phased QHA: (a) side view, (b) top view, (c) patterns, and (d) return loss.

where λ = the wavelength and m = 1, 2, 3, 4 ....; the ends of the helices must be short-circuited when m = an even integer and open when m = an odd integer.

The QHA can be considered a combination of two orthogonal, uncoupled bifilars fed in phase quadrature, where a bifilar is defined as a two-element helical antenna array. The top end of the four helical lines in self-phasing QHAs are connected to a coaxial balun, which is an acronym for a balanced to unbalanced transformer. The balun provides correct current paths between a balanced configuration such as the feeding coaxial cable and the unbalanced configuration such as the antenna. Baluns with different designs are used in QHAs [50, pp. 166–169, 54]. Figures 2.36(a) and 2.37(a) shows the geometry for a half-turn QHA fed by the coaxial feed line and a split-sheath type of balun located along the axis of the cylinder. This balun is designed such that the balanced outputs have equal amplitudes but are in antiphase (i.e., have a relative phase difference of 180°).

Two adjacent elements of the quadrafilar are connected to the center conductor which is one-half of the balun. The other two elements of the quadrafilar are connected to the outer conductor, which is the other half of the balun. Therefore two of the elements are at a 0° phase angle and the other two are at a 180° phase angle. Adjacent helical elements at the same phase are trimmed to different lengths. One element is made shorter than the ideal resonant length to the shorting plane of the balun to make it capacitive at the desired frequency and is cut such that the current leads by 45°. The other element in this pair is cut longer than the correct resonant length to the shorting plane of the balun to make it inductive and causes the current to lag by 45°. The lengths of the other pair of adjacent elements, which are 180° out of phase with the first pair, are adjusted the same way. The four elements of the QHA connected to the balun are now fed with equal amplitudes and with relative phases in the counter clockwise direction of (0°+45°), (0°–45°), (180°+45° = –135°), and (180°–45° = –225°). The θ and φ components of the radiated electric field of the combined bifilars have the same amplitude but are in-phase quadrature to generate RHCP. The normalized input impedance of the longer, inductive bifilar is $Z_L/Z_0 = 1+j$. Similarly the input impedance of the smaller, capacitive bifilar is $Z_S/Z_0 = 1-j$. The normalized admittances of the two bifilars are given by $y_L = 1/1+j$ and $y_S = 1/1-j$. Since the two bifilars are fed in parallel, their combined admittance is the sum of $y_L + y_S = 1$, and the QHA is matched in impedance to the coaxial feed line [55]. To obtain an elevated omnidirectional antenna pattern, the sign of the relative phase difference between the helix elements and the direction of the helix winding occur in the opposite direction. The feeding direction of the quadrature phasing determines the end-fire direction of the antenna. A relative phase increment of +90° in the clockwise direction between the individual helical elements produces an RHCP omnidirectional pattern if the helices are wound in a left-hand sense. Note that in a conventional end-fire helix, the helix has to be wound like a right-hand screw to obtain RHCP. Since the QHA operates in the backfire mode, with the main radiation directed towards the feed point and away from the shorted end, the helix has to be wound as a left-hand screw in addition to the correct phasing on the four helical wires [53, 54].

Since a CP signal changes its sense of polarization after a reflection, the wires of the QHA need to be wound in a left-hand sense so that upon reflection from either the ground plane in an externally fed QHA or the short circuit termination at the bottom end in a self-phased QHA, the signal radiated from the top (feed) end of the antenna is RHCP.

2.3.1.2 Computational Design of a Self-Phased QHA

Kilgus [56] was able to provide approximate, closed-form analytical formulas for the radiation patterns of ¼-, ½-, and a l-turn QHA by approximating the QHA as a combination of a loop and a dipole. However, ensuring a more accurate design of the QHA for GNSS applications, which includes estimating both the amplitude and phase for the θ and φ polarization components as well as the effects introduced by the central dielectric core or by the ground plane as in an externally phased QHA requires analysis using more advanced electromagnetic computational modeling methods. Method of moments (MOM) codes such as NEC-4 are useful for the analysis of QHAs that use a core made from a very low-dielectric constant such as foam or an air core. Accurate design of a QHA with a dielectric core with higher-dielectric constant than foam requires using commercial computational codes listed in Table 2.2, which are the same as those used for designing microstrip antennas.

Using these electromagnetic codes will require segmenting both the four helical and radial elements of the QHA into finite segments, which are fractions of a wavelength at the desired design frequency. Figure 2.38(a) shows the segmentation of a QHA for analysis using a MOM code. The geometrical equations for the QHA described below can be used to assist in this segmentation. The geometry of each of the QHA can be described in terms a helical coordinate system as shown in Figure 2.38(b).

The coordinate system shown in Figure 2.38 is defined by the following parameters: the pitch angle ψ of the helix given by $\cot\psi = 2\pi b/P = \tau\, b$ where $\tau = 2\pi/P$, where P is the pitch distance of the helix or the spacing between helical turn measured center to center, b = the radius of the helix to the center line of the wire

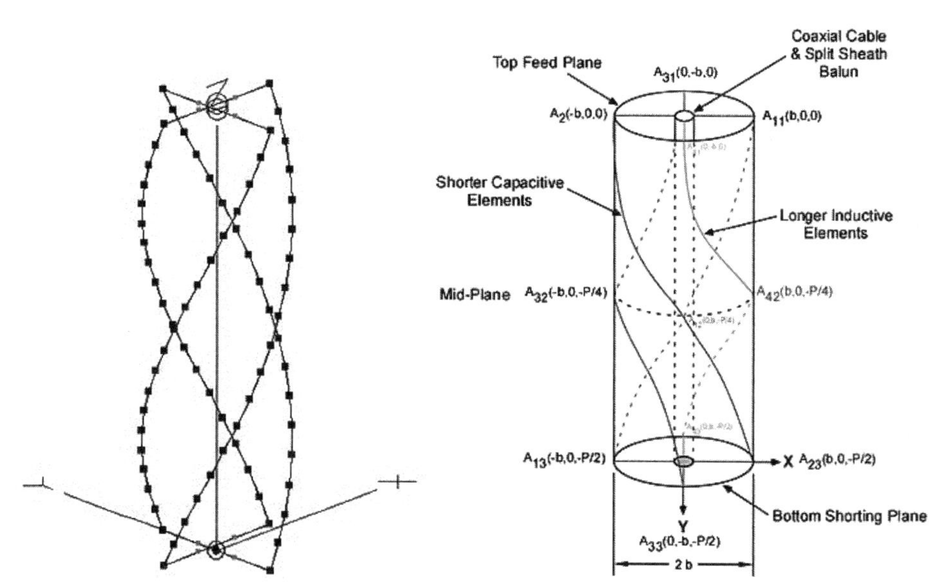

(a) (b)

Figure 2.38 Electromagnetic computational model of a self-phasing QHA. (Courtesy of S. R. Best.)

as shown in the figure, and a = radius of the antenna wire. The total axial length of the helix is $L_{ax} = N\,P$, where N is the total number of turns (or a fraction of a turn) in the helix. Any point at an axial distance z on each of the four helical wires originating from the feed region at the points A1, A2, A3, and A4 at the top of the balun can be represented by the following four equations:

$$A_1 = \hat{x}b\cos(\tau z) - \hat{y}b\sin(\tau z) + \hat{z}z \tag{2.53}$$

$$A_2 = -\hat{x}b\cos(\tau z) + \hat{y}b\sin(\tau z) + \hat{z}z \tag{2.54}$$

$$A_3 = b\hat{x}\sin(\tau z) + \hat{y}b\cos(\tau z) + \hat{z}z \tag{2.55}$$

$$A_4 = -b\hat{x}\sin(\tau z) - \hat{y}b\cos(\tau z) + \hat{z}z \tag{2.56}$$

In the above equations, $\hat{x}$, $\hat{y}$, and $\hat{z}$ represent the unit vectors along the three principal axes of the Cartesian coordinate system. Notice that helical wires originating from A1 and A2 form one pair of bifilars and those originating from A3 and A4 represent the orthogonal bifilar of the QHA. From the above equations it can be seen that for a half-turn QHA shown in Figure 2.38(b) the coordinates of the these four points at the feed plane $z = 0$ are:

$$A_1 = \hat{x}b,\ A_2 = -\hat{x}b,\ A_3 = \hat{y}b, \text{ and } A_4 = -\hat{y}b$$

Similarly, at the other end of the QHA away from the feed point, where $z = L_{ax} = -P/2$ for the half-turn QHA, the four helical wires are "shorted" to the outer sheath of the feeding coaxial balun. This is the shorting plane of the balun and is used as the reference plane for adjusting the lengths of the helical arms to obtain the quadrature phasing. The coordinates of the corresponding points of the four helical wires at this shorting plane represented by A'_1, A'_2, A'_3, and A'_4 are given by:

$$A_1' = -b\hat{x} - \hat{z}(P/2),\ A_2' = b\hat{x} - \hat{z}(P/2),\ A_3' = -b\hat{y} - \hat{z}(P/2),\ A_4' = b\hat{y} - \hat{z}(P/2)$$

Equations (2.53) through (2.56) can be applied to other QHAs with different fractional turns.

The elevation pattern or the cone aperture of the radiation pattern of the QHA can be controlled by adjusting the radius of the helix b, the pitch distance of the P, and the number of turns N. These three parameters can be optimized for providing a near-ideal RHCP pattern covering the upper hemisphere. For GPS applications typical QHA designs use from one-quarter to a full turn to keep the length of the antenna small. The optimum number of turns for a QHA used in some GPS and GLONASS antenna systems is between 1.5 to 3 turns. Increasing the length of a QHA by increasing the number of turns does not have much impact on the gain, but improves the beam shape and attenuates the undesired backward antenna lobe in the downward direction, thereby improving its resistance to multipath from ground reflections. Increasing the length of the antenna but maintaining the same number of turns reduces the front-to-back ratio; it will also increase the beamwidth and reduce the phase center variations at low-elevation angles. The diameter of the QHA plays an important role in determining the bandwidth with the sleeker style QHAs having a narrower bandwidth while those with more bandwidth require a larger diameter. The pitch angle, with the other parameters adjusted to be optimal, controls the beamwidth of the antenna. Higher pitch angles yield wider beamwidth at lower gain, whereas the lower pitch angles tend to yield narrower beamwidth at higher gain. The number of turns in itself is not a good indicator of the performance of the QHA.

Tranquilla and Best [57] by using MOM have provided detailed design information of the QHA with a foam (air) core specifically designed for GPS applications. They have considered both the amplitude and phase response of three different designs of the QHA summarized in Table 2.4. Of special importance is information they provide on the phase center stability of the QHA that strongly affects the use of carrier phase data that is important in differential GPS systems. The three antennas analyzed were one-half turn and one-half wavelength long measured along the curved helical segments including the radial section connecting to the central coaxial feed.

Figure 2.39(a) shows the computed RHCP antenna patterns for these three antenna designs. While all three QHAs provide good upper hemispheric gain coverage needed for acquiring visible GPS satellites, only one the shortest of these three with $L_{ax} = 0.2\ \lambda$ has a suppressed backlobe that is need to mitigate ground multipath. Figure 2.40(b) shows the phase patterns of the linear polarized θ and φ components in a fixed observation plane; the phase data in this referenced paper [68] has been presented in linear polarization rather than in CP to maintain the strict definition of "antenna phase center"; the authors claim that phase center for CP is not a definable quantity although this is the polarization of interest for GNSS applications.

Figure 2.40(a) shows the computed phase center location for the φ and θ linear polarizations for the half-wavelength QHAs. The phase center is measured in terms of a length L from the coordinate origin and an angle ψ as measured from the axis X of the QHA as shown in Figure 2.40(b). The improved phase stability

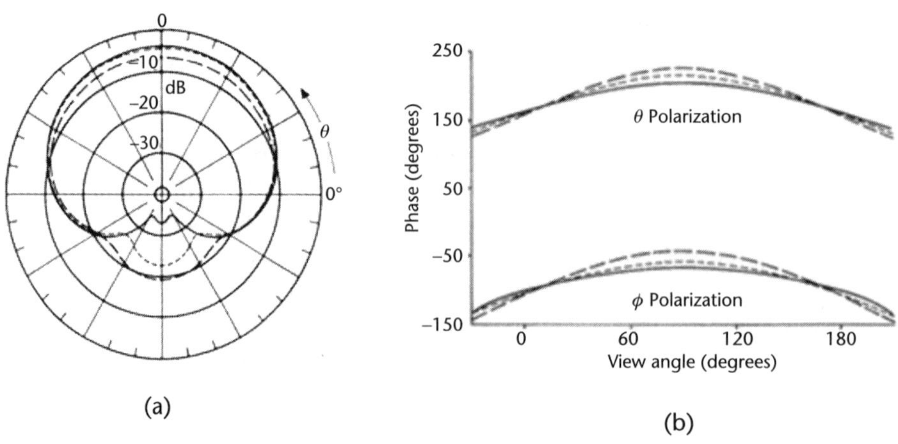

Figure 2.39 Calculated radiation pattern of a self-phased QHA. (From [68]. ©1990 IEEE.)

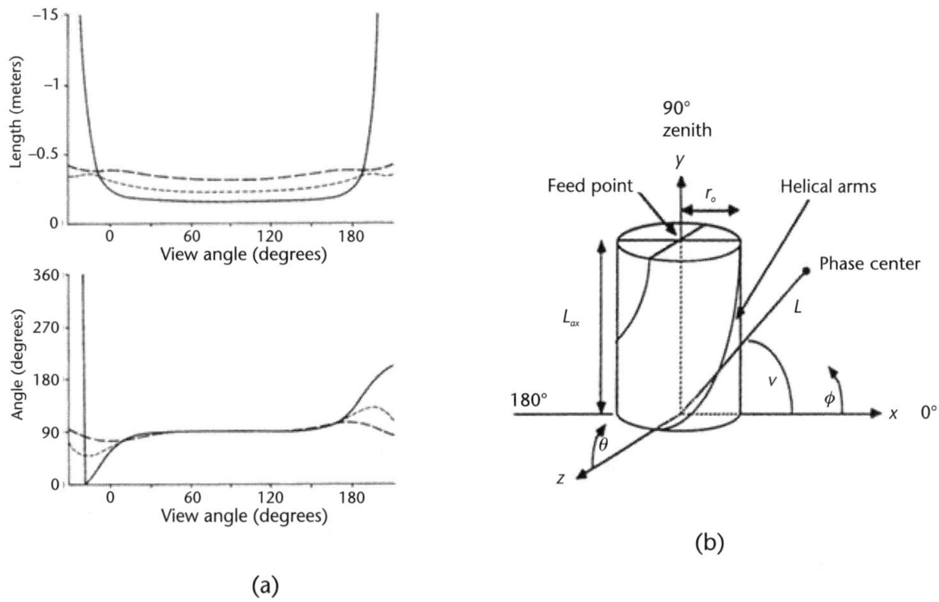

Figure 2.40 Computed phase center of self-phased QHA.(From [57]. ©1990 IEEE.)

provided by these longer and slimmer QHAs is due to the reduced movement in the phase center at low-elevation angles. This is attributed to the reduced separation distances between helical arms when viewed from near broadside to the antenna, which minimizes the interference effects between the field contributions from the opposing arms of the QHA

Hussain and Rengarajan [58] have conducted investigations using MOM techniques on a half-turn QHA with shaped ground planes specifically designed for GPS applications to examine their effects on the antenna backlobes in the radiation pattern as well as the impact on phase center stability caused by diffraction and mutual coupling from the edges of the ground plane. Both a flat ground plane $1\lambda \times 1\lambda$ and a bent ground plane with their edges bent downwards and away from the antenna at a 20° inclination were considered. The ground plane was placed at various distances from the base of the antenna and its effects on the antenna pattern and phase center were examined.

QHAs designed for GPS applications have a narrow bandwidth of between 3% to 5% for a return loss of –10 dB VSWR = 2:1; this allows operation only over a single GNSS frequency band. As stated earlier, the natural bandwidth of the QHA is proportional to the diameter of the QHA; hence, the smaller the antenna the smaller its bandwidth. The return loss of a QHA used in the GPS L_1 band was shown in Figure 2.37(d). Notice that these measurements indicate antenna resonance only at L_1 whereas the return loss in the L_2 band for this antenna is poor.

2.3.1.3 Effects of Dielectric Core on a Self-Phased QHA

The cylindrical core used in the QHA is generally made either from foam ($\varepsilon_r \sim 1.07$) or a hollow, thin-walled dielectric tube made from a low-dielectric constant material such as Teflon or from periodic dielectric spacers. In some designs the

Table 2.4 Design Parameters for Three GPS Quadrafilar Helix Antennas Designed by Tranquilla and Best for Operation at 1.5754 GHz (L_1 Band)

Antenna Number	*Axial Length (L_{ax}) Expressed in Wavelength λ*	*Radius of the QHA (b) in cm and in wavelength*
Antenna 1	0.20	1.60 (or 0.084 λ)
Antenna 2	0.27	1.39 (or 0.072 λ)
Antenna 3	0.35	1.04 (or 0.054 λ)

From: [57].

thin walled inner tube supporting the helical structure can be made from a high-temperature molded plastic. High-dielectric constant cores, such as ceramic cores, have been used for reducing the size of the QHA [59, 60]. The presence of the dielectric core alters the phase velocity of the RF current path along the helical windings and can affect the radiation pattern of the antenna [53]. Therefore, the pitch angle and the length of the helix need to be adjusted to compensate for these effects. In the uncompensated case, dielectric loading can cause the peak to occur at a lower-elevation angle and cause a significant loss in the peak gain. A reduction in length of the helix and the pitch angle effectively shortens the electrical path length and compensates for the phase error. Foo [53] has compensated for these dielectric loading effects by varying the pitch angle and reducing the overall length of the helix. Leisten [59] has used a cylindrical core made from very-high-permittivity ceramic material to design very-small-sized, lightweight QHAs that have been developed by Sarantel for use in GPS handsets. Liu and Chu [60] have recently designed an extremely compact triple-band, dielectrically loaded QHA intended also for handset applications. Using the high-dielectric core coupled with the antenna topology in both designs helps to insulate the antenna from interacting with the person carrying the handset and also with the surrounding environment. It does so by confining the near field to be closer to the antenna. This is an important type of antenna used in many GNSS handset applications and will be discussed in greater detail in Chapter 4. MOM techniques described in the previous section for analysis and design of air core or foam core QHAs cannot be used for designing QHAs with cores made from higher-dielectric constants. A transmission line method (TLM) such as Microstripes listed in Table 2.2 has been used by both Leisten [61] and Liu [60] for designing QHAs with ceramic cores. One major disadvantage in using ceramic cores for QHAs is the drastic reduction in bandwidth. Their bandwidth is so narrow that it allows them to receive only C/A code signals in the GPS L_1 band but not the P(Y) code or M code signals in the GPS L_1 or L_2 bands, requiring either 20-MHz or 24-MHz bandwidths or even the L_5 code, which requires a 20-MHz bandwidth.

2.3.1.4 Printed Self-Phased QHA

The four helical elements of the array of the QHA are made from either conducting wires (called WQHA for wire QHA) as shown earlier in Figure 2.36(a) or from metallic tapes using low-cost, printed circuit technology (PQHAs or printed QHAs). Extensive design work on PQHAs has been conducted by Sharaiha and coworkers [62]. The construction of a PQHA using low-cost, printed circuit tapes is illustrated in Figure 2.41(a) and (b) [55, 62]. The substrate used is a thin, flexible Kapton film (dielectric constant = 3.2 and about 50 μm in thickness) with electrodeposited copper cladding on its surface; four equally spaced, sloping-straight-line copper strips of width *w* are etched on its surface. A strip of width *w* is equivalent to a circular wire of diameter $4a$, where *a* is the radius of a circular wire. The angle of each strip relative to the base line Δ is ψ = pitch angle of the individual helical antenna

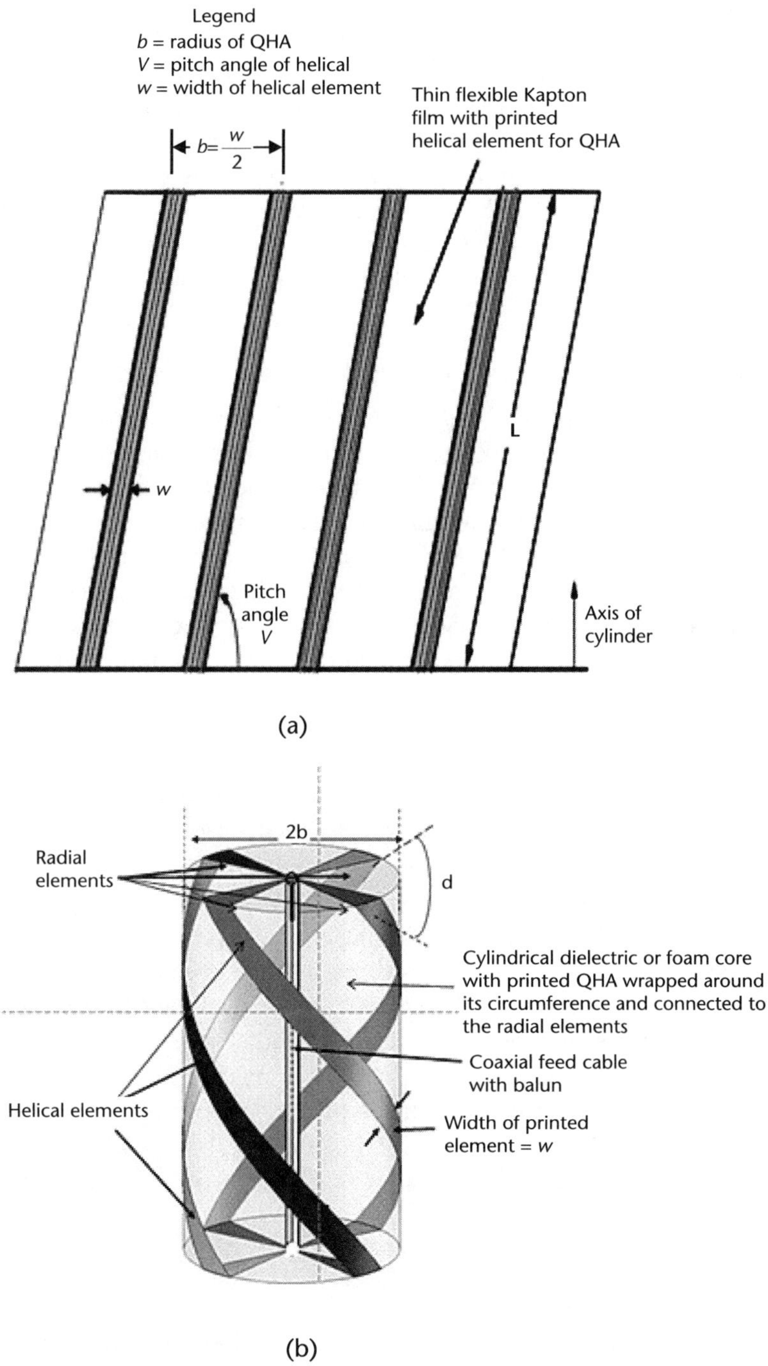

Figure 2.41 Construction of a printed self-phased QHA with a dielectric core.

elements of the QHA. The etched Kapton film containing the four lines are wrapped around a cylindrical core generally made from Rohacell foam whose circumference is equal to four times the interelement spacing d between the helical elements. The axis of the cylinder core is normal to the baseline Δ. The feeding arrangement for a self-phased QHA using tapes as the helical elements is shown in Figure 2.41(b).

2.3.1.5 Baluns for the QHA

Each of the two bifilars in the QHA operates as a balanced antenna when fed from a coaxial line through a balun. The function of the balun is to distribute equal currents but of opposite phase from the feed coaxial line to the two arms of the antenna. Bricker and Rickert [54] and Kumar [50] have proposed three different balun designs that can be used with the self-phased QHAs fed from the top. These are the folded balun, the split-sheath balun, and the infinite balun. The folded and split-sheath baluns are both fed from the center by coaxial cables mounted along the axis of the bifilar. A split-sheath balun is shown in Figures 2.36 and 2.37.

2.3.1.6 Printed QHA with Infinite Microstrip Balun Feeds

A new type of printed half-wavelength QHA consisting of microstrip fed helical elements with four opposite radial strips has been designed by Shumaker et al. [62]. Infinite microstrip baluns whose design is described in the paper is used to feed the helical elements. This design obviates the need for conventional folded or split-sheath baluns and reduces the physical size and weight of the antenna and makes it easy to mass produce. Experiments show that this antenna has a 3-dB bandwidth of more than 145° and a very impressive front-to-back ratio of 20 dB. This is a major improvement over QHAs with a conventional coaxial balun feeds described earlier.

2.3.2 Externally Phased QHA

The second type of QHA used in GNSS is the externally phased QHA shown in Figure 2.36(b) [50, 53, 54]. In this design the four antenna arms of the QHA are fed from the bottom of the antenna through an external broadband phasing network fed against a ground plane. Figure 2.36(b) shows a typical phasing network for the QHA consisting of 180° hybrid ring coupler and two 90° branch-line type couplers. The 180° hybrid ring coupler provides two outputs that are equal in amplitude but with relative phases of 0° and 180°, respectively. Each of these two outputs are in turn connected to the inputs of a 90° branch-line coupler to provide two outputs that have the same amplitude but with a relative phase difference of 90°. The feed network provides the four arms with equal amplitudes and with a relative phase of 0°, –90°, –180°, and –270°, resulting in RHCP end-fire radiation in the direction of winding. The external phasing method described above provides three advantages:

1. By eliminating the need for the narrowband balun, it allows a 50-ohm impedance match to the receiver over a much broader bandwidth
2. It allows the use of open-ended, equal-length helical elements and therefore allows the antenna designer to precisely tune the length of the helical ele-

ments by trimming the open end of each element or by varying the shape of each element.

3. It allows a dual-band design of the QHA by placing a resonant trap load at an appropriate position along the length of each open-ended helical element.

A dual-band QHA that operates in both the GPS L1 and L2 band has been designed on this principle [64]. Operation over dual-band designs is obtained by placing a resonant trap filter tuned to 1.5754 GHz (the center frequency of the GPS L1 band) at an appropriate position along the length of each open ended helical element. The bandstop trap filter acts as a high-impedance "open" circuit in the GPS L_1 band but as a virtual short circuit in the GPS L_2 band, thus allowing the antenna currents to flow along the full length of helical element so as to achieve resonance in the lower L_2 band.

2.3.2.1 Dual-Band and Triple-Band QHAs for GPS

A major disadvantage of the self-phased QHA as pointed out earlier was its narrow bandwidth indicated by Figure 2.37(c), which shows its measured return loss; hence, it is generally only designed to operate over only over a single GPS frequency band, generally the L_1 frequency band of the GPS system. Other QHA designs have been proposed that allow dual-band operation over the GPS L_1 and L_2 bands. One proposed dual-band design [51, pp. 455–462; 57] uses two antennas, each operating at a different band, placed either in a piggyback configuration with the L_1 antenna on top of the L_2 antenna or by enclosing one antenna inside a the other, with the L_1 antenna inside the L_2 antenna or by rotational offset with the L_1 and L_2 arms interleaved. Results of measurements performed on the first two proposed designs of dual-band QHAs have been provided by Tranquilla and Best [57]. However, neither of these proposed dual-band designs performs as well as the single-band version discussed earlier. Both designs show a reduction in gain in the upper hemisphere at both the bands, an increase in the backlobe level, and a poorer phase center performance at off-boresight angles when compared to the single-band design. In addition both designs appear physically clumsy and structurally suspect. There is no data provided on the third design of rotationally offset interleaved antennas. As mentioned in Section 2.3.1.3, a triple-band piggyback of ceramic-loaded QHAs for operation over the GPS L_1, L_2, and L_5 bands, has been designed and built by Liu [60]; their design consists of a piggyback configuration of two separate antennas—a dielectric-loaded QHA (DQHA) at the top and a dielectric-loaded octa-filar helix (DOHA) at the bottom. The two antennas work in combination to operate in the GPS L_1, L_2, and L_5 frequency bands. However, the bandwidth of these QHAs is extremely narrow because of ceramic loading—especially in the L_1 band where only C/A code operation over a 2-MHz bandwidth seems possible from the data that has been provided.

A dual-band printed quadrifilar helix (PQHA) fed from the bottom with an external phasing network has been designed by Lamensdorf and Smolinski [64] by using trap filters placed at an appropriate distance along the length of the four helical elements of the PQHA. A picture of the dual-band PQHA is shown in Figure 2.42.

This antenna can provide a 24-MHz bandwidth in both the GPS L_1 and L_2 bands, which allows it to receive the P(Y) code and M code GPS signals. The antenna design is also physically more robust and provides good performance in both bands than the coaxial or piggyback dual-band versions mentioned earlier. The four helical arms are excited with sequential phase variation of 0°, –90°, –180°, and –270° in a counterclockwise manner to generate RHCP. The external phasing network is similar to that shown earlier in Figure 2.36(b). The trap filter consists of a parallel resonant LC circuit and is used as a switch to tune the antenna for resonance at two different resonant frequencies. The trap filter acts as an open circuit offering high impedance to current flow in the antenna at its parallel resonance frequency, which is selected to be 1.5754 GHz, the higher of the two selected frequencies for the QHA. The location of the trap filter in each helical arm is selected so as to provide resonance at 1.5754 GHz. At 1.2276 GHz, the lower of the two selected frequencies, the trap filter has low reactive impedance, allowing an almost unimpeded current flow in the antenna. The length of the antenna beyond the filter is compensated for the presence of the trap filter is tuned to achieve resonance at 1.2276 GHz, the center frequency of the L2 band offering dual-band operation within a single integrated QHA. The four helical lines of the PQHA were etched on the thin copper-clad Mylar tape at the correct pitch angle. The tape was then wrapped around a cylinder made from Rohacell foam with an outer diameter of 0.36". Discrete microstrip L and C components were used in building the resonant trap filter that was soldered across a small gap in each of the four arms of the PQHA. The PQHA was mounted on a small circular ground plane with a diameter of 0.3 wavelengths at the lowest GPS frequency used in these measurements. The performance of the QHA was optimized using MOM techniques using the NEC-4 MOM code. Figure 2.43 shows the measured reflection loss S_{11} in each of the two GPS bands. The bandwidth is 5% in the L_1 band and 2.2% in the L_2 band. Figure 2.44 shows the measured elevation plane pattern radiation pattern for elevation angles from zenith down to –120° over a 24-MHz bandwidth at frequencies of 1.563, 1.575, and 1.587 GHz in the L_1 band. Figure 2.45 shows the corresponding measured elevation plane radiation patterns GHz in the GPS L_2 band at frequencies

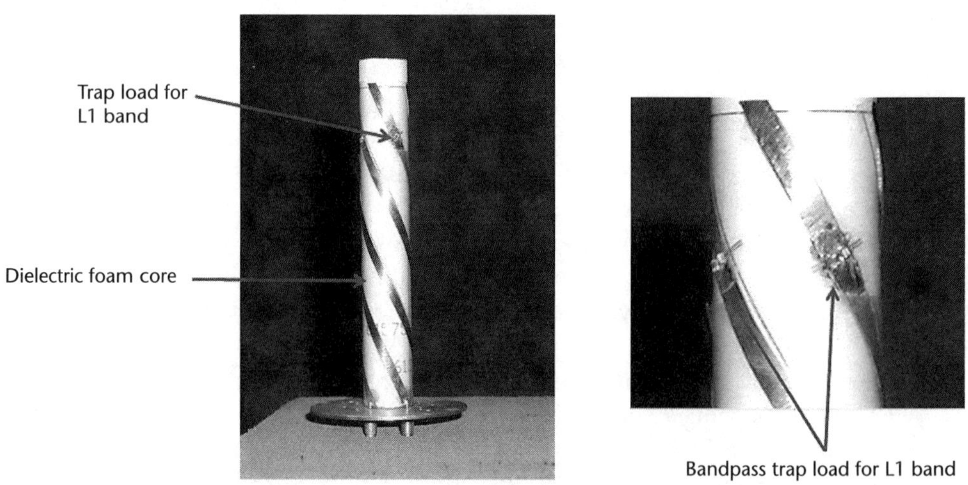

Figure 2.42 GPS dual-band externally phased QHA. (*From:* [64]. Courtesy of D. Lamensdorf.)

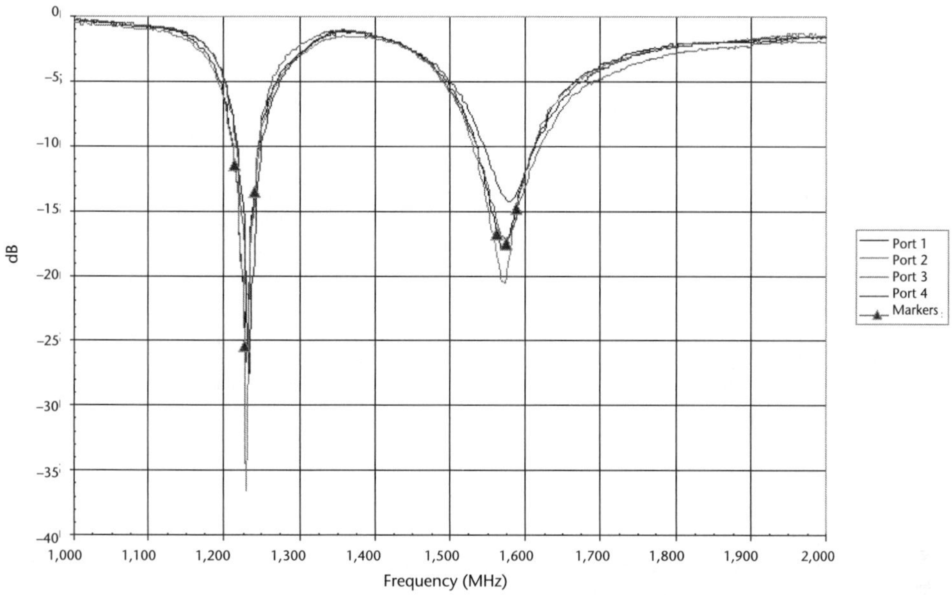

Figure 2.43 Measured return loss of a GPS dual-band QHA. (*From:* [64]. Courtesy of D. Lamensdorf.)

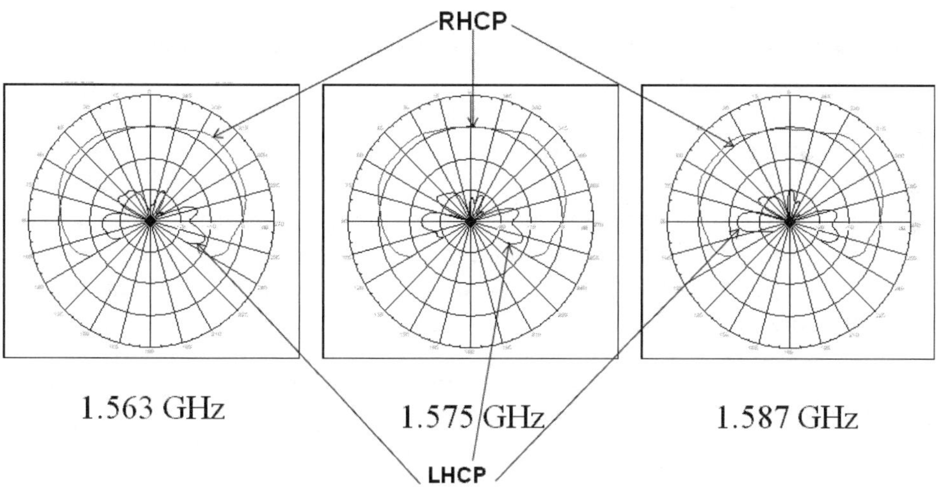

Figure 2.44 Measured elevation pattern in the GPS L_1 band of a dual-band QHA with trap filter. (*From:* [64]. Courtesy of D. Lamensdorf.)

of 1.215, 1. 227, and 1.239 GHz. Unfortunately, antenna patterns were not measured by the authors below −120 degrees; hence no information is available on the antenna backlobes. The phase of the antenna was also not measured so there is no information to determine its phase center stability.

2.3.3 A Summary of the Advantages and Disadvantages of QHAs for GNSS

Self-phasing QHAs, shown in Figures 2.36(a) and 2.37 are very popular for GNSS because of the many advantages that they provide. They are light in weight, small in

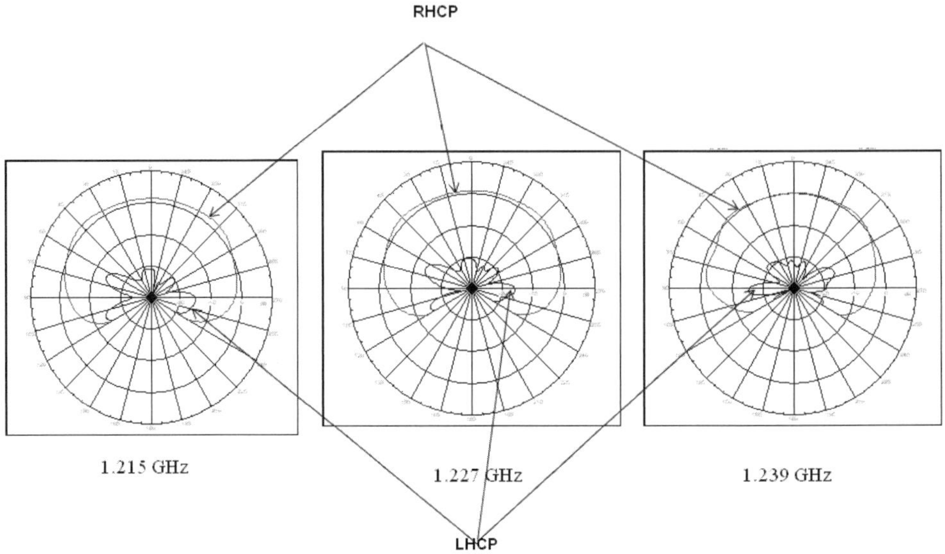

Figure 2.45 Measured elevation pattern in the GPS L_2 band of a dual-band QHA with trap filter. (*From:* [64]. Courtesy of D. Lamensdorf.)

size, and inexpensive to build especially if photolithographic techniques are used in its construction, as illustrated in Figure 2.41. Their biggest advantage is that they do not need a ground plane, which makes them very compact. Yet another advantage provided by a self-phased QHA is its shaped-conical radiation patterns with very broad antenna beamwidth of 100° to 180° as shown in Figure 2.37(c). This makes the self-phased QHA ideally suited for receiving signals from multiple GPS satellites visible to the antenna in the upper hemisphere. The ability to acquire and track nearly all of the visible satellites enables the QHA to provide excellent PDOP. The pattern can be shaped in elevation through small changes to its geometrical parameters. Their size can also be reduced by using a dielectric core made from a high permittivity material such as a ceramic, although this may also reduce the bandwidth.

Self-phased QHAs also have several disadvantages. One principal disadvantage is that its cylindrical shape and longer length makes it unsuitable for avionics applications, where because of ARINC 743 size requirements a very-low-profile antenna of height no greater than 0.75" is required to reduce drag. Another major disadvantage is that it also needs a coaxial balun, which makes its bandwidth narrow as shown in Figure 2.37. Therefore, it is generally only able to cover a single GNSS frequency band, handicapping its use in the multiple frequency bands used in modernized GPS or Galileo. Piggybacked and coaxially enclosed versions of dual-band QHAs have been proposed but do not work very well [57]. Its phase center can also vary with elevation angle since its radiating surface is distributed over a longer length. There can also be some asymmetry in azimuth caused by the uneven lengths of the adjacent helical elements required to generate quadrature phasing. Another disadvantage is that when the QHA is used with baluns made from coaxial lines, it generally has a poor front-to-back ratio as shown in Figure 4.39(c); this can make the self-phased QHA more susceptible to multipath. More recently, however, a modified design using infinite baluns made from microstrip lines has greatly improved the front-to-back ratios to 20 dB [63]. The use of small ground planes with a self-phased QHA is optional and can be used when there is a need to screen out multipath from ground reflections. The use of such a ground plane can, however, cause phase center excursions [60].

Externally phased dual-band QHAs provide greater bandwidths and when used with trap-loaded filters can perform over the GPS L_1 and L_2 frequency bands [64]. The disadvantage in this design is that it needs a small ground plane and also an external phasing network, making it heavier and bulkier than a self-phased QHA.

2.4 Hexafilar Slot Antenna for GPS

Antenna engineers from Garmin International have published a total of eight papers between 1996 and 1998 describing several different types of resonant cylindrical slot antennas that have been developed for GPS applications. These antennas are modifications of a conventional self-phased QHAs and use printed cylindrical slots in their construction. One of these new designs is a microstrip-fed, printed cylindrical slot antenna designed by Ho et al. [65]. References to several other types of resonant slot QHA designs developed by Garmin International are also given in Ho's paper. This antenna consists of six quarter-wavelength resonant slots, each of which is rolled half-turn around a cylindrical laminate. The slot antennas maintain the required phase relation to each other needed to provide a RHCP pattern with near hemispheric coverage. Experimental results indicate a 3-dB beamwidth of more than 120° and a front-to-back ratio of more than 15 dB provide good resistance to multipath signals from the ground. Tests conducted on this antenna with the Garmin GPS 90^{TM} receiver have confirmed its ability to track satellites at very-low-elevation angles for providing excellent PDOP. Figure 2.46(a) and (b) show the cross-sectional diagram of this antenna and its measured radiation pattern.

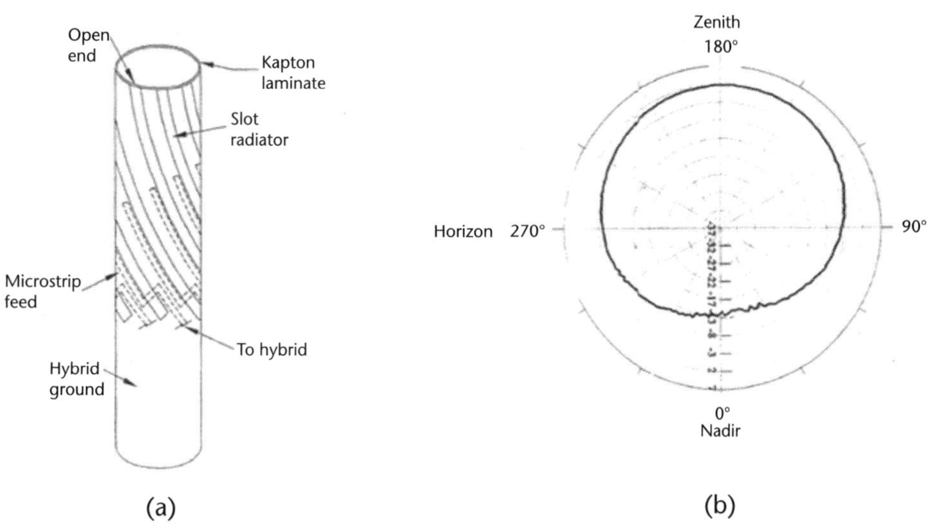

Figure 2.46 GPS hexafilar slot radiator: (a) cross section, and (b) measured radiation pattern (From [65]. ©1998 IEEE.)

2.5 Planar and Drooping Bow-Tie Turnstile Antennas for GNSS

Bow tie antennas are planar versions of the biconical antenna, which is well known for its broadband properties. RHCP needed for GNSS can be obtained from a pair of crossed bow-tie dipoles placed in the *x-y* plane over a ground plane and fed with equal amplitudes and in phase quadrature, with phases of 0° and –90°, respectively. The crossed dipoles are generally fed in shunt with either a split-sheath coaxial balun or a dual-folded balun; such a shunt connection will require two dipoles of different lengths with complex conjugate impedances needed to obtain the required 90 difference in phase. A physical arrangement of these crossed conjugate dipoles is called a turnstile configuration and it exploits the impedance properties of the dipole to obtain the required 90° phase difference. The impedance of the shorter dipole is capacitive, and its current will have a positive phase relative to its length at resonance. The longer dipole will have an inductive resonance and has a current with a negative phase relative to its condition at resonance. The relative phase quadrature needed to generate the CP can be obtained by optimizing the relative lengths of the crossed pair of dipoles. The dipoles need to be placed at an optimized height over a ground plane so that the LHCP signal they generate in the downward -*z* direction will upon reflection from the ground plane change polarization and add to the direct signal propagating upwards in the +*z* direction. Several factors such as optimizing the lengths of the dipoles, the height above the ground plane, and the finite bandwidth of the balun will place restrictions on taking full advantage of the maximum bandwidth capability provided by the bow-tie antenna geometry. Pattern coverage in the elevation plane can also be "tailored" by drooping the bow ties down towards the horizon to provide the gain needed for receiving signals from the low-elevation GNSS satellites. Planar as well as drooping crossed bow-tie dipole antennas on top of both planar and conical ground planes have been investigated by Kaswarra et al. [66] and Makarov [67] for modernized GPS and Galileo at frequencies ranging from 1165 to 1300 MHz. This covers the L_5 and L_2 bands of modernized GPS as well as the neighboring Galileo bands from 1250 to 1300 MHz. The geometry of these antennas is shown in Figure 2.47. The planar and drooping dipole versions of their antenna placed over conical ground planes are shown in Figure 2.48. Their investigation indicates that the best polarization bandwidth is obtained by a drooping bow-tie turnstile placed over a planar ground plane. The dimensions of the antenna were optimized using the HFSS finite element method (FEM) solver for an elevation angle $\theta = 60°$ (elevation of 30°) and for azimuth angles $\varphi = 0°$, 600, 90°, and 135°. A flare angle of 75° for the bow tie and a drooping angle of 30° provided the best performance with the antenna placed on top of a planar ground plane at a height $h/\lambda_0 = 0.26$ where h is the height above the planar ground plane and λ_0 is the wavelength corresponding to a frequency of 1.2 GHz, the center frequency of the frequency band that was needed for these applications band. The diameter of the planar ground plane was set to 0.8 λ_0 and for the conical ground plane the height to diameter ratio was set to 1.4. Optimum polarization performance was obtained when $l_y = 0.25\ \lambda_0$ and $l_x = 0.13\ \lambda_0$, where l_y and l_x are the half-lengths of the dipole. The measured return loss for the antenna is shown in Figure 2.48. The use of cavity-backed planar microstrip bow-tie dipoles for Galileo

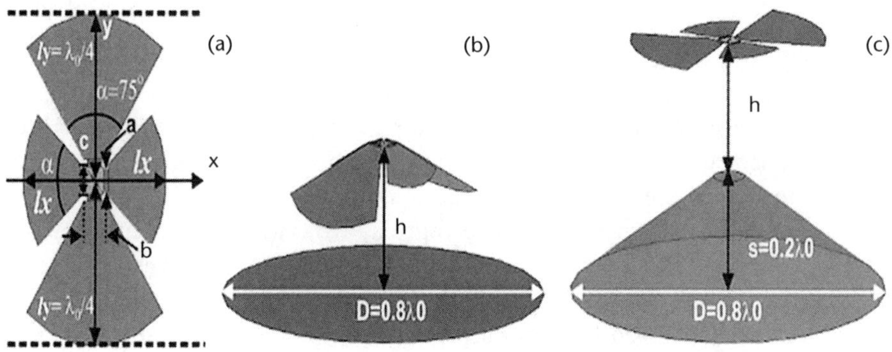

Figure 2.47 Geometry of planar and drooping crossed bow-tie dipole antennas for GPS. (From [66]. ©2007 IEEE.)

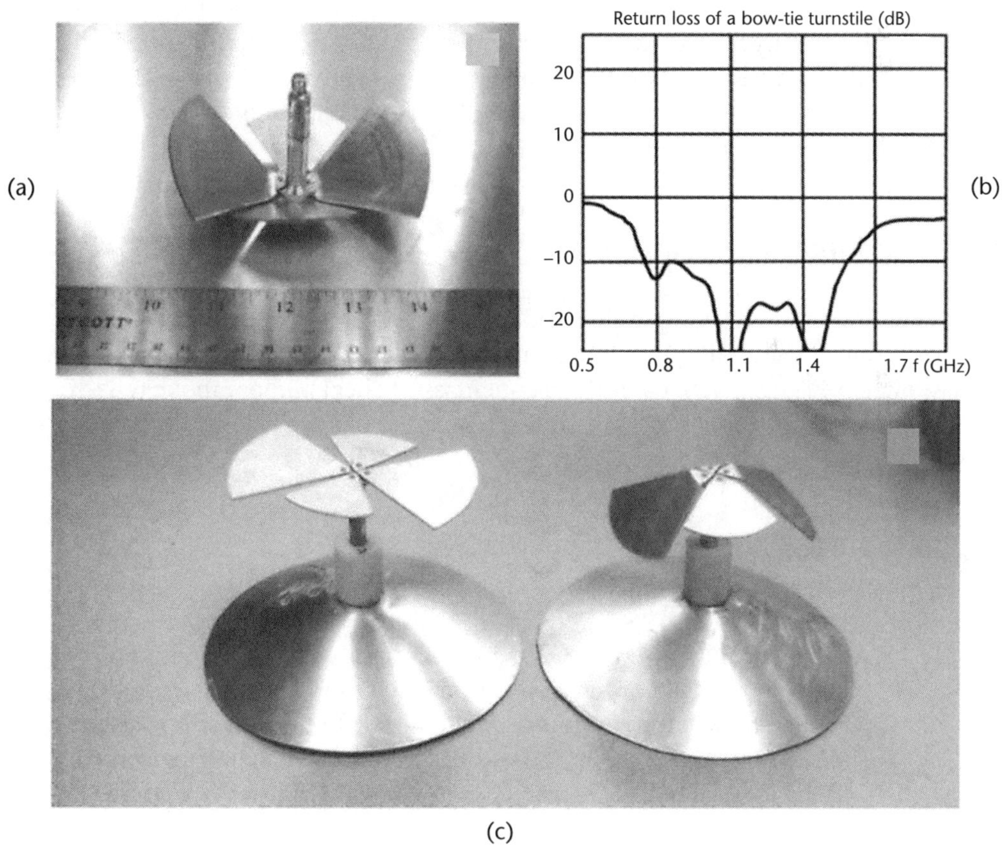

Figure 2.48 Picture of planar and drooping crossed bow-tie dipole antennas for GPS. (From [66]. ©2007 IEEE.)

has been mentioned by Granger et al. [68]; however, no detailed description has been provided in the paper of either the design or the measured performance of this antenna.

2.5.1 End-Fire Array of Turnstile Antennas for Multipath Limitation

An end-fire array of three identical turnstile dipole antennas has been used by for limiting multipath effects on a GPS receiver by Counselman [69]. This array directs the radiation towards zenith while suppressing backlobes to reduce multipath signals reflected off the ground. It obviates the need for a ground plane to suppress the antenna backlobes. The array is quite complex and consists of three antenna elements spaced one-third of a wavelength apart at the center frequency of the respective GPS bands. Each array element consists of separate turnstile antenna elements covering the L_1 and L_2 bands with 90° hybrids to generate RHCP. The turnstile elements are crossed horizontal dipoles; they are linear in shape and do not have bow-tie shapes to make them broadband as in the turnstile antennas considered in the previous section. The L_1 and L_2 arrays are connected to a GPS receiver through a L_1- L_2 duplexer. The array was tested at both the GPS L_1 and L_2 frequency bands and the multipath reject performance was compared against that of a planar metal ground plane. The array is approximately 0.1m in diameter, is 0.4m in height, and resembles a vertical post. Tests indicate that the three-element array was able to reject the multipath about 5 dB better than a 0.5m diameter metal ground plane. When the number of elements in the array was increased to five elements the improvement in multipath rejection was 5 dB better than obtained by a 0.9m diameter metal ground plane.

2.6 Directional GNSS Antennas

The design of three types of directional antennas used for GNSS applications: helical, conical spiral, and parabolic reflector antennas are described in this book and shown in Figure 2.1. Two of these antennas— helical and parabolic reflector—will be discussed in this chapter. Conical spiral antenna design along with other types of multiband spiral antennas are discussed in Chapter 4.

2.6.1 Helical Antennas

Two types of circularly polarized helical antennas are used in GNSS:

1. Cylindrical axial mode helix antenna with wide bandwidth and relatively high gain in the boresight direction [66] is shown in Figure 2.49(a);
2. Hemispherical low-profile helix antenna with a broad CP beam for providing coverage over a larger part of the upper hemisphere similar to a FRPA [70, 71], which has lower gain than a helix, is shown in Figure 2.49(b).

The hemispherical helix antenna, however, has an impedance bandwidth of only 6% as compared to the axial mode helix, which can be optimized to have a bandwidth of 34% to cover the entire GNSS spectrum if necessary. The hemispherical helix can only cover a single GNSS band and has normally been designed to cover just the GPS L_1 band. It has been used as an element in a four-element GPS beamforming antenna array for providing some mitigation of multipath and interference [71].

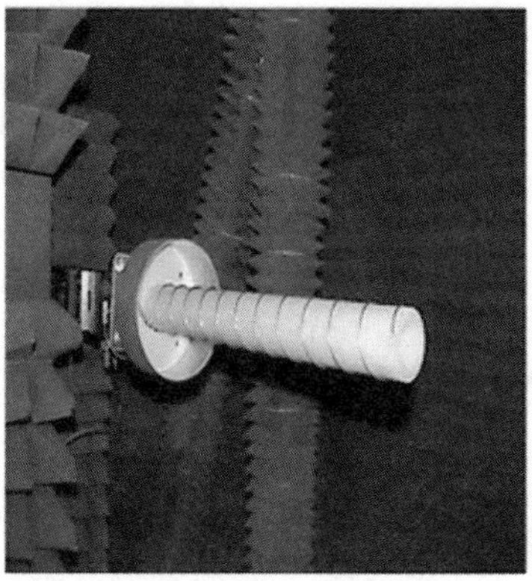

(a)

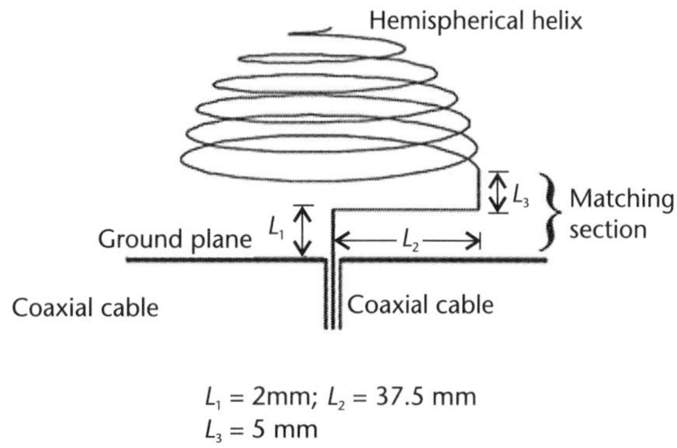

(b)

Figure 2.49 (a) GPS cylindrical axial mode antenna, and (b) hemispherical helix antenna. (Figure 2.49 (b), from [70]. ©IEEE 2005.)

2.6.1.1 Cylindrical Axial Mode Helix

The cylindrical, monofilar helical antenna, made from either a single wire or metalized tape, can provide a CP pattern with a relatively high gain over a fairly wide bandwidth. Figure 2.49(a) shows a picture of an axial mode helix for applications in the modernized GPS covering the L_5, L_2, and L_1 frequency bands [44]. It has been used as a stand-alone transmitting antenna or more frequently as an element in a transmitting antenna array in GNSS satellites. The narrow beamwidth and larger physical size and weight of the axial mode helix antenna generally precludes it from being used as a receiving antenna that can be integrated into GNSS receivers but it is popular as a stand-alone transmitting antenna in many tests and laboratory measurements. An interesting recent application for the axial mode helix has been as a transmit antenna in indoor navigational systems using GPS-like pseudolites (PLs) that are capable of providing centimeter accuracy [4]. A microstrip patch antenna cannot be used in such applications because its beam is too wide and can create serious multipath from interaction with the surrounding environment. In contrast, the beamwidth of the axial mode helix can be tailored to the desired narrower beamwidth by adjusting its length, diameter, and other parameters; hence, it is the preferred antenna of choice for such measurements.

2.6.1.1.1 Helical Antenna Arrays Used in GPS Satellites

Helical antennas with either cylindrical or conical cup ground planes are also very popular as transmitting elements in antenna arrays used in several GPS navigational satellites. There are four different types of satellite antennas: (1) Block I, (2) Block II, Block IIA, (3) Block IIR-A, and (4) Block IIR-B and Block IIR-M [1–3]. Figure 2.50 shows helical antennas used on the GPS IIR-M satellites built by Lockheed Martin Co. [1]. Figure 2.51 shows the helix antennas used in the Block IIA satellites by Rockwell Collins [2]. A summary of the antenna arrays of axial mode helical elements used as transmitting antennas in the GPS satellites—Block I, Block IIA, and Block IIR—are described by Aparicio et al. [3]. The bandwidths of the elements in these earlier satellites range from 1200 to 1600 MHz and hence do not include the L_5 signal. Each array contains twelve RHCP helical elements arranged in two concentric circles, with four elements in the inner circle and the remaining eight elements in the outer circle [72]. The earlier Block I satellites used a cylindrical cup ground plane but the later systems, the Block IIA and IIR, have used conical cup ground planes. The PCV of these satellite antenna arrays have also been studied by Schmid et al. [72] from the vast amount of data that they have collected. The PCV for the Block IIR-A, Block I and Block II and IIA satellite antennas was between ±5 mm, whereas for the Block IIR-B the PCV was almost double this value of nearly ±10 mm.

2.6.1.1.2 Design of Cylindrical Axial Mode Helix Antennas

The axial mode helix is a traveling wave antenna and consists of two parts: the director, which is the main helical portion, and the exciter/ground plane, which is the feed portion for the antenna. The helix can be made either from a single wire, a thin tube, or a flat strip conductor on a thin dielectric tape; these are wound around a supporting dielectric cylinder. Tapes are commonly used in satellite antennas, as shown in Figures 2.50 and 2.51, because of their lighter weight. The polarization is determined by the direction of the windings; thus a right-hand screw winding of the helix wire produces the required right-hand polarization needed for GNSS. The direction of the right screw winding for the helical wire is determined by pointing the thumb of the right hand along the desired direction of propagation and by wrapping the four other fingers in the direction of the wire winding. The geometry of the axial mode helix antenna with either a cylindrical or conical cup ground plane is shown in Figure 2.52. The performance of the helical portion is controlled by the various geometrical parameters of the helix antenna listed in Table 2.4. The early "classic" design models for this antenna [74] provided high gain but over a relatively narrower bandwidth. They assumed that the radius of the helical wire would have an insignificant effect on antenna performance; this assumption

Figure 2.50 GPS IIR-M satellite antenna assembly. (Printed with permission from Lockheed Martin Co.)

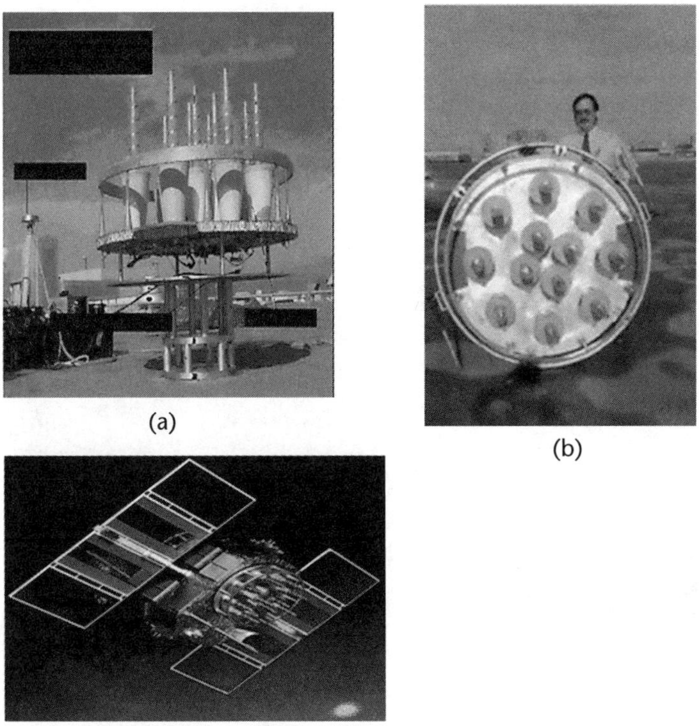

(a) (b) (c)

Figure 2.51 GPS Block IIA satellite antenna: (a) side view, (b) front view, and (c) array on satellite. (Printed with permission from Dr. G. L. Mader and Boeing Co.)

turned out to be erroneous in achieving the maximum possible bandwidth. New techniques for designing uniform diameter helical antennas have been proposed by Djordjevic [75]; these take into account the influence of the radius of the wire for achieving the maximum possible bandwidth and they provide a lower peak gain but over a much wider bandwidth than the classical designs. The trade-off between gain and bandwidth in these two designs are explained in Figure 11 of the paper by Djordjevic [75]. Using these new wideband design techniques the performance of the helical antenna can be optimized to provide good gain over the entire GNSS spectrum covering the GPS, Galileo, and GLONASS systems with frequencies ranging from 1.145 GHz (the lower-band edge of Galileo E5 band) to 1.615 GHz (the upper-band edge of the GLONASS L_1 band) for a total bandwidth of 34%. Several other parameters of the helix can also be varied to achieve even further improvements in the pattern and gain bandwidth. For example, the diameter of the cylinder can be tapered or made nonuniform, the pitch of the helical winding can be changed along the axis, or the diameter of the helical wire or the width of the helical flat strip conductor can be varied [75]. The return loss of a uniform diameter helix as a function of frequency can be improved simply by adding two additional tapered turns at the free end of the helix as in the satellite antennas for the Block IIR-M satellites shown in Figure 2.50.

The exciter portion of the helical antenna consists of a ground plane with either a cylindrical or a truncated conical cup at the base of the helix as shown in Figure 2.52. These improve the gain, front-to-back ratio and sidelobe level, thereby

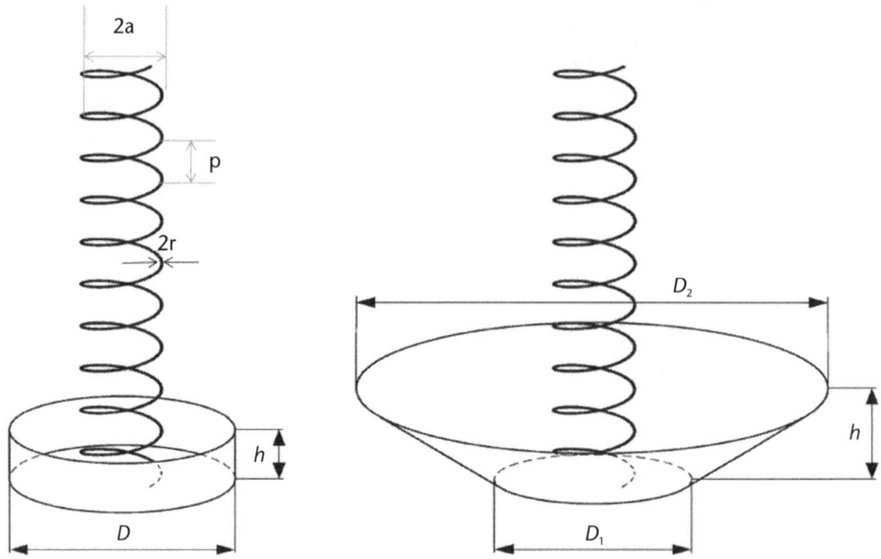

Figure 2.52 Axial mode helix antenna with cylindrical and truncated conical cup ground plane. (From [75]. ©2006 IEEE.)

reducing the susceptibility of the antenna to multipath. Inside this exciter portion the helix wire is connected to a coaxial connector mounted on the ground plane. The helical antenna has an input resistance R_{in} = 141 C/λ_C ohms where C = the circumference and λ_C = the wavelength at the midband frequency. A wideband impedance matching section is needed to match the helix to standard 50-ohm coaxial connector. This can be designed by using several methods such as by bringing the last quarter turn of the helix wire or tape parallel to the ground plane in a gradual manner and inserting a microstrip matching transformer between the coaxial feed and the beginning of the helix wire [76] or alternatively by using a vertical-profiled metal plate at this junction [75]. The optimum diameter D of a cylindrical cup used with the ground plane is $D = 1\ \lambda_C$ and the height of cup is $h = 0.25\ \lambda_C$. The cup can increase the antenna gain by 1.4 dB compared to a planar ground plane. The optimal dimensions for the truncated conical cup are as follows: the diameter at the base of the cone is $D_1 = 0.75\ \lambda_C$; the diameter at the apex of the cone $D_2 = 2.5\ \lambda_C$; the height $h = 0.5\ \lambda_C$. The cone suppresses the sidelobes near the horizon and also the back radiation but does not appreciably increase the gain in the boresight direction. Near the exciter feed region the current on the helix attenuates smoothly to a minimum, while the current over the remaining length of the helix is relatively uniform, allowing the current on the helix to be approximated as being relatively uniform while computing its radiation pattern.

2.6.1.2 Design and Measured Performance of an Axial Mode Helical Antenna for Modernized GPS

A picture of a helical antenna with a cylindrical cup ground plane built by the MITRE Corporation [44] is shown in Figure 2.49(a). The design parameters for this antenna are listed in Table 2.5. This antenna is capable of operating in the L_5, L_2, and L_1 bands of the modernized GPS system and has a 31% bandwidth; it can also

Table 2.5 Design Parameters with Dimensions of the Axial Mode Helix Antenna with Cylindrical Cup Ground Plane Shown in Figure 2.49(a)

Parameter	*Symbol*	*Dimension of Helix Antenna in Figure 2.48(a)*	*Normalized Dimension Where Applicable*
Radius of the helix	a	1.4"	$a/\lambda c = 0.163$
Helix circumference	$C = 2\pi a$	8.796"	$C/\lambda c = 1.024$
Helix pitch	p	2.214"	$P/\lambda c = 0.258$
Pitch angle	$\alpha = \arctan\left(\frac{p}{C}\right)$	14.13°	
Number of turns	N	11 full turns plus 0.75 turn	
Length of helix	$L = P\,N$	26.01"	$L/\lambda c = 3.03$
Radius of wire	r	0.05175"	$r/C = 0.0058$
Midband design frequency	$f_c = 1/2\,(f_{max} + f_{min})$ $f_{max} = 1.587$ GHz $f_{min} = 1.164$ GHz	$f_c = 1.3755$ GHz	
Bandwidth	Δ	428 MHz (31.1%)	
Wavelength at midband frequency	λc	8.586"	
Diameter of cylindrical cup	D	7.88"	$D/\lambda c = 0.917$
Height of cylindrical cup	h	3.94"	$h/\lambda c = 0.458$

be used in the neighboring Galileo E5 and E6 bands from 1145 to 1.300 GHz. Its use thus far has been as a transmitting antenna to evaluate the fidelity of M code reception of GPS receiving antennas through correlation measurements in the L_1 and L_2 bands [44]. The radiation patterns and gain of this antenna measured at 1.176, 1.227, and 1.575 GHz, and the center band frequencies of the L_5, L_2, and L_1 bands are shown in Figure 2.53. The measured pattern shows good performance at the two lower frequencies but there is some degradation in gain performance and broadening of beamwidth at 1.5754 GHz. The antenna was designed using formulas from classical narrowband design mentioned earlier without accounting for the effects of the diameter of the helical wire on the bandwidth. It would require only a small variation in its design to increase its bandwidth to the 34% needed to cover the entire GNSS spectrum. The design methods recommended by Djordjevic [75] can be used as discussed below for making such an improvement. The classical design recommends the following guidelines for varying the antenna parameters listed in Table 2.5 for achieving optimum performance [74]:

- Normalized circumference C/λ_C should be between $0.71 < C/\lambda_C < 1.20$, where C is the circumference and λ_C is the wavelength at the midband frequency;
- Pitch angle α be between $12° < \alpha < 15°$, where α = arc tan (p/C);
- Normalized pitch or spacing between turns p/λ_C should be between $0.15 < p/\lambda_C < 0.3$.

The number of turns N should be between $4 < N < 15$. These recommendations are based on the assumption that the antenna performance is independent of the

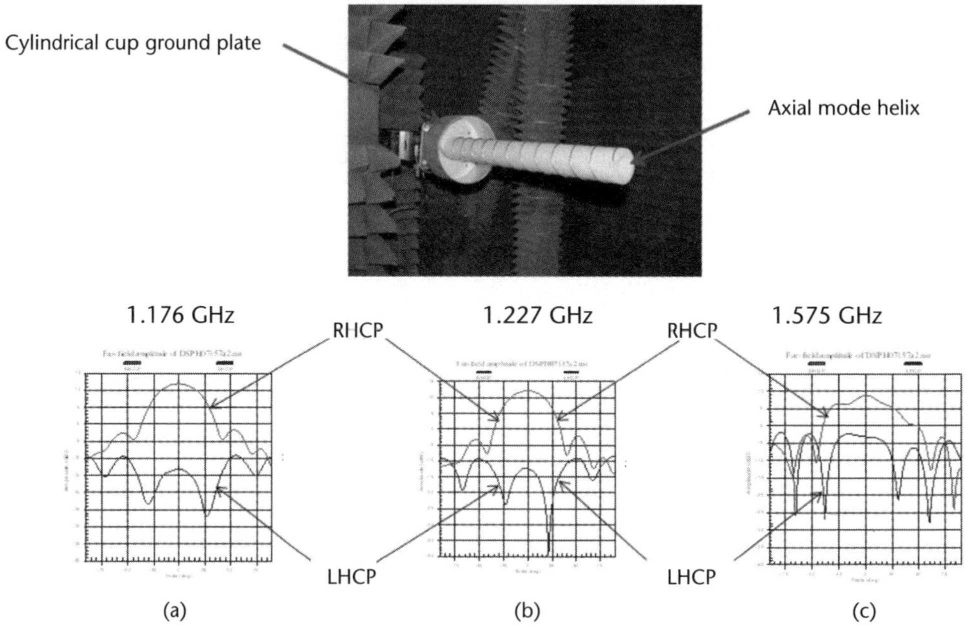

Figure 2.53 GPS cylindrical axial mode helix antenna with cylindrical cup cone and its measured patterns at (a) 1.176 GHz, (b) 1.227, and (c) 1.575 GHz.

radius of the wire. Other parameters needed for completing the design and building the antenna are the total length of the antenna $L = Np$ and the total length of the antenna wire $L_W = N\sqrt{p^2 + C^2}$ where $\sqrt{p^2 + C^2}$ is the length of the wire between each turn. The broadband design [75] takes into account the wire radius as well as the total length "L" of the helix. (The optimum values for these same parameters can be obtained from Figures 4 and 6 and the desired bandwidth, axial ratio, and maximum gain can be determined from Figures 3, 5, and 7, respectively, provided by Djordjevic [75].) The antenna shown in Figure 2.49(a) has a normalized wire radius $r/C = 0.0059$, where r = radius of the helical wire. Increasing the diameter of this helix to 1.574" from its current radius of 1.4" and the reducing the pitch angle to 13° from its current value of 14° would increase the bandwidth to the 34% needed to cover the GNSS spectrum. The length L can be adjusted to achieve a higher gain.

Empirical formulas for calculating the maximum gain, the optimum radius of the helix, the half-power beamwidth (HPBW), and the optimum axial ratio are also available. Emerson [73] has proposed a simple empirical expression for the maximum gain G_{max} (dBic) based on numerical modeling of the helix as a function of the total length of the helix L normalized to the wavelength λ_c; this formula gives results that compare well with measurements:

$$G_{\max}(\text{dBiC}) = 10.25 + 1.22\left(\frac{L}{\lambda_C}\right) - 0.0726\left(\frac{L}{\lambda_C}\right)^2 \tag{2.57}$$

The normalized optimum radius of the helix for maximum gain can be obtained from the wideband theory of Djordjevic [75]

$$\frac{a}{\lambda_C} = 0.2025 - 0.0079\left(\frac{L}{\lambda_C}\right) - 0.0726\left(\frac{L}{\lambda_C}\right)^2 \tag{2.58}$$

The empirical equation for the half-power beamwidth HPBW is [74; pp. 12-11]

$$\text{HPBW} = \frac{61.5\left(\frac{2N}{N+5}\right)^{0.6}}{\left(\frac{2\pi a}{\lambda}\right)^{\frac{\sqrt{N}}{4}}\left(\frac{Np}{\lambda}\right)^{0.7}}\left(\frac{\tan\alpha}{\tan 12.5°}\right)^{\frac{\sqrt{N}}{4}} \text{(degrees)} \tag{2.59}$$

The polarization of the normal mode helix is in general elliptical with an axial ratio given by

$$AR = \left|\frac{E_\theta}{E_\varnothing}\right| = \frac{2p\lambda_C}{C^2} \tag{2.60}$$

The antenna will be circularly polarized if AR = 1 is satisfied. This condition is satisfied when the circumference of the helix and the pitch or spacing between the turns are related as

$C = \sqrt{2p\lambda_c}$, which in turn be used to find the pitch angle of the helix as

$$\tan(\alpha) = \frac{p}{C} = \sqrt{\frac{p}{2\pi_c}} \tag{2.61}$$

The far-field radiation pattern of the axial mode helical antenna can be determined by considering the N turn helix as an array of N identical elements with an interelement spacing equal to p. The normalized field pattern is then obtained by multiplying the pattern of one turn of the helix by the array factor. The result is

$$E(\theta) = sin\left(\frac{\pi}{2N}\right)cos(\theta)\frac{sin\left(N\frac{\psi}{2}\right)}{sin\left(\frac{\psi}{2}\right)} \tag{2.62}$$

In the above equation cos(θ) is the element pattern and $\frac{\sin\left(N\frac{\psi}{2}\right)}{\sin\left(\frac{\psi}{2}\right)}$ is the array factor of a uniform array of N equally spaced elements $\psi = \frac{2\pi}{\lambda_c}p\cos\theta + \alpha$. Here α is the phase shift between successive elements and is given as $\alpha = -2\pi - \frac{\pi}{N}$

2.6.2 Helibowl Multipath Limiting Antenna

The Helibowl is a bowl-shaped reflector antenna that is excited by a helical antenna element located at the center of the bowl [77]. The Helibowl was originally developed by Don Spitmesser at the Jet Propulsion Laboratory in Pasadena, California [77] by modifying a low sidelobe "helicone" antenna concept first proposed by K. Carver [78]. It provides high and uniform gain over the desired range of elevation angles above a certain masking angle followed by a sharp cutoff in gain below the masking angle. The Helibowl is used in conjunction with a phased array to track low-elevation satellites and has been extensively investigated at Ohio University [79, 80]. Because of its low sidelobe and backlobe levels and its ability to limit multipath, this antenna is used frequently in DGPS ground reference stations.

2.6.3 Hemispherical Helix Antenna and Its Use in a GPS Beamforming Array

The axial mode helix described in the previous section provides a good axial ratio but over a narrow beamwidth because of its relatively high gain. Its radiation pattern is not that of a "classic" FRPA and will not be optimum for receiving signals simultaneously from four or more GPS satellites spread over a wide range of viewing angles in the upper hemisphere. A five-turn, printed hemispherical helical antenna for GPS applications has been designed by Zhang and Hui [70] and is shown in Figure 2.49(b); it provides a much broader, CP antenna pattern for acquisition of multiple GNSS satellites. It has a low profile that allows a stable mechanical installation above a ground plane and an impedance bandwidth of 6% with a return loss of greater than 20 dB. It also provides a 3-dB axial ratio bandwidth of between 6% to 7% and has a gain of 9 dB at its center frequency. The simulated patterns show good CP over a ± 45° range of elevation angles around zenith, but also show some asymmetry with variation of azimuth angle since the patterns measured at orthogonal azimuth angles of 0° and 90° are different at certain elevations. An input impedance matching network for the GPS L_1 band is provided by a three-element matching network also shown in this figure; the return loss it provides varies from –10.7 to –27.7 dB across the band.

The hemispherical helix (a wire version) has been used as an antenna element in a four-element GPS beamforming antenna array to mitigate multipath and interference effects [71].

Figure 2.54(a) shows a picture of the four-element helical antenna array. The radius of each hemispherical antenna element in the array is 37.5 mm and the wire radius is 0.5 mm. The height of the array is not given. Measured beam patterns of the four antenna lobes generated by the array are shown in Figure 2.54(b). The elements are arranged in a square lattice separated at a distance of 0.7 wavelengths at 1.5754 GHz along both the x and y axes. An inexpensive, analog Butler Matrix feeding network made from four branch-line quadrature hybrids is used in the array. The array can generate four spatial beams along two orthogonal directions; the beam directions are at azimuth angles of $\varphi = 45°$, 135°, 225°, and 315°, respectively and at an elevation angle $\theta \sim 25°$. The half-power beamwidth of each beam is ~ 44° and the gain is greater than 17 dB; the axial ratios of the beams are between 1.5 to 2.2 dB.

2.6.4 High-Gain Reflector Antennas for Monitoring GNSS Signals

Large reflector antennas providing very high gain are increasingly being used both in the United States and in Europe for monitoring the quality and spectral purity

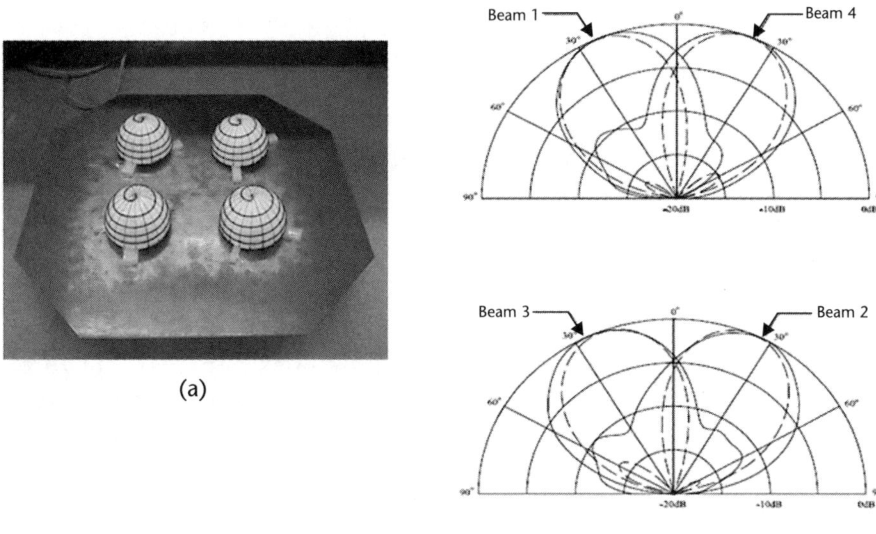

(a) (b)

Figure 2.54 GPS beamforming antenna array of four hemispherical helical elements with measured beam patterns. (From [71]. ©2005 *Radio Science.*)

of signals transmitted by recently launched satellites. These antennas also are being used for identifying signal interference and for characterizing potential signal anomalies and errors caused by the satellite hardware. The received signal power spectral density of GNSS spread spectrum signals is approximately 20 dB below the thermal noise floor. The baseband signal bandwidth is intentionally spread over a wide bandwidth and appears as noise; the original signal is recovered through a despreading process in the receiver that can provide between 10 to 60 dB of processing gain. Before this despreading the chips of the C/A and P(Y) codes are not discernible. The high gain of these large reflectors when coupled with new "dithered sampling frequency signal processing techniques" provides a positive signal-to-noise ratio (SNR) and allows both the individual chips and the navigation bits to be demodulated without performing the despreading procedure.

These types of monitoring investigations started in the early 1990s and were first used to diagnose an anomaly observed in satellite vehicle number (SVN) 19 [5]. It has since been used to monitor signals from recently launched satellites such as the Galileo satellites (GIOVE-A & B) [7] and the Compass M1 satellite [6, 81]. The largest of the reflector antennas used for this purpose is the 110m diameter Robert C Byrd Telescope antenna at Green Bank, West Virginia, which provides a gain at L band of about 70 dB [5]. Several of these large reflector antennas currently being used for these monitoring investigations are described below.

The GNSS Monitor Station (SGMS) at Stanford University, which uses two reflector antennas, [6, 81] has been very active in this field for the past few years. The smaller of these two is a front-fed parabolic reflector antenna 1.8m in diameter . Its estimated gain, assuming an aperture efficiency of 50%, is 23.3 dBic at 1.1 GHz and 26.75 dBic at 1.6 GHz. Since the maximum gain of a typical FRPA is around 1 to 3 dBic, this reflector antenna can provide a boost in gain (or in SNR) of nearly

20 to 23 dB across the GNSS band over a FRPA. The estimated HPWB is 8.85° at 1.1 GHz and around 6.08° at 1.6 GHz. The second dish at SGMS is even larger: a 45.7m (150 feet) reflector antenna called the SRI Stanford "Big" dish [81]. This antenna has a 50 dB of gain and a beamwidth of 0.25 degrees with an efficiency of 35% and it operates from 1.2 to 1.6 GHz.

Figure 2.1 shows a Cassegrain dual-reflector antenna, with a 30m diameter primary parabolic reflector and a 4m diameter hyperbolic subreflector is being used by the German Aerospace Center (DLR) in Weilheim, Germany [7]. Its gain is estimated to be 50 dBic with a HPWB of 0.5° [7]. The antenna has a broadband feed covering 1.1 to 1.6 GHz and a position accuracy of 0.001° and has been used for analysis of signals from the Galileo satellites. A parabolic reflector antenna 5.1m in diameter is currently being acquired by Alcatel Alexia Space Italia (AAS-I) for their NAVCOM laboratory [8]. This antenna is also a front-fed parabolic reflector with a diameter of 5.1m. Its gain is estimated by AAS-I to be 32 dBic at 1.1 GHz with a half-power beamwidth of 4°; its gain at 1.6 GHz is 36.9 dBic with a half-power beamwidth of 2.75°. Its G/T is 11.5 dB/K pointed at zenith and is 5.7 dB/K pointed at the horizon. Other notable examples of reflector used for GNSS monitoring antennas are a 2.4m antenna at the Toulouse Space Center (CST) in France and two dishes of 12 and 7m at Monitoring Earth Station in Leeheim, Germany [81]. Several other large radio telescopes in Europe are also being used in GNSS monitoring.

Most of these antennas are fed by broad L band feeds, allowing them to monitor signals transmitted by all current GNSS satellites including GPS, Galileo, GLONASS, and Compass. These reflector antennas can be one of two types: either a front-fed parabolic reflector or a Cassegrain dual reflector consisting of primary reflector and a subreflector. A detailed description of the design of these single- and dual-reflector (Cassegrain) antenna systems is beyond the scope of this book but is well described in two recent books [82, 83]. The last of these references [83] is devoted exclusively to a description of very large reflector antennas similar to those used in GNSS monitoring stations.

The gain G of these parabolic reflector antennas is determined mainly by the size of the aperture of the primary reflector and the illumination amplitude taper across the aperture to reduce the sidelobe level. The gain G can be calculated from:

$$G = \frac{\varepsilon \pi^2 D^2}{\lambda^2}$$

where

D = diameter of the reflector

λ = wavelength

ε = aperture efficiency, which is a function of several factors such as spillover loss from the f.eed, amplitude taper across the aperture to reduce sidelobe level, blockage from struts supporting the feed antenna, random surface errors, and so forth.

The efficiency can vary from 35% for a low-efficiency feed to 70% for better reflector designs. The HPWB in degrees for a uniform aperture taper across the primary reflector can be calculated from:

$$HPBW = 58.44\left(\frac{\lambda}{D}\right)$$

For the 1.8m Stanford reflector the HPBW is 8.85° at 1.1 GHz and 6.08° at 1.6 GHz. If for the purpose of reducing the level of the antenna sidelobes the taper across the aperture of the reflector were to be made quadratic instead of uniform, it would cause a decrease the gain and a corresponding increase in the beamwidth. Thus at a frequency of 1.6 GHz if the main reflector has an edge illumination of –10 dB relative to the center to decrease its sidelobe level its beamwidth can increase to 6.8° but its sidelobe level will decrease from –17.6 dB for a uniform illumination to –22.3 dB.

2.7 Beamforming Antenna Arrays

A beamsteering antenna array can steer four or more multiple high-gain beams towards GNSS satellites to receive only signals of interest (SOI) while simultaneously mitigating signals not of interest (SNOI) from either multipath or interference; it is able to differentiate between these two types of signals by using their spatial separation. Digital beamforming antenna arrays are frequently used in wireless communications for improving signal quality.

The beamsteering principles of this type of antenna array is best illustrated by first considering a 3-D array of N antenna elements whose geometry is shown in Figure 2.55.

The azimuth angle is measured counter clockwise from the X axis and an angle θ measured from the Z axis; $\theta = 90°$ represents the horizon. The location of the element n in the array is $\vec{r}_n = x_n\hat{x} + y_n\hat{y} + z_n\hat{z}$ where (x_n, y_n, z_n) can be considered as the coordinates of the phase center of the nth antenna element in the array. One of the centrally located antenna elements in this array, $n = 1$, is designated as the "reference" element and is located at the origin. The location of this reference element is $\vec{r}_1 = (0)\hat{x} + (0)\hat{y} + (0)\hat{z}$ is the wave number of a signal—the SOI—arriving at the array from the satellite at azimuth ϕ_S and at an angle θ_s measured from zenith and is represented by

$$\hat{k}_s = \beta\left(\cos\varphi_S \sin\theta_S\hat{x} + \sin\varphi_S \sin\theta_S\hat{y} + \cos\theta_S\hat{z}\right)$$

here $\beta = \dfrac{2\pi}{\lambda_0} = \dfrac{2\pi f_0}{c}$ and c is the velocity of light = 3×10^8 meters per second. The difference in phase between the satellite signal incident on an element n in the array and the reference element 1 at the origin is

$$\Delta\psi_n = \beta \cdot \Delta d_n = \beta\vec{k}_s \cdot (\vec{r}_n - \vec{r}_1) = \beta\left(x_n \cos\varphi_s \sin\theta_s + y_n \sin\varphi_S \sin\theta_s + z_n \cos\theta_s\right)$$

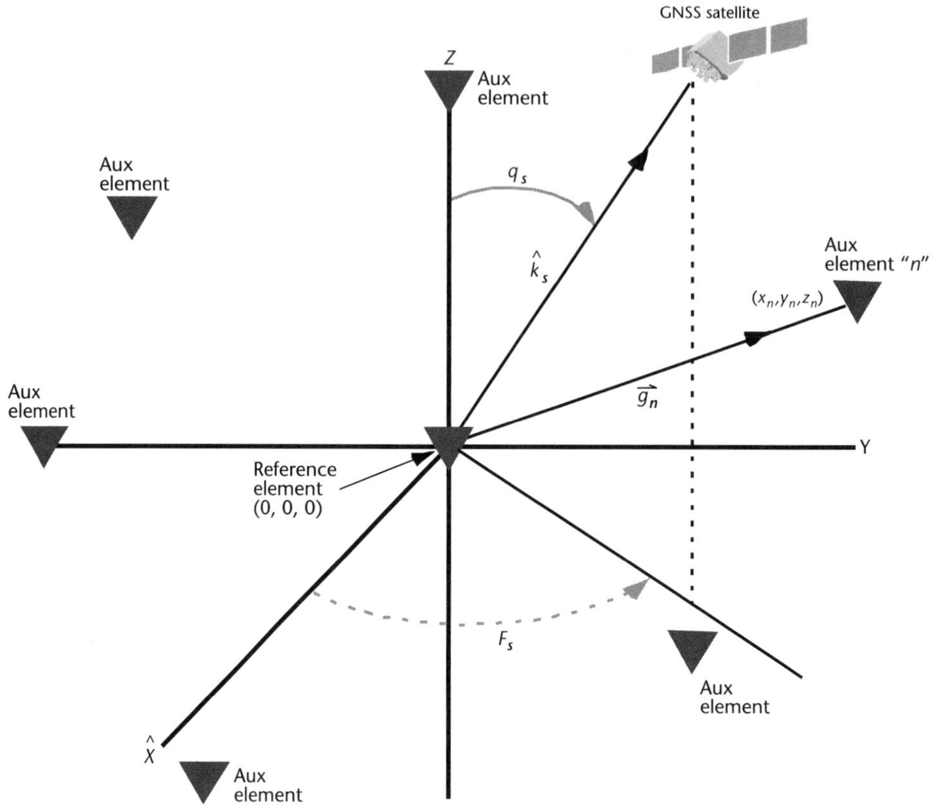

Figure 2.55 3-D representation of a GNSS antenna array.

The signals received by all N elements in the array compared to the signal in the reference element 1 can then be represented in terms of a steering vector $S(\theta_S, \phi_S)$, which describes the phase shifts of the array elements needed to steer the main beam of the antenna array in the direction (θ_S, ϕ_S) of the SOI from the satellite

$$S(\theta_s, \varphi_S)^T \left[1, s_1(\theta_s, \varphi_s), \ldots, s_n(\theta_s, \varphi_S)\right]$$
$$s_n(\theta_s, \varphi_s) = e^{-\Delta\psi_n}$$

It is also assumed that incident signal and each element in the array are copolarized so that there is no polarization loss on reception. The modulation of the incident satellite signal will be represented by its baseband complex envelope $a(t)$. The signal $s(t)$ at the output terminal of the array manifold will then be

$$s(t) = a(t) \sum_{n=1}^{n=N} g_{na}(\theta_s, \varphi_s, f_0) e^{-j\Delta\psi_N}$$

$g_{na}(\theta_S, \phi_S, f_0)$ is the gain of the nth array element along a direction specified by (θ_S, ϕ_S). In an actual array the gain $g_{na}(\theta_S, \phi_S, f_0)$ may not be the same for all elements in the array. It is necessary to accurately account for the differences in the phase and amplitude weight contributions of the individual antenna elements through calibration if optimum results are to be obtained from beam forming discussed in this section. However, to simplify the analysis we will make a number of initial assumptions that rarely, if ever, occur in a real antenna array. We will ignore mutual coupling effects between the antenna elements in the array and also the diffraction effects from the ground plane or the airborne platform on which the array is mounted. We will assume that all antenna elements in the array have identical radiation patterns. Furthermore, we will also assume that all array elements are RHCP with the same axial ratio. The gain of each of the N antenna elements for the principal RHCP under these simplifying assumptions will be represented as $g_a(\theta_S, \phi_S, f_0)$. Hence,

$$s(t) = a(t)\sum_{n=1}^{n=N} g_{na}(\theta_s, \varphi_s, f_0) e^{-j\Delta\psi_N} = a(t) g_a(\theta_S, \varphi_S, f_0) f(\theta_S, \varphi_S)$$

The term $f(\theta_S, \phi_S)$ is called the array factor; it determines the ratio of the received signal available at the array output to the signal measured at the central reference element of the array.

To steer the main beam pattern of the array, each antenna element in the array is assigned complex filter weight w_n that has both a magnitude and a phase component associated with it. The magnitude of weighting on the array elements is varied only if it is desired to reduce the sidelobes of the array pattern, otherwise it can be the same for all elements. The weight vector is represented by:

$$w^T = [w_1, w_2, \ldots, w_n]$$

In the vector notation the weighted sum of the total input signal received by the array is then represented by

$$u(t) = w^{T_v}$$

where

$$v^T = [v_1, v_2, \ldots, v_N]$$

The output of the array manifold from the signal received from the GNSS satellite at (θ_S, ϕ_S) is

$$u_S = w^{T_S}$$

where

$$s^T = [s_1, s_2, \ldots, s_N]$$

The received power at the array output

$$P_S(\theta_s, \varphi_s) = E\left|\sum_{n=1}^{n=N} w_n a(t) g_a(\theta_s, \varphi_s, f_0) e^{-j\Delta\psi_N}\right|^2$$

where E| | is the expectation. By adjusting the complex weights on each of the N elements it is possible to do "beamforming," which is directing the main beam of the array manifold in the direction (θ_S, ϕ_S) of the incoming signal to maximize the signal received by the antenna array from any visible GNSS satellite. In a real-world antenna array, the complex weights w_n on each array element can also be optimized if necessary to compensate for the presence of mutual coupling or any change in the gain or polarization state of the antenna element in the array through careful precalibration.

If the positions of the GNSS satellites relative to the antenna elements of the array are known it would be possible to steer the main beam of the array in these directions. This also requires that the position and the orientation of the array is also known. The adaptive beamforming antenna is therefore designed as an integrated INS/GNSS unit. The outputs provided by the INS unit are the position and attitude of the array relative to the earth, which is generally obtained from the inertial measurement unit (IMU). Using this information the beamsteering unit in the array can calculate the directions of the satellites relative to the array and select the appropriate complex weights for each element. A wide range of different beamforming algorithms have been proposed in literature; these are based on using a priori information to discriminate between the SOIs, which are the satellite signals, and the SNOIs, and also the correlation properties of the GNSS signals.

Figure 2.56 shows a schematic diagram of a generic beamforming GNSS antenna array.

The beamforming array of the type described above provides several advantages:

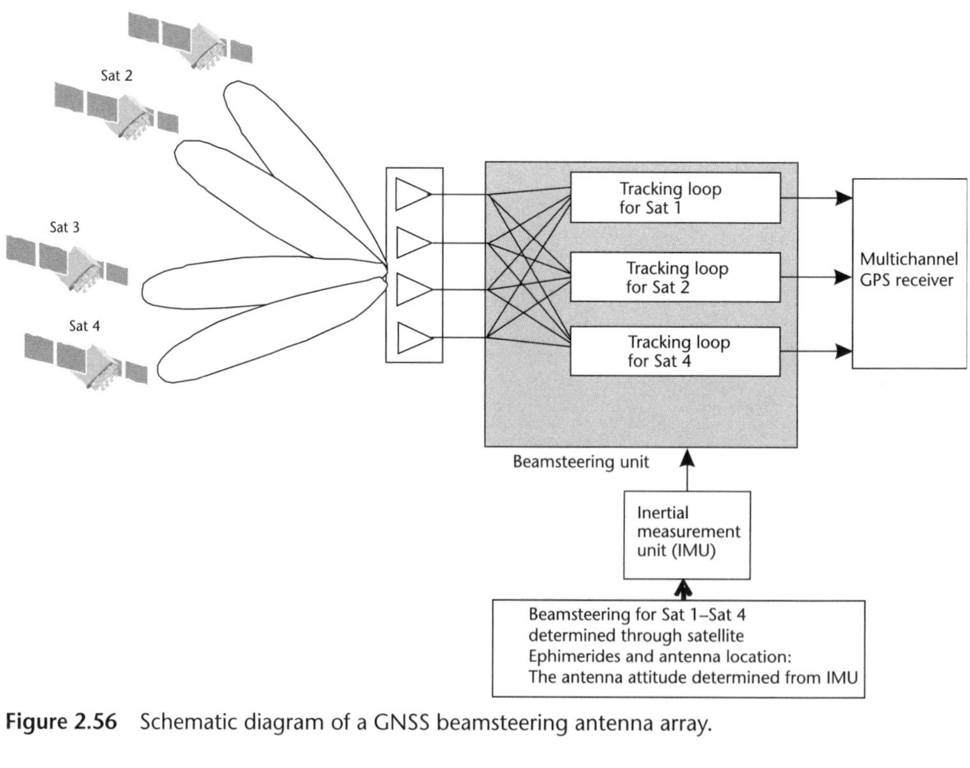

Figure 2.56 Schematic diagram of a GNSS beamsteering antenna array.

- Under ideal conditions the array can improve the satellite signal by a factor of log(N), where N is the total number of antenna elements in the array.
- Since each beamformer is directed towards a specific satellite, it avoids phase discontinuities in the received GNSS signal that can occur in a single beamformer array thereby allowing more accurate carrier and phase differential tracking techniques.
- Beamforming antenna arrays have also been investigated recently by several authors as a method for reducing multipath effects in GNSS systems [10]. They use the spatial separation between the satellite signals and the multipath signals to vary the antenna array pattern so as to minimize multipath signals while enhancing the satellite signals.

Despite the many advantages that they provide, beamforming antenna array also has some significant disadvantages:

- This type of antenna array can become complex, large, and expensive since it needs a separate beamformer for each GNSS satellite and a minimum of four beamformers are needed for the array to function
- The spatial environment of a highly dynamic vehicle can change rapidly resulting in large variations in attitude and directions of SOI. This requires that the complex weights of the array elements need to be constantly calculated and updated; but every change in the weight vector can result in an abrupt phase shift to the signal delivered to the receiver, which could create problems for the tracking loop.
- Another disadvantage is that each beamformer output needs to be passed independently to the GNSS receiver. This means that it cannot be used with most current GNSS receivers without making significant design changes although a software-defined radio is a good candidate for accomplishing this task.

(a) (b)

Figure 2.57 Two high-gain GPS beamforming antenna arrays (HAGR) built by NAVSYS. (a) Beamforming array of 100 elements, and (b) beamforming array of 16 elements. (Figures provided courtesy of A. Brown of NAVSYS Corp.)

A beamforming array built by the NAVSYS Corporation has recorded signal improvements of between 10 to 20 dB depending on the number of elements in the array [9, 10]; a picture of a very large and a smaller beamforming GNSS antenna arrays built by NAVSYS is shown in Figure 2.57. Various computer and hardware simulations have been performed by Konovaltsev et al. [84] to demonstrate to the efficacy of various beamforming algorithms. The beamforming is conducted in the digital domain and requires downconversion of the L band GNSS signals in the receiver front end.

References

[1] Lockheed Martin Space Systems Co. 80215-8500, Littleton, CO. (Permission for pictures shown in the book was granted by Stephen O. Tatum.)

[2] Mader, G. L., and F. M. Czopek, "Calibrating Antenna Phase Center (The Block IIA Satellite)," *GPS World,* May 2002. (Permission for picture granted by Gerald L. Mader.)

[3] Aparicio, M., et al., "GPS Satellites and Payload," in *Global Positioning System: Theory and Application,* Vol. 1, B. Parkinson and J.J. Spilker (eds.), Washington DC: American Institute of Aeronautics and Astronautics, 1996, pp. 234–239.

[4] Kee, C. et. al, "Development of Indoor Navigation Using Asynchronous Pseudolites," Proceedings ION-GNSS 2000, Salt Lake City, Utah, September 19–22, 2000, pp. 1038–1045.

[5] Pini, M., and D. M. Akos, "Analysis of GNSS Signals Using the Robert C. Byrd Green Bank Telescope," in *Satellite Communications and Navigation Systems,* E. Del Re and M. Ruggieri (eds.), Springer U.S.A., 2008, pp. 283–290.

[6] Gao, G., et al., "Compass–M-1 Broadcast Codes in E2, E5b, and E6 Frequency Bands," *IEEE Journal of Selected Topics in Signal Processing,* Vol. 3, No.4, August 2009, pp. 509–612.

[7] Yerker, S., S. Tholert, and M. Meurer, "Concept for Interference Detection and Analysis on Navigation Satellites," Proceedings European Navigation Conference, Naples, 2009.

[8] Gotta, M., et.al., "GNSS Signal Monitoring Laboratory Facility at AAS-I Premises—An Overview," Proceedings 2nd Workshop on GNSS Signals & Signal Processing, GNSS Signals 2007, European Space Research & Technology Center (ESTEC), 2007.

[9] Brown, A., and R. Siva, "A GPS Digital Phased Array Antenna and Receiver," Proceedings of IEEE Phased Array Symposium, Dana Point, CA, May 2000, pp. 1–4.

[10] Brown, A.K., and B. Mathews, "GPS Multipath Mitigation Using a Three Dimensional Phased Array," Proceedings 2005 ION GNSS Conference, Long Beach, CA, September 2005, pp. 1–8.

[11] Konovaltsev, A., F. Antreich, and A. Hornbostel, "Performance Assessment of Antenna Array Algorithm For Multipath and Interference Mitigation," 2nd Workshop on GNSS Signals & Signal Processing, GNSS Signals 2007, April 24, 2007, pp. 24–28.

[12] Malmstrom, J., Robust Navigation with GPS/INS and Adaptive Beamforming, Scientific Report FOI-R- 0848, SE, Swedish Defence Research Agency–System Technology Division, April 2003.

[13] Anon, Aeronautical Radio Inc., "ARINC Characteristic 743A GPS/ GLONASS Sensor Prepared by the Airlines Electronic Engineering Committee," March 16, 1992.

[14] Van Dierendonck, A. J., "GPS Receivers," in *Global Positioning System: Theory and Application,* Vol. 1, B. Parkinson and J. J. Spilker (eds.), Washington DC: American Institute of Aeronautics and Astronautics, 1996, pp. 340–341.

[15] SC-159 RTCA Inc., Minimum Operational Performance Standards for Global Navigation Satellite System (GNSS) Airborne Antenna Equipment, RTCA DO-228, October 20, 1995.

[16] Betz, J. W., "Effect of Linear Time-Invariant Distortions on RNSS Code Tracking," Proceedings of the Institute of Navigation Technical Meeting, ION NTM 2002, pp. 1–12.

[17] Schaffrath, M., and P. Salonen, "Characterization of Synthetic Fabrics as GPS Antenna Substrates," *Proceedings of the Second IASTED International Conference on Communications, Internet, and Information Technology,* 2003, pp. 167–172.

[18] Waterhouse, R. B., *Microstrip Patch Antennas: A Designers Guide,* Boston: Kluwer Academic Publishers, 2003.

[19] Jackson, D. R., J. T. Williams, and A. K. Bhattacharya, et al., "Microstrip Patch Antenna Designs That Do Not Excite Surface Waves"; *IEEE Transactions on Antennas and Propagation,* Vol. 41, No. 8, August 1993, pp. 1026–1037.

[20] Boccia, L., G. Amendola, and G. Di Massa, et al., "Shorted Annular Patch Antenna for Multipath Rejection in GPS Based Attitude Determination," *Microwave Optical Technology Letters,* January 2001, pp. 47–51.

[21] Garg, R., P. Bhartia, and I. Bahl, et al., *Microstrip Antenna Design Handbook,* Norwood, MA: Artech House, 2001.

[22] James, J. R., and P.S. Hall, *Handbook of Microstrip Antennas*, Volumes 1 and 2, IEE Electromagnetic Series 28, Peter Peregrinus Ltd., 1989.

[23] Bancroft, R., *Microstrip and Printed Antenna Design,* Noble Publishing Company, 2004.

[24] Pozar, D. M., S. Duffy, and N. Herscovici, "A Comparison of Commercial Software Packages for Microstrip Antenna Analysis," *Proceedings 2000 IEEE Antenna & Propagation Society International Symposium,* Vol. 1, July 16–21, 2000, pp. 152–155.

[25] Sainati, R. A., *CAD of Microstrip Antenna for Wireless Applications,* Norwood, MA: Artech House, 1996.

[26] Hammerstad, E. O., "Equations for Microstrip Circuit Design," *Proceedings 5th European Microwave Conference,* September 1975, pp. 268–272.

[27] Schneider, M. V., "Microstrip Dispersion," *Proceedings IEEE,* Vol. 60, No.1, January 1972, pp. 144–146.

[28] Waterhouse, R. (ed.), *Printed Antennas for Wireless Communications,* Hoboken, NJ: John Wiley and Sons, Ltd., 2007.

[29] Stutzman, W. L., *Polarization in Electromagnetic Systems,* Norwood, MA: Artech House, 1993.

[30] Kumar, G., and K. P. Ray, *Broadband Microstrip Antennas,* Norwood, MA: Artech House, 2003.

[31] Rama Rao, B., et al., "Compact Co-Planar Dual-Band Microstrip Patch Antennas for Modernized GPS," *Proceedings 2009 ION/GNSS Conference,* Savannah, GA, 2009.

[32] Yang, F, and Y. Rahmat-Samii, "A Single Layer, Dual-Band Circularly Polarized Microstrip Antenna for GPS Applications," *Proceedings 2002 IEEE International Antenna and Propagation Society Symposium,* Volume 4, June 16–20, 2002, pp. 720–723.

[33] Doust, E. G., M. Clenet, and Y. Hemmati, et al., "An Aperture-Coupled Circularly Polarized Stacked Microstrip Antenna for GPS Frequency Bands L_1, L_2 and L_5," *Proceedings IEEE Antenna & Propagation Society Symposium,* July 2008, pp. 1–4.

[34] Dong, W., J. T. Williams, and D. R. Jackson, et al., "Phase and Group Delays for Circularly Polarized GPS Microstrip Antennas," *Proceedings ION 63rd Annual Meeting,* Cambridge, MA, April 23–25, 2007, pp. 545–554.

[35] Krantz, E., S. Riley, and P. Large, "The Design and Performance of the Zephyr Geodetic Antenna," *Proceedings 2001 ION/GPS Conference,* Salt Lake City, UT, September 11–14, 2001, pp. 1942–1951.

[36] Lennen, G., W. Hand, and B. Westfall, GPS Receiver with N-Point Symmetrical Feed Double-Frequency Patch Antenna, United States Patent #5,515,957, USPTO, 1996.

[37] Pozar, D. M., and S.M. Duffy, "A Dual-Band Circularly Polarized Aperture-Coupled Stacked Microstrip Antenna for Global Positioning Satellite," *IEEE Transactions on Antennas and Propagation,* Vol. 45, No.11, November 1997, pp. 1618–1625.

[38] Wirola, L, I. Kontola, and J. Syrjarrine, "The Effect of the Antenna Phase Response on the Ambiguity Resolution," *Proceedings 2008 IEEE/ION Position Location and Navigation Symposium (PLANS),* May 5–8, 2008, pp. 606–615.

[39] Toko America Inc., GPS Dielectric Patch Antenna; DAK 1575 MS50T, May 2002, www.tokoamerica.com.

[40] Guo, Y.-X., K.-M. Luk, and K.-F. Lee, "Analysis and Design of L-Probe Proximity-Fed Patch Antenna," *IEEE Transactions on Antennas and Propagation,* Vol. 49, No. 2, February 2001, pp. 145–149.

[41] Zhou, Y., C.-C. Chen, and J. L. Volakis, "Dual-Band Proximity Stacked Patch Antennas for Tri-Band GPS Applications," *IEEE Transactions on Antennas and Propagation,* Vol. 55, No.1, January 2007, pp. 220–223.

[42] Iwasaki, H., H. Sawada, and K. Kawabata, "A Circularly Polarized Microstrip Antenna Using a Singly-Fed Proximity Coupled Feed," *Proceedings of the 1992 International Symposium on Antennas and Propagation (ISAP),* September 1992, pp. 797–800.

[43] Basilio, L. L., et.al., "A New Planar Dual-Band GPS Antenna Designed for Reduced Susceptibility to Low–Angle Multipath," *IEEE Transactions on Antennas and Propagation,* Vol. 55, No. 8, August 2007, pp. 2358–2366.

[44] Rama Rao, B., et.al., "Characterizing the Effects of Mutual Coupling on the Performance of a Miniaturized GPS Adaptive Antenna Array," *Proceedings of the 2000 ION/GNSS Conference,* Salt Lake City, UT, September 2000.

[45] Jedlicka, R. P., et.al., "Measured Mutual Coupling Between Microstrip Antennas," *IEEE Transactions on Antennas and Propagation,* Vol. AP 29, 1981, pp. 147–149.

[46] Garcia-Garcia, Q.,"Radiated Cross-Polar Levels and Mutual Coupling in Patch Radiators," *International Journal of RF and Microwave Computer Aided Engineering* (CAE), Vol. 10, 2000, pp. 343–352.

[47] Ngai, E. C., and D. J. Blejer, "Mutual Coupling Analysis of Small GPS Adaptive Arrays," *Proceedings IEEE Antenna & Propagation Society International Symposium,* Vol. 4, 2001, pp. 38–41.

[48] Murphy, T., P. Geren, and T. Pankaskie, "GPS Antenna Group Delay Variation Induced Errors in a GNSS Based Precision Approach and Landing Systems," *Proceedings 2007 ION GNSS Conference, 20th International Meeting of the Satellite Division,* Ft. Worth, TX, September 2007, pp. 2974–2989.

[49] Van Dierendonck, A. J., and A. J. Erlandson, "RTCA Airborne GPS Antenna Testing and Analysis for a New Minimum Operational Performance Standards (MOPS)," *Proceedings of the 2007 ION National Technical Meeting,* pp. 399–408.

[50] Kumar, A., *Fixed and Mobile Terminal Antennas,* Norwood, MA: Artech House, 1991, pp. 183–193.

[51] Fujimoto, K., and J. R. James (eds.), *Mobile Antenna Systems Handbook*, Norwood, MA: Artech House, 1994, pp. 535–536 and 455–462.

[52] Vidmar, M., "A DIY Receiver for GPS and GLONASS Satellites, Part 3b, Quadrifilar Backfire Helix Antenna," *VHF Communications,* Vol. 26, No. 4, Winter 1994, pp. 197–200.

[53] Foo, S., "A Quadrifilar Helical Antenna for Low Elevation GPS Applications," *Microwave Journal,* Vol. 41, No. 1, January 1998, pp. 178–184.

[54] Bricker, R. W., "A Shaped Beam Antenna For Satellite Data Communications," *IEEE Antenna and Propagation Society Symposium Digest,* October 1976, pp. 121–126.

[55] Vu, T. B., "Smart Antenna System Architecture and Hardware Implementation," in *Handbook of Antennas in Wireless Communications,* L. C. Godara (ed.), Boca Raton, FL: CRC Press, 2002, pp. 22-19–22-21.

[56] Kilgu,s C. C., "Resonant Quadrifilar Helix," *IEEE Transactions on Antennas and Propagation,* Vol. AP-17, May 1969, pp. 349–351.

[57] Tranquila, J. M., and S. R. Best, "A Study of the Quadrifilar Helix Antenna for Global Positioning System (GPS) Applications," *IEEE Transactions on Antennas and Propagation,* Vol. 38, No. 10, October 1990, pp. 1545–1550.

[58] Hussein, Z. A., and S. R. Rengarajan, "Ground Plane Effects on Quadrifilar Helix Antenna Phase Center and Radiation Characteristics for GPS Applications," *IEEE Antennas and Propagation Society Symposium,* Vol. 3, June 25–28, 1991, pp. 1594–1597.

[59] Leisten, O., et al., "Miniaturised Dielectrically Loaded Quadrifilar Antennas for Global Positioning Systems (GPS)," *Electronic Letters,* Vol. 37, No. 22, October 25, 2001, pp 1321–1322.

[60] Liu, S., and Q.-X. Chu, "A Novel Dielectrically Loaded Antenna for Triband GPS Applications," *Proceedings 2008 38th European Microwave Conference (EuMc),* 2008, pp. 1759–1762.

[61] Leisten O. P., J. C. Vardaxoglu, and Z. Agboraw, "Simulating the Dielectric-Loaded Quadrifilar Helix Antenna Using Bruye-Force TLM Approach," *Annular Review of Progress in Electromagnetics,* Vol. 1, 1999, pp. 479–483.

[62] Sharaiha, A., and C. Terret, "Analysis of Quadrifilar Resonant Printed Helical Antenna for Mobile Communications," *IEE Proceedings–H,* Vol. 140, No.4, August 1990, pp. 269–273.

[63] Shumaker, P. K., H. O. Chien, and K. B. Smith, "A New GPS Quadrifilar Helix Antenna," *IEEE Antennas & Propagation Society Symposium Digest,* Vol. 3, July 21–26 1996, pp. 1966–1969.

[64] Lamensdorf, D., and M. A. Smolinski, "Dual-Band Quadrifilar Helix Antenna," *IEEE Antennas and Propagation Society Symposium,* Vol. 3, June 16–21 2002, pp. 488–491.

[65] Ho, C. H., D. W. Huang, and B. Wu, "A New GPS Hexafilar Slot Antenna," *1998 IEEE APS Conference on Antennas and Propagation For Wireless Communications,* 1998, pp. 137–140.

[66] Kaswarra, B., et.al., "Circular Polarization Bandwidth for a Bowtie Turnstile Over a Ground Plane," *2007 IEEE Antenna and Propagation Society International Symposium,* pp. 3546–3547.

[67] Makarov, S. N., et.al., "Circular Polarization Bandwidth of a Planar Turnstile Antenna," *2007 IEEE Antennas and Propagation Society International Symposium,* pp. 2646–2649.

[68] Granger, R., P. Readman, and S. Simpson, "The Development of a Professional Antenna for Galileo," *2006 ION GNSS Conference,* September 2006, Fort Worth, TX, pp. 799–806.

[69] Counselman, C.C., "Multipath Rejecting GPS Antenna," *Proceedings of the IEEE,* Vol. 87, No.1, January 1999; pp. 86–97.

[70] Zhang, Y., and H. T. Hui, "A Printed Hemispherical Helical Antenna for GPS Receivers," *IEEE Microwave and Wireless Components Letters,* Vol. 15, No. 1, January 2005, pp. 10–12.

[71] Hui, H. T., et.al., "A Small Hemispherical Helical Antenna Array for Two-Dimensional GPS Beam-Forming", *Radio Science,* Vol. 40, RS1006, 2005, pp. 1–5.

[72] Schmid, R., M. Rothacher, D. Thaller, and R. Steigenberger, "Absolute Phase Center Corrections of Satellite and Receiver Antennas," *GPS Solutions,* Vol. 9, 2005, pp. 283–293.

[73] Emerson, D. T., "The Gain of Axial Mode Helix Antenna," *Antenna Compendium ARRL,* Vol. 4, 1995, pp. 64–68.

[74] King, H. E., and J. L. Wong, "Helical Antennas," in *Antenna Engineering Handbook, Fourth Edition,* J. L. Volakis (ed.), New York: McGraw-Hill, 2007, pp. 12-1–12-26.

[75] Djordjevic, A. R., et al., "Optimization of Helical Antennas," *IEEE Antennas & Propagation Magazine,* Vol. 48, No. 6, December 2006, pp. 107–115.

[76] Kraus, J. D.," A 50 Ohm Input Impedance for Helical Beam Antennas," *IEEE Transactions on Antennas & Propagation,* Vol. AP-25, November 1997, p. 913.

[77] Cheng, C.-Y., Numerical Electromagnetic Modeling of a Small-Aperture Helical-Fed Reflector Antenna, Masters thesis, Ohio University, August 1998.

[78] Carver, K., "The Helicone–A Circularly Polarized Antenna with Low Side lobe Level," *Proceedings of the IEEE,* Vol. AP-55, No. 4, April 1967, pp. 559.

[79] Braasch, M., "Optimum Antenna Design for DGPS Ground Reference Station," *Proceedings ION-GPS 1994,* Salt Lake City, UT, September 1994, pp. 1291–1297.

[80] Surathu, M., Investigating Global Positioning System Helibowl Antenna Performance Sensitivity with Variation in Design Parameters, M.S. thesis, Ohio University, June 1999.

[81] Lo. S., et al., "GNSS Album, Images and Spectral Signatures of New GNSS Signals," *Inside GNSS,* May/June 2006, pp. 46–56.

[82] Imbriale, W. A., "Reflector Antennas," in *Modern Antenna Handbook,* C. A. Balanis (ed.), Hoboken, NJ: John Wiley & Sons, 2008, pp. 201–262.

[83] Imbriale, W. A., *Large Antennas of the Deep Space Network,* Hoboken, NJ: John Wiley & Sons, 2006.

[84] Konovaltsev, A., F. Antreich, and A. Hornbostel, "Performance Assessment of Antenna Array Algorithm For Multipath and Interference Mitigation," *2nd Workshop on GNSS Signals & Signal Processing, GNSS Signals 2007,* April, 24 2007, pp. 24–28.

CHAPTER 3

Multiband, Handset, and Active GNSS Antennas

Basrur Rama Rao

3.1 Introduction

This chapter describes multiband, handset, and active antennas used in GNSS. The first two categories are recent additions to the better known GNSS antennas that were described in Chapter 2. They are becoming increasingly important and necessary for meeting the demands of new global navigational systems as well as other consumer market driven applications. In this chapter we will be considering some of the very smallest of GPS antennas currently in use, which are miniaturized antennas internally mounted inside handsets. One of the smallest of these is a dielectric chip antenna made by Pulse (Model W3010) with a volume of just 5.6 cubic millimeters (mm), it is 3.2 mm wide, 1.6 mm long, and 1.1 mm high. The surface area of this tiny antenna is 3×10^{-9} smaller than Stanford University's Big Dish reflector antenna used for monitoring GPS satellite signals, described in Chapter 2 .The diameter of this giant reflector was 45.7 feet. The size difference between these two antennas is a dramatic illustration of the wide variety and sizes of antennas currently being used for many different GPS applications.

Active antennas, the third category of antennas considered in this chapter, have been used for a long time in navigation systems. These active antennas, which are rapidly replacing conventional passive antennas because of their compactness and versatility, have electronic components such as bandpass prefilters and low-noise amplifiers that are directly integrated into the antenna package. This makes it difficult to separate the contribution of the passive antenna from those of other active electronic components associated within the package because of the lack of access to the terminals of the passive antenna. This chapter will describe new metrics and measurements for the evaluation of the performance of active antennas that are different from those used for testing conventional passive antennas.

3.2 Multiband GNSS Antennas

3.2.1 Current and New Satellite Navigations Systems

For the past 20 years most of the antennas used for global navigation satellite systems were designed to serve either the legacy U.S. GPS system (L_1 and L_2 bands) or the GLObal NAvigation Satellite System (GLONASS) system (G1/L1 and G2/L2 bands) launched by the former USSR. A few dual-band antennas were also designed to cover the frequency bands of both these systems. However, this current scenario is expected to undergo a very significant change within the next 5 years:

- GPS is being modernized by the United States;
- The Russian Federation has also been "modernizing" and replenishing the GLONASS system with "healthy" new satellites (e.g., the GLONASS "M" and GLONASS "K") to restore the system to a full 24-satellite constellation [1,2];
- In addition to the L1 and L2 bands currently available, a third frequency band (L3) from 1201.743 to 1212.250 GHz is likely to be added [1].
- Two entirely new global navigational systems are being introduced: "Galileo" by the European Union and "Compass" by China [3, 4].

The deployment of these two latest GNSS systems is already underway and some satellites have been launched in both systems. Information on specific frequency bands, carrier frequencies, and modulation schemes employed by these four new global satellite navigation systems are shown in Tables 3.1 and 3.2. Table 3.1 covers modernized GPS and Galileo and Table 4.2 covers GLONASS and Compass. Table 3.3 shows the expected number of satellites in the full constellation for each system, their orbits, and orbit radius. Figure 3.1(a) and (b) show a chart of the frequency bands for these four systems. When completed and operating at their full constellation strength (estimated to be around 2015), there will be a total of 113 satellites just from these four global navigation systems alone. They will be operating over 12 separate L-band frequencies and offering new modulations, data formats, and message protocols. Some of these modernized codes are expected to have lower noise, better resistance to multipath, and will offer several new advantages as described in Section 3.2.2. The bandwidth necessary to cover the full spectrum of GNSS frequencies extends from 1.164 to 1.6065 GHz, for a total bandwidth of 32%. The frequency band of GNSS can be further subdivided into two distinct frequency regimes: the first extending from 1.164 to 1.3 GHz and the second from 1.559 to 1.6065 GHz. A brief description of these new and existing global and regional GNSS systems is provided in Chapter 1.

More detailed descriptions of these systems have been provided by Hoffman-Wellenhof et al. [5], Hein [6], and Rizos [7]. In addition to these four global systems there are two other regional navigation systems: QZSS (Japan) and IRNSS (India) [5–7] and also several other space-based augmentation systems (SBASs) [7, 8 (see listing in Table 14.1)]. SBAS are extra satellites added to the GNSS constellations for transmitting signals that provide enhanced accuracy, availability, and integrity for civilian aviation users.

Table 3.1 Modernized GPS and Galileo

GNSS	*Frequency Bands*	*Carrier Frequency (MHz)*	*Bandwidth (MHz)*	*Modulation*
GPS III	L_1 : [C/A, P(Y), and M	1575.42	1563–1587	C/A: BPSK P(Y): BPSK; M: BOC (10,5)
GPS III	L_1 : C/A, P(Y), M L_2C:Civilian	GNSS	1215–1239	C/A: BPSK P(Y): BPSK; M: BOC(10,5)
GPS III	L_5	1176.45	1164–1188	BPSK
Galileo	E5A	1176.45	1164–1191.75	BPSK R-10
Galileo	E5B	1207.14	1191.75–1215.0	BPSK R-10
Galileo	E2 – L1 – E1	1575.42	1559–1592	L1P: BOCcos(15,2.5) L1F: BOCcos(1,1)
Galileo	E4 / E6	1278.75	1258–1300	BPSK–R5

Table 3.2 GLONASS and Compass

GNSS	*Frequency Bands*	*Carrier Frequency (MHz)*	*Bandwidth (MHz)*	*Modulation*
GLONASS	L1 (or G1)	1602.0	1598.0625–1606.5	FDMA $f_1^p =$ 1602 + 0.5625.p MHz where p denotes the channel number specific to the satellite: $-7 \leq p \leq +8$ for current constellation $-7 \leq p \leq +13$ for full constellation [1, 2]
GLONASS	L2 (or G2)	1246.0	1242.9376–1249.5	FDMA $f_2^p =$ 1246 + 0.4375.p MHz $-7 \leq p \leq +8$ for current constellation $-7 \leq p \leq +13$ for full constellation [1, 2]
*				
Compass	B1	1575.42	N/A	BOC (14, 2)
Compass	B3	1268.52	N/A	QPSK (10) [3, 4]
Compass	B2	1191.795	N/A	Alt BOC (15, 10)

FDMA = frequency division multiple access;
*BeiDow Phase III signal plan.

The four most prominent international SBASs are WAAS (United States), EGNOS (EU), MSAS (Japanese SBAS), and GAGAN (Indian SBAS) [6, 7]. There are also at least three other commercial SBASs: carrier-phase differential GPS (CDGPS), Star Fire, and OmniSTAR [8]. The imminent emergence of multiple satellite navigational systems will bring about a fundamental change in the way future GNSS measurements are likely to be performed so as to take full advantage of the many benefits that they can offer. Some of these advantages (described in greater detail in Section 4.2.2) may include improved precision in positioning and greater reliability, availability, and integrity.

The mixing of signals from multiple navigation systems has also some drawbacks such as RF compatibility, interoperability, and issues with national security

Table 3.3 Expected Number of Satellites in the Full Constellation for Each System, Their Orbits, and Orbit Radius

GNSS	*Modernized GPS (US)*	*Galileo (EU)*	*GLONASS (Russian Federation)*	*Compass (China)*
Number of SV	24 (Medium Earth Orbit [MEO])	30 (Walker MEO)	24 (MEO)	35 (27 MEO, 5 GEO, 3 IGSO)
Orbits	6	3	3	5
Nominal altitude (km)	20,180	23,223	19,100	21,500 and 35,786
Orbit period (sidereal days)	1/2	10/17	8/17	7/13
Repeat period of apparent orbit (sidereal days)	1	10	8	7 and 1
MEO inclination	55°	56°	64.8°	55°

MEO = medium Earth orbit (MEO), IGSO = inclined geosynchronous satellite orbit.

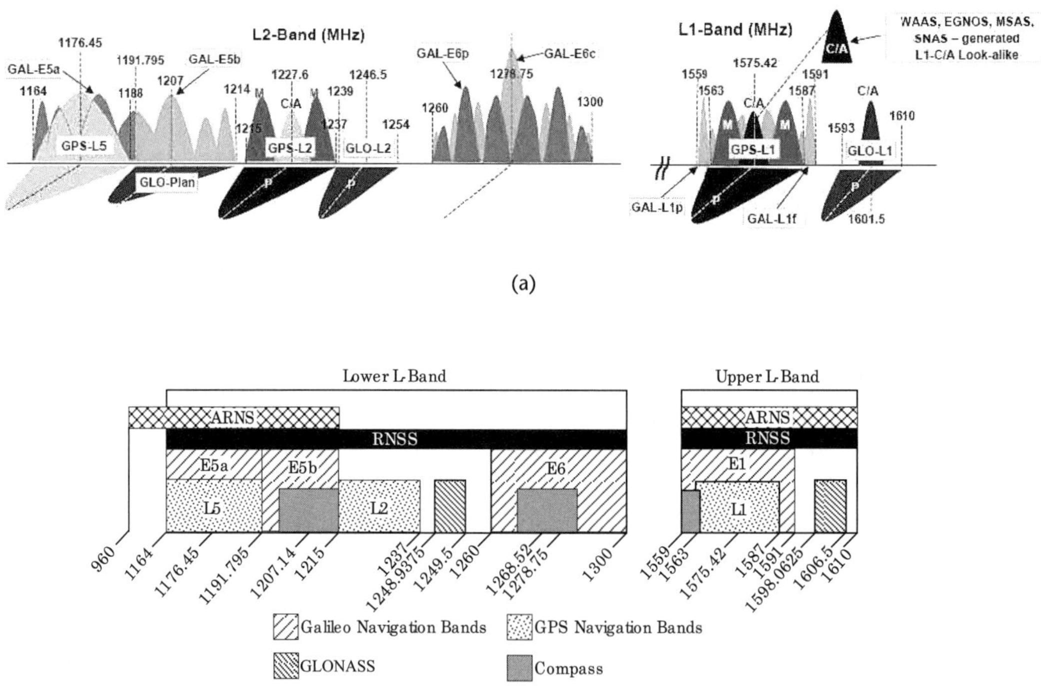

(b)

Figure 3.1 GNSS bands. (a) GPS, Galileo, GLONASS, (b) Compass. [(a) provided by Dr. Johnson Wang, WEO Corporation.]

that may need to be resolved. Some of these problems have been discussed by Hein et al. [6, Part 2]. Since these systems use closely spaced frequencies in the L band, signals from one GNSS system can be considered as a potential source of interference to a receiver designed to process signals from another system. Multipath effects, receiver antenna phase center variations, and receiver-hardware-induced delays can also be different for the different frequencies requiring careful calibration.

Several new designs for multiband/multisystem GNSS receivers have been proposed using the latest technology using ASICs, FPGAs, and DSPs; some of these receivers are capable of operating from 1.1 to 1.6 GHz [9, 10]. The January 2010 survey of GNSS receivers reported in *GPS World* [11] indicates that most high-end manufacturers are already offering multiband system of systems (SOS) GNSS receivers capable of acquiring and tracking signals from GPS, Galileo, GLONASS, Compass, and various SBASs. An entirely new genre of wideband antennas capable of operating between 1.164 to 1.606 GHz will be soon needed for use with these new GNSS receivers. Modernized GPS alone will require triple-band antennas. Galileo may require triple- and even quadruple-band antennas for the optimum use of all frequency bands that they offer to multiple users. As with wideband GNSS receivers, a 2010 GNSS antenna survey conducted by *GPS World* [12] has indicated that several of the major GNSS antenna manufacturers are already offering geodetic-grade, multiband antennas that cover this entire GNSS band. The design and performance characteristics of five of these commercial multiband antennas developed since 2006 as well as six other noncommercial wideband GNSS antennas that have been proposed recently are covered in this chapter.

3.2.2 Advantages Offered by Multiple GNSS

Several new applications are likely to emerge when the additional GNSS frequencies and satellite constellations offered by these new navigational systems are combined with those of current systems. The use of dual frequencies for correcting propagation errors caused by the ionosphere is an already well-established procedure, but more recent applications for exploiting additional frequencies can include some of the following:

- It will allow implementation of three-carrier (TCAR) and multicarrier (MCAR) ambiguity resolution techniques to obtain correct integer values in carrier phase observations for achieving the highest precision in pseudorange that is possible. TCAR and MCAR processing methods for ambiguity resolution have received much attention [5, 13–15]. The three signal bands (L_1, L_2, and L_5) of modernized GPS as well as L1, E5a, and E6 signal bands of Galileo can be used for TCAR ambiguity resolution. The benefits expected to emerge from these applications include ambiguity resolution over long distances, fixing correct and highly reliable integer solutions within much shorter time periods, and enabling RTK positioning in urban areas where signal obstruction is a problem. These techniques have been described by Vollath [13], Feng and Rizos [14], Hoffman-Wellenhof et al. [5, pp. 210–214], and by many other authors. The use of these techniques can enable greater precision in applications such as geodesy, surveying, avionics, mapping, machine guidance in agriculture, mining, and the construction industry. It has been shown by Verhagen [15] that the probability of achieving correct integer estimation in ambiguity resolution, referred to as the "success rate," can consistently reach levels of 0.999 (with maximum being 1.0) by combining the signals from both GPS and Galileo as compared to using signals from only a single

GNSS system. The recent cooperation between GPS and Galileo in terms of a common receiver for both systems may also hold future possibilities for increased use of multicarrier ambiguity resolution [16]. However, the prospect of extending this concept to all GNSSs/RNSSs appears to be daunting at the moment because of interoperability issues alluded to earlier.

- It will improve satellite availability in challenging environments degraded by blockage from buildings, shadowing from foliage canopies, and similar "urban canyon" types of environments such deep river valleys, open-pit mines, and so forth [22]. Satellite signals lost in such environments due to poor availability can be replaced by signals from satellites in other constellations that are not subject to such blockage [6, Part 2,17]. This will enable the sensitivity of GNSS receivers in an indoor environment to be enhanced by allowing acquisition and tracking of weak GNSS signals. This is accomplished through a new signal processing technique called collective detection [17, 18], which uses a collective detection strategy to generate frequency and Doppler correlograms by combining satellite signals from different constellations. This allows the receiver location to be tracked with greater sensitivity and in a shorter time. The increase in the number of satellites will ease the constraints in separation distance between the reference and rover receivers in RTK systems. A key limitation in existing RTK systems is the service distance between the reference station and the roving user receiver due to the impact of distant-dependent biases. The accuracy generally decreases as a function of distance from the reference station at about a 1-cm-per-km rate, which currently places maximum limits on the baseline distance. The use of TCAR and MCAR from the availability of new GNSS frequencies can extend this distance to a few hundreds of kilometers and can redefine future regional RTK services on both a regional and global basis [19].
- There will be less vulnerability to interference and jamming from narrowband sources since several alternate frequencies will be available to users over a much wider band to defeat such efforts, or if not, make it vastly more difficult [20]. However, the wideband, noise-jamming threat still remains and can be mitigated only by using adaptive antenna arrays described in Chapter 4.
- Interoperability between various navigational systems will improve integrity, reliability, and continuity through a higher redundancy of observations with lower position dilutions of precision (PDOP). It has already been predicted that interoperability between GPS, Galileo, and SBAS will significantly improve aviation safety and security and help to achieve CAT-1 navigation performance during all phases of flight [21]. Two recent lengthy studies conducted by a working group set up under a 2004 EU–U.S. agreement have also confirmed the substantial benefits to user equipment that is obtained by combining two systems, such as GPS III and Galileo, using a dual-constellation receiver. According to this research the benefits were particularly helpful in partially obscured environments, where buildings, trees, or terrain blocked large portions of the sky [22]. In all the cases studied, the combination of GPS with Galileo led to noteworthy improvement in performance as compared to that of a single system operating alone.

The announcement made on July 30, 2010 on collaboration of between the United States and the EU on the use of their respective GNSSs will provide a fresh impetus to these efforts [16]. The advantage provided by these new systems has even prompted a recent communiqué on the national space policy of the United States, issued on June 28, 2010. This has the following statement: "Foreign positioning, navigation, and timing (PNT) services may be used to augment and strengthen the resilience of GPS" [23].

3.3 Current Technology for Multiband GNSS Antennas

The design of currently available multiband GNSS antennas that cover the entire GNSS frequency band from 1.1 to 1.6 GHz is described in this chapter. Five are commercial antennas listed in Table 3.4; they are also in the list of antennas covered in the 2010 antenna survey conducted by *GPS World* [12]. These five antennas were selected for this book because good descriptions of their designs were available, obtained from personal communications with the various authors and from conference publications, white-papers, patents, and brochures published by their respective antenna manufacturers. The reference sources that were used for collecting the data for each of these antennas are also provided in Table 3.4. However, Table 3.4 does not include many other commercial multiband antennas that are currently available. All the commercial antennas included here were introduced to the market since 2006 and use different design concepts for meeting the demanding performance requirements expected from geodetic-grade antennas (described in Section 3.4). These antenna designs also show significant differences in their weight and size. The smaller and lighter designs are more attractive for mobile DGPS (Rover) or user receivers (UR) user applications; the heavier ones may be more suitable for reference receivers (RR).

In addition to these five commercial multiband antennas, eight other noncommercial multiband GNSS antenna designs that have been developed recently were also selected and are listed in Table 3.5. Five of these cover the entire GNSS band from 1.1 to 1.6 GHz. These noncommercial antennas have not yet found their place into the commercial marketplace but show considerable promise for future applications. We also describe three other triple-band antenna designs which have a more limited bandwidth and were developed specifically for the L_1, L_2, and L_5

Table 3.4 List of Commercial Multiband GNSS Antennas

List Number	*Company*	*Model Number*	*Frequency Band*	*Year Introduced and References*
1	NovAtel Inc. and Leica Geosystems	GNSS 750Leica AR25	1.15–1.61 GHz	2008; [24, 25]
2	Trimble	Zephyr Geodetic 2	L1/L2/L5/G1/G2/ E1/ E5ab/E6	2006; [26–28,76,77]
3	Wang Electro-Optic Corporation (WEO)	Universal GNSS Antenna Series 250	1.10–2.0 GHz	2009; [29–31]
4	NovAtel Inc.	GPS–704X	1.15–1.65 GHz	2006; [32–36]
5	Roke Manor Research Ltd.	Tri-G07 Tri-G08	1.150–1.650 GHz 1.105–1.630 GHz	2007; [37–40], 2008

Table 3.5 List of Noncommercial Multiband GNSS Antennas

List Number	*Source*	*Antenna Design*	*Frequency Band (GHz)*	*Year Introduced and References*
1	Mardel/ ESA ©2008 Institute of Electrical and Electronics Engineers (IEEE)	Crossed dipole on AMC ground plane	1.0–1.596	2008 [41]
2	Physical Sciences Inc. ©2009 IEEE	Turnstile drooping dipole on two-dimensional (2-D) wideband choke-ring ground plane	1.1–1.6	2009 [42]
3	Electronics and Telecommunications Research Institute (ETRI)/Korea ©2008 IEEE	Conical spiral	1.164–1.610	2008[43]
4	European Conference on Antennas and Propagation (EuCAP) 2007	Microstrip patch with parasitic coupling	1.16–1.64	2007 [44]
5	©2008 IEEE	Miniaturized spiral antenna array	1.1–1.6 GHz	2008 [45]
6	©2008 IEEE, etc.	Stacked microstrip antennas	GPS triple band: L_1, L_2 and L_5 only	2008 to 2006 [46–49]
7	MITRE: U.S. Patent 6,856,287, February 15, 2005	Four element , inverted L Antenna array with trap load filter	GPS triple band: L_1, L_2 and L_5 only	2005 [50]
8	Liu and Chu (China); ©2008 EuMA; 38th European Microwave Conference (2008)	Stacked, ceramic loaded, quadrifilar helix, and octafilar helix	GPS triple band: L_1, L_2, and L_5 only	2008 [142]

AMC = artificial magnetic conductor.

bands of modernized GPS. These antennas can also cover the E5a and E1-L1 bands of Galileo. This list includes different designs for very compact, stacked microstrip patch antennas; a low-profile, four-element array of trap filter loaded inverted L type antennas, and a compact, triple-band QHA. One possible application for this very compact and low-profile triple-band, microstrip antennas could be in avionics for improving integrity [21]; the triple-band QHA [142] is suitable for handsets and is described in greater detail under handset antennas. An entirely separate group of multiband antennas that combine GNSS with wireless communications and are primarily meant for use in handsets is discussed separately later in this chapter.

3.4 Performance Requirements for Geodetic-Grade Multiband GNSS Antennas

Most of the five commercial antennas listed in Table 3.4 are geodetic-grade antennas capable of providing very high precision in GNSS measurements for applications such as surveying and mapping [51], geodynamics including accurate measurements of tectonic motion (crustal dynamics) [52, 53], and monitoring of engineering structures, remote sensing, and determination of attitude [51]. In some of these geodynamic measurements the baseline between markers occupied by different geodetic systems may need to be measured with accuracies of better than 2 mm for baselines up to 450 km in length [52, 53]. These antennas are also used in baseline determinations in DGPS and RTK systems with an RR and with the "rover" or mobile UR; some of these reference stations can be networked as in netted RTK (NRTK) [51, 54–55]. The antenna and instrumentation issues associated with such precision surveying and mapping systems have been well explained in several recent books that specifically deal with satellite geodesy (e.g., by D. A. Grejner-Brezinska [56], Hoffman-Wellenhof et al. [57], Seeber [51], and Langley [58]).

In addition to meeting the gain and axial ratio requirements expected from all GNSS antennas(and described in Chapters 1 and 2), these wideband geodetic grade antennas are required to satisfy three key additional key requirements.

3.4.1 Phase Center Stability Requirements and Calibration of Multiband GNSS Antennas

Bias errors introduced by phase center variations in a geodetic-grade antenna have long been recognized as one of the factors limiting the accuracy of high-precision GNSS measurements; hence, geodetic grade antennas are required to keep such bias errors to a minimum level.

The stability and repeatability of the phase center of the antenna with respect to variation in elevation and azimuth angle of the received satellite signal and also its frequency are important factors for the achieving millimeter precision in GNSS range measurements. Two specific requirements are associated with the phase center variations of these geodetic-grade antennas: minimum PCV and PCO relative to the ARP with respect to elevation (above a specified low cutoff angle), azimuth angles (over a complete circumferential circle), and as a function of frequency. These various terms are defined in Chapter 1. Techniques for measuring phase centers are discussed in much greater detail in Chapter 6.

The conventional definition of ARP as defined by the International GNSS Service (IGS) is the point where the vertical antenna axis of symmetry meets with the bottom of the antenna mount (BA). As indicated by Hoffman-Wellenhof [5, pp. 149] and by Mader [60], these PCV variations of some GNSS antennas can be as large as 1 to 2 cm in the horizontal direction and up to 10 cm in the vertical direction in certain antennas. To correct these bias errors the PCV and PCO of all GNSS antennas used in precision measurements need to be calibrated taking into consideration their specified orientations. This data is needed to connect the GPS measurements to physical monuments. However, several geodetic-grade GNSS antennas, such as NovAtel's 704 [32, Table 6] and Trimble Zephyr Geodetic GNSS 2 antenna [26] have been designed to keep these phase center movements to less than 2 mm according to the manufacturer's brochures. Recent measured data provided by Wang Electro-Opto Corporation (WEO) also indicates that their Universal GNSS Spiral Mode microstrip antenna also has a PCV of less than 3 mm over the whole GNSS band from 1.1 to 1.6 GHz for all elevation and azimuth angles in the upper hemisphere.

Displacements of the PCV and PCO with respect to frequency are generally unavoidable. For example, PCV can be different for the GPS L_1 and L_2 bands as indicated from measurements made by Mader [60], Rothacher et al. [62], and

Gorres [63]. This shows a need for PCV calibration of multiband geodetic antennas across the various GNSS frequency bands if these frequencies are to be used for high-precision measurements. For these high-precision applications even correcting for the mean phase center is not sufficient for ensuring the required accuracy and phase center variations with angle of arrival of the satellite signals and frequency needs to be fully accounted for. These errors become particularly important when different antenna types and receivers are combined during DGPS and RTK measurements. Accurate calibration of the phase centers of each antenna becomes particularly important especially during height measurements [51, pp. 366–368); 57, pp. 162–163)].

Several independent antenna calibration programs have been undertaken to measure such antenna bias errors. PCV calibration techniques are described in greater detail in Chapter 6. The U.S. National Geodetic Survey (NGS) has calibrated many geodetic antennas and provided corrections along the horizontal "North" (X) and "East" (Y) coordinates and along the perpendicular Height (Z) coordinate, but this calibration data so far has been only for the GPS L_1 and L_2 bands [60,6 1]. These calibration tables provide the average or mean spatial separation between the ARP and the electrical phase centers provided by the NGS-ant.cal table; it is normally included in commercial GPS software.

Antenna calibration over a wide frequency band can be difficult and tedious since it would require a large number of antenna phase measurements over several different azimuth and elevation angles to cover almost the entire upper hemisphere at the many different frequencies covered by these multiband systems. PVC and PCO data for some of these geodetic antennas covering the entire GNSS band from 1.15 to 1.65 GHz has been provided only recently by Becker et al. [64]. Calibration for two of the geodetic antennas listed in Table 3.4—the Leica AR25 and the Trimble Zephyr Model 2—across the GNSS band from 1.15 to 1.64 can be found in this paper. NovAtel has also provided extensive phase center data for their 704X geodetic-grade antenna, for both the horizontal and vertical planes at several different frequencies covering the entire GNSS band [31]. A limited amount of PCV information has also been provided by the various antenna manufacturers such as Trimble [27], but only for various selected frequencies, mostly for the modernized GPS system. As mentioned earlier, WEO [29] has conducted an extensive calibration of the PCV of their universal GNSS SME antenna over the entire GNSS band from 1.1 to 1.6 GHz; this information is provided later in this chapter.

3.4.1.1 Mixing of Antennas in DGPS and RTK Measurements

One main driver for the above requirements is the needs of RTK differential applications where mixed receiver and antenna types are likely to be used [57, 65]. Phase center errors introduced by antennas can be minimized by using antennas of the same make and model at the end of a baseline and orienting both antennas towards the same direction—for example, magnetic north. This data is sometimes included in the commercial GPS software. However, if a mixture of antennas of different make or model are used in a baseline or in a network, then the data processing software, such as RINEX [51, pp. 281–283; 66], will need to know the relative positions of the phase centers of the antennas so that appropriate corrections can be made. Investigations have been conducted by Alemu [65] on the effects of mixing different types of geodetic antennas and receivers in precision DGPS measurements. His measurements on mixing antennas indicate great variations are observed in the height vector components than on the horizontal vector components. These errors can be ±1–2 mm for the horizontal component, but can be as large as ±2–3 cm for the vertical error component when antenna calibration correction factors are removed from the rover station data. He also observed that certain combinations of antennas, such as Leica-to-Ashtech combinations, can produce large errors in the vertical component—sometimes as large as ±2–4 cm.

3.4.2 Multipath Rejection Ratio and FBR

Multipath rejection ratio (MPR), as illustrated in Figure 3.2, is a particularly critical requirement for high-quality geodetic grade antennas since multipath can cause significant errors in both pseudorange and carrier phase measurements as discussed in Chapter 1. The MPR is defined as [67, 68]

$$MPR = \frac{G_{RHCP}(\theta)}{G_{RHCP}(180-\theta)+G_{LHCP}(180-\theta)}$$

The angle θ is the zenith angle or 90° minus the elevation angle. When the antenna has negligible gain in the backward direction towards the ground in either polarization, then the denominator in the above equation is zero at θ=0° and MPR becomes infinity, indicating that the antenna has perfect multipath rejection capability.

An alternative definition used for evaluating multipath rejection by the antenna is the "up/down ratio" or the "front-to-back ratio" given by [67, 68].

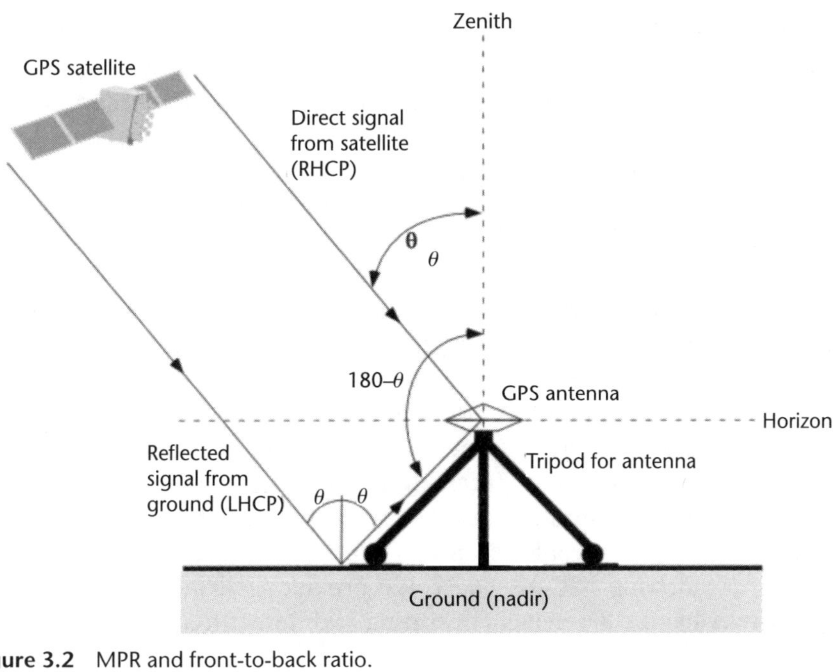

Figure 3.2 MPR and front-to-back ratio.

$$\frac{Up}{Down}Ratio = \frac{G_{RHCP}(\theta)}{G_{LHCP}(180 - \theta)}$$

In the above two equations, G_{RHCP} and G_{LHCP} are the RHCP and LHRP gains (i.e., the antenna gain of principal-polarized and cross-polarized components) of the radiated field from the antenna and $\theta = 90°$ elevation angle in degrees; $\theta = 0°$ corresponds to Zenith, and $(180 - \theta)$ is pointed directly towards the ground. For well-designed geodetic grade antennas, $G_{RHCP}\,(180 - \theta)$ is zero when $\theta = 0$; hence the MPR and up/down ratio are the same at this antenna pointing angle. The cross-polarization level especially in the antenna backlobes pointed towards the ground needs to be low to provide good a MPR; multipath from ground reflections are LHCP. Note these ratios also need to be defined for different azimuth angles covering a complete 360° circle.

The gain below the horizon should be as low as possible for maintaining a good front-to-back ratio or MPR so as to avoid reception of multipath signals. Rover antennas are typically mounted close to the ground on a handheld pole. This means they need a good MPR, a low-axial ratio, and decent gain roll-off at low elevations to minimize reception of multipath reflections off the ground. Most of these rover antennas are mounted on pillars, poles, tripods (with tribrach), or even bipods to allow easy portability to speed up the surveying task. Hence ground bounce multipath is a prime concern in such measurements. The front-to-back ratio should be a minimum to allow maximum suppression of multipath for achieving the highest precision possible from these measurements. In NovAtel's 704X, the front-to-back ratio varies between 32 to 27 dB. [31, Table 4].

3.4.3 Size and Weight Requirements for Rover Antennas Used in DGPS and RTK

Rover antennas have to comply with restrictions imposed on their weight and size so as to be compatible for use with newer mobile DGPS receivers. Receivers today are light in weight, small in size, and consume very low power; several are multifrequency receivers with multiple channels and capable of viewing all in view satellites. The wideband antennas to be used with such receivers need to have similar weight size restrictions to be compatible; some of these receivers were listed in January 2010 issue of *GPS World* [11].

A geodetic quality antenna used with the RR has a less stringent requirement on its weight and size since it is permanently stationed at a known, carefully surveyed location that is selected to be far from buildings and trees. Criteria for selecting a suitable site location for these antennas are discussed in Chapter 7. The primary source of multipath for these antennas is from GPS signals reflected off the ground. Being stationary, this antenna can use heavier and larger size choke-ring ground planes needed for suppressing ground multipath across a wide frequency band. These RR antennas can also be used with high-end, expensive multiband receivers that may require a lot of power to operate. When coupled with a wideband receiver, a wideband RR antenna could provide DGPS service to a large number of narrower band UR receivers operating at different frequencies within its service sector. In contrast, a geodetic-grade wideband antenna used in a mobile UR system poses a different type of technical challenge for the antenna designer.

Since these antennas need to be portable, they have to be lighter in weight, smaller in size, and lower in cost. They must also meet most other requirements expected of geodetic antennas, sometimes in less rigorous terms. They may also need to be used only with lower-cost and less power-hungry receivers to allow mobile battery pack operation. These mobile UR antennas need to be compatible for use with modern GNSS receivers used in surveying field equipment that can weigh less than 1 kg and is some cases only a few hundred grams and be smaller than 1000 cubic cm in volume and consume a power of less than 5 watts to as low as 1 watt [51, p. 248; 57 p.145]. The weight requirement is a particularly difficult requirement that only some of the new multiband antennas listed in Table 4.4 can meet. Since the use of heavy choke-ring ground planes are excluded for UR due to their weight and cost, the UR antenna may in some designs potentially experience larger phase center variations compared to the antenna used for RR.

Antennas in this second category are more likely to provide coverage only over specific bands of a system and take advantage of frequency overlap between two different GNSS systems. Some of these antennas may cover only the L_1, L_2, and L_5 bands of modernized GPS and the E5A, E5B, and the E2-L1-E1 bands of the Galileo system. Some of the older antennas cover the legacy GPS; GLONASS systems also fall within this category. Using different multiband antennas at the stationary RR and the mobile UR receivers due to the restrictions imposed by size, weight, and power can cause errors due to difference in the PCV variations between antennas making accurate calibration an imperative requirement for maintaining precision.

The five commercial multiband antennas listed in Table 4.4 can be divided into two basic groups: the first group uses wideband antennas mounted on an assortment of specially designed multiband ground planes to help in the suppression of multipath. The second group of wideband geodetic-grade antennas is based on design variations of conducting spirals or spiral slots; some of these spiral antennas come with their own specially designed ground planes that are integrated directly into the antenna structure.

3.5 Geodetic-Grade Multiband GNSS Antennas Using New Ground Plane Technology

Surface waves diffract from the edges of the ground plane and create antenna backlobes, increasing susceptibility of the GNSS antenna to ground bounce multipath. They also create phase and amplitude ripples in the elevation plane of the main antenna pattern caused by "multipath speckle interference" between the space wave radiated by the antenna and the secondary diffracted wave caused by ground plane edge diffraction, which can lead to phase errors in GNSS measurements. Chapter 5 provides a more detailed theoretical overview of these diffraction effects, including some measured antenna patterns. Conventional ground planes cannot mitigate such diffraction effects over the entire GNSS frequency band from 1.1 to 1.6 GHz, requiring new ground plane designs to achieve such suppression. Here we describe the design of a selected group of ground planes that have been specially designed for multipath mitigation in wideband geodetic grade GNSS antennas. These include:

- 2-D cutoff as well as noncutoff choke-ring ground planes using frequency selective surfaces (FSSs) [42];
- Three-dimensional (3-D) conical choke-ring ground planes [24];
- Resistivity tapered ground planes [26–28];
- AMC ground planes, which are also sometimes called electromagnetic bandgap (EBG) ground planes because of their ability to suppress surface waves over a limited frequency band [41].

While they provide good electrical performance over a wide band, choke-ring ground planes for DGPS applications suffer from three distinct disadvantages: weight, size, and cost. This puts them at a disadvantage for their use with mobile URs, limiting their role primarily to the stationary RR. Resistivity tapered and AMC ground planes offer several attractive advantages for use with URs. They are lighter, smaller, and have a lower profile, making them more suitable for UR applications in DGPS systems. Resistivity tapered ground planes have the widest bandwidth sufficient to cover the entire GNSS band from 1.1 to 1.6 GHz. They also appear to provide a subcentimeter level of phase center stability and have been used by Trimble for their Zephyr geodetic antennas [26].

Ground planes using AMC are of relatively recent origin; they are specially designed FSSs consisting of a periodic array of specially shaped metal patches that are etched on to the surface of a grounded dielectric substrate much like a printed circuit board. In some designs each metal patch is connected at its center to the ground plane underneath it through a via connection (an electrically conducting grounding stub). The magnetic field is zero at the surface over a limited frequency band, which is the reason for calling them "artificial magnetic conductors." Their reflection phase is also zero over this frequency band. This allows the antenna operating at this frequency band to be placed in close proximity to the AMC surface, making for a very-low-profile antenna.

The periodic array of metal-dielectric patches in an AMC surface can be represented as a spatially distributed network of capacitances and inductances; this also allows this surface to specifically be tuned to act as a stopband filter to prevent transverse electric (TE) and transverse-magnetic (TM) surface wave propagation over a certain frequency band. With proper design this stopband can be made to coincide with the AMC frequency band and thus help in mitigating multipath caused by ground-plane edge diffraction. This is the reason why they are sometimes also called EBG ground planes.

AMC/EBG ground planes are being investigated of late as ground planes for multipath mitigation in precision antennas used in the GPS and Galileo systems [41, 69, 70, 71]. They are light and low cost since they can be built using printed circuit technology. Thus they provide a major advantage over the traditional choke-ring designs that require deep grooves to be hogged into metal surfaces. A detailed discussion of their design has been provided by Sievenpiper [72], who first proposed this concept.

Various types of AMC surfaces for improving the performance of antennas, including GPS antennas, have been investigated by Rahmat Samii and Yang [73. However, the bandwidth of AMC ground planes in some of the designs appears to be more limited than obtained with resistive ground dual-band planes. Early versions were designed to cover only the legacy GPS L_1 and L_2 bands. The design of a dual-band AMC ground plane designed specifically for the GPS L_1 and L_2 bands are provided in Chapter 5 along with its measured elevation plane antenna patterns to demonstrate its multipath suppression capabilities. Later in this chapter we investigate a relatively broader-band design of one such AMC ground plane that has designed to cover GPS, Galileo, and Compass [41]; however, its bandwidth is not sufficient to cover the GLONASS L1 (or G1) band.

3.5.1 Wideband 2D GNSS Cutoff and Noncutoff Corrugated Ground Planes

Corrugated ground planes are a very popular method to mitigate multipath. To suppress or "cut off" surface waves propagation, the depth of the corrugation d in a corrugated choke-ring ground plane using a "traditional" design needs to satisfy the following two relationships:

$d > \frac{\lambda_L}{4}$ where λ_L is the wavelength of the lowest frequency signal, or the longest wavelength in the frequency band

$d < \frac{\lambda_H}{2}$ where λ_H is the wavelength of the highest frequency signal, or the shortest wavelength in the frequency band. This type of choke ring is called a cutoff choke ring [74].

Each corrugation in this textured electromagnetic surface performs like a quarter-wave transmission line, where the short circuit at the bottom of the corrugation is transformed into an open circuit at the top surface. This provides a high-impedance boundary condition to the waves propagating at the top surface of the ground plane whose electric field is perpendicular to the grooves, but low-impedance for electric fields parallel to the grooves. To operate in the GPS L_1/L_2 bands a dual-band choke ring using this design will need a corrugation depth d greater than 61.7 mm but less than 95.7 mm. The larger dimension corresponds to a quarter-wavelength at 1.215 GHz—the lowest frequency in the L_2 band with a 24-MHz bandwidth needed for M-code signal modulation. The smaller dimension corresponds to a half-wavelength at 1.567 GHz, the highest frequency in the L_1 band. Typically the depth of the choke rings needed to cover the L_1 and L_2 bands is adjusted to be between 61 to 64 mm [42]. These dimensions are based on the assumption that the tooth of each corrugation is very thin compared to the wavelength. Dual-band corrugated ground planes operating in the L_1 and L_2 bands use either tapered corrugations or alternate corrugation depths for adjacent grooved rings to meet the requirement of achieving good suppression of surface waves at both bands [74].

If the choke rings were also designed for triple-band operation to cover the GPS L_5 band, then the depth of the choke rings would need to be even deeper; at least 64.43 mm corresponding to a quarter-wavelength at 1.164 GHz, the lowest frequency in the L_5 band (24-MHz bandwidth). The dimensions of the corrugations will now need to be adjusted to cover the L_5, L_2, and L_1bands of modernized

GPS. The design of these triple-band choke rings gets even more complex, requiring further optimization in the depth of the corrugations and its spacing. An optimized design for this type of cutoff 2-D ground plane that works over the entire GNSS band from 1.1 to 1.6 GHz has been obtained by Scire´-Scappuzo and Makarov [42]. The performance of different types of cutoff choke-ring ground planes used with a wideband, droopy turnstile dipole is discussed in Table III of their paper. These consist from two to up to five rings with ring depths of between 65 to 69 mm but with different ring diameters, groove width, and tooth thickness.

A new type of 2-D, wideband choke-ring ground plane using a noncutoff, FSS, corrugated surface has also been investigated by Sciré-Scappuzzo and Makarov [42]. This ground plane uses corrugations that have much shallower depth than those used in the conventional cutoff ground plane described above but is still able to get good multipath limitation over the entire GNSS band from 1.15 to 1.64 GHz. This type of noncutoff ground plane is designed in such a way that the line of sight (LOS) signal from the broadband, droopy turnstile bow-tie antenna used with this type of ground plane and the surface wave propagating above the ground plane cancel each other out at the rim of the ground plane, as illustrated in Figure 3.3(d). This condition is achieved when the depth and width of the corrugations and the outer radius of the ground plane satisfy the condition:

$$\sqrt{1+\left(\frac{g.\tan k_0 d}{g+t}\right)^2}-1=\frac{n\pi}{k_o R}$$

R = distance from the antenna to the rim of the ground plane

$k_0 = \frac{2\pi}{\lambda_0}$ where λ_0 = wavelength

d = depth of the corrugation

g = width of the corrugation

t = thickness of the each tooth in the corrugation

These dimensions are illustrated in Figure 3.3(b). The above cancellation condition between the LOS signal from the antenna and surface wave signal on the ground is rather incongruous as it is strictly valid only at a single frequency whose wavelength is given by λ_0, as implied in the above equation. A perfect cancellation requires a relative phase shift of 180° between these two signals at this particular frequency. It is therefore hardly applicable over the extended frequency range from 1.1 to 1.6 GHz covering the entire GNSS band. Surprisingly, according to the authors an average phase shift across the band of 90° is optimum for such wideband operation. No satisfactory explanation is provided for this anomaly, which appears to have been derived from the results of computer simulations. The maximum shift that provides the best cancellation or multipath suppression is 150° at the maximum frequency of 1.6 GHz.

A ground plane with this new design has been optimized through computer simulations to operate over the GNSS band from 1.15 to 1.64 GHz when used with the droopy turnstile bow-tie antenna. The optimized ground plane has a slightly conical shape with a semiflare angle of 85° and has the following dimensions:

diameter $2R = 380$ mm, $d = 33$ mm, $t = 4$ mm, $g = 6$ mm. A picture of this ground plane and its dimensions are shown in Figure 3.3(c) and (d).

Figure 3.4(a) shows the comparison of measured and simulated RHCP and LHCP elevation plane antenna patterns of the wideband turnstile antenna on the noncutoff ground plane measured from 1.15 to 1.60 GHz at frequency increments of 0.05 GHz. The top figure shows a measurement of 0° azimuth angle. The measured results show that the LHCP gain at zenith is a little higher than the result predicted from simulations.

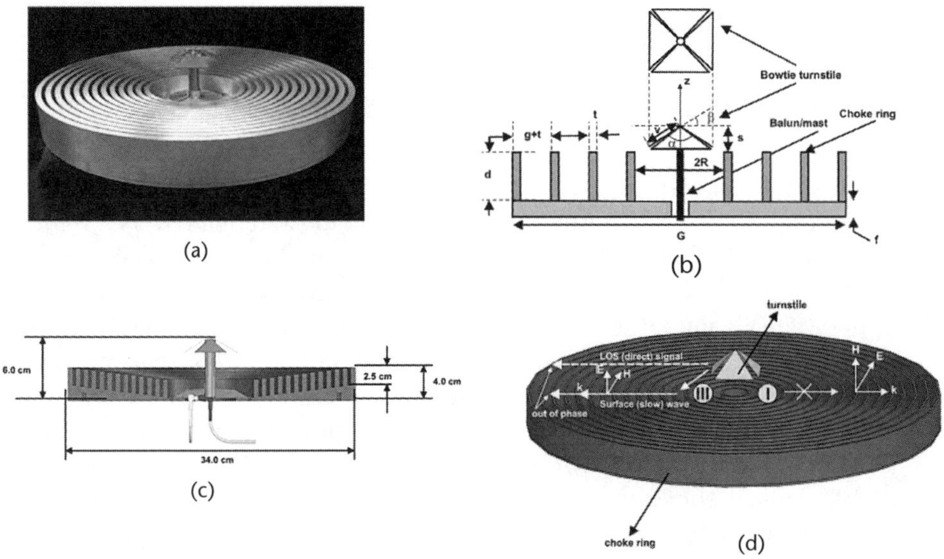

Figure 3.3 Wideband GNSS noncutoff FSS ground plane and drooping turnstile antenna. (©2009 IEEE [42].)

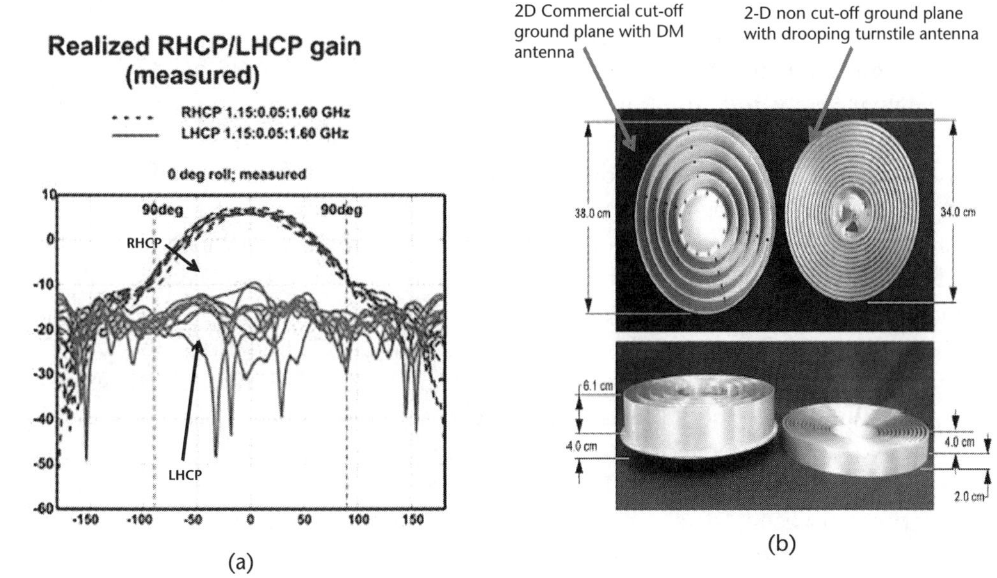

Figure 3.4 Measured elevation plane patterns of noncutoff ground plane and size comparison with commercial cutoff ground planes. (©2009 IEEE [42].)

Figure 3.4(b) shows a comparison of the new noncutoff and a traditional design cutoff ground planes. The new design provides a much better axial ratio and hence higher RHCP gain in the upper hemisphere; it also provides better multipath suppression through much lower antenna backlobes than a standard cutoff ground plane. Its diameter is also smaller and the relatively shallow corrugation reduces the volume. The FSS noncutoff ground plane is less than 40% smaller in volume than the classic 2-D choke-ring ground plane with deeper corrugations. The disadvantage is that it is also heavier since the adjacent grooved rings are closer and the width of the teeth is also thicker. The weight of the cutoff ground plane is 19 lbs from information provided by the author; this is even heavier than the NovAtel GNSS 750 3-D ground plane described in the next section, which weighs 16.75 lbs.

3.5.2 NovAtel GNSS-750 and Leica AR25 Wideband GNSS Antenna with Hemispherical Choke-Ring Ground Plane

These two antennas are identical since Leica Geosystems produces this antenna under a license from NovAtel. This geodetic-grade antenna covers a frequency range from 1176 to 1610 MHz and supports GPS, Galileo, GLONASS, and Compass as well as SBAS, Omnistar, and CDGPS. A picture of this antenna is shown in Figure 3.5(a); some of its antenna patterns measured by the MITRE Corporation at the three center-band frequencies of the modernized GPS frequency bands: 1.176, 1.227, 1.575, and 1.616 GHz (the highest GLONASS frequency) are shown in Figure 3.5(b). It uses a compact, ultrawideband antenna element made by Dorne-Margolin/EDO Corp. (Model Number DMC-146-24-1) on a specially designed 3-D conical choke-ring ground plane [24]. This ground plane is composed of a 3-D array of coaxial grooved slots and is in the shape of a pyramid. It is claimed that a ground plane of this design allows better tracking of signals from low-elevation-angle satellites and also improves multipath rejection than the traditional 2-D choke-ring ground planes.

A multidepth choke-ring design first proposed by Blankney et al. [75] has been used in this design for achieving multiband performance. The antenna has a 380-mm diameter, is 200 mm high, and weighs 7.6 kg (16.75 lbs). This may be somewhat heavy and large for certain mobile UR applications. Information on the specific depth and spacing of the different concentric groove rings comprising this ground plane has not been provided by NovAtel. The size and weight is contributed almost entirely by the large ground plane since the wideband antenna located at its center on top of the ground plane is small in size: just 5.19 inch diameter and 1.55 inch height. The EDO antenna by itself is also light, just 0.22 kg (0.5 lbs) and comes with a broadband LNA with a gain of 43 dB with a noise figure of 2 dB. The antenna when combined with the 3-D choke-ring ground plane provides good phase center stability with an offset of less than 2 mm and has excellent multipath suppression across the GNSS band, as seen from patterns shown in Figure 3.5. Extensive measurements of the PCVs of the Leica AR 25 has been made by Becker, Zeimetz, and Schonemann [64] over the GNSS band from 1.18 to 1.6 GHz and for elevation between 0° to 90°; the measured PCV is well within ±5 mm. Information on the brochure for this antenna provided by NovAtel indicates that this antenna meets all of their performance requirements for gain, axial ratio, front-to-back ra-

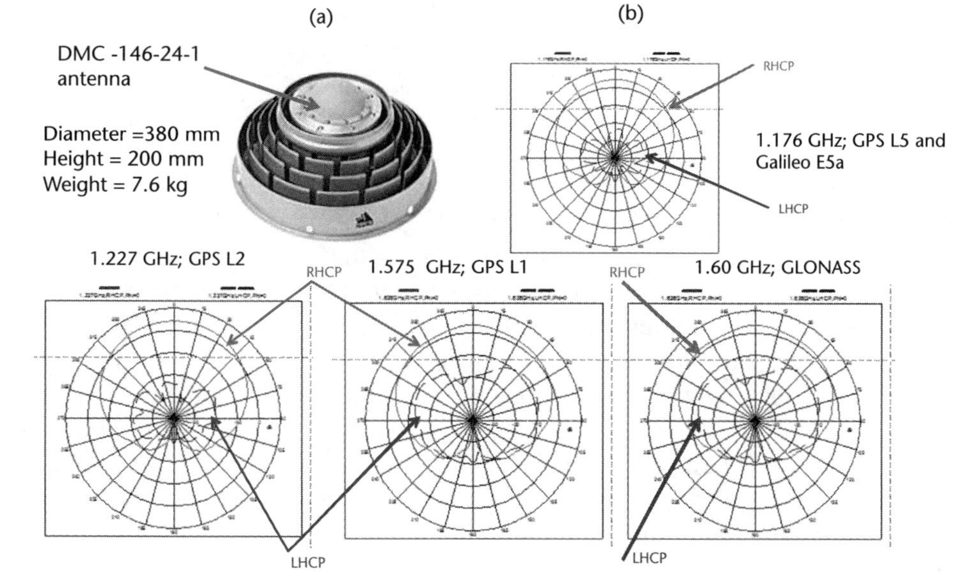

Figure 3.5 NovAtel's GNSS 750/Leica AR25 multiband antenna and 3-D conical choke-ring ground plane with measured antenna patterns. (Antenna photo courtesy of Pat Fenton of NovAtel.)

tio, and gain rolloff with elevation as well as PCV limits in the *X*, *Y*, and *Z* planes expected of geodetic-grade antennas.

3.5.3 Trimble's Zephyr Geodetic 2 GNSS Antenna with Resistivity Tapered Ground Plane

The design of this antenna is based on two U. S. patents issued to the Trimble Corporation [76, 77]. A 2001 ION/GNSS conference paper also provides a detailed description of its design [27]. A picture of the Zephyr Geodetic 2 GNSS antenna is shown in Figure 3.6(a) mounted on a pole and also on a tribrach. The Zephyr geodetic antenna has a 34.3-cm diameter and a 7.6-cm maximum depth. Its weight

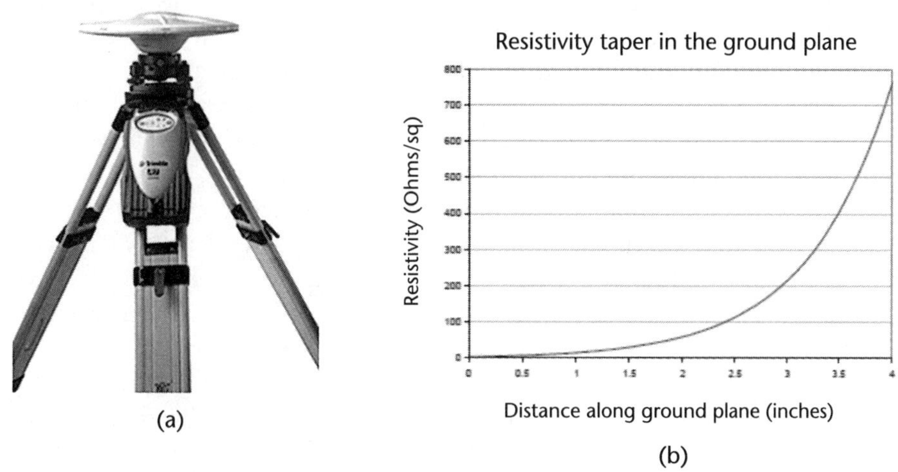

Figure 3.6 Trimble Zephyr Geodetic 2 Antenna. (Courtesy of E. Krantz, Trimble Corporation.)

is 3 lbs. The small size and light weight of this antenna allows portability needed for the roving RTK user in DGPS. The antenna has sufficient bandwidth to receive signals from modernized GPS, Galileo, GLONASS, and Compass satellites. The unique design feature of this antenna is the resistivity tapered broadband ground plane that Trimble describes as a "stealth" ground plane since it uses resistive card technology similar to what is used for reducing the radar cross section of stealth aircraft. According to their 2001 paper [27], the resistivity of their ground plane varies from zero (i.e., perfect conductor) at the center of the ground plane to 800 ohms/square at the outer perimeter of the ground plane. The resistivity taper showing the gradient in the surface resistivity (in ohms per square) as a function of distance from the center of the ground plane is shown in Figure 3.6(b) [27]. The tapering of surface resistivity prevents the electromagnetic fields radiated by the antenna from being diffracted from the edge of the ground plane. This ground plane design has a very wide bandwidth that is more than sufficient to cover the entire GNSS band. The surface of the ground plane is protected against weather effects by a radome cover, as shown in Figure 3.6(a).

A similar type of resistivity tapered ground plane for GPS applications was built around the same time by the MITRE Corporation [78] and is described in Chapter 5. The measured elevation plane patterns of a microstrip antenna placed at the center of this resistive ground plane are also shown in Chapter 5 to demonstrate the suppression in the antenna backlobes achieved by this types of resistive ground plane design.

The 2001 ION/GNSS conference paper by Trimble [26] states that a dual-band L_1/L_2 microstrip antenna with a four-point feed for each band was used to achieve good pattern symmetry in the azimuth plane and also a low circular polarization (CP) axial ratio in the radiation pattern needed to reduce the electrical phase center error. This is based on their patent [77]. It is not clear if later models of this type of the Zephyr geodetic antenna made by Trimble still use the same type of microstrip antenna. The antenna is placed at the center of a resistivity tapered ground plane to reduce antenna backlobes for mitigating multipath over the entire GNSS band. The radiation patterns of a microstrip antenna measured at three representative frequencies in the L_1, L_2, and L_3 bands of modernized GPS as measured with their resistive tapered ground plane are shown in Figure 3.7. These measured patterns were provided for this book by Eric Krantz of the Trimble Corp. The antenna is able to achieve submillimeter phase center repeatability when the microstrip antenna feed design is combined with the ground plane. The antenna uses a low-profile radome that covers the entire antenna element and the ground plane assembly, thereby sealing and protecting the special resistive ground plane against adverse weather conditions as shown in Figure 3.6(a). According to data provided in Trimble's information brochure [26], the phase center accuracy is better than 2 mm and phase center repeatability is less than 1 mm. Independent PCV measurements of this antenna made by Becker et al. [64] over a frequency band from 1.18 to 1.6 GHz and elevation angles varying from 90° to 0° have also confirmed that PCV of this antenna is in the range of +2.5 to –1 mm. The combined antenna/ground plane assembly is lighter and smaller than choke-ring ground planes described earlier. The axial ratio is 2 dB at zenith. An even smaller Zephyr antenna also sold by Trimble has a 6-in diameter and a maximum depth of 2.25 inches. It weighs only

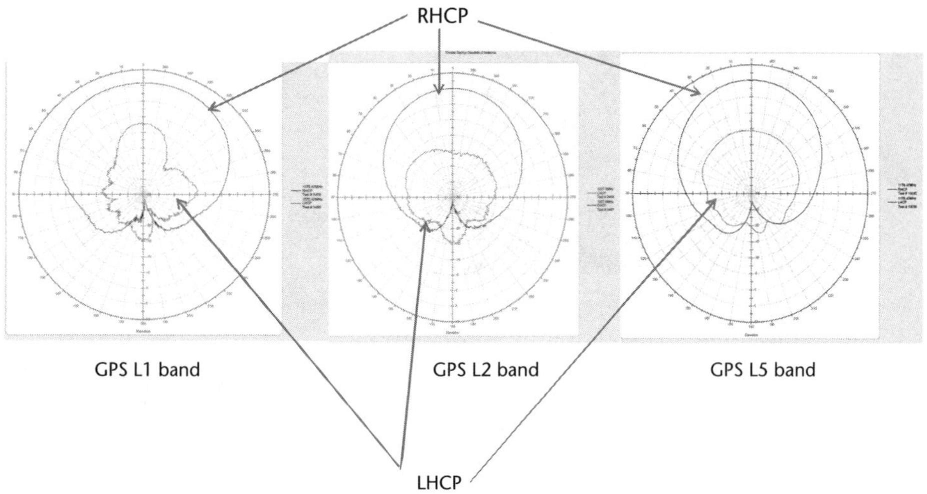

Figure 3.7 Antenna patterns of Trimble Zephyr Geodetic 2 Antenna. (Courtesy of E. Krantz, Trimble Corp.)

1 lb and also shows submillimeter phase center repeatability [28]. Both antennas come with integrated LNAs with a gain of 50 dB.

3.5.4 Wideband and Low-Profile GNSS Antenna on an AMC/EBG Ground Plane

A crossed dipole antenna on a wideband AMC/EBG ground plane covering the GPS, Galileo, and Compass frequency bands from 1.16 to 1.59 GHz has recently been designed and tested [41]. The unique feature of this antenna design is the 2-D periodic structure directly below the antenna known as AMC because it is designed to make the tangential magnetic go to zero at the surface over a certain frequency band. The AMC ground plane is designed to perform as a resonant LC structure; an ideal AMC behaves as a perfect magnetic conductor (as opposed to a perfect electric conductor where the tangential electric field is zero) and exhibits a reflectivity of +1 where the reflected wave is in phase with the incident wave. These properties of the AMC ground plane allow a GNSS antenna such as a crossed dipole to be placed very close to the surface of the AMC ground plane, allowing for low-profile geometry with this design. These AMC surfaces can be made thin and also light by using low-cost, printed-circuit technology.

The AMC ground plane is also designed to present a high impedance to prevent propagation of unwanted surface waves of both polarizations and for all propagation directions within a certain frequency band. Hence such surfaces are called EBG ground planes because they display frequency bandgap characteristics to the propagation of surface waves. To be useful for GNSS, the high-impedance surfaces are needed to provide simultaneously both AMC as well as EBG characteristics—the first for allowing antennas to be mounted close to the surface of the ground plane and the second to suppress surface wave propagation so as to prevent diffraction from the edges of the ground plane.

The top surface of this antenna, shown in Figure 3.8(a), consists of two planar crossed linear dipoles that are excited in phase quadrature to receive RHCP signals.

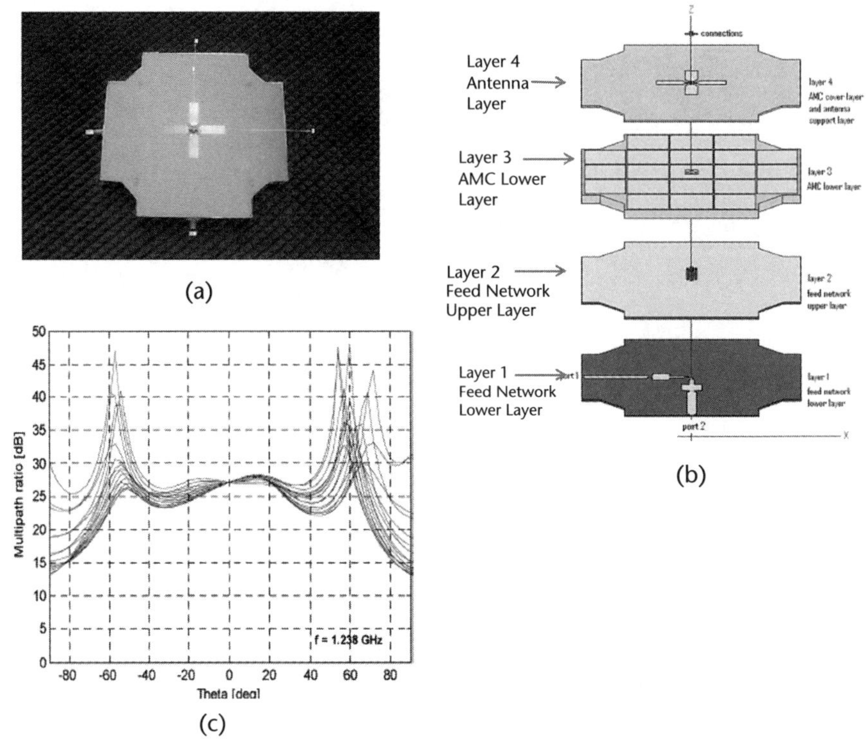

Figure 3.8 Multiband GNSS crossed dipole antenna on an AMC ground plane. (a) Top view of the antenna, (b) cross-sectional view, and (c) MPR measured at 1.238 GHz. (©2008 IEEE.)

The total antenna structure of this multiband GNSS antenna is comprised of four separate layers shown in Figure 3.8 (b). The first layer at the top consists of two linearly polarized crossed dipoles that are printed on top of a dielectric substrate; the dipoles are 151 mm long along the X and Y directions with a chamfer added in the 45° plane for improving the circular symmetry of the pattern. The crossed dipole antenna with its underlying dielectric substrate layer is placed over a specially designed AMC surface that acts as a reflecting ground plane for the antenna. This is the second layer of the whole antenna assembly. The frequency selective surface used in this design is a 2-D array of conducting patches printed on top of a thick dielectric substrate. The array is designed to operate as a high-impedance surface at a specific band of frequencies; within this frequency band the electric field has a node at the surface while the magnetic field has an antinode. At frequencies much lower than the resonance frequency the reflection phase is $+\pi$, and the surface will behave as a metal surface. As the frequency increases the reflection phase decreases and eventually becomes zero. At this resonance frequency the periodic surface acts as a perfect magnetic conductor (PMC) point. The periodic surface presents higher impedance than free space at this frequency. Above the resonance frequency the reflection phase returns to $-\pi$. The bandwidth over which the AMC surface continues to act as a PMC surface is determined by the frequency range where the reflection phase remains within $\pm\pi/2$ of the PMC point where the phase is zero. For the AMC periodic surface built for this antenna the PMC point was at 1.37 GHz with a 25% bandwidth; the PMC point is a little higher than the initial design goal 1.34 GHz. Its bandwidth appears to be sufficient to cover GPS, Galileo, and Compass frequency bands but not wide enough to cover the higher GLONASS frequency band. This AMC surface that was designed for this antenna does not need a via shorting connection of each patch to the underlying ground plane as in some other AMC/EBG designs described by Sievenpiper [72]; the omission of shorting connections to the ground plane reduces its cost. However, very little information is provided in this paper on the actual design of this vialess periodic surface used in the AMC ground plane and additional information on it could not be obtained from the authors. Fortunately, there have been several other recent papers that discuss the design of similar types of AMC/EBG impedance surfaces that use square conducting patches on dielectric substrates without using via connections to the ground plane underneath. These papers may prove useful in understanding their design and performance [83–85].

The use of an AMC ground plane allows the crossed dipoles to be placed much closer to reflecting surface, making for a lower profile and lighter antenna. Because the surface presents high impedance with a zero reflection phase at resonance, the antenna is not shorted out by its proximity to the surface but instead has high-radiation efficiency since the reflected energy from the AMC ground plane is combined with the radiated energy. In a conventional design the dipole antenna would need to be placed a quarter wavelength above a conducting ground plane to prevent phase cancellation from the reflected signal. The antenna is just 12.43 mm as compared to nearly 50 mm needed for a classical cavity-backed, cross-dipole and 95 mm if a corrugated ground plane were to be used, which is a 4- to 7.6-fold reduction in thickness according to the authors. The third layer in the antenna assembly is a conducting ground plane that serves as a backing for the AMC periodic surface. This metal ground plane also acts as the upper ground plane for the fourth dielectric layer at the very bottom, which carries the self-balanced stripline feed network for the crossed dipole antenna at the top. The dipoles on the top surface are phased with equal amplitude and with a relative phase difference of 90° to generate RHCP. They are connected to the stripline feed network through four coaxial cables.

Elevation plane patterns of this antenna measured at several different frequencies are given in [41].This antenna has an excellent multipath rejection capability; The up/down ratio at a frequency of 1.238 GHz is shown in Figure 3.8(c). The phase center of this antenna has been calculated from its 3-D pattern. The phase center remains within a sphere of 13 mm for each of the frequency bands in the Galileo system. The antenna has good gain and polarization purity across the frequency band. The weight of this antenna has not been provided in the paper but is expected to be much lighter than either of the choke-ring antennas described previously.

3.6 GNSS Geodetic-Grade Antennas Based on Spiral Antenna Technology

The next group of multiband geodetic-grade GNSS antennas use some form of planar or conical spiral geometries. Spiral antennas are made from either conducting spiral wires or spiral slots cut into a metal conductor. A slot cut into a thin metal ground plane is the electromagnetic Babinet complement of a spiral conductor. Both of these spiral shapes are well known for their ability to provide a wide impedance bandwidth and good RHCP gain and axial ratio radiation patterns extending over an ultrawide frequency band. Both have been widely used in antennas for radar and communications when wide bandwidth performance is needed. Hence it is not surprising that they have been used for building wideband GNSS antennas as well. These antennas belong to a group called frequency-independent antennas; this independence is however limited to a frequency band that is determined by the size of the antenna, especially when it comes to setting the low frequency limit.

3.6.1 Introduction to the Design of Spiral Antennas

A brief description of the design of spiral antennas is provided to help better understand the performance of more intricate spiral antenna designs to be described later. Figure 3.9(a) shows a diagram that represents the top-view cross section of the RHCP, four-arm Archimedean spiral antenna [79]. Four spiral arms are typically used for GNSS antennas to achieve better phase symmetry in the azimuth plane. A self-complementary spiral is one in which the width of the spiral conductor is the same as the interelement spacing between the conducting spirals. A spiral antenna of this design can theoretically have an input impedance of 188 ohms over a wide frequency band. The outer diameter of the antenna determines the lowest frequency of operation for the antenna; if D is the outside diameter of the spiral, the lowest wavelength $= \pi D$. The highest frequency is determined by the diameter of the innermost feed region where the four spirals originate. The radiation from the antenna is bidirectional; hence to prevent a 3-dB loss in gain and to prevent undesirable backlobes, it needs to be backed by a metal cavity whose depth is a quarter wavelength at its lowest frequency of operation or alternatively at its midband frequency as a compromise. The polarization of the antenna is determined by the direction in which the spiral conductor are wound; thus the polarization is RHCP when the fingers of the right hand placed on the antenna surface point in the direction of the

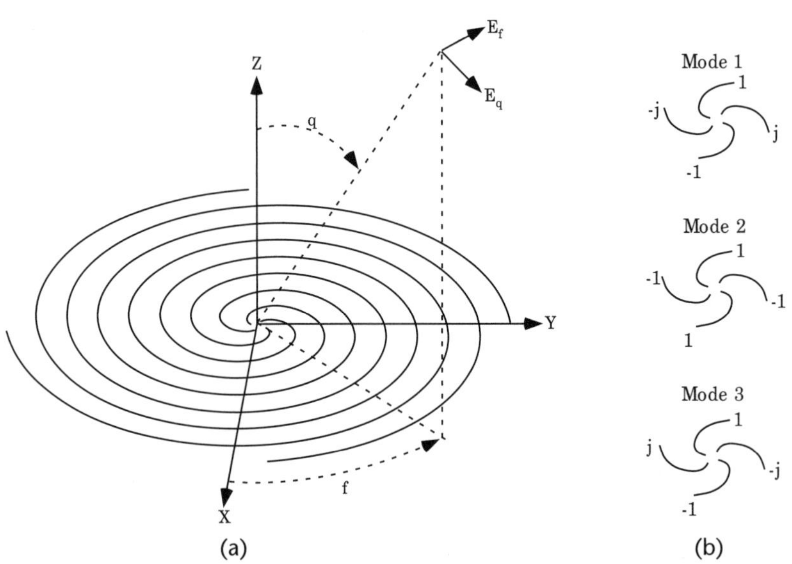

Figure 3.9 Geometry of four-arm Archimedean spiral antenna.

spiral windings; the direction of propagation of the RHCP signal is determined by the thumb, which is pointed upwards from the plane of the spiral. Similarly, if the fingers of the left hand are pointed in the direction of the spiral winding, the direction of propagation and the polarization of the backward radiation are determined by the thumb, which in this case is pointed downwards from the plane of the spiral. Hence the wave traveling towards the bottom end of the cavity in which the antenna is mounted is LHCP; upon reflection it becomes RHCP and adds in phase with the signal radiated upwards by the antenna when the bottom of the cavity is a quarter of a wavelength from the plane of the antenna. This imposes a limit on the maximum bandwidth of this antenna of edge diffraction effects that can cause multipath.

Special ground planes may sometimes be used with certain spiral GNSS antennas for suppression. This is not shown in Figure 4.9(a), but is discussed separately later.

The relative phases of the feed arrangement for the four arms are also shown in Figure 3.9(b); the amplitude of excitation for the four arms is equal. The equation in planar cylindrical coordinates (r, φ) for the geometry of one of the four arms of this spiral is given by φ. The other three arms are identical, except that they are rotated sequentially by degrees in a clockwise direction. Their respective radii can be represented as:

$$r_2 = r_0\left(\phi - \frac{\pi}{2}\right), r_3 = r_0\left(\phi - \pi\right) \text{ and } r_4 = r_0\left(\phi - \frac{3\pi}{2}\right)$$

where π is the radius at the start. The input terminals are numbered in the counterclockwise direction and their relative phases for RHCP radiation is as shown in the figure. A spiral antenna with N arms has (N-1) different modes of radiation. The four arm spiral has therefore three modes of radiation depending on the phase combinations of the spiral arms. These are represented by

$$A_1\left(1,-j,-1,j\right),\ A_1\left(1,-1,1,-1\right) \text{ and } A_3\left(1,j,-1,-j\right)$$

where $A_n(I_1, I_2, I_3, I_4)$ is the current vector notation for the feed points of this antenna. These are called, respectively, mode 1, mode 2, and mode 3. For GNSS applications, mode 1 and mode 3 are the most important in determining the polarization purity of the RHCP-radiated signal since mode 3 also has a narrow LHCP pattern at zenith. Hence a good understanding of their properties is essential for understanding the performance of the two spiral antennas considered in this section. This understanding is also essential in the following section that describes the universal spiral mode microstrip (SMM) antenna developed by the WEO Corporation, which also uses Archimedean spiral geometry. The current distribution for these modes has been analyzed by Nakano [79] and by Kraft [80]. The computed current distributions and the radiation patterns for modes 1 and 3 are shown in Nakano's paper [79] and help explain the critical performance parameters of this antenna. Mode 1 is the desired mode of radiation and generates a single RHCP broad beam pointed towards zenith. For mode 1, the currents in the spiral of the spiral antenna become "in phase" in a region of the antenna where the circumference is a wavelength;

hence this mode generates RHCP radiation in the forward direction. Due to this radiation, the current on the antenna decays exponentially starting from the feed point and proceeding towards the end of the four spiral arms as shown in Figure 2 of Nakano's paper [79].The current distribution for mode 3 has two components:

1. A transmission-line current component that shows little decay as it approaches the ends of the four spiral wires since it does not radiate;
2. An active-mode component which, like in mode 1, radiates as it proceeds towards the end of the spiral wire and decays as its approaches the end of the spiral wire.

Consequently, because of its slowly decaying transmission-line component, the current for mode 3 is still strong when it reaches the end of the spiral and can be reflected back from the end of the spiral wire, thereby setting up a standing wave. When these currents pass the region where the circumference of the antenna is one wavelength it generates a circularly polarized signal that has the opposite rotational sense to that generated by mode 1. Mode 3 is detrimental to the polarization purity of the antenna since it is generates a narrow LHCP beam directed towards zenith in addition to a split-beam RHCP pattern as shown in Figure 3 of [79]; it becomes excited due to imperfections in feed excitation network [80]. Poor axial ratio can result from second-pass radiation from the currents reflected from the ends of each spiral arm, which travel in reverse direction to the main currents and are hence polarized LHCP, and therefore it generates an LHCP radiation that degrades axial ratio, reduces bandwidth, and produces backlobes resulting in increased multipath. Every effort needs to be made while designing the GNSS antenna to suppress the current from being reflected from the ends of the spiral winding.

Four multiband geodetic GNSS antennas using various types of spiral geometries have so far been developed:

1. A cavity-backed Archimedean spiral antenna with its own specially designed ground plane developed by Roke Manor Research (U.K.) [37, 38];
2. SMM Universal GNSS antenna developed by WEO Corporation (U.S.A.) [29–31];
3. Pin Wheel Spiral Slot Array Antenna (model number GPS-704X) developed by NovAtel Inc. (Canada) [34–36];
4. Wideband GNSS conical spiral antenna developed by S. Tun et al. (The Electronics and Telecommunications Research Institute, Korea) [43].

Each of these antennas will be described in detail next.

3.6.2 Roke Manor's GNSS Multiband Cavity-Backed Archimedean Four-Arm Spiral Antenna with a Special Ground Plane

Figure 3.10(a) shows the wideband GNSS antenna built by Roke Manor Research Ltd. It is a cavity-backed, four-arm spiral antenna designed to operate from 1.1 to 1.6 GHz. It measures 125 mm in diameter and has a cavity that is 55 mm deep. To suppress these reflections, at the end of each spiral arm in the Roke Manor antenna is a short length of narrow line wound more tightly than the rest of the antenna

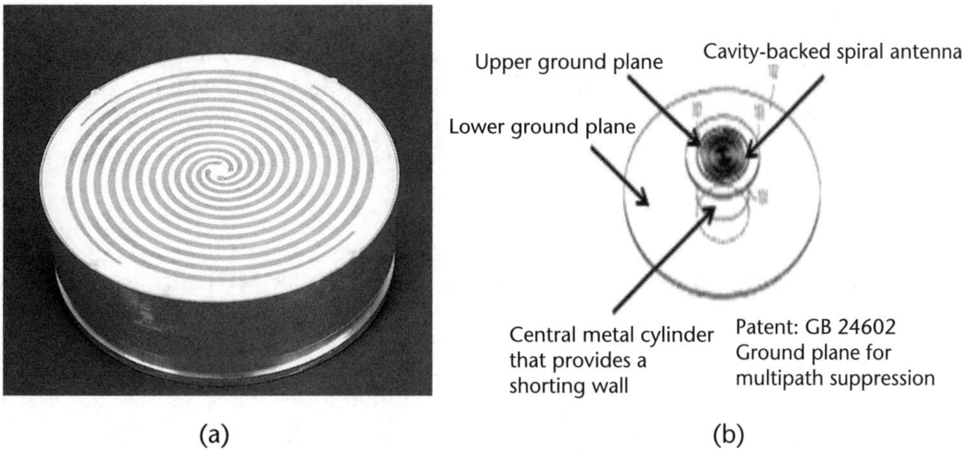

Figure 3.10 Wideband GNSS Archimedean spiral antenna and special ground plane by Roke Manor Research Ltd. (Image of antenna and copy of patent provided courtesy of James Horne, Roke Manor Ltd.)

and terminated in an open circuit. The Archimedean spiral itself is only 105 mm across; hence the lowest frequency is approximately 0.989 GHz. Sufficient circular symmetry is obtained by using three turns per conductor, resulting in an axial ratio of less than 3 dB in a 120° sector of the main beam; the RHCP to LHCP isolation at zenith is 30 dB. The 55-mm-deep cavity is a quarter of a wavelength at 1.36 GHz. The bottom of this cavity is made from a printed circuit board that contains the phase forming network consisting of a 90° hybrid and two baluns. The balun is necessary for matching the impedance of the antenna that has an input impedance of approximately 188 ohms to a 50-ohm coaxial connector. The printed circuit board at the bottom of the cavity also carries a LNA, interference rejection filters, and the power supply. Radiation patterns of this antenna measured at four different frequencies: 1189 GHz (Galileo E5a and E5b), 1228 GHz (GPS L_2), 1279 GHz (Galileo E6), and 1575 GHz (GPS L_1 and Galileo E1-L1-E2) [38]. PCV data for this antenna have not been provided in the paper.

A special ground plane has been designed for this antenna to suppress multipath [39, 40]. A picture of this ground plane is shown on in Figure 3.10(b). This ground plane has the shape of a "stretched" C and consists of a large lower ground plane and an upper smaller diameter ground plane. The two ground planes are separated by a cylinder. The surface of this cylinder provides a circumferential vertical metal wall that separates the two ground planes; the height of this central cylinder is approximately one-fifth of a wavelength at the central design frequency. The operational principle of this ground plane is outlined by Granger in his patent application and is also in a paper presented at an ION GNSS conference [39]. The underlying design is that multipath wave diffracts around the leading edge of the lower ground plane, creating a vertically polarized surface wave. The surface wave propagates towards the antenna until it encounters the upper ground plane. Together with the lower ground plane, these two metal ground planes form a kind of a parallel plate waveguide that continues to guide the wave towards the antenna.

Finally the weave meets the wall forming the central metal cylinder and is reflected back from the way it came. It is likely there is a secondary reflection from the upper ground plane but simulations show that it is relatively insignificant. A number of candidate designs of this ground plane were simulated using Ansoft HFSS v11. With the addition of the ground plane the height of the antenna increases to 113 mm.

3.6.3 Ultrawideband GNSS Antenna Based on SMM Technology by WEO

A low-profile, ultrawideband RHCP antenna covering all GNSS bands from 1.1 to 1.64 GHz based on their patented Spiral Mode Microstrip technology has been designed by WEO [29–31]. The WEO antenna design combines the best features of two different designs: the ultrawideband properties of an Archimedean spiral with the low-profile characteristics of a planar microstrip antenna. A conventional planar Archimedean spiral antenna, despite its well-know wideband properties, has two drawbacks for GNSS applications.

Its radiation is bidirectional and hence it needs a metal cavity backing placed underneath the antenna to reflect this downward radiated energy upwards to avoid backlobes pointed towards the ground, which can contribute to multipath. This design was adopted by the Roke Manor GNSS antenna described in the previous section. The cavity also prevents a 3-dB loss in gain by preserving the energy radiated into the lower hemisphere. Since the cavity backing needs to be a quarter wavelength deep, it is unsuitable for a low-profile shape needed for certain GNSS applications. The cavity backing also prevents taking full advantage of the truly broadband frequency-independent performance provided by the spiral antenna. WEO's patented SMM design [30] obviates the need for a cavity backing by using a microstrip spiral with a ground plane placed at a small distance (i.e., thickness of the dielectric substrate) below the plane of the antenna.

Radiation from higher-order modes is generated by the currents reflected from the tip of the spiral wires as explained in the introduction; this can degrade axial ratio and bandwidth. WEO is able to avoid this through another an innovative design modification. It prevents the generation of higher-order phase modes (modes 2 and 3 discussed earlier) and also the negative first-order "Minus 1" mode by placing a ring of absorbing material at the truncated edges of the spiral outside the radiation zone to allow absorption of the power in the higher-order modes, thus avoiding a degradation in the radiation patterns especially at the off-boresight angles. It also improves the RHCP axial ratio at zenith. In one such design employed by WEO [31], a .5 inch wide ring of microwave-absorbing material, thick enough to be fit the gap between the spiral and the ground plane, is placed half inside the truncated spiral and half outside.

WEO's Universal GNSS antenna, shown in Figure 3.11, is a four-arm, self-complementary Archimedean microstrip spiral antenna [29]. The antenna is 11.5 cm in diameter and is 3.0 cm in height. The top half of the antenna contains the planar microstrip spiral etched on a thin dielectric substrate with metal backing that acts as the ground plane for the antenna. The antenna is fed by a broadband tapered coaxial balun and feed network located in the bottom portion of the antenna. The WEO antenna has a very wide-impedance bandwidth with a 2:1 standing wave ratio (SWR) extending from 1 to 2 GHz. WEO has also provided results of several

Figure 3.11 WEO's Universal Spiral Mode Microstrip (SMM) GNSS-101 antenna. (Courtesy of Johnson Wang, WEO Corp.)

measurements on their antenna for inclusion in this book. The PCV with frequency is shown in Figure 3.12(a). The measured axial ratio versus the angle Theta = 90° – elevation angle is shown in Figure 3.12(b). These results show a low PCV and good RHCP axial ratio across the entire GNSS band from 1.1 to 1.64 GHz. The measured antenna patterns for several important frequencies in the GNSS band were also provided by WEO and are shown in Figure 3.13. The measured PCV in the azimuth and elevation planes are shown in Figure 3.14.

3.6.4 NovAtel's GPS-704X Wideband GNSS Pinwheel Antenna

The design information for the GPS-704X antenna was provided by NovAtel Inc. Waldemar Kunysz is the inventor of their patented Pinwheel™ array antennas [32, 34]. Several U.S. patents on different types of pinwheel antennas obtained by Kunysz are listed in [33–36]. The original wideband pinwheel antenna is shown in Figure 3.15(a); the latest improved version of this design, which includes fractal loop configurations at the end of each spiral slot for improving the bandwidth, is shown in Figure 3.15(b). Finally, a picture of the entire antenna enclosed inside a radome is shown in Figure 3.15(c). Since this design does not use a choke ring there is a considerable reduction in weight compared to the previously discussed GNSS-750

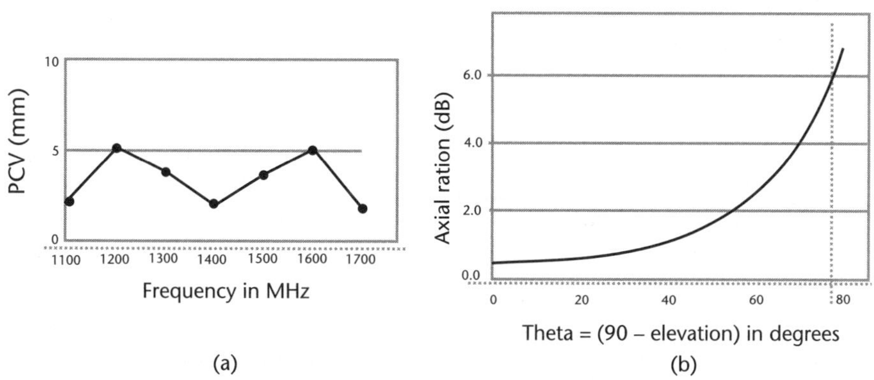

Figure 3.12 (a) PCV and (b) axial ratio of WEO's Universal Spiral Mode (SMM) GNSS-101 antenna. (Courtesy of Johnson Wang, WEO Corp.)

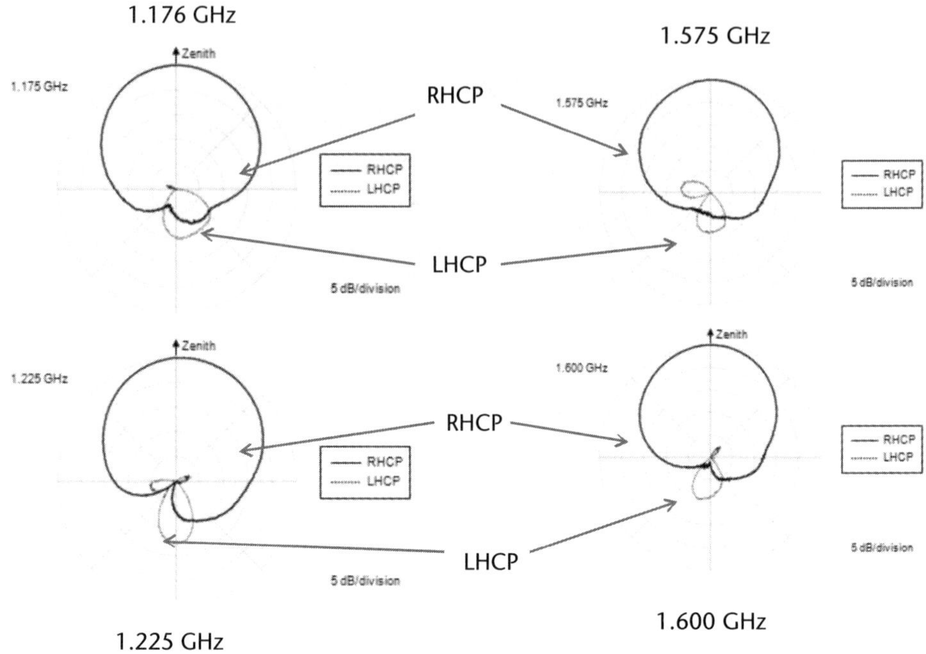

Figure 3.13 Measured antenna patterns of WEO–GNSS-101 Spiral Mode Microstrip antenna at four GNSS frequencies. (Courtesy of Johnson Wang, WEO Corp.)

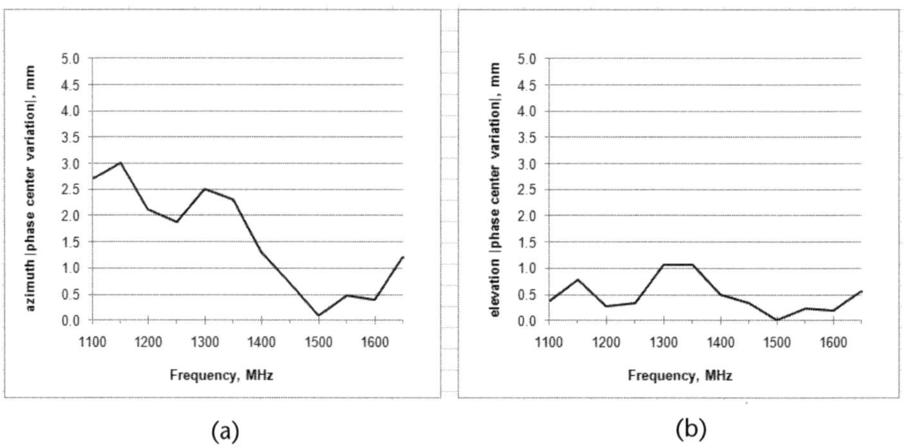

Figure 3.14 WEO Universal SMM GNSS-101 antenna: calculated PCV in (a) azimuth and (b) elevation planes. (Courtesy of Johnson Wang, WEO Corp.)

3-D choke-ring design; the weight of the 704X antenna is 468 grams compared to 7.6 kg for the GNSS 750. The diffraction backlobes are suppressed by using EBG techniques around the rim of these pinwheel antenna arrays in the 704X design.

The 704X is a fixed-beam, 12-element phased array of aperture-coupled spiral slots printed on a PC board. The slots are curved counterclockwise to receive RHCP. The spatial difference between two consecutive spiral arms is 30°. The feeding network for array used in the new design of this antenna is a leaky wave, multiple-turn spiral microstrip transmission line that is located on the lower surface of the substrate underneath the multislot array. The spiral shape of this feed line has a

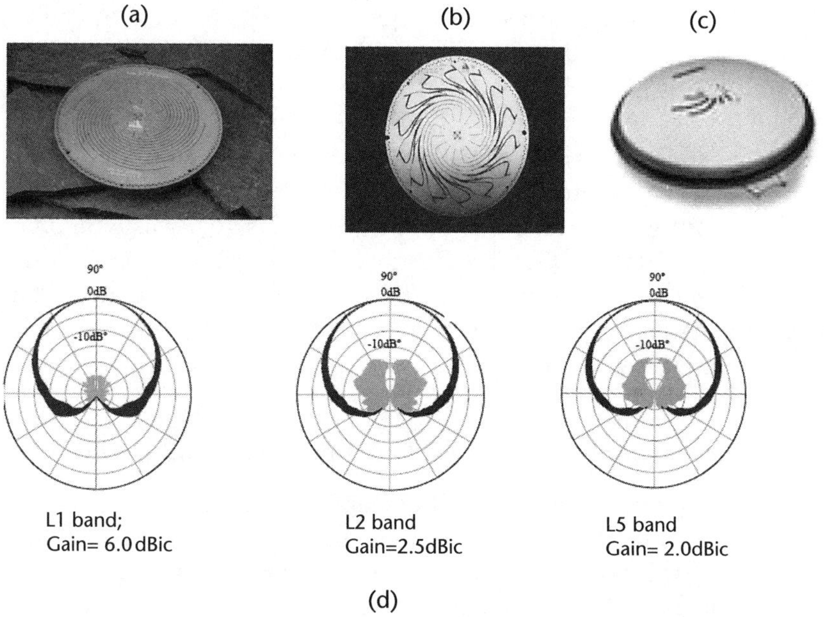

Figure 3.15 NovAtel's model 704X Pinwheel Antenna and measured antenna patterns at three bands of modernized GPS. (Courtesy of Pat Fenton of NovAtel.)

larger bandwidth than the circular feeding structure that was used in the previous design. There is sequential coupling of energy between this transmission line and the slot array through the aperture openings of the slots. Since the spatial difference between the each of two consecutive spiral arms is 30°, the electrical phase length of the feeding network is also set to 30°. This feed arrangement ensures that the correct relative phasing and equal amplitude between the spiral arms necessary to maintain RHCP over a wide bandwidth. It also ensures a stable phase center at all frequencies of interest. Its bandwidth extends from 1.15 to 1.7 GHz with a common phase center across the entire band [32, 34]. A receiver may be connected to end the leaky wave feed line through either a LNA or a connector to receive the RHCP signals from the slot array. To reduce edge diffraction effects from the edges, a surface wave suppression region consisting of multitude of openings encloses the slotted array. In the latest design modification of this antenna, shown in Figure 3.15(b), the slots are designed to initially begin as spirals but then flare out into fractal loop configurations. Each loop is connected to an adjacent loop of another slot. The composite radiating structure obtained in this fashion creates many open paths according to the antenna designer and opens up the bandwidth and also provides a common phase center for the new GNSS frequencies. The fractal loops reduces the current distribution at the ends of the arms by increasing its impedance. This results in a smoother amplitude and phase variation across the azimuth plane. The antenna was designed using high frequency structure simulator (HFSS)–a 3-D electromagnetic design software available from Ansoft. This computer code was also referred to earlier in Chapter 2 while discussing the design of microstrip antennas.

Extensive measurements have been made by NovAtel to measure the key performance parameters of this antenna at all the frequency bands of GPS III, Galileo, and GLONASS. The measured phase center movement for this antenna is less than 2 mm, which is the design goal set by NovAtel for all the frequency bands ranging from 1.15 to 1.7 GHz. The gain of this antenna meets the minimum required gain of –2 dBic at an elevation of 15°. The peak gain at an elevation of 90° varies from +7 dBic at 1608 GHz (GLONASS-L1 band) to a low of +2.7 dBic at 1176 GHz (GPS L_5). The antenna also meets the axial ratio and the front-to-back ratio goals set by NovAtel. For the front-to-back ratio this goal was a minimum of 25 dB at GPS L_2 and 30 dB at GPS L_1. The goal was to achieve an axial ratio not to exceed 3 dB at an elevation of 45°. The 704X meets both requirements handily. The measured radiation patterns for this antenna at the L_1, L_2, and L_5 bands of GPS III are shown in Figure 3.15(d).

3.6.5 Wideband Conical Spiral Antenna for GPS and Galileo

We have up until now discussed planar, wideband spiral antennas, but spirals can also be designed to fit over a conical surface. The advantage of the conically shaped spiral is that is produces a single-lobe radiation pattern with a maximum directed along the axis of the cone, much like the helix antenna discussed in Chapter 2. Planar spiral antennas on the other hand, with the sole exception of WEO's SMM antenna, all produce bidirectional, double-lobe radiation patterns, where one lobe is directed towards the ground that is undesirable from multipath considerations. Planar spiral antennas can also result in a gain loss of 3 dB unless prevented by either a reflecting cavity or a suitably placed conducting ground plane below the antenna. A conical spiral avoids this problem with a unidirectional main lobe, but the antenna geometry now becomes relatively more complex and it loses its advantage of being low profile.

A bifilar conical spiral that covers the GPS and Galileo frequency bands has been designed and built by Yun et al. [43]. A cross-sectional diagram of this antenna and its geometrical parameters from which its performance can be determined are shown in Figure 3.16(a); a picture of this antenna is shown in Figure 3.16(b). The conical spiral is a two arm (bifilar) symmetrical structure that is excited at its apex. It produces backfire radiation when cone half angles are 15° or less. It is fed by a balanced coaxial twin wire line that is brought along its axis, with a wire connected to each arm of the conical spiral. The metallic shield of the twin wire coaxial line has little effect on the radiation pattern if its diameter is kept small. This antenna uses a balun along with a 1:4 RF transformer to match the feed line to the radiating mode of the antenna.

The frequency band of operation of a conical antenna is determined by its physical size [81]. The upper frequency of operation is determined by the diameter of the truncated section at the vertex of the cone as shown in Figure 4.16(a). It can be estimated approximately from: $\frac{\lambda_U}{4} \sim d$ where d = diameter of the top truncated section of the cone and λ_U is the wavelength at this highest frequency. Similarly the lowest frequency of operation is determined by the diameter of the base of the cone and can be determined approximately from $\frac{3\lambda_L}{8} \sim D$, where λ_L = wavelength at the lowest frequency, and D is the diameter at the base of the cone. The bandwidth of the structure is defined as the ratio of the base $D/2$ to the radius of the apex $d/2$.

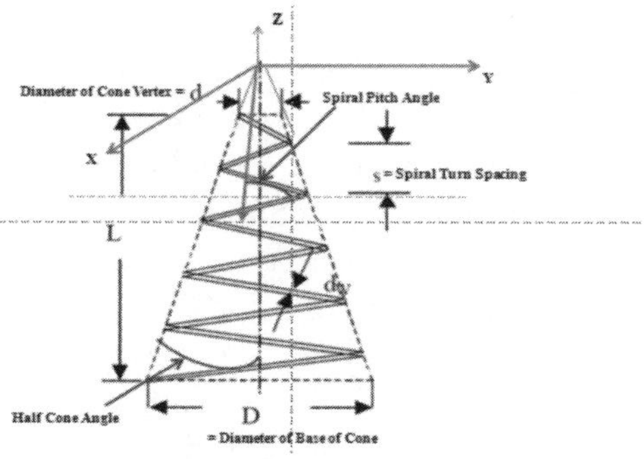

(a)

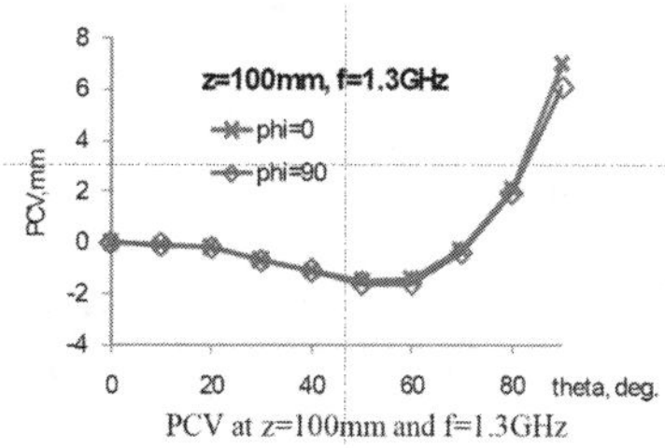

(c)

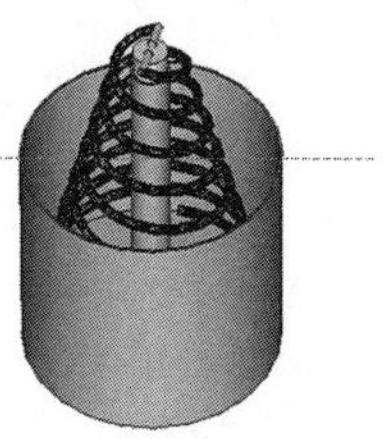

(b)

ANTENNA PERFORMANCE SUMMARY

PARAMETER		Specification	Analysis	Tested
Freq.	E5 & E6	1164-1300 MHz	1164-1300	1164-1300
	L1	1559-1610 MHz	1559-1610	1559-1610
PCV		< 5 mm	4	4.3
Gain	El=90°	> 3 dBi	5.7	5.9
	El=10°	> -3 dBi	-3.6	-3.5
	Below 0°	> 10 dBc	10	11.0
GDV		< 18 ns	2.1	17.0
VSWR		< 2:1	1.3	1.6

(d)

Figure 3.16 Bifilar conical spiral antenna for GPS and Galileo. (Courtesy of Sohyeun Yun.)

The other geometrical parameters that determine the performance of the conical spiral antenna are its axial length *L*, the spacing between turns *s*, the spiral pitch angle α, and the number of turns and the diameter of the wire *dw*. The spiral pitch angle α is angle between the radius along the cone to the edge of the antenna wire and the tangent to that edge. The half-cone angle and the spiral pitch angle influence the antenna pattern that is obtained. The smaller the value of the half-cone angle the larger the front-to-back ratio; similarly an increase in the spiral pitch angle results in a more uniform lobe. The parameters *s*, *d*, *D*, and α were adjusted to provide a backward mode of radiation (of the first T_{-1} type described in [43] in the frequency band from 1.1 to 1.6 GHz. The following design parameters were obtained at the middle of this frequency band: s = 30 mm, α = 10°; and D = 56 mm. The information provided in Yun's paper [43] for the design of this antenna is still very sparse; unfortunately, not all of the design parameters, such as the cone half angle at the apex or the length of the cone in this design, are available from the paper.

The gain, antenna patterns, PCV, and group delay variation of this antenna are from 1164 to 1300 MHz (Galileo E5 and E6 bands) and also between 1559 to 1610 MHz; the GPS L_1 and Galileo E1 and GLONASS (G1 or L1) bands have been measured. Figure 3.16 (c) shows the measured PCV versus Theta = 90°, elevation angle in degrees. The table in Figure 3.16 (d) shows a summary of performance achieved by this antenna. The PCV is <5 mm and the group delay variation (GDV) is less than 18 ns from these measurements. The peak gain is about 5.9 dBic with a gain drop of about 9.4 dB between zenith and a masking angle of 10° in elevation.

A conical spiral antenna similar to the antenna described above but designed primarily for the GPS L_1 and L_2 bands has been built and tested by Padros [82]. This antenna is also bifilar and the radius of the antenna wires is fixed. Principal and cross-polarized radiation patterns measured in the L_1 and L_2 bands are provided but other information of interest to the GNSS community, such as the measured gain, PCV, or group delay for this antenna has not been provided.

The antenna as designed by Yun [43] is strictly not frequency-independent since the radius of the bifilar wires is fixed instead of the radius changing with the azimuth angle as required by an equiangular conical spiral antenna first suggested by Dyson and Mayes [81]. An equiangular conical spiral for Galileo has been built by Granger et al. [38]; however, except for a picture of this antenna shown in Figure 4 of their paper, there is little information provided either on its design features or its performance. It has been pointed in their paper that a conical spiral with four arms fed in quadrature can provide an even better axial ratio over a much wider range of elevation angles than two-arm conical spirals but with greater complexity in the feeding arrangement [38]. Conical spirals, in general, are capable of covering very large bandwidths when the apex half-cone angle is about 15°. The gain can be increased by mounting the conical spiral above a ground plane and the front-to-back ratio can be improved further by placing absorber material at the base of the antenna.

3.7 Microstrip and Other Wideband Antennas Designs for GNSS

3.7.1 Wideband GNSS Microstrip Antenna Covering 1.164 to 1.610 GHz

A new, low-cost microstrip antenna suitable for high-precision GNSS applications and capable of operating over the entire band from 1.164 to 1.610 GHz has been designed by Popugaev et al. [44]. A picture of this antenna is shown in Figure 3.17(a). It consists of a broadband-matched, square microstrip patch antenna element with a four-probe feed, which is surrounded by 12 wedge-shaped parasitic elements arranged circumferentially around it. The patch antenna is a square patch antenna that is 50 mm wide and 30 mm high. It uses a dielectric substrate with very low relative dielectric constant that is almost comparable to air. Its large size provides a wider bandwidth. The four-probe feeds have a relative phasing of 0°, 90°, 180°, and 270° so as to generate RHCP. The patch element has an input impedance of approximately 200 ohms at 1.61 GHz. Hence, each probe feed needs a triangular shape transition connecting the 50-ohm feed line to the patch. The patch antenna is located at the center of a ground plane that is 150 mm in diameter. This antenna is capable of providing a gain of +3 dBic and a stable phase center over the GNSS frequency band from 1.16 to 1.64 GHz. The front-to-back ratio is better than 30 dB, thus ensuring good multipath rejection. The measured antenna patterns for RHCP and LHCP at 1.16 and 1.61 GHz. for this antenna at elevation angles of zenith and at an elevation of 15° are also shown in the lower right-hand side of Figure 3.17(a). In these measured patterns the LHCP level, also shown in this figure, is 15 dB lower than the RHCP level. The suitability of this antenna for high-precision geodetic measurements has been demonstrated through successful RTK applications.

(a)

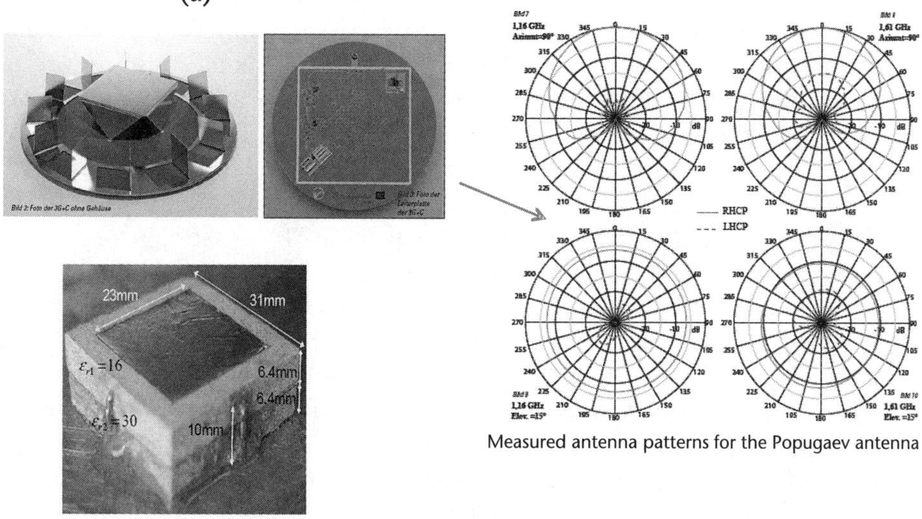

(b)

Figure 3.17 (a) Wideband GNSS microstrip antenna by Popugaev. (From [44]. Courtesy of Popugaev.) (b) Triple band GPS microstrip antenna by Zhou et al. (From [47]. ©2007 IEEE.)

Several types of microstrip antennas have been designed recently for reception of the L_1, L_2, and L_5 band signals of the modernized GPS system; a brief description of five of these designs is provided in this chapter. Two of these antenna designs, despite being designed for the L_5 band (whose wavelength is ~10 inches at the midband) have very small aperture sizes and low profiles that are just a fraction of the wavelength at L_5. A very compact, triple-band, proximity-fed, stacked microstrip patch antenna for the GPS L_1, L_2, and L_5 has also been designed by Zhou et al. [47]. The aperture size of this antenna, even smaller than the one described above and shown in Figure 4.17(b), is 1.2-inches square or $\frac{\lambda}{8} \times \frac{\lambda}{8}$ in terms of wavelength λ in the L_5 band.

The size reduction in this antenna is has also been achieved by using a two-layer stack of very high dielectric constant substrates, each of which is 31 mm square. The top substrate layer has a dielectric constant of 16 and has thickness of 6.4 mm. The bottom substrate has a dielectric constant of 30 and also has thickness of 6.4 mm. Both are made from HI K dielectrics made by Emerson & Cummings. To achieve resonance over the required frequency bands, the size of the top- and bottom-square-shaped conducting patches on each substrate layer were adjusted to be 23 and 20 mm, respectively. The stacked patches were fed by two L-shaped probes that are end fed by coaxial feed lines connected to a quadrature hybrid to achieve RHCP. This feed arrangement is similar to that described earlier in Chapter 2. The horizontal section of the L-shaped probe is placed close to the top patch, and the vertical and horizontal lengths are adjusted for an optimum match across all three frequency bands.

Zhou et al. [46] have also designed such a compact and low-profile, triple-band antenna using a two-layer dielectric substrate with a novel feed configuration consisting of four F-shaped probes. The antenna has an aperture size of just 1.5 inches by 1.5 inches, or only about $\frac{\lambda}{7} \times \frac{\lambda}{7}$, and has a thickness of just $\frac{\lambda}{13}$ in terms of wavelength λ in the L_5 band. The small size is achieved by using 38-mm, square-shaped dielectric substrates that have very high dielectric constants. The top substrate has a thickness of 11 mm and uses a ceramic substrate with a dielectric constant of 25. The bottom substrate is 9 mm thick and has a dielectric constant of 10.2. The antenna is fed by four F-shaped conductor sections placed at the center of each of the four sides of the square-shaped, two-layer substrates. Each F-shaped section is connected to a 50-ohm coaxial cable and fed in quadrature (i.e., with equal amplitude signals but with a sequential 90° phase shift in the counterclockwise direction to generate RHCP radiation). The inner and outer widths of each tapered F section and its length are adjusted to achieve resonance control across all three frequency bands. This antenna has a boresight gain of about 2 dBic in the L_5 and L_2 bands and about 0 dBic in the L_1 band.

Because of their low profile and small size, both of the antenna designs described above could be used as a triple-band FRPA for airborne and other modernized GPS applications as well as in multielement arrays used for multipath suppression, beam forming, and so forth.

Two other types of triple-band stacked microstrip antennas covering the GPS L_1, L_2, and L_5 bands have also been designed. The design used by Doust [48] consists of two stacked patches with the larger patch at the top to increase the bandwidth. The two patches are excited by a crossed slot with tuning stubs and fed by an aperture-coupled feed network (shown in Figure 2.22). This type of feeding arrangement results in the ends of the slots being fed with equal amplitudes but with phases of 0°, 90°, 180°, and 270°, respectively, to produce RHCP over a wide band from the two stacked patch antennas. The larger of the stacked patches was placed on the top to get the higher bandwidth offered by the thicker dielectric substrate. The antenna design was optimized by using the Ansoft HFSS code. Both of the substrates were made from Roger's RT Duroid 6002 with a dielectric constant of 2.94. The thickness of each substrate was 3.05 mm. The width of the top and bottom square-shape patch was 61 and 43 mm, respectively. The dielectric substrate used for the feed lines for the crossed slots was Roger's TMM10i with a dielectric constant of 9.8 and a thickness of 1.9 mm. The crossed slots were 52 mm long and 6 mm wide with tuning stubs that were 41 mm long. This antenna has a peak boresight RHCP gain that varies between +2.5 to –1.5 dBic.

The triple band for modernized GPS has also designed by Targonski et al. [49]. This is also a two dielectric layer stacked microstrip but with a symmetrical four-port feed network with a probe coupled to the lower patch. This type of four-port feed configuration is discussed at length in Chapter 2. A crossed slot is cut into the lower patch to facilitate stronger coupling to the top patch. The patch antenna is about 3.5 inches square and uses a 0.32-inch thick Roger's Duroid 4003 substrate material. Three different prototype designs have been considered. The height of the microstrip antenna above the ground plane in these three designs is 1.7 inches, 1.2 inches, and 0.65 inches, respectively. The tallest design with the 1.7-inch height provides the widest bandwidth that virtually covers the entire GNSS from 1.16 to 1.6 GHz when measured in terms of a –10-dB return loss. The lowest height of 0.65 inches provides adequate triple-band coverage with 24-MHz bandwidth in each band.

The fifth is a low-cost, low-profile, triple-band antenna also for GPS L_1, L_2, and L_5 bands by Rama Rao et al. [50]. This design does not need any dielectric substrates or extensive machining or circuit etching in its construction thus reducing the cost, especially when compared to designs where machining of ceramic substrates are needed. This design consists of a RHCP, four-element, trap-loaded inverted-L antenna array. The four L-shaped elements are arranged at 90° intervals around the sides of a square-shaped foam material. The foam layer provides mechanical support for the elements of the array. An RF trap filter designed for the 1.5754-GHz band is inserted at an appropriate distance from the feed end into the horizontal portion of each of the inverted L-shaped elements of the array. This parallel resonance trap filter consists of a lumped element 2.2-picofarad capacitor in parallel with a 2.8 nanohenry, high-Q inductor. This trap-load filter presents high impedance in the L_1 band causing the section of the antenna between the feed and the trap filter to resonate in the L_1 band. The currents at the other two frequency bands proceed beyond the trapped filter to the end of each L-shaped element and cause each element to resonate in the other two bands, L_2 and L_5. The Q of its resonance is low enough to accommodate both bands. The four elements of the array were excited with equal amplitudes but with relative phase differences of 0°, –90°, –180°, and –270°. The feed network consisted of a low-cost and compact microstrip feed network consisting of 180° "rat race" hybrid and two compact, surface-mounted 90° hybrids. The return loss, impedance characteristics, and measured antenna patterns in the three GPS bands are shown in [50]. The antenna provides 24-MHz bandwidth in each of the three GPS bands; it has a relatively low profile of 0.87 inches. However, since no dielectric substrates are used in its construction, its lateral dimensions as designed are 4.77 inches square. These dimensions can be reduced further with a more simulations. The antenna was designed using both Ansoft HFSS and the NEC (MOM code.

A compact, triple-band, ceramic-loaded, stacked QHA has been proposed and built by Liu and Chu [86]. This consists of a dielectrically-loaded quadrifilar helix antenna (DQHA) mounted "piggyback" on top of a dielectrically-loaded octafilar helix antenna (DOHA). This antenna is discussed in Section 3.11.

3.8 GNSS Antennas for Handset

3.8.1 Techniques for Locating Wireless Devices

GPS receiver chip-sets and antennas, now an integral part of mobile phones, smart phones, personal digital assistants (PDAs), personal navigation equipment, and several other similar devices; this has generated a vast market for new applications. These include location-based services (LBS) and navigational services for providing position-related information to customers, remote tracking, fleet and traffic management, and so forth [51, pp. 371–372]. The demand for such services is expected to grow to 500 million or more by 2011. The installation of GPS in mobile phones has also become mandatory as a result of the Wireless E 911 Automatic Location Identification (ALI) directive (94-102), Phase 2, issued by Federal Communications Commission (FCC) [87]. This FCC directive sets a stringent requirement for the location of all mobile 911 caller locations; the required accuracy is within 50 mm for 67% of the calls and within 150 mm for 95% of the calls for wireless handset-based solutions. A similar type of emergency and rescue regulation for cell phones in Europe, E112, has also been issued by EU telecomm; this system will use navigational signals provided by Galileo [88]. The regulation for E112 is less stringent than for assisted GPS (A-GPS); it imposes only a "best effort" requirement to the European cell phone operators [88].

There are currently three techniques available for the location of wireless devices: network based positioning, stand-alone GPS, and A-GPS. While network based positioning can determine the location faster than the other two GPS techniques it does not provide the accuracy needed for meeting FCC E911 requirements. Stand-alone GPS has problems receiving weak GPS signals, especially when operating indoors or in a dense urban environment (urban canyons) where signal levels can drop down to –150 or even –160 dBm compared to levels of between –123 to –130 dBm for typical unobstructed outdoor environments [89, Table 6.14; 90, Table 1]. A conventional receiver can nominally only acquire signals above –130 dBm [90, p. 141]; hence, signals at a lower level can prevent a GPS receiver from being fully functional or require more processing gain (longer integration) for successful acquisition. As a result many wireless carriers now look to A-GPS, which combines the benefits of the location accuracy afforded by GPS with the reliability provided by wireless network-based positioning for meeting the FCC location criteria. A-GPS is able to achieve two goals:

1. It allows location of an emergency caller accurately even when the received signals are very weak with expected attenuations as high as 20 to 30 dB, as mentioned earlier;
2. A-GPS also decreases time-to-first-fix (TTFF) to allow faster rescue in an emergency situation.

The GPS receiver may take up to several minutes from a cold start to download the satellite almanac and a further 30 seconds per satellite for calculating each satellites ephemeris. This is not acceptable when an emergency call is being made. A much faster acquisition of GPS signals is needed for summoning immediate help. A-GPS is able to accomplish this by providing the GPS receiver of the 911 caller with assistance data needed to speed up the acquisition process and make the assisted cold start transform into almost a "hot start." It also allows the receiver to use less processing power saving on battery life. Hence, A-GPS is able to achieve its two difficult goals of being able to receive GPS signals virtually anywhere cellular phone service is available and also very quickly with the required accuracy and sensitivity that is needed.

3.8.2 Fundamentals of A-GPS

A brief description of A-GPS is given here since it has assumed such an important role in the design of antennas needed for mobile phones; more detailed descriptions can be found in two recent books [89, 90] and in several journal articles [88, 91]. Figure 3.18 shows a schematic outline of an A-GPS system.

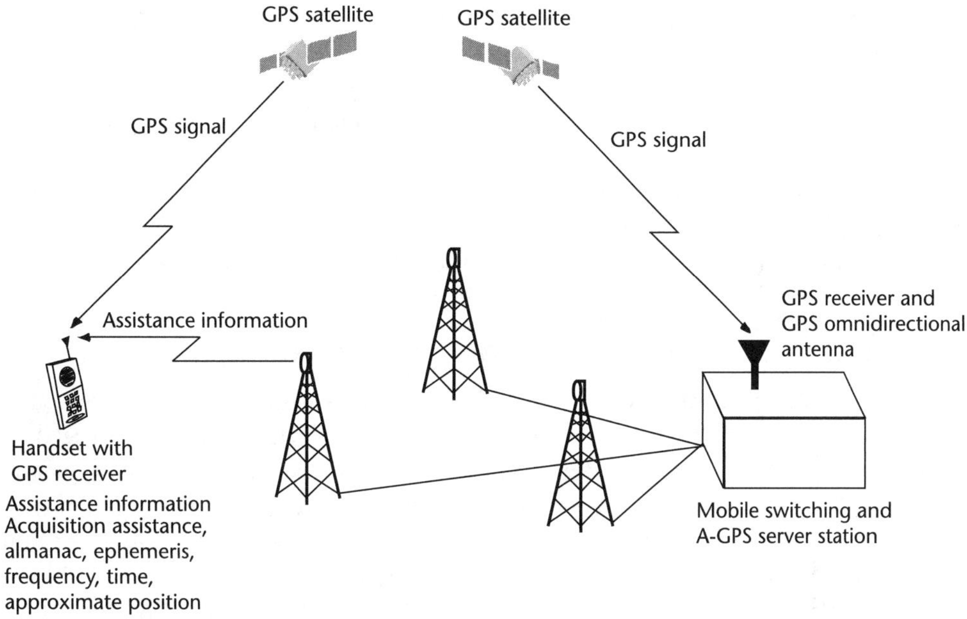

Figure 3.18 Concept of the A-GPS system.

An A-GPS system consists of the following three components:

- A wireless device (mobile phone) that contains the partial GPS receiver making the emergency call. It is called a partial receiver since it is handicapped by weak signal reception from a nonoptimal environment.
- An A-GPS server associated with the network base station with a fully functioning "reference" GPS receiver that can receive the same satellite signals due to its proximity with the partial GPS receiver in the mobile wireless device that is making the emergency call. The reference GPS receiver is on all the time and collecting and processing data from all the visible satellites and computing its own navigational solutions. Signals from at least three satellites are required as a minimum for determining the precise location.
- A wireless network infrastructure to allow sharing of information between the 911 caller (or UR) and the base station of the wireless network consisting of a RR and the A-GPS server.

The network through the A-GPS server sends the mobile wireless device 911 caller with key information needed for acquisition and tracking of the GPS signals. This allows the partial receiver to avoid spending time to extract all the required data from the satellite signals since most of it is supplied by the network, thus decreasing TTFF. This information provided by the network could include reference time, reference frequency, satellite Doppler and Doppler rate, almanac, and/or ephemeris. This information enables the partial receiver to enhance its sensitivity and reduces the search space needed for the acquisition process. Technical details on A-GPS, including two types of A-GPS called MS-assisted and MS-based GPS, can be obtained from several books and reference articles [88–91].

3.9 Design of Handset Antennas: Requirements and Technical Challenges

Some of the key requirements and technical challenges in designing a suitable GPS antenna for a modern mobile handset are listed below. These requirements can often conflict with one another because of the opposing limits imposed by size, performance, cost, and cosmetics. This makes the design antennas of modern mobile handsets an extremely complex and technical challenging problem [92, 93].

3.9.1 Size, Appearance, Cost, Weight, and Functionality

The size of mobile telephones has been rapidly shrinking due to increased emphasis on extreme compactness and aesthetic considerations that are being driven by a fiercely competitive market. Typical cell phones used these days typically are 40 to 50 mm wide and 100 to110 mm long. In addition they have become ultrathin, with some being just 0.5 inches thick. The average volume of handset phones is estimated to have shrunk from an average volume of 131 cubic cm in 2004 to just 55 cubic cm in some current designs [94]. These dimensions are for the entire handset, which includes not only just the GPS antenna but other antennas and components needed to serve several functions expected from mobile phones. These can include a liquid crystal display (LCD) screen, the battery, keyboard, speaker, camera, and so forth. It is common to have two to four antennas in a phone. Table 3.6 lists the most commonly used frequency bands that current mobile phones may need to carry [95].

For example, a GSM-enabled cell phone these days in the mid- to high-price range may support five different cellular bands—(GSM 2G or 3G), WLAN, Bluetooth, digital TV (B-H)—in addition to GPS that is mandatory for meeting the FCC E 911 requirement. In a recent paper, Vainikainen [96] has identified a total of 20 antenna elements that may be needed in future mobile phones if this trend continues. Cost and size limitations require a single antenna to cover multiple frequency

Table 3.6 Frequency Bands of a Selected Group of Commercial Mobil Communications and WLAN Systems

System	*Frequency Range (MHz)*	*Power Level*
(2G) GSM	880–960; 1710–1880; 1850–1910	33 dBm
DCS 1800/DCS 1900	1710–1880	30 dBm
(2G) PCS	810–956;1429–1501;1850–1990	30 dBm
3G GSM network (WCDMA)	1920–2170	
UMTS 3G	1920–2175	
4G (multicarrier CDMA; TDMA)	2000–8000	
US-DVB-H	1670–1675	
Bluetooth	1710–2170	4 dBm
WLAN 802.11b/g	2400–2483	20 dBm

From: [95].

GSM = Global System for Mobile Communications; DCS = digital communications system; PCS = personal communication system; WCDMA = wideband code division multiple access; UMTS = Universal Mobile Telecommunications System; CDMA = code division multiple access; TDMA = time division multiple access; US-DVB-H = U.S. Digital Video Broadcasting—Handheld; WLAN = wireless local area network.

bands needed for different functions; antennas that cover four to five different cellular and WLAN frequency bands plus GPS are becoming quite common in cell phones.

For aesthetic considerations externally mounted antennas, such as retractable whip antennas and stubby helices that were previously used for cell phones, have almost all been replaced by compact antennas with complex mechanical shapes that are embedded directly inside the handset and are invisible to the user. The superior performance and intrinsic isolation provided by some of these externally mounted antennas has been sacrificed in the interest of cosmetics, smaller size, and reduced chances of breakage provided by internally mounted antennas that vastly increases the technical challenges involved in their design. Considerable size reduction has, however, been achieved by using a variety of design techniques. Mobile phones generally use several multilayered printed circuit boards (PCBs), some of which have metal coatings; these metalized PCBs can act as a ground plane for a wide variety of compact, internally mounted printed circuit antennas that include microstrip patch antennas, small printed monopoles on dielectric chips, isolated magnetic dipoles, planar inverted F antennas (PIFAs) and its many design variants such as the folded inverted conformal antenna (FICA), and balanced dipoles [97–101] are some antenna designs that are currently popular. Fractal geometries have also been applied for expanding bandwidth [104, 105]. A very good description of a large variety of small printed circuit antennas especially suitable for internal mounting in current day wireless systems has recently been described by Waterhouse [99].

Unfortunately, antennas, unlike other electronic components, do not follow Moore's law for size reduction; instead the size and performance are limited more by the wavelength of the signal.

One of the miniaturization techniques used for reducing the size include loading the antenna with high permittivity (and permeability materials) as substrates such as ceramics. The use of a high dielectric constant material as a substrate can reduces the size of the antenna roughly by a factor of $\frac{1}{\sqrt{\varepsilon_r}}$ where ε_r is the relative dielectric constant of the material although other geometrical factors such as thickness, width, and location are also determining factors. The use of ceramics can also increase both weight and cost. More recently low-temperature cofired ceramic (LTCC) substrates, with dielectric constants ranging from 4 to 10, have been widely used for building compact GPS antennas in cell phones; these substrates allow printed circuit antennas to be deposited directly on their surfaces [102,103,106–108]. The addition of capacitive and inductive loads, meandering and folding of printed circuit antennas, and the use of shorting pins and slots are also techniques being used to reduce antenna size by making the antenna resonate at a lower frequency [97, 99]. The design of these antennas and others will be described in greater detail later.

These miniaturized antennas are generally placed all along the periphery of the cell phone, generally either at the top or along the sides or bottom. GPS antennas are normally at the top and antennas used for cellular frequency bands are at the bottom or along the sides where they may be most impacted by the presence of the hand of the user. Modern cell phones also come in various form factors such as clamshell, bar, swivel, slider, and flip shape [92]. The selected location for the various antennas depends on the form factor of the phone. These antennas are also very light and some weigh only a few milligrams. The price is also steadily decreasing even as functionality is being increased with new systems being added each year to keep up with market demands.

3.9.2 Influence of Receiver Chassis on Bandwidth of Cell Phone Antennas

Decreasing the size and height of the antenna can also degrade antenna bandwidth and its radiation efficiency. According to the Chu-Harrington limit, the impedance bandwidth is inversely proportional to the minimum quality factor $Q_{\min}$, which is given by [100]:

$$Q_{\min} = \frac{1+3(k_0R)^2}{(k_0r)^3\left[1+(k_0R)^2\right]}$$

R is the radius of the minimum sphere that completely encloses the antenna. The impedance bandwidth, which is defined by BW, is related to the actual Q, which is generally much higher than $Q_{\min}$ and is given by

$$BW = \frac{2(f_u - f_L)}{f_u + f_L} \times 100\% = \frac{VSWR - 1}{Q\sqrt{VSWR - 1}}$$

f_U and f_L are the upper and lower edge frequencies and VSWR is the voltage standing wave ratio. For a well-matched impedance bandwidth over the required operating frequency band VSWR = 2, which corresponds to a return loss of 9.8 dB. A return loss of 6 dB is often considered adequate for an antenna in a mobile phone due to the small size limitations and other stringent constraints placed on it. An internally mounted handset antenna couples to all components located in its immediate vicinity. Hence an accurate determination of the actual bandwidth of such an antenna can only be estimated through an analysis performed on a 3-D model that takes into account the mutual coupling between the antenna and the chassis, the metalized PCB underneath the antenna, as well as all other surrounding electronic components. The effects of the interaction between an internal antenna and its surroundings have been studied by a large number of authors [102–106].

The chassis now becomes part of the radiating structure of the antenna and can support a dipole-type mode. The analysis using such a model shows that by maximizing the coupling the between the antenna and its chassis and surroundings the bandwidth can be made nearly five times larger than the minimum bandwidth indicated by the above equations. For example, Lindbergh [92] has indicated that while the bandwidth of an isolated microstrip patch antenna on a large ("infinite") ground plane is typically only 1% to 3%, but its bandwidth can increase to 10% when placed inside a standard cell phone chassis due to coupling between the patch antenna and its surrounding environment. A PIFA antennas in sleek cell phones need to be placed at very low heights above a ground plane (or equivalent PCB) that is needed, which can restrict their bandwidth when used in cell phones.

3.9.3 Polarization and Radiation Pattern Requirements of GPS Cell Phone Antennas

While RHCP is the optimum polarization for a GPS antenna allowing maximum reception of signals from the satellites in a clear environment, it may not be the best polarization for an antenna inside a cell phone due to the different environment in which it operates. Most cell phones are used in urban settings that are dominated by heavy multipath caused by reflections from tall surrounding buildings and structures. The polarization of the GPS signal changes to LHCP after suffering the first such reflection. While a RHCP antenna with a low axial ratio provides good rejection against such multipath in a clear environment, the polarization of signals received in a dense urban environment is often poorly defined. This is from reflections and obstruction suffered by the direct signals from satellites. In such an urban environment, the beamwidth of a cell phone antenna and its pointing direction often are more important than its pristine state of RHCP polarization. A linearly polarized antenna with a nearly omnidirectional (figure-eight) pattern is able to receive both RHCP and LHCP signals with an associated 3 dB of loss for each CP state. This can improve satellite availability since its radiation pattern can uniformly cover the entire azimuth plane down to very low elevation angles, but with slight degradation in C/N_0 resulting from the 3-dB reflection loss. In contrast, an RHCP antenna with a good axial ratio can, however, can cause considerable attenuation of an LHCP signal, which can reduce satellite availability in urban environments. These conclusions have been validated by Pathak et al. [107]. The results of his study have shown that in a multipath dominated environment, a linearly polarized GPS antenna often performs better if not the same as an RHCP antenna. He made comparison tests between a linearly polarized antenna and an RHCP antenna by measuring their CNRs and PDOP. In one example, an RHCP antenna that was tested acquired five satellites with an average C/N_0 of 45 dB; the linearly polarized antenna, on the other hand, was able to acquire six satellites with an average C/N_0 of 41 dB. These results indicate that the linear antenna in this case provided better satellite coverage than an RHCP antenna. The PDOP for both antennas that were tested was between 0.8 and 2.5 most of the time. As a result of these findings most cell phone antenna manufacturers have now opted to use linear polarization for their GPS antennas since it reduces complexity of the feed network and reduces both cost and size and allows the use of highly compact GPS antenna designs.

Due to the many different ways in which the handset can be gripped by the user there is no specific "up" or a "slant down" direction for a handset antenna. This requires that the GPS antenna be designed to have an omnidirectional antenna pattern to allow acquisition of satellites when the phone is held at any arbitrary orientation. A linearly polarized antenna, as mentioned earlier, can provide such a broad antenna pattern but generally with a shallow null at zenith. It is thus able to acquire satellites appearing at low elevation angles and also obtain early signal acquisition of rising satellites.

3.9.4 Isolation Techniques to Mitigate EMI/EMC Effects on GPS Antennas in Cell Phones

As discussed in Chapter 1, the GPS receiver is highly susceptible to out-of-band interference signals that are coupled into the antenna from adjacent transmitters inside the cell phone. Table 3.6 shows the frequencies of the various other L band systems in a cell phone, which include cellular, Bluetooth, and WLAN whose frequencies vary from 900 MHz to 2 GHz—all of which are close to the GPS band. They transmit at power levels range from 33 dBm for 2G GSM to 4 dBm for Bluetooth as indicated in this table. Hence mutual coupling or "cross-talk" between the GPS antenna and antennas used by the other wireless systems in a handset can raise a number of potential issues. These include the following:

- Transmit power signal from an adjacent antenna for one of these wireless systems could saturate the GPS receiver and generate what is known as "blocking";
- High output noise levels from the transmitters could elevate the noise threshold of the GPS receiver, degrading C/N_0;
- Reduce antenna efficiency from direct leakage of received power into an adjacent system.

High surface currents can be excited by an internally mounted antenna on the surface of a metalized PCB ground plane that is also likely to be shared by several other antennas. Adequate isolation between an antenna and other nearby conducting elements and antennas becomes necessary to allow their coexistence within a small space. GPS antennas are particularly susceptible to such interference. Provisions need to be taken in installation and in testing to ensure that there is adequate isolation and selectivity to avoid interference between various L band systems to mitigate electromagnetic interference (EMI) and electromagnetic compatibility (EMC). Coupling mechanisms between antennas could occur either through conduction or radiation. One prime conductive coupling mechanism are currents set up in the PCB ground planes [97,108]. Radiation coupling can occur through interaction between the near fields of closely spaced antennas. A good antenna design should include effective measures to suppress or minimize such conductive and radiation coupling, and therefore antenna designs that minimize dependence on the ground plane are important for cell phones. Suppression of currents on the PCB ground plane is also important for minimizing interaction between the user and the antenna. PIFAs [97, 130], shown in Figure 3.19, are popular for mobile phones but they can generate strong currents over an extended surface of their ground plane, creating a potential for interference with adjacent antennas and the user; there will be more discussion about the PIFA design later. In an effort to improve the bandwidth and to minimize cross-talk coupling caused by currents generated on the ground plane, some antenna manufacturers have chosen to provide "ground clearance" by removing a portion of the metal layer on the PCB boards and chassis directly below the antenna that serves as a ground plane [97].

Some antenna manufacturers also specify "keep out areas" around the antenna to minimize its interaction with its surrounding environment. These restrictions can affect the true volume requirements of the antenna in a space-restricted

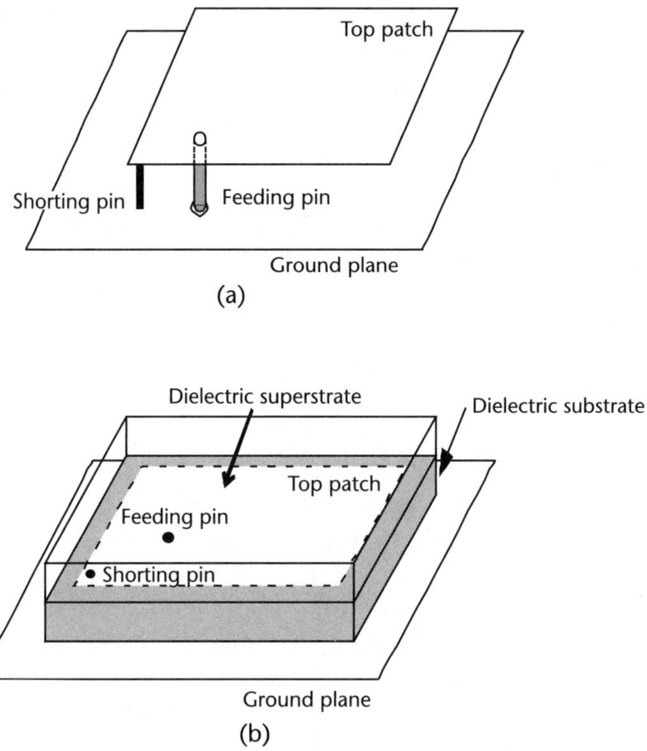

Figure 3.19 PIFA (a) without dielectric loading, and (b) with dielectric loading using substrate and superstrate.

environment and these designs reduce the susceptibility to detuning from the hand and the body of the user. Removing a portion of the ground plane also improves the bandwidth by increasing the effective height of the antenna above the ground plane. If a significant amount of ground plane is removed under a PIFA antenna it begins to resemble a folded monopole or an inverted L antenna (ILA); when a shunt stub is attached for matching purposes it is called an inverted F antenna (IFA). Diagrams illustrating ILA and IFA antennas are shown in Figure 3.20.

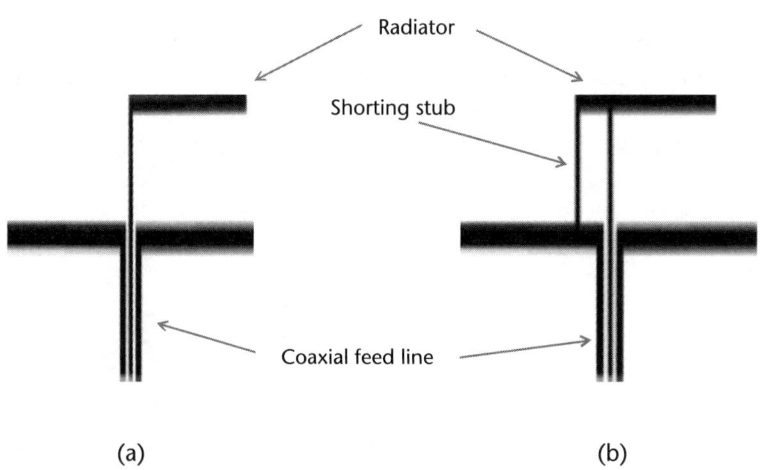

Figure 3.20 (a) Inverted L, and (b) inverted F antennas used in handsets.

Some antenna designs use parasitic coupling structures to modify the antenna radiation to minimize interaction with the user. Ceramic chip antennas, most commonly used for GPS antennas in handsets, help in alleviating this coupling to adjacent antennas as well as coupling to the hand and body of the user by confining and intensifying the near zone fields to a small region in the immediate vicinity of the antenna. This can reduce the bandwidth that is acceptable for GPS antennas operating in the C/A code of the L_1 band; unfortunately, the penalty is an increase in weight and cost. Some antenna designs that have been developed specifically to improve isolation; the design of four recent GPS handset antennas that achieve this will be described in greater detail later in this chapter. Ethertronics has recently developed a patented GPS antenna design called an Isolated Magnetic Dipole (IMD) [133–136], a folded G-shaped slot antenna above a ground plane that reduces the currents generated on the ground plane that is placed below the antenna. A diagram of this antenna is shown in Figure 3.21.

The IMD antenna greatly reduces coupling between the antenna and its adjacent components, thereby reducing interference. Balanced antennas fed from baluns, such as the balanced planar dipole design being used by Antenova [137] shown in Figure 3.22, also greatly improve isolation by reducing such coupling. Ceramic cores have been used by Sarantel [138, 140, 141] for building a compact GPS quadrifilar helix antennas shown in Figure 3.23. The ceramic core confines the near field to a narrow region around the antenna reducing susceptibility to detuning from the hand and body of the user. Although this antenna is generally mounted externally to the chassis of the cell phone but some new designs of this antenna also allow internal mounting.

Selectivity is defined as the out-of-band rejection capability of the antenna. From this perspective a narrowband antenna can be a blessing in terms of minimizing interantenna interference. Bandpass filters can be inserted to improve selectivity but their insertion loss can degrade the noise figure of the antenna; minimizing the bandwidth through dielectric loading also improves selectivity.

3.9.5 User Effects on Handset Antennas

Loading effects on handset antennas can be caused by the users hand as well as the body and head, which can all be considered as dielectrics with a high permittivity as well as high loss because of their water content. Recent studies have shown that in printed L band antennas typically used in handsets approximately 50% of the energy radiated by an antenna is absorbed by the user [110]. In addition, the presence of a human hand in close proximity to a handset antenna operating at 1800

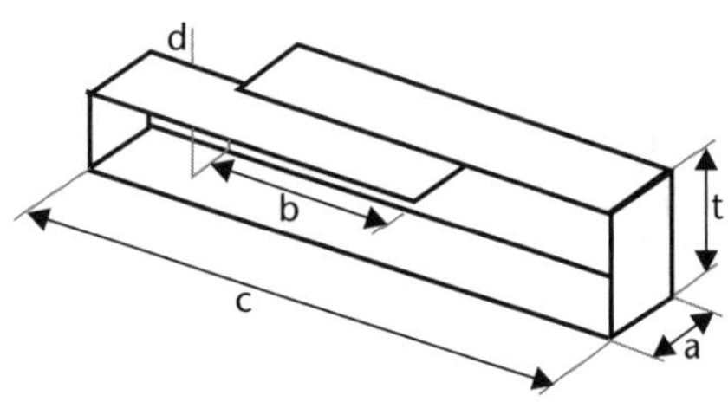

Figure 3.21 GPS Isolated Magnetic Dipole by Ethertronics. (Outline of antenna derived from [136]).

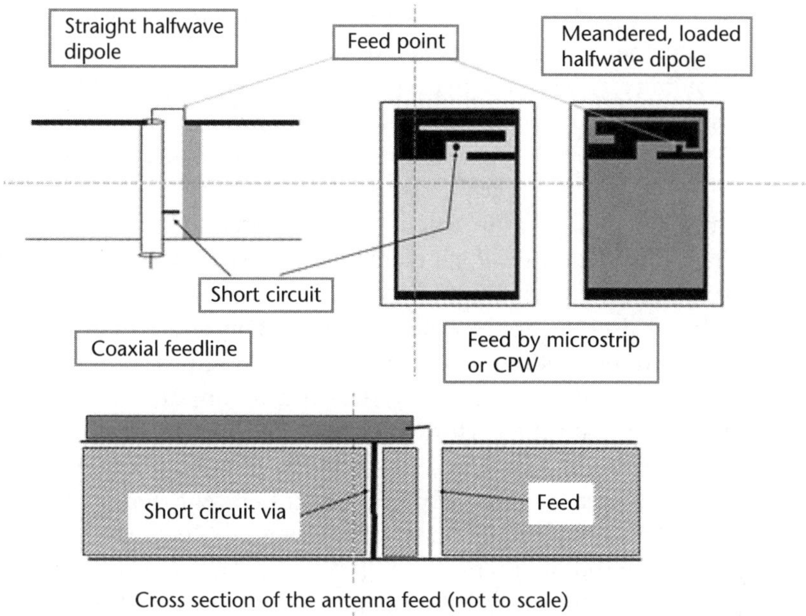

Figure 3.22 Internally mounted balanced dipole handset antennas by Antenova Ltd. for GPS and Galileo receiver. (Coutrtesy of Prof. S. P. Kingsley.)

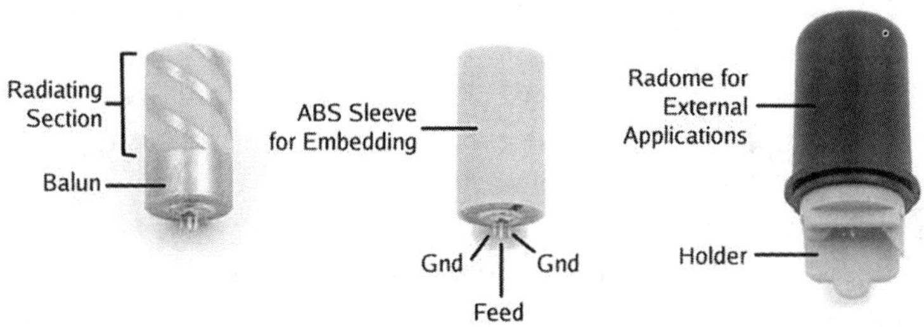

Figure 3.23 Sarantel's Geo-Helix for GPS handset. (Courtesy of Dr. O. Leisten of Sarantel Inc.)

MHz can reduce its efficiency by 10% [110]. The same magnitude of absorption and loss in efficiency can also be expected to occur at GPS frequencies because it is close enough to 1800 MHz. User interaction can change the antenna impedance and detune the matching network; in addition to degrading antenna efficiency due to absorption losses it can cause blockage and shadowing and alter the radiation pattern, affecting satellite availability. The bandwidth of a small GPS antenna operating in the C/A code of the L_1 band is very narrow because of its small size—often just 5 MHz wide. Hence, detuning caused by user interactions unless properly accounted for in the design can pull antenna right out of its predesigned tuning band. Several recent studies, using either suitable electromagnetic computer models or phantom models, have been conducted to evaluate the effects of the human hand and body on the performance of handset antennas that operate at L band cellular frequencies [111–114]. Most of the studies on user effects on handset antennas have been conducted in the cellular communications frequency bands ranging from 800 to 2100 MHz; however, as mentioned earlier, their proximity to the GPS C/A code frequency makes these study results relevant for evaluating similar effects on GPS antenna performance. The hand, especially the thumb and fingers, because of their proximity to the antenna have the most effect [112–114]. The effects were demonstrated in a most dramatic manner by Steve Jobs during his frustrated efforts to get the new iPhone 4 to work during a marketing pitch to customers of Apple in June 2010. His incorrect grip of the iPhone has been diagnosed as the cause of this problem; the remedy was instructions issued later by Apple on the correct ways to hold the phone. The use of supplementary plastic and rubber bumpers also mitigated these effects.

Table 3.7 lists the approximate dielectric constant and conductivity of the human hand and head as well as material used in the construction of the phone. These values are provided by Futter et al. [106] and are important in the understanding of user interaction with the performance of the cell phone antenna.

The body also acts as a large absorbing medium and can degrade efficiency but has a lesser impact on antenna performance—except that it can prevent reception of some visible GPS satellites due to shadowing or blockage by the body of the user. The results of hand interaction with four different types of PIFA antennas, including dual-band, triband, and multiband PIFAs, covering frequency ranges 880 to 960 MHz and 1710 to 1990 MHz have been studied by Huang and Boyle [112]. Although GPS band was specifically not covered as part of this study, the results are still relevant because of its proximity to these two frequency bands. They examined user interaction when the antennas were held in six different handgrip positions by 63 volunteers. The antennas were placed in the top part of the cell phone at its outer periphery.

This is generally the position selected for installation of GPS antennas in cell phones. This study reveals the extent to which the input impedance of the antenna can change relative to hand positions and the extent to which antenna designs can influence such changes. The results of these studies conducted by Huang and Boyle [112] show three distinctive categories of user interaction. The majority of users held the phone with the thumb and finger on the side faces; the resulting interaction was minor causing a minor-to-medium impedance mismatch. However, a severe interaction resulted when a user placed either the hand or finger directly on the antenna plate, requiring a corrective measure to get the antenna rematched.

Table 3.7 Dielectric Constant of Human Hand and Head and Other Phone Materials

Human Body Components	*Relative Dielectric Constant*	*Conductivity; mho*
Muscle (hand)	53.44	1.385
Skin (hand)	38.75	1.22
Bones (hand)	11.73	0.289
Head (skull of phantom model)	3.7	—
Head (interior of phantom head)	39	1.42
Phone materials	*Relative Dielectric Constant*	*Conductivity; mho*
PCB material	4.5	0.07
Housing parts	3.5	0.025
Display glass	3.0	0.01
Antenna dielectric	~2.5	0.0045

From: Futter, et al. [106]

In the low-frequency range (around 880 to 960 MHz) the user interaction resulted in the antenna becoming inductively loaded; at higher frequencies (1710 to 1990 MHz) somewhat closer to the GPS band, there were different effects depending on antenna design.

Pelosi and his colleagues have conducted a large number of studies [113, 114] to evaluate user influence on performance of handset antennas in the UMTS frequency bands I, II, and V covering a frequency range from 824 to 2170 MHz overlapping the GNSS band, which extends from 1.1 to 1.61 GHz. Several PIFAs at four different location on the cell phone—at the top left and right, bottom left and right, and also at the very top and bottom of the cell phone were studied. Detailed CAD models of hand phantoms including dimensions based on a previous anthropometric study were generated. The dielectric properties were chosen to represent an average of several tissues. Electromagnetic analysis to evaluate interaction of the hand with the antenna was conducted using a finite difference time domain (FTTD) code. A -6-dB return loss was used to define an impedance match. Six different hand grip positions were investigated consisting of different palm-handset gaps and index finger locations. These are illustrated in Figure 3.24(a).

Figure 3.24(b) shows the location of the phones in the antenna. Handgrip positions 1 to 3 were considered "firm" and 4 to 6 were considered "soft." The location of the index finger was varied relative to its proximity to the PIFA antenna. The absorption loss, mismatch loss, and minimum isolation for different receive and transmit antenna pair configurations for UMTS band transmit frequency between 1920 and 1980 MHz and receive frequency between 2110 and 2170 MHz is presented in these papers. The amount of presented data is too large and complex to summarize concisely for this book. Interested readers should refer to the original published paper for details. The result of this study has again confirmed that the way a mobile phone is held is important in determining the absorption loss, detuning, and the antenna isolation.

Narrowband antennas are less subject to mutual coupling and provide better isolation in a cell phone environment. Although the frequencies used for the hand interaction studies in these two previous studies are either lower or higher than for GPS, they are still in the L-band frequency range. Hence, these results of these previous investigations provide a clue of likely effects of hand positions on the

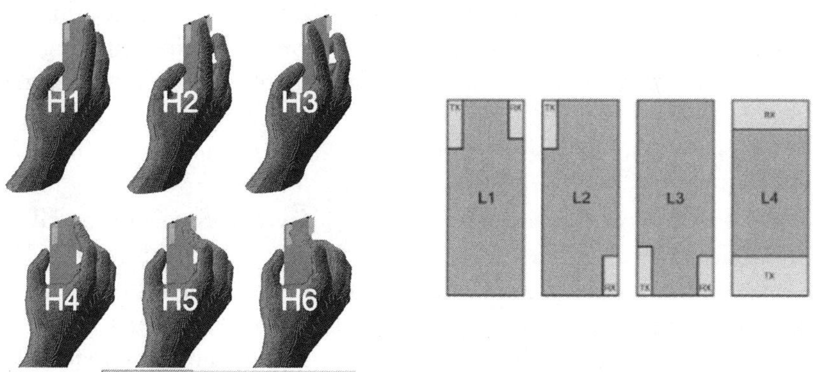

(a) (b)

Figures 3.24 (a) Six selected hand positions and (b) location of antennas in cell phone for evaluation of cell phone performance. (Figures courtesy of Prof. M. Pelosi [113].)

performance of GPS antennas; since the GPS frequencies are lower than UMTS the hand effects are likely to be even greater. Methods used for minimizing effects of the user's hand on antenna performance is the same as those for obtaining better isolation between an internal antenna and its surrounding environment that were described in the earlier section. A unique solution suggested recently is to enclose the handset in a flexible metallo-dielectric electromagnetic band gap (MEBG) surface suggested by Gousettis [115].

3.10 Representative GPS Antenna Designs Used in Handsets

As mentioned earlier, antennas covering many different frequency bands are needed to serve the multiple functions demanded from cell phone antennas. Several recent books have described the design of antennas used in handsets for wireless communications, Bluetooth, and WLAN frequency bands [97–101 MHz]; these include quad-band and penta-band antennas covering frequency bands from 800 to 2100 MHz. The discussion in this book will be confined solely to a selected group of GPS handset antenna designs from the many that are currently available. Some of these commercially available antennas developed for handsets are listed in Table 3.8, which also provides information on their size, gain, efficiency, bandwidth, and weight (if available). These antennas were selected for the compactness of their size and for the novelty of their designs. Some other handset antenna designs obtained from various journal publications are also described below including antennas using LTCC, fractals, and so forth; these journal articles provide a more detailed description of these antenna designs that are often not available for commercially sold antennas because of proprietary restrictions. A few dual-band antenna designs that combine GPS and Bluetooth that are being sold commercially will also be covered in this list.

3.10.1 Dielectric Chip GPS Handset Antennas Using Ceramic and LTCC Substrates

Dielectric chip antennas are some of the smallest and most frequently used GPS antennas in handsets. They generally consist of a ceramic base with a dielectric constant > than 5 to allow a compact size (the "chip"); on this chip are printed metallic circuits of different shapes that constitute the radiating structure. These are generally printed on the top surface of the chip but sometimes on both the top and bottom surfaces which are then connected by vias. The antenna element is often encapsulated inside the body of the ceramic chip. Some examples of these printed circuits on the top surface of the chip are shown in Figure 3.25(c) and (d). They are designed as surface mounting devices (SMDs) that allows them to be monolithically integrated directly into the printed circuit boards. There are a large number antenna designs in this category that are being sold commercially, including end-fed, ceramic-loaded microstrip patch antennas discussed in Chapter 2, small multiband printed monopoles of various shapes, dielectric loaded ILAs, IFAs, and PIFAs. Sketches of some of these are shown in Figures 3.19 and 3.20.

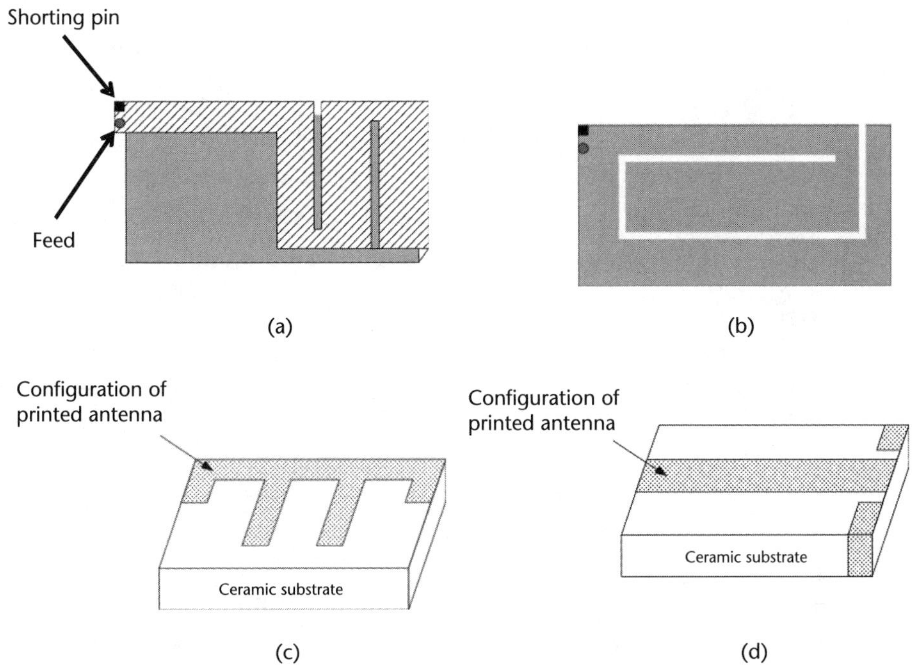

Figure 3.25 Design variants of the top plate of a PIFA for multiband operation: (a) meander line top plate, (b) top plate with bent slot, (c) dielectric chip antenna configuration 1, and (d) dielectric chip antenna configuration 2.

One technique currently being used to reduce the size of GPS dielectric chip antennas is to increase its electrical surface current length is by changing the shape of the radiating element by utilizing meander or helical line structures. Some examples are shown in Figure 3.25(a) and (b) for a PIFA-type design.

One of the smallest GPS antennas used in handsets is a dielectric chip antenna made by Pulse (model number W 3010) [116]; its performance data is listed in column 2 of Table 3.8. This antenna is only 3.2 mm long, 1.6 mm wide, and is 1.1 mm high; the ground removal area is 10.80 × 6.25 mm and the polarization of this antenna is linear. Design details of this antenna are proprietary to Pulse, but permission has been obtained to show a picture of this antenna and its measured maximum and average gain in Figure 3.26.

Pulse also makes a dual-band GPS/Bluetooth Ceramic Single-Feed Antenna Model W3056 that operates at 1575 ±10 MHz and from 2400 to 2480 MHz. The efficiency in the GPS band is 75% with a peak gain of –1.25 dBi (linear polarization) and is 80% in the Bluetooth band with a –1.55-dBi gain at the band edges [117].

The characteristics of a ceramic-loaded GPS microstrip antenna made for handsets by MAXTENA (Model Number MPA 1575D1) is shown in column 5 in Table 3.8. More information is available from www.maxtena.com. The dimensions of this antenna are 10 mm × 10 mm × 4mm; the polarization of this antenna in contrast to other antennas shown in this table (except the QHA by Sarantel) is RHCP. A similar type but larger size antenna made by Toko-America is discussed in Chapter 2.

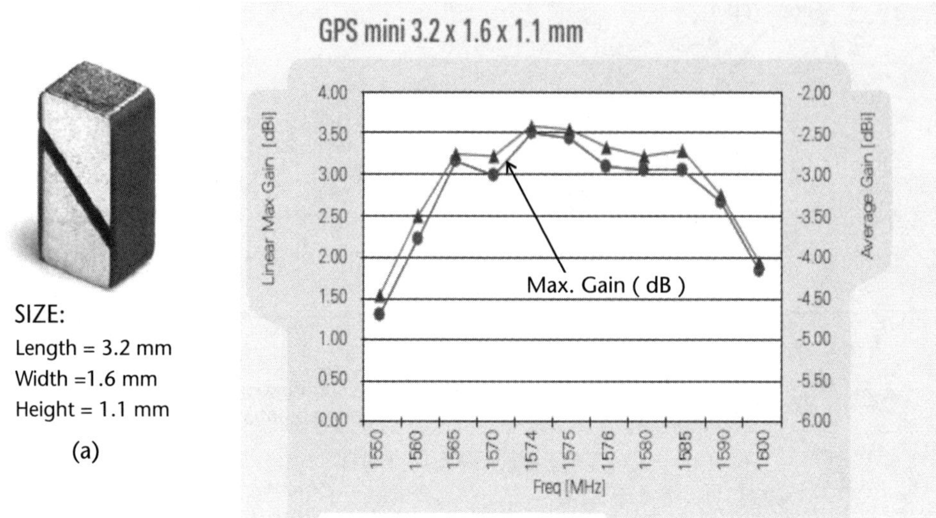

Max and Average Gain

(b)

Figure 3.26 Pulse W 3011 GPS ceramic chip antenna. (Courtesy of Aaron Logan, Pulse, Vancouver, WA.)

Table 3.8 Representative List of Commercially Available GPS Handset Antennas

Company and Model of Antenna	*Pulse W3010*	*Ethertronics Savi: M 210110*	*Yageo 3216*	*Maxtena MPA 1575D*	*Antenova/ RadioNova A 10137-R*	*Sarantel-UK; Geohelix P2*
Antenna design	Ceramic chip-FIFA	Isolated magnetic dipole	PIFA mode—ceramic chip	Microstrip patch	Coplanar balanced dipole	Ceramic-loaded quadrafilar helix
Size (mm) length/ width/height	3.2 × 1.6 × 1.1	2.0 × 2.0 × 1.8	3.2 × 1.6 × 1.2	10 × 10 × 4	9 × 20.1 × 1.6 (reference board size)	10 mm diameter; 17.75 mm length
Polarization	Linear	Linear	Linear	RHCP	Linear/RHCP	RHCP
Bandwidth (MHz)	1565–1585	1560–1610	1535–1620	1570–1580	1560–1590	1573–1577
Gain: RHCP inear	0.4 dBic, 3.4 dBi	−2 dBic, 1 dBi	4.32 dBic, 7.32 dBi	3.0 dBic, 6 dBi	−0.2 dBic, +2.8 dBi	−2.8 dBic, +0.2 dBi
Efficiency	85%	65%	81.7%	50%	60%	45%
Weight (grams)	0.033	0.2	N/A	N/A	< 1.	7.

3.10.1.1 LTCC Technology for GPS Handset Antennas

Multilayer LTCCs are increasingly being used for building miniaturized and surface mountable GPS and wireless technology antennas used in handsets [118, 119]. The dielectric constant of commercial LTCC materials can range from 3.9 to as high as 80 and their dissipation factors (DF) are low in the order 10^{-3}. Another factor of importance in LTCC is the frequency stability of the resonant structure, determined by the temperature coefficient of the resonance frequency, τ_F, which for LTCC ceramics is < 10 ppm/degree centigrade. LTCC technology makes it possible to integrate and build different types of electronic components directly into the multilayer ceramic boards, allowing greater multifunctionality and performance while simultaneously allowing miniaturization needed for handheld devices operating in the frequency range from 1 to 4 GHz.

LTCC allows parallel processing where layers of complicated electronics structures are fabricated separately and then brought together for a final lamination and cofiring process leading to high yield, fast turnaround, and lower cost. Hence, for GPS applications the entire receiver front end including antenna, filter, diplexer, and the mixer can all be integrated into a single package, greatly reducing size and cost.

LTCC provides precise thickness control of the dielectric layer as well as allow the printing of fine conducting lines on their surfaces with precise geometries through photo processing. These printed circuits can be of different shapes (e.g., for antennas they could be linear lines, square patches, meander lines, spirals, or helices).

Recently, several types of LTCC handset antennas for GPS alone or for multiband antennas that combine GPS with cellular or WLAN systems have been built. Three examples of such antennas published in recent journal articles are described next to illustrate the size reduction and performance achieved. Kim et al. [120] have provided design and measurements of a dual-band LTCC chip antenna operating in the GPS and PCS bands. The volume of this LTCC chip antenna is 9 × 15 × 1.2 mm dual-layer, meander-type radiating patches have been designed for dual-band operation with via holes and a height of 0.3 mm connecting the upper and lower layer antennas. The lower layer is tuned for the lower GPS frequency and the upper layer for the higher PCS frequency. The LTCC ceramic used has a dielectric constant of 7.8. The gain of the antenna at GPS and PCS is 2.5 and 2.8 dBi, respectively.

A compact GPS chip antenna with a volume of 10 × 10 × 2 mm has been designed by Chung et al. [121]. This antenna consists of a meander line that is imprinted on to the surface of an LTCC substrate made from calcium borosilicate glass (CBS) with a dielectric constant of 6 and an LOS tangent of $\tan \delta = 0.0014$. The polarization is linear and several antenna patterns, which are omnidirectional, are also shown in this paper; however, the gain achieved from these antennas is not mentioned.

A compact GPS helical LTCC antenna has been designed by Tao and Qin [122]. The volume of this antenna is 10 × 3 × 0.77 mm and the size of the ground plane required is 40 × 20 mm The LTCC substrate has a dielectric constant of 27, but other specific details on the substrate have not been provided. The antenna operates at 1.575 GHz and has a –10-dB return loss bandwidth of 60 MHz (1.65 to –1.60 GHz). The polarization is linear and the measured patterns for this antenna shown in the paper are omnidirectional with gains varying from –1.2 dBi to –1.316 dB.

In addition to the above LTCC antennas, information on which was obtained from published journal articles, a number of U.S. companies are also selling very compact GPS antennas made from LTCC material. Two such examples are Yageo's Phycomp brand of LTCC antennas, which cover GPS, wireless communication, and Bluetooth frequency bands (www.phycomp.components.com), and WP

Wireless (www.wp-wireless.com), which sells similar antennas. Details on the performance characteristics of these antennas can be obtained from the company brochures available on these Web sites.

3.10.1.2 GPS Fractal Microstrip Handset Antennas

Microstrip antennas suffer from narrow bandwidth. Recently microstrip and other types of antennas capable of providing much wider bandwidths needed for wireless communications as well as GPS in handsets and other devices have been designed by application of fractal geometries. Two important characteristics that define fractal geometry are "self-similarity" (where small regions of the geometry duplicate the overall geometry on a reduced scale) and "self-affinity" (where small regions of the geometry not identical to the overall geometry but are skewed or distorted and have differing scale factors). Based on these two underlying characteristics, fractals can have a geometrical form which are self similar and include multiple copies of themselves within their structure; these are created through successive iterations at several scales. The fractal antenna thus includes many characteristic sizes to allow multifrequency operation. Second, fractal antennas have highly convoluted and irregular shapes that can have sharp edges, corners, and discontinuities that can enhance the radiation from these antennas. A detailed discussion of fractal antennas is beyond the scope of this book but is well described in two recent articles [123, 124].

A good description of microstrip antennas using fractal geometry for wireless and other frequency bands has been provided by Anguera et al. [125]. Fractal shapes frequently used for increasing the bandwidth include the Sierpenski gasket, the Koch island or Koch snowflake fractal, the Minkowski Like Prefractal (MLPF) and space-filling curves such as the Hilbert fractal curve [123, 124]. Figure 4.27 shows the shapes obtained from different iterations using the first two fractals—the Sierpinski gasket shown in Figure 4.27(a) and the snowflake gasket shown in Figure 4.27(b). The application of fractals in antenna engineering is not without controversy. Best [129] in particular has questioned the benefits offered by fractals in increasing bandwidth. He has used the Sierepinki gasket fractal as one example to illustrate his concerns through simulations.

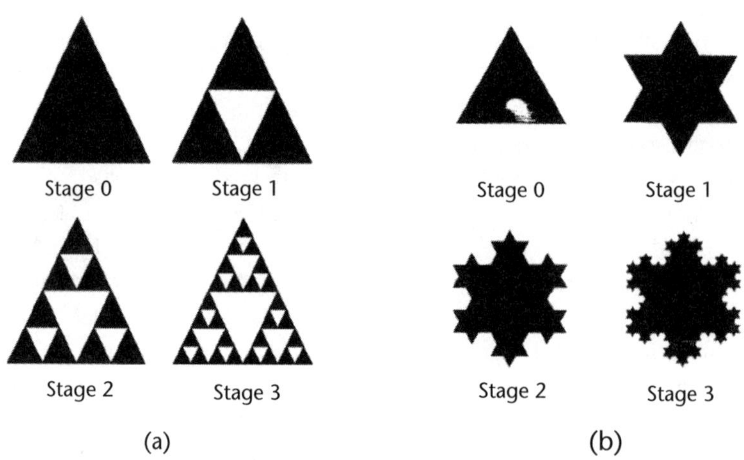

Figure 3.27 Fractal geometry applied to antennas: (a) Sierpinski gasket, and (b) Koch Island.

Currently there is only one GPS fractal antenna available commercially that is small enough to be suitable for handsets: the Fractus Geo-find model # FR05-S1-E-0-103. The size of this antenna is 10 × 10 × 0.9 mm. It is linearly polarized and a gain of 1 dBi and an efficiency of 70%. Its bandwidth is limited to the GPS C/A code; the narrow bandwidth is surprising since one of the expected benefits from the fractal design is an ultrawide bandwidth. Details on this antenna are available from www.fractus.com.

Several other wideband fractal antennas covering GPS, cellular, Bluetooth, and WLAN frequency bands have been published recently. The size of these antennas may not be small enough for being mounted inside a handset but may be suitable for mounting outside the handset and for other larger personal communications devices. The design of bow-tie fractal antennas, based on the third-order Sierpinski gasket, has been described by Gelman et al. [126]. This antenna operates at four different L-band frequencies: 915, 1575, 1750, and 2150 MHz and covers the GPS, cellular, and Bluetooth frequencies. This paper also compares standard bow-tie antennas over this band of frequencies. The dielectric substrate had a dielectric constant of 3.38 and a thickness of 1.5 mm. Ali and Jalal [127] have described the design of a microstrip "fractal" patch antenna based on the MLPF geometry; this antenna can provide wide, dual-band coverage from 1.48 to1.61 GHz and again from 1.64 to 2.5 GHz and has sufficient bandwidth to cover GPS, 3G-PCS, WLAN (IEEE 802.11), and so forth. The application of other fractal geometries for wireless communications antennas has also been described in a recent paper by Singh et al. [128].

3.10.2 Basic PIFA and Design Variations of a PIFA

Figure 3.19(a) shows a sketch of the PIFA antenna in its most basic form; it is one of the most popular antennas used in handsets due to their low cost and compact size. It can also be adapted through many design variations to operate over multiple frequency bands such as cellular, Bluetooth, WLAN, and GPS. Some of these designs have been described in great detail by Hall et al. [130] and have also been covered in many other recent books [97–100, 104]. For example, its size can be reduced by using high-permittivity ceramic substrates and sometimes also by using a superstrate placed on top of the metal plate as shown in Figure 3.19(b). The addition of the substrate/superstrate combination reduces the size but also increases both weight and cost. It can also degrade the gain, efficiency, and the bandwidth of the PIFA. The length of a PIFA can also be reduced by using a meander line to increase the path of current flow as a replacement for top plate and dual-band operation can be achieved by using slots cut to resonate at the selected frequency, as illustrated in Figure 3.25.

In its simplest form, as shown in Figure 3.19(a), the PIFA consists of a rectangular metal plate (or patch) of width W and length Lpositioned above a ground plane, which in the handset could be a metalized PCB. The PIFA has a short-circuit pin connecting the patch to the ground plane at one corner of the patch. The patch is connected to a feed pin (the center conductor of a coax cable) placed close to the shorting pin. The shorting pin adds a shunt inductance with a magnitude determined by its thickness and its proximity to the feed point. The impedance of the PIFA can be adjusted by varying the distance between the feed pin and the shorting pin used for the ground return. The impedance becomes smaller as the feed pin is moved closer to the sorting pin; the 50-ohm impedance point can sometimes be located very close to the shorting pin. The frequency of a PIFA can be determined approximately by the formula

$$f_0 = \frac{c_0}{4\sqrt{\varepsilon_{eff}}\,(W + L)}$$

In the above formula ε_{eff} is the "effective" dielectric constant of the substrate or the dielectric constant of the substrate suitably modified by the geometrical shape and dimensions of the antenna. This is similar to that of the effective dielectric constant of a microstrip antenna considered in Chapter 2. W and L are the width and length of the conducting patch, respectively.

Other modifications of the location and shape of the ground connection and feed pin and their impact on the resonant frequency of a PIFA have been described by Gustrau and Manteuffel [104]. If a superstrate is placed above the patch in addition to the substrate, the formula for the resonant frequency gets too complicated to be explained by a simple equation. The PIFA has a geometry that is similar to an end-fed microstrip patch but with a shorting pin added. When the top plate of a PIFA is reduced to the thin wire, the geometry of this antenna changes to that of an IFA shown in Figure 3.20(a), or to an ILA when the shorting pin is removed, as shown in Figure 3.20(b). The top plate has also been modified to generate a new antenna shape called a folded inverted conformal antenna (FICA) where generally three different resonant modes that each uses the entire antenna volume of the antenna have been used for obtaining a multiband frequency response [97].

Yageo/Taiwan has recently introduced one of the smallest ceramic chip antennas operating in the PIFA mode (part number: CAN4311 712 15 158 3K). It is 3.2 mm long, 1.6 mm wide, and 1.2 mm high. Its performance features are listed in column 4 of Table 3.8. More information about this antenna can be obtained from the product data sheet [131] available from their Web site (www.yageo.com).

3.10.2.1 GPS/Bluetooth Handset Antenna Using a PIFA-IFA Combination

A triple-band antenna that combines the 800 and 1900 MHz wireless cellular frequency bands with the GPS L_1 C/A code frequency band at 1575 MHz has been built and measured by Li and Rahmat-Samii [132]. The configuration of this antenna is shown in Figure 3.28. It consists of a dual-band PIFA with a dielectric substrate covering the two wireless cellular bands. A J-shaped slot was used to achieve dual-band tuning of the top plate of the PIFA as shown in this figure. The size of the top plate or patch"is 26 × 34 mm. The size of the ground plane is 98 × 34 mm. The substrate has a dielectric constant of 2.25 and is 10 mm thick. A GPS IFA is mounted on the side of the dielectric substrate. Although the IFA is normally linearly polarized it produces RHCP in this type of antenna configuration due to its coupling to the top patch that performs as an RHCP microstrip patch antenna. The location of the feed and ground connections are selected to optimize performance in all three bands. The distance between the feed and the ground was fixed at 6 mm for the PIFA and 2.5 mm for the IFA. The size of this antenna as designed may be

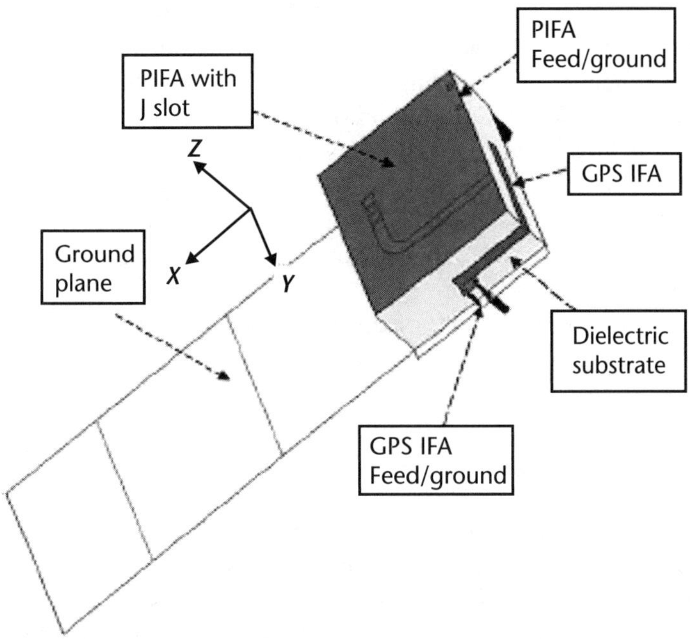

Figure 3.28 GPS and Bluetooth PIFA/IFA integrated antenna. (©2005 IEEE.)

slightly large for integration in to small handset since the dielectric constant of the substrate was just 2.25 and the size could have been reduced with a higher permittivity dielectric, but the design of this antenna does show how multiband operation could be achieved with a PIFA-IFA combination.

3.10.3 Isolated Magnetic Dipole for GPS Handsets by Ethertronics

The isolated magnetic dipole is a very compact, low-profile GPS antenna developed by Ethertronics using their patented design [133]. The performance characteristics of this antenna obtained from the product data sheet [134] are listed in column 3 of Table 3.8. As seen from this table the size of these antennas can be very small—the smallest being only 2 × 2 × 1.8 mm for a total volume of just 7.2 cubic mm. This design provides excellent isolation from its surrounding environment since the fields are confined to the immediate vicinity of the antenna and current flow on the ground plane underneath the antenna is restricted, unlike in some PIFA antenna designs. This is explained in great detail in several of the product brochures issued by this company. In addition, modifications to the antenna can be made to allow dual-band operation and small antennas operating at both GPS and Bluetooth frequencies are available [135].

A description of this antenna has been provided in a recent journal article [136]. An outline of this antenna is shown in Figure 3.21. It was also derived from reference [136]. The antenna is a stretched G-shaped metal structure when looked at from the side. It consists of three layers; the top and second layers overlap each other and are both connected to the bottom layer, but not directly to each other. The bottom layer is also connected to the ground plane and the "tongue" in the top layer is the feed point for the antenna. At resonance a cylindrical current loop is formed by the current that travels back and forth around the three layers. The overall structure can be considered as a capacitive loaded inductive loop. This generates a magnetic field along the axis of the loop that is the main mechanism for the radiation. The overall reduction in size is achieved by a capacitive loading. The electric field is highly confined between the two overlapping capacitor plates at the top as shown in Figure 3.21, which results in high isolation and reduces coupling to surrounding metal objects. Because of this restriction in current flow on the bottom layer (which acts as the ground plane) the isolation provided by the isolated magnetic dipole has been shown to be superior to that of a PIFA antenna [108]. The resonance frequency f of this antenna is given by the following equation:

$$f = \frac{1}{2\pi\sqrt{\frac{\varepsilon\mu cbt}{d}}}$$

where c, b, t, and d are the respective dimensions of the antenna component indicated in Figure 3.21, are, respectively, the permittivity and permeability. Ethertronics has also developed a dual-band version of this antenna in a slightly larger size that covers GPS and Bluetooth [135]. The model number is M 442100 and operates at 1.5754 GHz with an average gain of -2.6dBi and at 2.4 to 2.5 GHz with an average gain of –2.4 dBi. The polarization in both bands is linear and the size of the antenna is $14 \times 4 \times 1.3$ mm with a weight of 0.2 grams.

3.10.4 Coplanar Balanced Dipole Antenna for Handsets by Antenova Ltd.

A compact and coplanar balanced dipole antenna that operates in the GPS L_1 and the Galileo E5a frequency bands has been developed by Antenova Ltd. This new design goes by its marketing name as GPS Radionova and is protected by U.K. and international patents. Technical details of this antenna were published in a recent conference paper [137]. In this design both radiating arms of the dipole are coplanar and are located one on each side of a PCB substrate. The dipole is a fed transmission line balun that is derived from what is known as a Pawsey stub, which is integrated into the feed circuit. A diagram of this stub is shown in Figure 3.22.

The two conducting arms are connected together by a short-circuit conductor whose function is to limit excitation to a region that is confined to the feed line and the antenna so that arrangements in the remainder of the circuit of the device does not affect the balance of antenna circuit. To keep the design compact the radiating arms of the dipole have been meandered, as shown in Figure 3.22. Although the radiating arms of the dipole may overlap one another it has been found that by displacing the arms with respect to each other they do not follow exactly the same path in their respective planes. This provides good efficiency. Likewise it has been found that making the arms have different lengths leads to improved bandwidth.

The antenna can be mounted either vertically since it is only 7 mm in height or it can be co-planar with the PCB in a handset. Like other dipole antennas the radiation pattern depends on the size of the PCB; the pattern is dipolar and gives good coverage of the upper hemisphere and is particularly suited for acquiring low elevation satellites. The performance characteristics of this antenna are listed in column 6 of Table 3.8. Since it is a balanced dipole, the resonant frequency is not very susceptible to the size of the main PCB, nor is it affected by its location on the PCB. The efficiency is 70% when located at the edge of the handset but is lower if it is placed towards the center. Comparison tests conducted against a RHCP microstrip patch antenna showed an average C/N_0 that was 2 dB lower for this antenna than for the patch but it also allowed acquisition of a larger number of satellites and reduced the TTFF.

3.10.5 Miniaturized Ceramic Loaded QHA for Handsets

Three different types of compact ceramic-loaded GPS QHAs have been developed for use in handsets. These include the GeoHelix –P2 developed by Sarantel in the United Kingdom [138], a compact quadrifilar antenna (CQA) developed by Maxtena, Inc. in the United States [139], and a novel, piggyback style of triple-band covering the L_1, L_2, and L_5 bands of GPS developed in China [86]. The design of a QHA, including a dielectric-loaded QHA, was described in great detail in Chapter 2.

In these three antennas a ceramic material with a high dielectric constant is used as a core element inside the QHA to reduce its size. Sarantel was one of the first companies to develop a compact, ceramic-loaded QHA for commercial GPS handsets. Its design has been explained in several papers published by Sarantel [140]. The early models of this antenna were externally mounted outside the handset; however, the later versions like the GeoHelix P2 (shown in Figure 3.23) also allow internal mounting within the housing case of the handset, as explained in their product information sheet. This antenna is much like other QHAs described earlier in Chapter 2, except for the presence of a ceramic core located along its longitudinal axis. By decreasing the guide wavelength, this ceramic core reduces the size of the antenna to a size small enough to be suitable for handsets. Like other QHAs this antenna is constructed from two quadrature-phased bifilar helical loops. A sleeve balun is integrated into antenna structure at the base. It ensures that a balanced feed signal is projected to the backfire tip of the antenna. The balun acts like an RF choke so that all currents remain within the radiating structure of the antenna and there is minimum interaction with other components of the handset.

The ceramic core also acts like a capacitor confining the near-field energy close to the antenna and the twisted loops that constitute the antenna also constrain the magnetic fields tightly around the antenna. All three of these design features combine to ensure that there is high isolation between the antenna and other components within the handset chassis, as well with the hand of the user, or when the handset is held close to the head. Performance parameters of this antenna are listed in column 7 of Table 3.8. Laser and imaging technology has been used to ensure a tight mechanical tolerance required for ensuring good antenna performance [141].

Maxtena Inc. also makes a compact QHA with a ceramic core (model number M 1575CQA) suitable for handsets. This antenna has a diameter of 10 mm and is 20 mm long. Polarization is RHCP with a 0.2-dB axial ratio and an efficiency of 38%. The operational frequency is 1575.42 ±10 MHz. The beam width is 140° along both axes. This antenna has feed board at the base containing the balun and three LTCC hybrid couplers needed for obtaining 0°, 90°, 180°, and 270° degrees of relative phasing to the arms of the QHA. The front-to-back ratio is 12 dB.

3.10.5.1 Miniaturized Ceramic-Loaded Triple-Band GPS QHA for Handsets

A triple-band QHA, which covers the GPS L_1, L_2, and L_5 frequency bands, has been designed and built by Liu and Chu [86] for application in handsets. These frequency bands can also cover some bands of the Galileo system. The antenna is 17 mm long with an aperture (diameter) of 17 mm. The triple-band behavior is obtained from a piggyback operation of two antennas, one stacked coaxially on top of the other; the bottom antenna is a DOHA and the top antenna is a DQHA. The dielectric constant of the ceramic core for the top antenna was 39 and for the bottom antenna was 21. Figure 3.29(a) shows the physical configuration of this antenna and its simulated antenna patterns at center frequencies of the three frequency bands. The antenna has a gain of –2.5, –1, and –0.5 dBic at 1176, 1227, and 1575 MHz, respectively. Figure 3.29(b) shows these simulated radiation patterns. The ceramic loading assures low interaction with other objects in its proximity much like that offered by the handset QHA that have been developed by Sarantel and Maxtena. The return loss for this antenna shown in the conference paper has a rather narrow bandwidth in all three frequency bands.

3.11 Active GNSS Antennas and the Use of the G/T Ratio as Its Performance Metric

An "active GNSS antenna" in its most basic form consists of a passive antenna in series with an active section; a representative cross section of such an antenna is

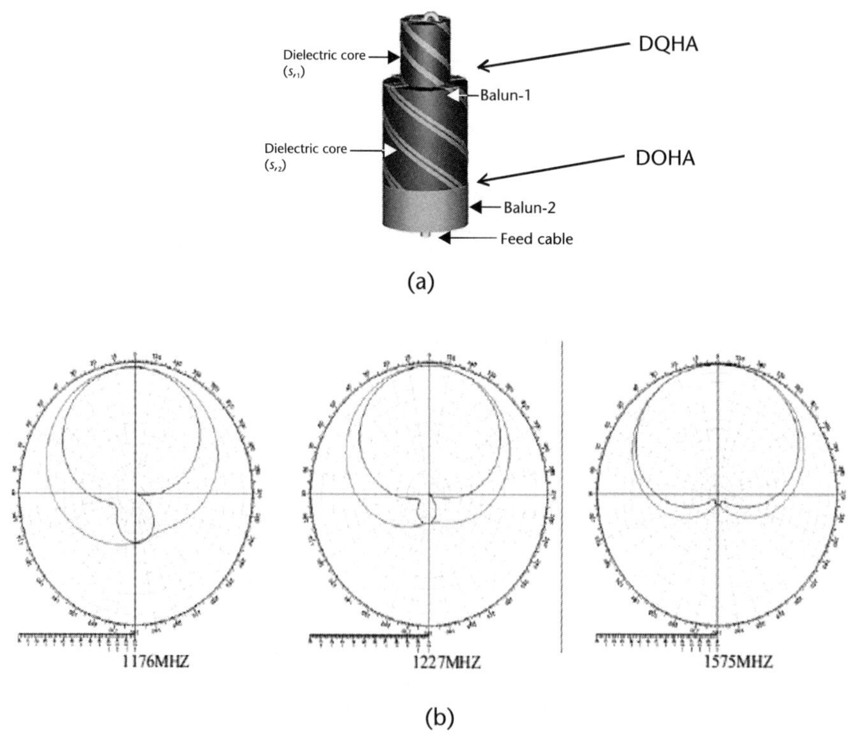

Figure 3.29 (a) and (b) GPS triple-band ceramic-loaded QHA in piggyback configuration by Liu and Chu (L_1, L_2, and L_5 bands). (Figure courtesy of Professor Qing-Xin Chu.)

shown in Figure 3.30. The active section may, at a minimum, consist of a chain of several electronic components:

- A very short length of coaxial cable;
- A bandpass filter located immediately after the antenna to filter off unwanted signals, which is closely followed by an LNA to amplify the weak GNSS satellite signal.

The LNA being an active device also needs a bias-T, which precedes the LNA and provides it with direct current (DC) power from an external DC power supply. In some cases this DC power required for the LNA is provided directly from the colocated GNSS receiver through the coaxial cable connecting the receiver to the active antenna.

In rare instances, some active antenna assemblies may use a second filter following the LNA to provide additional filtering before the IF and digitization stages of the receiver, although this is less likely in order to minimize insertion loss because the LNA serves as a buffer and minimizes an increase in noise level. The filter generally used in a GNSS receiver is either a surface acoustic wave (SAW) filter, which provides very sharp filtering or lumped element filters made from capacitors and inductors. This prefilter, which is placed before the LNA, introduces insertion loss that degrades the net gain available from the LNA and will also add noise. The LNAs provide high gains in the order of tens of decibels and can have low-noise figures ranging from just 0.5 dB for the latest super-low-noise LNAs using high-electron mobility transistor (HEMT) devices to as high as 4 dB for less expensive devices. Commercially available active GNSS antennas typically have only one con-

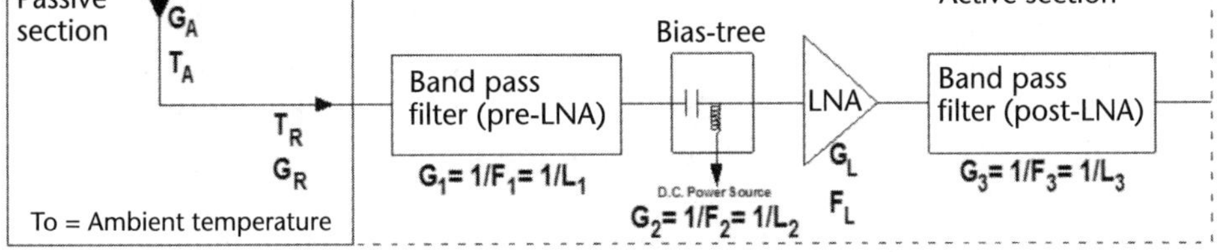

Effective noise temperature:
$$T_R = (F_1 - 1)\,T_0 + \frac{(F_2 - 1)T_0}{G_1} + \frac{(F_L - 1)T_0}{G_1 G_2} + \frac{(F_3 - 1)T_0}{G_1 G_2 G_L}$$

Net gain:
$$G_R = G_L - L_1 - L_2 - L_3$$

Figure 3:30 G/T measurements for an active GNSS antenna.

nector interface, generally a SMA coaxial connector, which provides the output for the amplified and filtered GNSS signals.

It is likely that more complex multiband antennas covering several GNSS systems could perhaps also have a variety of other components included inside the antenna package besides those mentioned above. Components such as diplexers, switches, and additional filters may also be directly integrated inside the antenna package to allow greater miniaturization, lower manufacturing costs, and improved versatility. The need for such active antennas is likely to increase as multifrequency and multisystem receivers used in conjunction with wideband antennas increase in popularity. These types of receivers will allow switching between the various bands to acquire, track, and process signals from different constellations, sometimes called system of systems (SOS) receivers.

It is impossible in active antennas of the type described above to separate the antenna function from those of the other electronic components that are integrated directly into the package of an active antenna. New performance metrics are needed in the evaluation of these active antennas. The G/T ratio has been suggested as the most appropriate figure of merit for evaluating their performance [143–145]. This is the ratio of the gain G of the antenna in dBic to the overall noise temperature of the receiving system expressed in degrees Kelvin. The unit of this figure of merit is dBic/K (degrees Kelvin).

$$G/T = \{G(dBic) - 10\log_{10}(T°K)\} \quad (dBic/K)$$

Figure 3.30 shows a schematic diagram of a basic active GNSS antenna divided into a passive and an active section. Each component in the active section is represented by either its gain G and by its noise figure F. If the component is a passive and lossy two-port system such as a coaxial cable, bandpass filter, or a bias-tee, it is represented as having a gain

$$G = 1/L \text{ with } L > 1$$

where L is the insertion loss and its equivalent noise temperature is represented by $T_e = (L - 1)\ T_0$ or by a noise figure $F = L = 1/G$. In other words, the noise figure is simply the inverse of the gain. T_0 is the ambient temperature, and is normally assumed to be 290°K. If the component in this section is active, such as an LNA, it will be represented by a gain G_L and by its noise figure F_L; the noise figure of an LNA is generally provided as part of its component specifications. The effective input noise temperature T_R and the net gain G_R of the active section that is connected to the passive antenna can be calculated using Frii's formula [89, pp. 133–140, 147].

In this formulation, the contribution to the effective temperature T_R from each component in the active section of the antenna is calculated as follows: the ambient temperature T_0 is multiplied by the noise figure of that component minus 1 and then divided by the gains of the earlier components in the active section. The LNA, which has a large gain, amplifies both the signal and the noise inputs to components in the chain. The effective temperature T_R calculated in this manner is given below:

$$T_R = (F_1 - 1)T_0 + \frac{(F_2 - 1)T_0}{G_1} + \frac{(F_L - 1)T_0}{G_1G_2} + \frac{(F_3 - 1)T_0}{G_1G_2G_L} \tag{3.1}$$

Alternatively, the above equation could also be written as

$$T_R (L_1 - 1)T_0 + (L_2 - 1)T_0 + (F_L - 1)L_1L_2T_0 + \frac{(L_3 - 1)L_1L_2}{G_L} \tag{3.2}$$

The net effective gain G_R of the active section is given by

$$G_R = G_L - L_1 - L_2 - L_3 \tag{3.3}$$

Note that the large gain of the LNA, which is in the denominator, can minimize the effects of noise sources that come after it. Hence, it is advisable to locate the LNA as close as possible to the output terminal of the antenna; it is generally placed immediately after the preselector filter with the shortest length of coaxial cable that is possible.

As shown in Figure 3.30, the active section of the antenna, represented by its net gain G_R and its effective input noise temperature T_R, is in series with the passive antenna section and represented by its gain $G_A(\theta, \phi, f)$ and by the effective antenna temperature T_A. The effective antenna temperature T_A may need some explanation: it is not literally the actual temperature of the antenna but is instead the temperature of a resistor that would produce the same thermal noise power of its environment that is coupled into the antenna. The radiative noise sources that contribute to it can be grouped into three categories: sky noise, man-made noise, and background noise due to terrestrial objects [146, 147]. The sky noise component, which is due to cosmic, galactic, and solar noise, has a typical noise temperature of around 10°K at GNSS frequencies (i.e., at L band). Man-made noise caused by electric motors, transformers, switching circuits, and so forth may count for 20 to 25K of the back-ground noise when the antenna is used in an urban center. The warm earth, buildings, trees, vehicles, and other objects near the antenna are black-body radiators of RF noise with equivalent temperature approximately equal to the ambient temperature of the environment (typically in the vicinity of around 290° K). While the noise sources described are extrinsic to the antenna, their total effect on the antenna noise temperature is determined by weighting each noise source by the antenna's radiation pattern, which can be represented by the following equation:

$$T_A(f) = \frac{\int_0^{2\pi} \int_0^{\pi} G(\theta,\varphi)T(\theta,\varphi)\sin(\theta)\,d\theta\,d\varphi}{\int_0^{2\pi} \int_0^{\pi} G(\theta,\varphi)\sin(\theta)\,d\theta\,d\varphi}$$

In the above equation $T(\theta, \phi)$ is the effective temperature profile of the various radiative noise sources and $G(\theta, \phi)$ is the gain pattern of the GNSS antenna. Since the gain pattern of a GNSS antenna is nearly omnidirectional (i.e., $G(\theta, \phi)$

~ constant), all noise sources around it appear to be about uniformly weighted by the antenna pattern with no specific directional preference. For GPS antennas used outdoors, the value generally used for T_A= 130° K [89, p. 135]. For airborne GPS antennas, T_A varies between 75° to 100° K [143, 144] when the antenna is mounted outdoors and pointed straight up towards zenith; for antennas mounted inside an anechoic chamber whose absorbing material can be at room temperature (23°C or so), the antenna noise temperature can be 300° K.

3.12 G/T Measurement of an Active GNSS Antenna

The G/T of an active GNSS antenna could be measured by making precision gain (or loss) as well as noise measurements of each individual component of the active section of the antenna, such as the filters, LNA, bias tee, and also components of the passive antenna section, as shown in Figure 3.30. However, to do such measurements on an active GNSS antenna, access is needed to the input and output terminals of each individual component of both the active and passive sections of the antenna; this is not possible since all components are integrated into a PC board and then packaged and covered over by a radome. The only port available for measurements is the SMA output connector located at the base of the antenna.

For very large reflector antennas and other large-size antennas such as arrays, a method commonly used for measuring G/T is by measuring the output noise level when the main beam of the antenna is pointed directly towards a radio star with a known flux density (such as Cassiopeia A, Cygnus A, Taurus, and even the moon) and then comparing it to the noise power received when the antenna is pointed away from the source towards the "cold sky." The flux density of some of these radio stars has been provided by Guidice and Castelli [148]. In the first measurement the received power is comprised of the system noise temperature plus the radio source's known flux density. In the second measurement, the power received by the antenna is comprised of only the system noise temperature. The ratio of these two measurements, noise plus radio source flux density divided by the system noise is referred as the Y factor from which the G/T can be estimated by using the formula:

$$\frac{G}{T} = \frac{8\pi k(Y-1)}{S\lambda^2}$$

In this equation, k = Boltzmann's constant = 1.38×10^{-23} J/K, S is the spectral power density of radio source (W/m^2/Hz), and λ is the operating wavelength. For small-size GNSS antennas with low gain this Y factor method using radio stars is not feasible since their flux density from the radio star is too small compared to the total noise power received to allow accurate measurement [149, 150]. Hence, new techniques are needed for measuring G/T of small-size and low-gain GNSS antennas.

A new method for measuring the G/T ratio of active GNSS antennas proposed by R. J. Erlandson is outlined below; more details are available in RTCA /DO-301 [143, 144]. The other method for measuring the G/T ratio of a small active antenna proposed by Tejedor et al. [145] was also studied in detail but found to be less accurate and unsuitable for GNSS active antennas. It does not properly account for antenna polarization and antenna temperature issues, which need many additional corrections to be made; hence, it will not be considered here.

3.12.1 Technique for G/T Measurement of Active GPS Antennas

Two separate measurements are needed in this method by R. J. Erlandson for measuring the G/T of an active GPS antenna [143, 144]. The first is measuring the total transducer gain, $G_A(\theta, \phi, f)$ which is the total integrated gain of the entire active antenna:

$$G_{TTG}(\varphi,\theta,f) = G_A(\varphi,\theta,f) * G_R(f) \tag{3.4}$$

In the above equation, $G_A(\phi, \theta, f)$ is the gain of the passive antenna element that is a function of the polar coordinate angles (ϕ, θ) where ϕ is the azimuth angle, θ is the elevation angle, and is the frequency. $G_R(f)$ is the net gain of the active section of the antenna that is equal to the gain of the LNA minus the loss in the filters, bias-tee, and any other lossy two-port device inserted into the active section. The total gain $G_{TTG}(\phi, \theta, f)$ is obtained from conventional antenna pattern measurements in a closed antenna range minus the losses introduced by the two filters and the bias-tee. These antenna measurement techniques will be described in greater detail in Chapter 6. The gains can be normalized to a convenient angle and frequency such as the zenith where degrees and the frequency is the L_1 center band frequency of 1.5754 GHz.

The second measurement that is needed is the output power density, PSD_{OUT} (f)

$$PSD_{OUT}(f) = k * (T_A(f) + T_R(f)) * G_R(f) \tag{3.5}$$

k = Boltzmann constant = 1.38065×10^{-23} W/Hz/K

$T_R(f)$ = effective input noise temperature (K) of the active antenna section of the antenna

$T_R(f) = (F_R(f) - 1) * 290$

$T_A(f)$ = effective antenna temperature

Since the antenna temperature is a significant component in the sum with $T_R(f)$, the output noise power density measurements should be taken outdoors with the antenna mounted on the same ground plane as the one used during the measurements of the antenna radiation pattern in the anechoic chamber. The antenna should be positioned with its main beam directed upwards to replicate antenna pointing conditions during normal use of the antenna. The equipment needed for measuring the output power density include:

1. Either noise measurement receiver or spectrum analyzer (Agilent N8973A, HP-8970B, or equivalent;

2. Noise diode or a noise calibration source (HP 346 B or equivalent); RF bias-tee, and a DC power supply.

More details on performing these measurements are provided in Section 2.4.2.4 of [144].

The G/T ratio of the active antenna can now be determined from these two measurements by scaling the measured total transducer gain G_{TTG} given in (3.4) by the Boltzmann's constant k and then dividing it by the measured output noise power density PSD_{out} given by (3.5).

$$\frac{k * G_{TTG}(\varphi,\theta,f)}{PSD_{out}(f)} = \left[\frac{k * G_A(\varphi,\theta,f) * G_R(f)}{k * \left[T_A(f) + T_R(f)\right] * G_R(f)}\right] = \frac{G_A(\varphi,\theta,f)}{\left(T_A(f) + T_R(f)\right)}$$

In the denominator of the above equation, the definition of the input temperature of the combined passive and active antenna combination $T_{in}(f) = T_A(f) = T_A(f) + T_R(f)$ can be substituted, which yields the final result for the G/T ratio of the active GNSS antenna

$$\frac{k * G_{TTG}(\varphi,\theta,f)}{PSD_{out}(f)} = \left[\frac{G_A(\varphi,\theta,f)}{T_{in}(f)}\right]$$

References

[1] Revnivykh, S., "GLONASS: Status and Progress," *Proceedings ION GNSS Conference 2008*, September 17, 2008, Savannah, GA.

[2] Leick, A., *GPS Satellite Surveying*, Hoboken, NJ: John Wiley & Sons, 2004, pp. 86–90.

[3] Data provided by C. Hegarty, MITRE Corp.

[4] Cameron, A., "Compass Delegates Share Status–GNSS Design and Test Newsletter," *GPS World*, March 23, 2010.

[5] Hoffman-Wellenhof, B., H. Lichtenegger, and E. Wasle, *GNSS—Global Navigation Satellite Systems; GPS, GLONASS, Galileo and More*, Springer-Verlag/Wien 2008, pp. 309–345.

[6] Hein, G.W., J.A. Avila-Rodriguez, and S. Wallner, et al, "Envisioning a Future GNSS System of Systems," *Inside GNSS*, Part 1, January/February 2007, pp. 58–67 and Part 2 , March/April 2007, pp. 64–72.

[7] Rizos, C., "GPS, GNSS and the Future," in *Manual of Geospatial Science and Technology, Second Edition*, J. D. Bossler (ed.), Boca Raton, FL: CRC Press, 2010, pp. 259–284.

[8] Lachapelle, G., P. Heroux, and S. Ryan, "Servicing the GPS/GNSS User," in *Manual of Geospatial Science and Technology Second Edition*, J. D. Bossler (ed.), Boca Raton, FL: CRC Press, 2010, pp. 235–257.

[9] Hein, G. W., T. Pany, and S. Wallner, et al., "Platforms for a Future GNSS Receiver: A Discussion of ASIC, FPGA, and DSP Technologies," *Inside GNSS*, March 2006, pp. 56–62.

[10] Detratti, M., "Antennas and R.F. Front Ends for Multi-frequency GNSS Receivers," ACCORDE Technologies S. A. 2010/ SPACOMM tutorial, organized by Galileo Ready Advanced Mass Market Receiver (GRAMMAR) , June 13, 2010, pp. 1–30.

[11] "2010 Receiver Survey," *GPS World*, Vol. 21, No. 1, January 2010, pp. 35–57, www.gpsworld.com.

[12] "2010 Antenna Survey," *GPS World*, Vol. 21, No. 2, February 2010, pp. 36–49, www.gpsworld.com.

[13] Vollath, U., S. Birnbach, and H. Landau, et al., "Analysis of Three-Carrier Ambiguity Resolution (TCAR) Technique for Precise Relative Positioning in GNSS-2," *Proceedings of the ION-GPS*, 1998.

[14] Feng, Y., C. Rizos, and M. Higgins, "Multi-Carrier Ambiguity Resolution Methods and Performance Benefits for Regional RTK and PPP GNSS Positioning Services," 20th International Technical Meeting of the Satellite Division of the U.S. institute of Navigation, Fort Worth, TX, September 25–28, 2007, pp. 668–678.

[15] Verhagen, S., "How Will the New Frequencies in GPS and Galileo Affect Carrier Phase Ambiguity Resolution," *Inside GNSS*, GNSS Solutions, March 2006, pp. 24–25.

[16] "Combined Performance for GPS/Galileo Receivers," EU-US Cooperation on Satellite Navigation, Working Group C, July 19, 2010.

[17] Axelrad, P., J. Donna, and M. Mitchell; "Enhancing GNSS Acquisition by Combining Signals from Multiple Channels and Satellites," *Proceedings ION GNSS 2009 Conference,* Savannah, GA, September 25, 2009, pp. 2617–2628.

[18] Axelrad, P., D. James, and M. Megan, et al., "Collective Detection," *GPS World*, January 2010, pp. 58–64.

[19] Feng, Y., and C. Rizos, "Impact of Multiple Frequency GNSS Signals on Future Regional GNSS Services," *International Global Navigation Satellite Systems Society, IGSS Symposium 2007,* Sydney, Australia, December 2007.

[20] Last, D., "GNSS: The Present Imperfect," *Inside GNSS*, May 2010, pp. 60–64.

[21] Blomenhofer, H., W. Ehret, and A. Leonard, et al., "GNSS/Galileo Global and Regional Integrity Performance Analysis," *17th ION/ GNSS Conference 2004,* September 21–24, Long Beach, CA, pp. 2158–2168.

[22] "No Surprise: Study Confirms Dual-GNSS Value," *Inside GNSS*, September 2010, pp. 16–17, www.insidegnss.com, and also http://ec.europa.eu/enterprise/policies/satnav/documents/galileo/index_en.htm.

[23] *National Space Policy of the United States of America*, June 28, 2010, p. 5.

[24] NovAtel antenna brochure, *GNSS-750; Revolutionary GNSS Wideband Antenna Delivers Enhanced Accuracy*, www.novatel.com, and "Leica AR25; 4 Constellation GNSS Antenna," product brochure, Leica Geosystems AG, www.leica-geosystems.com.

[25] Kunysz, W., "A Three Dimensional Choke Ring Ground Plane Antenna," *Proceedings 2003 ION/GNSS Conference,* September 9–12 2003, Portland, OR, pp. 1883–1888.

[26] "Trimble GNSS Geodetic Antennas," antenna information brochure provided by Trimble Navigation Ltd., www.trimble.com.

[27] Krantz, E., S. Riley, and P. Large, "The Design and Performance of the Zephyr Geodetic Antenna," *Proceedings 14th International Technical Meeting, ION/GPS 2001 Conference,* Salt Lake City, UT, September 11–14, 2001, pp. 1942–1951.

[28] Krantz, E., S. Riley, and P. Large, *GPS Antenna Design and Performance Advancements: The Trimble Zephyr, Trimble White Paper*, Trimble Navigation Ltd., 2001, www.trimble.com.

[29] Wang, J. J. H., and D. J. Triplett, "High Performance Universal GNSS Antenna Based on SMM Antenna Technology," *Proceedings IEEE 2007 International Symposium on Microwaves, Antenna, Propagation, and EMC Technologies for Wireless Communications*, 2007 IEEE.

[30] Wang, J. J. H., Spiral-Mode Microstrip (SMM) Antennas and Associated Methods for Exciting, Extracting and Multiplexing the Various Spiral Modes, U.S. Patent Number 5,621,422, issued April 15, 1997.

[31] Wang, J. J. H., and V. K. Tripp, "Design of Multioctave Spiral-Mode Microstrip Antennas," *Proceedings IEEE Transactions on Antennas and Propagation*, Vol. 39, No. 3, March 1991.

[32] NovAtel Inc., *GPS-704X Antenna Design and Performance*, www.NovAtel.ca/Documents/Paper/GPS-704X White Paper.pdf.

[33] Kunysz, W., Slot Array Antenna with Reduced Edge Diffraction, U.S. Patent 6,452,560 B2, issued USPTO, September 17, 2002.

[34] Kunysz, W., "High Performance GPS Pinwheel Antenna," Proceedings ION GNSS Conference, 2000, September 12–22, Salt Lake City, UT, pp. 2506–2511.

[35] Kunysz, W., *Aperture Coupled Slot Array Antenna,* U.S. Patent 6,445,354.

[36] Kunysz, W., Leaky Wave Antenna with Radiating Structure Including Fractal Loops, NovAtel Inc., U.S. Patent No. 7,250,916 and EU Patent No. EP1905126, issued 2006-07-10.

[37] Roke Manor Research Ltd., data sheet, *Triple GNSS Geodetic Grade Antenna.*

[38] Granger, R., P. Readman, and S. Simpson, "The Development of a Professional Antenna for Galileo," *Proceedings 2008 ION GNSS 19th International Technical Meeting of the Satellite Division,* September 26–29, 2006, Fort Worth, TX, pp. 799–806.

[39] Granger, R., and S. Simpson, "An Analysis of Multipath Mitigation Techniques Suitable for Geodetic Antennas," *Proceedings 2008 ION GNSS Conference*, Vol. 3, Savannah, GA, pp. 1260–1270.

[40] Granger, R., Ground Plane, Roke Manor Research (GB), U.S. Patent # US2009289868 (A1), issued November 26, 2009.

[41] Baracco, J.-M., L. Salghetti-Drioli, and P. de Maagt, "AMC Low Profile Wideband Reference Antenna for GPS and Galileo Systems," *IEEE Transactions on Antennas and Propagation*, Vol. 56, No. 8, August 2008, pp. 2540–2547.

[42] Sciré-Scappuzo, F., and S. N. Makarov, "A Low-Multipath Wideband GPS Antenna with Cutoff or Non-Cutoff Corrugated Ground Plane," *IEEE Transactions on Antennas and Propagation*, Vol. 57, No. 1, January 2009, pp. 33–46.

[43] Yun, S., D. Chang, and I. Yom, "Wideband Receive Antennas of Sensor Stations for GPS/Galileo Satellite," *2008 ISSTA and 2008 ASMS Conference Proceedings*, IEEE, 2009.

[44] Popugaev, A. E., R. Wansch, and S. F. Urquijo, "A Novel High Performance Antenna for GNSS Applications," *Proceedings EuCAP 2007, the Second European Conference on Antennas and Propagation,* November 11–16 2007, pp. 1–5.

[45] Kasemodel, J. A., C.-C. Chen, and I. J. Gupta, et al., "Miniature Continuous Coverage Antenna Array for GNSS Receivers," *IEEE Antennas and Propagation Letters*, Vol. 7, 2008, pp. 592–595.

[46] Zhou, Y., S. Koulouridis, and G. Kiziltas, et al., "A Novel 1.5" Quadruple Antenna for the Tri-Band GPS Applications," *IEEE Antennas and Wireless Propagation Letters*, Vol. 5, 2006, pp. 224–227.

[47] Zhou, Y., C.-C. Chen, and J. L. Volakis, "Dual-Band Proximity-Fed Stacked Patch Antenna for Tri-Band GPS Applications," *IEEE Transactions on Antennas and Propagation*, Vol. 55, No. 1, January 2007, pp. 220–223.

[48] Doust, E. G., M. Clenet, and V. Hemmati, et al., "An Aperture Coupled Circularly Polarized Stacked Microstrip Antenna for GPS Frequency Bands L_1, L_2, and L_5," *Proceedings IEEE Antenna & Propagation Society Symposium*, July 2008, pp. 1–4. (More details on this antenna can also be found in E. G. Doust, *An Aperture-Coupled Stacked Microstrip Antenna for GPS Frequency Bands L_1, L_2, and L_5*, Contract Report DRDC Ottawa CR-29007-147, August 2007.)

[49] Targonski, S. D., E. C. Ngai, and D. J. Blejer, et al., "Tri-Band Patch Antenna for Small GPS Polarimetric Arrays," *2005 IEEE Antenna and Propagation Society Symposium*, pp. 430–433.

[50] Rama Rao, B., M. A. Smolinski, and C. C. Quach, et al., "Triple-Band GPS Trap-Loaded Inverted L Antenna Array," *Microwave and Optical Technology Letters*, Vol. 38, No. 1, July 5, 2003, pp. 35–37, and U..S. Patent 6,856,287, issued February 15, 2005.

[51] Seeber, G., *Satellite Geodesy*, Berlin: Walter De Gruyter GmbH & Co., 2003 pp. 356–383.

[52] Schupler, B. R., R. L. Allhouse, and T. A. Clark, "Signal Characteristics of GPS User Antennas," *Navigation: Journal of the Institute of Navigation*, Vol. 41, No. 3, Fall 1994, pp. 277–295.

[53] Schupler, B. R., and T. A. Clark, "How Different Antennas Affect the GPS Observable," *GPS World*, November/December 1991, pp. 32–36.

[54] Rizos, C., "Making Sense of GNSS Techniques," pp. 173–190, and "Carrying Out a GPS Surveying/Mapping Task," pp. 217–234, in *Manual of Geospatial Science and Technology*, J. D. Bossler (ed.), Boca Raton, FL: CRC Press, 2010.

[55] Rizos, C., D. Smith, S. Hilla, J. Evjen, W. Henning, "GPS Projects: Some Planning Issues," pp. 191-215, in Chapter 12 of *Manual of Geospatial Science and Technology*, (Editors: John D. Bossler), Boca Raton, Fl., CRC Press, ©2010

[56] Grejner-Brezezinska D.A., "GPS Instrumentation Issues," *Manual of Geospatial Science and Technology*, (Editors: John D. Bossler), Chapter 10, pp. 151-172; Boca Raton, Fl., CRC Press, ©2010

[57] Hoffman-Wellenhof B., H. Lichtenegger, J. Collins, *GPS – Theory and Practice*, Fifth Revised Edition, Springer-Verlag/Wien, ©2001

[58] Langley R.B., "GPS Receivers and the Observables," *GPS For Geodesy*, Editors: Teunissen P. J. G., A. Kleusberg, Second Edition,1998, Springer-Verlag Publishers, pp.154-155.

[59] Schupler B. R., T. A. Clark, "Characterizing the Behavior of Geodetic GPS Antennas," *GPS World*, February 2001, Vol. 12, Pt. 2, pp. 48 – 57.

[60] Mader G., "GPS Antenna Calibration at the National Geodetic Survey," *GPS Solutions*, Vol. 3, (1), 1999; pp. 50-58; also in http://www.grdl.noaa.gov/GRD/GPS/Projects/ANTCAL/Files/summary/html

[61] NGS-antcal, 2009, http://www.ngs.noaa.gov/ANTCAL/, accessed March 10, 2009

[62] Rothacher, M., "Comparison of Absolute and Relative Phase Center Variations," *GPS Solutions*, Vol. 4, No. 4, pp. 55–60, 2001.

[63] Gorres B., J. Campbell, and M. Becker, et al., "Absolute Calibration of GPS Antennas: Laboratory Results and Comparison with Field and Robot Techniques," *GPS Solutions*, Vol. 10, 2006, pp. 136–145.

[64] Becker, M., P. Zeimetz, and E. Schonemann, "Anechoic Chamber Calibration of Phase Center Variations For New and Existing GNSS Signals and Potential Impacts in IGS Processing," IGS Workshop, June 28– July 2, 2010, Newcastle Upon Tyne, England.

[65] Alemu, G. T., "Assessment on the Effects of Mixing Different Types of GPS Antennas and Receivers," master's of science thesis in geodesy, No. 3106, TRITA-GIT-EX 08-009, Royal Institute of Technology (KTH), 100 44, Stockholm, Sweden, September 2008.

[66] Gurtner, W., "RINEX: The Receiver Independent Exchange Format," *GPS World*, Vol. 5, No. 7, 1994, pp. 48–52.

[67] Moernaut, G. J. K., and D. Orban, "Basics of GPS Antennas," *GPS World*, Vol. 20, No. 2, 1 February 2009, pp. 42–48.

[68] Moernaut, G. J. K., and G.A.E. Vandenbosch, "The Shorted Annular Patch Antenna as a Multipath Suppressing Antenna," *Proceedings 2nd European Microwave Conference*, EuCAP 2007, pp. 782–787.

[69] McKinzie, W. E., R. Hurtado, and W. Klimczak, "Artificial Magnetic Conductor Technology Reduces Size and Weight for Precision GPS Antennas," *Proceedings of the Institute on Navigation*, National Technical Meeting, San Diego, CA, January 28–30, 2002, pp. 1–12.

[70] Baggen, R., M. Martinez-Vasquez, and J. Leiss, et al., "Low Profile Galileo Antenna Technology Using EBG Technology," *IEEE Transactions on Antennas and Propagation*, Vol. 56, No. 3, March 2008.

[71] Bao, X. L., G. Ruvio, and M. J. Amman, "Low-Profile Dual Frequency GPS Patch Antenna Enhanced with Dual Band EBG Structures," *Microwave & Optical Technology Letters*, Vol. 49, No. 11, November 2007, pp. 2630–2634.

[72] Sievenpiper, D. "Review of the Theory, Fabrication, and Application of High Impedance Ground Planes," in *Metamaterials: Physics and Engineering Explorations*, N. Engheta and R. K. Ziolkowski (eds.), Hoboken ,NJ: John Wiley & Sons, 2006, pp. 287–309.

[73] Rahmat-Samii, Y., and F. Yang., "Development of Complex Artificial Ground Planes in Antenna Engineering," in *Metamaterials: Physics and Engineering Explorations*, N. Engheta and R. K. Ziolkowski (eds.), Hoboken, NJ: John Wiley & Sons, 2006, pp. 313–346.

[74] Ashjaee, J., V. S. Fillipov, and D. V. Tatarnikov, et al., Dual Frequency Choke Ring Ground Planes, Topcon Positioning Systems Inc., San Jose, CA, U.S. Patent #6278407, issued August 21, 2001.

[75] Blankey, T.L., D. D. Connel, and J. Lambert, et al., Broad-Band Antenna Structure Having Frequency Independent, Low Loss Ground Plane, U.S. Patent #4,608,572, issued August 26, 1986.

[76] Westfall, B., *Antenna with R Card Ground Plane*, U.S. Patent #5,694,136, USPTO, 1997.

[77] Lennen, G., W. Hand, and B. Westfall, GPS Receiver with N-point Symmetrical Feed Double Frequency Patch Antenna, U.S. Patent #5,515,957, USPTO,1996.

[78] Rama Rao, B., and Solomon M. N., et al., "Research on GPS Antennas at MITRE," *Proceedings IEEE Position Location and Navigation Symposium (PLANS)*, Palm Springs, CA, April 1998, pp. 634–661.

[79] Nakano, H., "Research on Spiral and Helical Antennas at Hosei University," *IEEE Antenna & Propagation Society Newsletter*, June 1968, pp. 19–27,and also Nakano, H., Y. Yamasaki, and S. Hashimoto, "Numerical Analysis of 4-Arm Archimedean Spiral Antenna," *Electronics Letters*, Vol. 19, No. 3, February 3, 1983, pp. 78–89.

[80] Kraft, U. R., "Optimization of Circular Polarization Performance of the 4-Arm Planar Spiral Antenna with Non-Perfect Excitation Networks," *Proceedings IEE Microwaves, Optics and Antennas Pt. H*. Vol. 137, No. 1, February 1990, pp. 45–50.

[81] Dyson, J. D., and P. E. Mayes, "New Circularly-Polarized Frequency Independent Antennas with Conical Beam or Omnidirectional Patterns," *IRE Transactions on Antennas & Propagation*, July 1961, pp. 334–342.

[82] Padros, N., J. I. Ortigosa, and J. Baker, et al., "Comparative Study of High-Performance GPS Receiving Antenna Designs," *IEEE Transactions on Antennas and Propagation*, Vol. 45, No. 4, April 1997, pp. 698–706.

[83] Zhang, Y., M. Younis, and C. Fischer, et al., "Planar Artificial Magnetic Conductors and Patch Antennas," *IEEE Transactions on Antennas & Propagation*, Vol. 51, No. 10, October 2003, pp. 2704–2712.

[84] Feresidis, A. P., G. Gousettis, J. C. and Vardaxoglou, "Metallodielectric Arrays without Vias as Artificial Magnetic Conductors and Electromagnetic Band Gap Surfaces," *Proceedings 2004 IEEE Antenna & Propagation Symposium Digest*, Vol. 2, 2004, pp. 1159–1162.

[85] Gousettis, G., A. P. Feresidis, and J. C. Vardaxoglou, "Tailoring the AMC and EBG Characteristics of Periodic Metallic Arrays Printed on Grounded Dielectric Substrate," *IEEE Transactions on Antennas & Propagation*, Vol. 54, No. 1, January 2006, pp. 82–89.

[86] Liu, S., and Qing-Xin Chu, "A Novel Dielectrically Loaded Antenna for Tri-Band GPS Applications," *Proceedings of the 38th European Microwave Conference,* October 28, Amsterdam, Netherlands, pp. 1759–1762.

[87] Federal Communications Commission, CC. Docket Number 94-102, Report and Order, adopted June 12, 1996.

[88] Monnerat, M., R. Couty, and N. Vincent, et al., "The Assisted GNSS, Technology and Applications," *Proceedings ION GNSS 17th International Technical Meeting of the Satellite Division,* September 21–24, 2004, Long Beach, CA.

[89] Van Diggelen, F., *A-GPS: Assisted GPS, GNSS, and SBAS*, Norwood, MA: Artech House, 2009.

[90] Richton, B., G. Vannuci, and S. Wilkus, "Assisted GPS For Wireless Phone Location–Technology and Standards," in *Next Generation Wireless Networks*, S. Tekinay (ed.), Kluwer Academic Publishers, 2001, pp. 129–155.

[91] Bryant, R., "Assisted GPS—Using Cellular Telephone Networks for GPS Anywhere," *GPS World*, May 2005, pp. 40–46, www.gpsworld.com.

[92] Lindbergh, P., "Challenges in Mobile Phone Antenna Development," in *Design of Ultra Wideband Antenna Matching Networks via Simplified Real Frequency Techniques*, Berlin-Heidelberg: Springer-Verlag, pp. 45–66.

[93] Haddrel, T., M. Phocas, and N. Ricquier, "A New Approach to Cellphone GPS Antennas," *Proceedings of ON GNSS 21st International Technical Meeting of the Satellite Division,* September 16–19, 2008, Savannah, GA, pp. 2747–2754.

[94] Ethertronics, "Advances in High Performance Ceramic Antennas for Small-Form Factor, Multi-Technology Devices," White Paper, 2007, www.ethertronics.com.

[95] Ah You, D. M. K., and R. Emrick, "Frequency Bands for Military and Commercial Applications," in *Antenna Engineering Handbook*, Fourth Edition, (J. L. Volakis, ed.), New York: The McGraw Hill Companies, 2007.

[96] Vainikainen, P., J. Holopainen, and C. Icheln, et al., "More Than 20 Antenna Elements in Future Mobile Phones, Threat or Opportunity," *Proceedings European Conference on Antennas & Propagation*, EuCAP 2, pp. 2940–2943.

[97] Lindbergh, P., "Design Techniques for Internal Terminal Antennas," in *Design of Ultra Wideband Antenna Matching Networks via Simplified Real Frequency Techniques,* Berlin–Heidelberg: Springer-Verlag, pp. 67–91.

[98] Vardaxoglou, and J. R. James, "Mobile Handset Antennas" in *Antenna Engineering Handbook*, Fourth Edition, (J. L. Volakis, ed.), New York: The McGraw Hill Companies, 2007.

[99] Waterhouse, R. (ed.), "Small Printed Antennas for Wireless Systems—Part 2," in *Printed Antennas for Wireless Communications*, Chichester, West Sussex, England: John Wiley & Sons Ltd., 2007, pp. 197–296.

[100] Chen, Z. N., and Chia M. Y. W., *Broadband Planar Antennas: Design & Applications*, Chichester, West Sussex, England: John Wiley & Sons Ltd., 2006.

[101] Wong, K.-L., *Planar Antennas for Wireless Communications*, Hoboken, NJ: John Wiley & Sons, 2006.

[102] Kivekas, O., J. Ollikainen, and T. Lehtiniemi, et al., "Effect of the Chassis Length on the Bandwidth, SAR, and Efficiency of Internal Mobile Phone Antennas," *Microwave and Optical Technology Letters*, Vol. 36, March 20, 2003, pp. 457–462.

[103] Vainkainen, P. J., Ollikainen, and O. Kivekas, et al., "Resonator- Based Analysis of the Combination of Mobile Handset Antenna and Chassis," *IEEE Transactions on Antennas & Propagation*, Vol. 50, No. 10, October 2002, pp. 1413–1444.

[104] Gustrau, F., and D. Manteuffel, *EM Modeling of Antennas and RF Components For Wireless Communication Systems*, Berlin Heidelberg: Springer-Verlag, 2006.

[105] Manteuffel, M., A. Bahr, and D. Heberling, et al., "Design Considerations fFor Integrated Mobile Phone Antennas," *Proceedings Eleventh International Conference on Antennas & Propagation*, Vol. 1, Manchester U.K., April 2001, pp. 252–256.

[106] Futter, P., N. Chavannes, and R. Tay, et al., "Reliable Prediction of Mobile Phone Performance for Realistic In-Use Conditions Using FTTD Method," *IEEE Antennas & Propagation Magazine*, Vol. 50, No. 1, February 2008, pp. 87–94.

[107] Pathak, V., S. Thornwall, and M. Kriz, et al., "Mobile Handset Performance Comparison of a Linearly Polarized GPS Internal Antenna with a Circularly Polarized Antenna," *IEEE Antenna and Propagation Society International Symposium Digest,* Vol. 3, Part 3, June 22–27, 2003, pp. 666–669.

[108] Poilasne, G., S. Rowson, and L. Desclos, "Achieving Antenna Isolation Within Wireless Systems," *Microwaves & RF*, January 2003, pp. 74–80.

[109] Lindbergh, P., "Antennas for Mobile Wireless Communication," in *Design of Ultra Wideband Antenna Matching Networks via Simplified Real Frequency Techniques*, Berlin- Heidelberg: Springer-Verlag, pp. 39–44.

[110] Bernhard, J. T., "Printed Antennas in Packages," in *Printed Antennas for Wireless Communications*, Chichester, West Sussex, England: John Wiley & Sons Ltd., 2007, pp. 281–296.

[111] Pelosi, M., O. Franek, and M. B. Knudsen, et al., "Influence of Dielectric Loading on PIFA antennas in Close Proximity to User's Body," *Electronics Letters*, Vol. 45, No. 5, February 26, 2009, pp. 246–247.

[112] Huang, T.,and K. R. Boyle, "User Interaction Studies on Handset Antennas," *Proceedings European Conference on Antennas & Propagation*, EuCAP 2007, November 11–16, 2007, Edinburgh, U.K., p. 6.

[113] Pelosi, M., G. F. Pedersen, and M. B. Knudsen, "A Novel Paradigm for High Isolation in Multiple Antenna Systems with User's Influence," *Proceedings European Conference on Antennas & Propagation 2010*, EuCAP 2010, April 12–16, 2010, Barcelona, Spain, p. 5.

[114] Pelosi, M., O. Franek, and M. B. Knudsen, et al., "A Grip Study for Talk and Data Modes in Mobile Phones," *IEEE Transactions on Antennas & Propagation*, Vol. 57, No. 4, April 2009, pp. 856–865.

[115] Gousettis, G., J. C. Vardaxoglou, and A. P. Feresidis, "Handset Antenna Performance Using Flexible MEBG Structures," *Proceedings IEEE International Workshop on Antenna Technology: Small Antennas and Metamateterials*, 2005, pp. 55–58.

[116] Pulse Finland Oy, product data sheet for Pulse Part Number: W3011/W3011,1.575 GHz GPS Ceramic Chip Antenna, September 2008, www.pulseeng.com/antennas.

[117] Pulse Finland Oy, product data sheet or pulse part number W3056, Ceramic Single Feed GPS/BT Antenna, June 2010, www.pulseeng.com/antennas.

[118] Barnwell, P., W. Zhang, and J. Leibowitz et al., "An Investigation of the Properties of LTCC materials and Compatible Conductors for Their Use in Wireless Applications," *Proceedings 2000 International Symposium on Microelectronics IMAPS*, SPIE, Vol. 4339, 2000, pp. 659–664.

[119] Imanaka, Y., *Multilayered Low Temperature Cofired Ceramics (LTCC) Technology*, Springer U.S., 2005

[120] Kim,Y. -D., M. Kim, and H. Kim, et al., "Dual-band Chip Antenna Using LTCC Multilayer Technology for Mobile Communications Applications," *Microwave and Optical Technology Letters*, Vol. 45, 2005, pp. 271–273.

[121] Chung, Y.-T., J. Sun, and S. Wang, et al., "Compact Chip Antenna Applications in GPS by LTCC Processing," *Microwave and Optical Technology Letters*, Vol. 51, No. 4, April 2009, pp. 1017–1019.

[122] Tao, H., and Y. Qin, "Design of a 1.575 GHz Helical LTCC Chip Antenna for GPS Application, *Proceedings of PIERS 2010*, Xi'an, China, March 22–26, 2010.

[123] Werner, D. H., and S. Gangully, "An Overview of Fractal Antenna Engineering Research," *IEEE Antennas & Propagation Magazine*, Vol. 45, No. 1, February 2003, pp. 38–57.

[124] Puente, C., J. Romeau, and A. Cardama, "Fractal Shaped Antennas," in *Frontiers in Electromagnetics*, D. H. Werner and R. Mittra (eds.), IEEE Press, 2000.

[125] Anguera, J., C. Borja, and C. Puente, "Microstrip Fractal-Shaped Antennas–A Review," *Proceedings European Conference on Antennas & Propagation-EuCAP 2007*, 2007, p. 7.

[126] Gelman, Y., E. Ben-Ari, and R. Shavit, "Multi-Band Bowtie Antenna Based on Fractal Geometry," *Proceedings IEEE International Antennas & Propagation Symposium*, 2004; pp. 3441–3444.

[127] Ali, J. A., and A. S. A. Jalal, "A Miniaturized Multiband Minkowski-Like Pre-Fractal Patch Antenna for GPS and 3G, IMT-2000 Handsets," *Asian Journal of Information Technology*, Vol. 6, No. 5, 2007, pp. 584–588.

[128] Singh, K., V. Grewal, and R. Saxena, "Fractal Antennas: A Novel Miniaturization Technique for Wireless Communications," *International Journal of Recent Trends in Engineering*, Vol. 2, No. 5, November 2009.

[129] Best, S. R., "On the Significance of Self-Similar Fractal Geometry in Determining Multi-Band Behavior of the Sierpinski Gasket Antenna," *Antennas and Wireless Propagation Letters*, IEEE; Vol. 1, 2002, pp. 22–25.

[130] Hall, P. S., E. Lee, and C. T. P. Song, "Planar Inverted-F Antennas," in *Printed Antennas for Wireless Communications*, Chichester, West Sussex, England: John Wiley & Sons Ltd., 2007, pp.197–227.

[131] Yageo Taiwan, *3216 Multilayer Ceramic Chip Antenna (PIFA Mode) for GPS Application*, print date April 29, 2010, www.yageo.com.

[132] Li Zhan, and Y. Rahmat Samii, "Optimization of PIFA-IFA Combination in Handset Antenna Designs," *IEEE Transactions on Antennas and Propagation*, Vol. 53, No. 5, May 2005, pp. 1770–1778.

[133] Desclos, L., J. Shamblin, and G. Poilasne, et al., Low Profile, Built-Frequency Multi-Band, Magnetic Dipole Antenna, U.S. Patent #6,717,551 B1, issued April 6, 2004.

[134] Ethertronics, product data sheet for Savvi™ GPS 2x1 M210110 Embedded Ceramic Chip Antenna, www.ethertronics.com.

[135] Ethertronics, product data sheet for Savvi™ Embedded Ceramic GPS & Bluetooth Antenna, 1.575 and 2.4–2.5 GHz, part number M442100, www.ethertronics.com.

[136] Rowson, S., G. Poilasne, and L. Desclos, "Isolated Magnetic Dipole Antennas: Application to GPS," *Microwave and Optical Technology Letters*, Vol. 41, No. 6, June 20, 2004, pp. 449–451.

[137] Collins, B. S., D. Iellici, and S. P. Kingsley, et al., "Balanced Antennas for GPS and Galileo Receivers," *Proceedings 2007 Loughborough Antennas and Propagation Conference*, April 2–3, 2007, Loughborough, U.K., pp. 321–324.

[138] Sarantel Ltd., U.K., product specification, Geohelix-P2, Embeddable Passive GPS Antenna, www.srantel.com.

[139] Licul, S., J. Marks, and W. L. Stutzman, "The Compact Quadrifilar Antenna (CQA for Hand Held GPS and Satellite Radio Reception," *Antenna Systems Conference*, Denver, CO, September 26–27, 2007.

[140] Leisten, O., Y. Vardaxoglou, and T. Schmid, et al., "Miniature Dielectric -Loaded Personal Telephone Antennas with Low User Exposure," *Electronics Letters*, Vol. 34, No. 17, 1998, pp. 1628–1629.

[141] Leisten, O., J. Fierset, and I. et al., "Laser Assisted Manufacture for Performance Optimized, Dielectrically Loaded GPS Antennas for Mobile Telephones," *Proceedings of SPIE*, Vol. 4637, 2002, pp. 397–403.

[142] Liu, S., and Qing-Xin Chu, "A Novel Dielectrically Loaded Antenna for Tri-band GPS Applications," *Proceedings of the 38th European Microwave Conference*, Amsterdam, Netherlands, October 2008, pp. 1759–1762.

[143] Van Dierendonck, A. J., and R. J. Erlandson, "RTCA Airborne GPS Antenna Testing and Analysis of a New Antenna Minimum Operational Performance," *Proceedings of ION NTM*, January 2007, pp. 692–701.

[144] RTCA Inc., Minimum Operational Performance Standards for Global Navigation Satellite System (GNSS) Airborne Active Antenna Equipment for the L_1 Frequency Band, RTCA/DO-301, December 13, 2006.

[145] Tejedor, P., J. Barbero, and C. Martin; "On the Measurement of G/T of Active Antennas," *Proceedings of AMTA (Antenna Measurement Techniques Association)*, Boulder, CO, October 7–11, 1991, pp. 9-47–9-49.

[146] Morgan, W. L., and G. D. Gordon, *Communication Satellite Handbook*, New York: John Wiley & Sons, 1989, pp. 291–309.

[147] Estabrook, P., and W. Rafferty; "Mobile Satellite Vehicle Antenna: Noise Temperature and Receive G/T," *Proceedings 39th IEEE Vehicular Technology Conference*, Vol. 2, 1989, pp. 757–762.

[148] Guidice, D. A., and J. J. Castelli; "The Use of Extraterrestrial Radio Sources in the Measurement of Antenna Parameters," *IEEE Transactions on Aerospace & Electronic Systems*, Vol. AES -7, No. 2, March 1971, pp. 226–233.

[149] Dybdal, R. B., "G/T Measurement of Small Antennas," *Proceedings of AMTA (Antenna Measurement Techniques Association)*, 1997, pp. 37–42.

[150] Dybdal, R. B., "Measurement of Antennas with Integrated Electronics," *Proceedings of AMTA (Antenna Measurement Techniques Association)*, 1996, pp. 9–13.

CHAPTER 4

Adaptive GPS Antennas

Dr. Ronald L. Fante and Dr. Keith F. McDonald

4.1 Introduction

In recent years, the military and civilian communities have become increasingly reliant on GPS for accurate knowledge of geographical location as well as timing for synchronous operations, asset deployment, and communications. GPS receivers have been shown to operate with accuracy in benign environments and enjoy a wide variety of modern day applications. The high level of military and civilian dependence on GPS makes it a likely electronic warfare (EW) target. This effort concerns the effects and mitigation of a particular type of EW known as jamming. That is, in order to deny GPS operation, an adversary may use multiple jammers that radiate interference either over the entire GPS operating bandwidth or selected portions thereof. The GPS user must counter this threat to operate successfully. This can be done by using antenna arrays that employ space-time (or space-frequency) adaptive processing. Such signal processing allows the GPS receiver to place antenna pattern nulls in the directions of any interferers, while simultaneously attempting to preserve the desired signals from GPS satellites. In this chapter, we will explain this methodology. We begin with an extremely simple approach, and then proceed to more sophisticated techniques.

4.2 Two-Element Adaptive Array

4.2.1 Theory of Operation

The simplest adaptive antenna uses a power minimization technique to place nulls at the locations of any jammers that are interfering with the operation of the antenna. In order to illustrate this approach, suppose we have an antenna with a voltage gain $g_1(\theta)$ that is attempting to receive a signal from an angle θ_s, but is being jammed by an interferer at an angle θ_j. The interference can be mitigated by adding a second (auxiliary) antenna with a voltage gain $g_2(\theta)$, as shown in Figure 4.1, and then weighting and combining the outputs of the two antennas to cancel the jammer. If the jammer interference voltage is far larger than either the signal or thermal noise voltages, a technique to mitigate the jammer would be to choose the unknown weight w in Figure 4.1, such that it minimizes the output power.

Let us assume the signal is received from $\theta_s = 0$ and produces a voltage[1] $s(t)$ $\exp(-i\omega_0 t)$ on an isotropic antenna, where $\omega_0 = 2\pi f_0$, f_0 is the carrier frequency, t is time, and $s(t)$ is an envelope that varies slowly in comparison with f_0. A jammer is assumed to lie at $\theta = \theta_j$ and produce a voltage $j(t)\exp(-i\omega_0 t)$ on an isotropic antenna. Therefore, the total voltage v_1 at the terminals of the reference antenna (after downconversion by $\exp(i\omega_0 t)$ is

$$v_1(t) = g_1(0)s(t) + g_1\left(\theta_j\right)j(t) + \xi_1(t) \tag{4.1}$$

where $\xi_1(t)$ is the main-antenna receiver thermal noise. Likewise, the voltage at the terminals of the auxiliary antenna is

$$v_2(t) = g_2(0)s(t) + g_2\left(\theta_j\right)j\left(t - \frac{L}{c}\sin\theta_j\right)e^{ik_0 L\sin\theta_j} + \xi_2(t) \tag{4.2}$$

where L is the separation between antennas, c is the speed of light, $k_0 = \dfrac{\omega_0}{c}$, and $\xi_2(t)$ is the antenna 2 receiver thermal noise voltage. These voltages are then weighted and combined to form

$$u(t) = v_1(t) - wv_2(t) \tag{4.3}$$

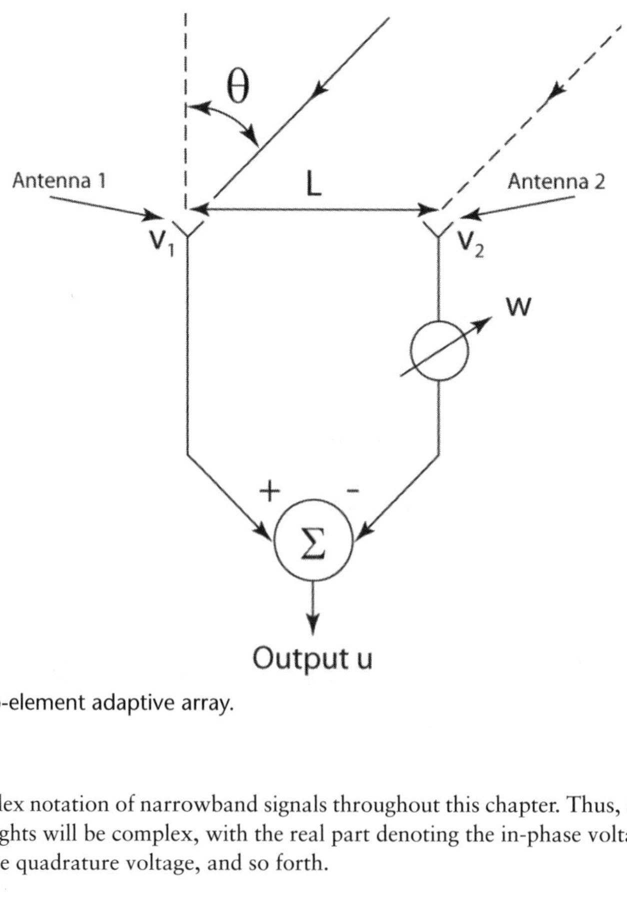

Figure 4.1 Two-element adaptive array.

1. We will use the complex notation of narrowband signals throughout this chapter. Thus, after downconversion, the antenna weights will be complex, with the real part denoting the in-phase voltage and the imaginary part denoting the quadrature voltage, and so forth.

We now choose w to minimize the mean output power $\left\langle |u|^2 \right\rangle$, where < > denotes an expectation. If we multiply $u(t)$ by its complex conjugate and average, we get

$$\left\langle |u|^2 \right\rangle = \left\langle |v_1|^2 \right\rangle - w\left\langle v_1^* v_2 \right\rangle - w^* \left\langle v_1 v_2^* \right\rangle + |w|^2 \left\langle |v_2|^2 \right\rangle \tag{4.4}$$

where the time dependence of u, v_1, and v_2 has been suppressed for brevity. In order to minimize $\left\langle |u|^2 \right\rangle$, we differentiate (4.4) with respect to w^* and set the result equal to zero, to get

$$w = \frac{\left\langle v_1 v_2^* \right\rangle}{\left\langle |v_2|^2 \right\rangle} \tag{4.5}$$

If (4.5) is used in (4.4), we find

$$\left\langle |u|^2 \right\rangle = \left\langle |v_1|^2 \right\rangle - \frac{\left|\left\langle v_1 v_2^* \right\rangle\right|^2}{\left\langle |v_2|^2 \right\rangle} \tag{4.6}$$

We will discuss $\left\langle |u|^2 \right\rangle$ in more detail in the next section.

If $v_1(k)$ is a sample $v_1(t)$ at $t = kT$, where T is the intersample period, then the averages in (4.5) can be estimated as

$$\left\langle v_1 v_2^* \right\rangle = \frac{1}{N_s} \sum_{k=1}^{N_s} v_1(k) v_2^*(k) \tag{4.7}$$

with similar expressions for $\left\langle |v_1|^2 \right\rangle$ and $\left\langle |v_2|^2 \right\rangle$. The number N_S of samples required will be discussed in Section 4.3.2. Once the weight in (4.5) has been estimated using (4.7), it can be applied to M samples (including those used for the estimate, if possible) and then periodically updated. This will also be discussed further in Section 4.4.

4.2.2 Effect of Bandwidth and Gain for Ideal Antennas

Consider the case where the interferer is a broadband noise jammer. Let us express the received jammer waveform $j(t)$ in terms of its Fourier transform $J(f)$

$$j(t) = \int_{-\infty}^{\infty} df \; H_B(f) J(f) e^{i2\pi ft} \tag{4.8}$$

where f is frequency, $J(f)$ is the Fourier transform of the jamming signal, and $H_B(f)$ is the system function of the front-end filter behind each antenna (the filter is used to limit the operating bandwidth). If $H_B(f)$ is an ideal, brickwall[2] filter of bandwidth B and $j(t)$ represents a noise jammer with power spectrum $\Phi_{jj}(f)$, then

$$\left\langle J(f)J^*(f')\right\rangle = \Phi_{jj}(f)\delta(f-f') \tag{4.9}$$

where is the Dirac delta function. We can use (4.8) and (4.9) to show that

$$\left\langle j(t)j^*(t+\tau)\right\rangle = \int_{-B/2}^{B/2} df\ \Phi_{jj}(f)e^{-i2\pi f\tau} \tag{4.10}$$

If the jammer power spectral density $\Phi_{jj}(f)$ is white (i.e., independent of frequency) over the operating bandwidth B, then it is readily shown that (4.10) gives

$$\left\langle j(t)j^*(t+\tau)\right\rangle = P_j\ \mathrm{sinc}(\pi B\tau) \tag{4.11}$$

where $P_j = \Phi_{jj}B$ is the jammer power received by an isotropic antenna and $\mathrm{sinc}(x) = \frac{\sin x}{x}$. If we use (4.11), and realize that the noises in the two channels are statistically independent both of each other and of the jammer voltage, we can use (4.1) and (4.2) to obtain all the expectations needed to evaluate (4.6) for a broadband jammer. These are

$$\left\langle |v_1|^2\right\rangle = G_1(\theta_j)P_j + \sigma_\xi^2 \tag{4.12}$$

$$\left\langle |v_2|^2\right\rangle = G_2(\theta_j)P_j + \sigma_\xi^2 \tag{4.13}$$

$$\left\langle v_1v_2^*\right\rangle = g_1(\theta_j)g_2^*(\theta_j)P_j\mathrm{sinc}\left(\frac{\pi \mathrm{BL}}{\mathrm{c}}\sin\theta_j\right)\exp\left(-ik_0L\sin\theta_j\right) \tag{4.14}$$

where σ_ξ^2 is the noise power, and G_1, G_2 are the antenna gains defined as $G_1(\theta) = |g_1(\theta)|^2$ and $G_2(\theta) = |g_2(\theta)|^2$. In writing (4.12) through (4.14), we have ignored the terms in (4.1) and (4.2) involving the signal because in realistic GPS systems, the signal power is much less (i.e., signal power << σ_ξ^2) than the thermal noise floor before processing. In fact, if the signal is above the thermal noise floor, and the jammer is absent, minimization of the output power will result in a null being placed in the direction of the signal, which is clearly not desirable. Thus, the power minimization would never be used in situations where the received signal is higher than the noise level.

If (4.12) through (4.14) are used in (4.6) and we assume that $G_2P_j >> \sigma_\xi^2$ (this is a reasonable assumption, because if the jammer was not strong compared with the noise, we would not need to bother canceling it), we obtain, approximately

2. That is, the frequency transfer function $H_B(f)$ of this filter is such that $H_B(f) = 1$ for $|f| < B/2$, and $H_B(f) = 0$, otherwise.

$$\left\langle |u|^2 \right\rangle = G_1 P_j \left[1 - \mathrm{sinc}^2 \psi_j\right] + \sigma_\xi^2 \left[\frac{G_1}{G_2} + 1\right] \tag{4.15}$$

where $\psi_j = \pi BLc^{-1}$, $G_1 = G_1(\theta_j)$, $G_2 = G_2(\theta_j)$. The first term on the right-hand side of (4.15) is the residual jammer power after adaptation and the second term is the noise residue (sometimes called noise carryover). Quite often the system bandwidth is such that $\frac{BL}{c} << 1$, so that (4.15) can be approximated by

$$\left\langle |u|^2 \right\rangle = \frac{1}{3} G_1 P_j \left(\frac{\pi BL}{c} \sin\theta_j\right)^2 + \sigma_\xi^2 \left(\frac{G_1}{G_2} + 1\right) \tag{4.16}$$

where the first term on the right-hand side of (4.16) represents the jammer residue after cancellation and the second term is the noise. There are two very important points to note from (4.16):

1. Bandwidth is an important factor. For zero bandwidth, the jammer residue vanishes, but for finite bandwidth, it increases as the square of bandwidth. Therefore, a simple space-only adaptive array of the type shown in Figure 4.1 is inadequate for broadband jamming.
2. The gain of the reference antenna at the jammer location ($\theta = \theta_j$) should not be much larger than the gain of the auxiliary at that angle. Otherwise, there can be a significant increase in the noise residue. Thus, the auxiliary antenna gain must be selected carefully.

Another point to be discussed but is not evident from (4.15) and (4.16), is the effect of the second antenna on the radiation pattern of the antenna array when the adaptive weight w is used. In order to derive this pattern, which is called the adapted radiation pattern, we note that a point source in the far field of the array at an arbitrary angle θ produces an array output $u(\theta)$ given by

$$u(\theta) = g_1(\theta) - w g_2(\theta) e^{ik_0 L \sin\theta} \tag{4.17}$$

If (4.13) and (4.14) are used in (4.5), we find for $\frac{\pi BL}{c} << 1$ (narrowband limit) that

$$w \simeq \frac{g_1(\theta_j)}{g_2(\theta_j)} e^{-ik_0 L \sin\theta_j} \tag{4.18}$$

where thermal noise is ignored. If (4.18) is used in (4.17), the result is

$$u(\theta) = g_1(\theta) - \frac{g_1(\theta_j)\, g_2(\theta)}{g_2(\theta_j)} e^{ik_0 L(\sin\theta - \sin\theta_j)} \tag{4.19}$$

Note that when $\theta = \theta_j$ (i.e., at the jammer location), $u(\theta_j) = 0$, so that there is a pattern null on the jammer. However, there are no guarantees on the value of $u(\theta)$

at the signal location $\theta = 0$. Because we have not constrained the response in the signal direction, we have just "taken our chances," and hoped that there would be no significant signal dropouts. However, as is evident from (4.19), it is possible to have a fade (i.e., deep loss) at the signal direction $\theta = 0$. This is the reason why more complex algorithms are often used.

Although we have illustrated power minimization for only one auxiliary antenna, the results are readily generalized to multiple auxiliary antennas. In the narrowband limit, it is found that N auxiliary antennas (a total of $N + 1$ antennas), can control the interference from N narrowband jammers provided the antenna geometry is properly chosen.

4.2.3 Effect of Dispersive Errors (Channel Mismatch)

In the preceding analysis, we implicitly assumed that the receivers and antennas of the two channels were perfectly matched across the frequency band of operation. In practice, because of manufacturing differences from receiver to receiver and antenna to antenna, this is not the case. This mismatch leads to a limitation on the amount of jammer cancellation that is possible. According to (4.16), if the antenna pair in Figure 4.1 had no dispersive errors, and was illuminated by a jammer at broadside ($\theta_j = 0$), the jammer residue (as given by the first term on the right-hand side) would vanish, independent of operating bandwidth B. We will now demonstrate that when dispersive errors are included, this is no longer the case. Suppose both antennas in Figure 4.1 now are isotropic (i.e., $g_1 = g_2 = 1$) and the reference channel (channel 1) has a frequency response $H_1(f)$ and the auxiliary (channel 2) has a different response $H_2(f)$. Then, if these antennas are illuminated by a broadband noise jammer located at broadside, the voltages v_1 and v_2 in Figure 4.1 are

$$v_n(t) = \int_{-B/2}^{B/2} df\, H_n(f) J(f) \exp(i2\pi f t) \tag{4.20}$$

where f is frequency, B is again the bandwidth of the front-end brickwall filter in each channel, $n = 1$ corresponds to antenna 1, $n = 2$ is for antenna 2, and $J(f)$ is the Fourier transform of the received jammer voltage envelope $j(t)$. In (4.20), for ease of presentation, we have neglected the noise in each channel, as well as any signal present.

The two channels are designed to be identical, but because of construction errors, they are actually different. Thus, although $H_n(f)$ is designed to ideally have a constant amplitude and a linear phase across the passband B, its actual response is

$$H_n(f) = \left[1 + \varepsilon_n(f)\right] e^{i\beta_0 f + i\delta_n(f)} \tag{4.21}$$

where β_0 is arbitrary and $\varepsilon_n(f)$ and $\delta_n(f)$ are the deviations from the desired amplitude and phase[3] for receiver n. The function $\delta_n(f)$ should not be confused with the

3. Usually, the phase $\beta_0 f + \delta_n(f)$ in (4.21) will not be directly measured. Instead, a quantity called the group delay, $T_g(f)$, is usually available, where: $T_g(f) = \frac{1}{2\pi}\frac{d(phase)}{df} = \beta_0 + \frac{1}{2\pi}\frac{d\delta_n}{df}$.

Dirac delta function $\delta(f)$ in (4.9). In general, unless a system is very poorly designed, we may assume $|\varepsilon_n| << 1$ and $|\delta_n| << 1$.

We now wish to calculate the average power after cancellation for a system with dispersive errors. If we use (4.9) and (4.20), it can be shown that

$$\left\langle v_n(t) v_m^*(t+\tau) \right\rangle = \int_{-B/2}^{B/2} df \; \Phi_{jj}(f) H_n(f) H_m^*(f) e^{-i2\pi f\tau} \tag{4.22}$$

where n and m can be either 1 or 2. For a white-noise jammer, $\Phi_{jj}(f)$ is independent of frequency, so (4.22) becomes

$$\left\langle j(t) j^*(t+\tau) \right\rangle = \int_{-B/2}^{B/2} df \; \Phi_{jj}(f) e^{-i2\pi f\tau} \tag{4.23}$$

We now assume the jammer is at broadside ($\theta_j = \theta$). Then, using (4.23) in (4.6) yields for the expected residual jammer power after cancellation

$$\left\langle |u|^2 \right\rangle = \frac{P_j}{B^2} \left[\int_{-B/2}^{B/2} df \; |H_1|^2 \int_{-B/2}^{B/2} df \; |H_2|^2 - \left| \int_{-B/2}^{B/2} df \; H_1 H_2^* \right|^2 \right] \tag{4.24}$$

where we have suppressed the frequency dependence of $H_1(f)$, $H_2(f)$, and have approximated the term

$$\frac{1}{B} \int_{-B/2}^{B/2} df \; |H_2|^2$$

that appears in the denominator of (4.24) by unity because $|H_2|^2 = (1+\varepsilon_2)^2 \simeq 1$, and $|\varepsilon_2| << 1$. Equation (4.24) demonstrates that because of channel mismatch, the jammer is not completely canceled.

When the jammer is located off boresight and system noise is included, (4.24) is replaced by

$$\left\langle |u|^2 \right\rangle = \frac{P_j}{B^2} \left[\int_{-B/2}^{B/2} df \; |H_1|^2 \int_{-B/2}^{B/2} df \; |H_2|^2 - \left| \int_{-B/2}^{B/2} df \; H_1 H_2^* e^{i\psi_j} \right|^2 \right] + 2\sigma_\xi^2 \tag{4.25}$$

where $\psi_j = 2\pi f L c^{-1} \sin\theta_j$ and is the noise in each channel. In the limit when $H_1 = H_2 = 1$, (4.25) reduces to (4.15) with $G_1 = G_2 = 1$.

There is a quantity known as cancellation ratio that is used to quantify channel mismatch. In order to calculate cancellation ratio, broadband white noise is injected into channels 1 and 2 and the output voltage is subtracted, giving

$$\Delta v(t) = \int_{-B/2}^{B/2} df \; J(f) \left[H_1(f) - H_2(f) \right] e^{i2\pi f t} \tag{4.26}$$

If we use (4.9), it is readily shown that

$$\left\langle |\Delta v|^2 \right\rangle = \frac{P_j}{B} \int_{-B/2}^{B/2} df \, |H_1 - H_2|^2 \tag{4.27}$$

The cancellation ratio is then defined as $CR = -10 \log_{10}(CC)$ where

$$CC = \frac{\left\langle |\Delta v|^2 \right\rangle}{\left[\left\langle |v_1|^2 \right\rangle \left\langle |v_2|^2 \right\rangle \right]^{1/2}} = \frac{\int_{-B/2}^{B/2} df \, |\mathrm{H}_1 - H_2|^2}{\left[\int_{-B/2}^{B/2} df \, |\mathrm{H}_1|^2 \int_{-B/2}^{B/2} df \, |\mathrm{H}_2|^2 \right]^{1/2}} \tag{4.28}$$

If we again employ the approximation used earlier, we have

$$CC = \frac{1}{B} \int_{-B/2}^{B/2} \mathrm{df} \, |\mathrm{H}_1(f) - \mathrm{H}_2(f)|^2 \tag{4.29}$$

If we compare (4.29) with (4.24), it is evident that, although the cancellation ratio yields a measure of channel mismatch, it is not exactly equal to the jammer residue after cancellation.

4.2.4 Effect of Multipath

When an adaptive antenna array is placed on an aircraft, or in the presence of other structures, the array must also cope with the effect of jammer multipath. Let us again assume that the jammer is located at antenna boresight ($\theta_j = \theta$) and that there is a scatterer of strength β near to the antenna. Then after downconversion, the interference voltages on antennas 1 and 2 are (ignoring noise)

$$v_1(t) = j(t) + \beta j(t - \tau_1) \, e^{i\omega_0 \tau_1} \tag{4.30a}$$

$$v_2(t) = j(t) + \beta j(t - \tau_2) \, e^{i\omega_0 \tau_2} \tag{4.30b}$$

where τ_1 and τ_2 are the differential delays between the multipath scatterer and antennas 1 and 2, respectively. The Fourier transforms of (4.30a) and (4.30b) are

$$V_1(f) = J(f)\left[1 + \beta_1 e^{-i2\pi f \tau_1}\right] \tag{4.31}$$

$$V_2(f) = J(f)\left[1 + \beta_2 e^{-i2\pi f \tau_2}\right] \tag{4.32}$$

where $\beta_1 = \beta \exp(i\omega_0 \tau_1)$ and $\beta_2 = \beta \exp(i\omega_0 \tau_2)$. In this case, we obtain

$$\left\langle v_1(t)v_2^*(t+\tau)\right\rangle = \int_{-B/2}^{B/2} df\, \Phi_{jj}(f)\tilde{H}_1(f)\tilde{H}_2^*(f)e^{-i2\pi f\tau} \tag{4.33}$$

where

$$\tilde{H}_1(f) = \left[1+\beta_1 e^{-i2\pi f\tau_1}\right] \tag{4.34}$$

$$\tilde{H}_2(f) = \left[1+\beta_2 e^{-i2\pi f\tau_2}\right]. \tag{4.35}$$

Upon comparing (4.33) with (4.22), we see that multipath produces the same type of effect as channel mismatch. In this case, the residual jammer power is

$$\left\langle |u|^2 \right\rangle = \frac{\dfrac{P_j}{B}\left[\int_{-B/2}^{B/2} df\, \left|\tilde{H}_1\right|^2 \int_{-B/2}^{B/2} df\, \left|\tilde{H}_2\right|^2 - \left|\int_{-B/2}^{B/2} df\, \tilde{H}_1\tilde{H}_2^*\right|^2\right]}{\int_{-B/2}^{B/2} df\, \left|\tilde{H}_2\right|^2} \tag{4.36}$$

Unlike channel mismatch, it is no longer acceptable to approximate the denominator by unity because $|\beta_2|$ may not be very small in comparison with unity.

When multipath and channel mismatch are both present, the total jammer residue power after cancellation is the sum of the results in (4.25) and (4.36).

It is now clear by examining (4.25) and (4.36) that a two-element adaptive antenna can cancel a narrowband jammer, but is increasingly ineffective as the jammer bandwidth is increased. This suggests that using only spatial (i.e., multiple antenna elements) degrees of freedom is insufficient to effectively cancel a broadband jammer plus jammer multipath, and temporal degrees of freedom are also required.

4.3 Space-Time Adaptive Array

4.3.1 Introduction

In the last section, we demonstrated that broadband jammers and their multipath cannot be totally cancelled using a space-only adaptive antenna, because a spatial array cannot match the dispersive and multipath effects. This suggests that we should add an adaptive finite impulse response (FIR) filter behind each antenna element, thus producing the configuration shown in Figure 4.2, called a space-time adaptive processor (STAP).

In Figure 4.2, the time taps are spaced by the intersampling delay (i.e., $T = \frac{1}{f_s}a$ where f_s = sampling rate) and the filter weights w_{nq} (n denotes antenna and q denotes time tap), where $n = 1 \ldots$ N, $q = 1 \ldots$ Q) are determined adaptively using algorithms that will be discussed later. It is readily seen that once the weights w_{nq} have been selected, the FIR filter behind antenna n produces a transfer function

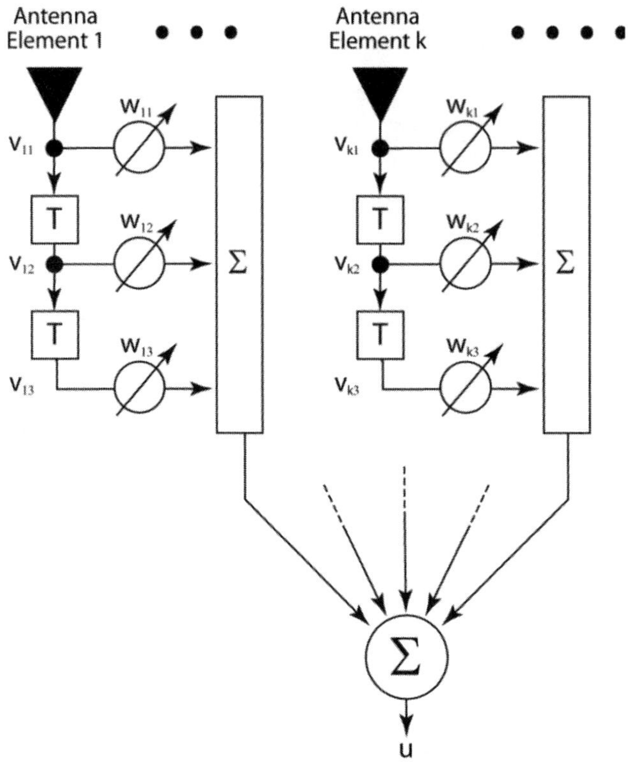

Figure 4.2 Space-time adaptive array.

$$\tilde{H}_n(f) = \sum_{q=1}^{Q} w_{nq}\, e^{i2\pi(q-1)fT} \tag{4.37}$$

which can be used to compensate for the frequency variation produced by multipath, channel dispersion, and so forth.

The configuration in Figure 4.2 is not the only option to minimize the effect of jamming. A second architecture is suggested by our result in the last section that it is easier to cancel narrowband jammers than broadband ones. If the total band B can be segmented into Q subbands, it should be possible to cancel the interference separately in each subband. This decimation of the system bandwidth B can be accomplished using the discrete Fourier transform, and the resulting architecture is shown in Figure 4.3. This type of processor is called a space-frequency adaptive processor (SFAP).

In the sections to follow, we will concentrate on the space-time adaptive processor, but the space-frequency adaptive processor can achieve essentially the same results[4]

4. For a more complete discussion of the space-frequency adaptive processor, please refer to the paper in the Bibliography by Fante and Vaccaro published in the *IEEE Transactions on Aerospace and Electronic Systems*, Vol. AES-36, 2000, pp. 549–564.

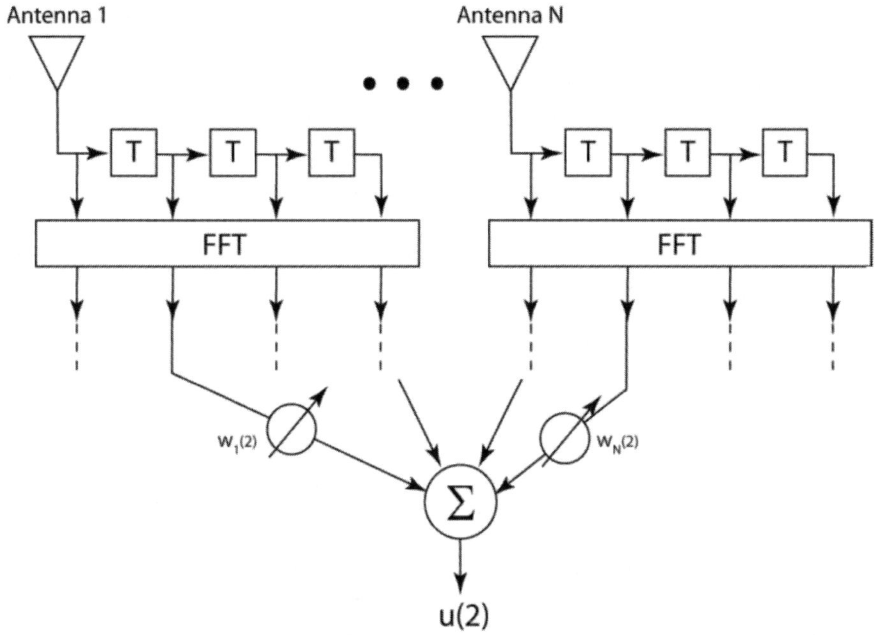

Figure 4.3 Space-frequency adaptive array.

4.3.2 Calculating the Adaptive Weights

In order to discuss how the adaptive weights are computed, let us refer to Figure 4.4, and assume there are N isotropic antenna elements located at positions (x_n, y_n, z_n) where $n = 1, 2, \ldots N$ and the angles (θ, Φ) shown in Figure 4.4 denote the direction to a GPS satellite of interest. The origin in Figure 4.4 is chosen as the geometric center of the array. Also, in the discussion to follow, mutual coupling between antennas will be ignored.

Let us first discuss a constrained optimization approach. We desire to minimize the jammer interference while simultaneously pointing a receive beam in the

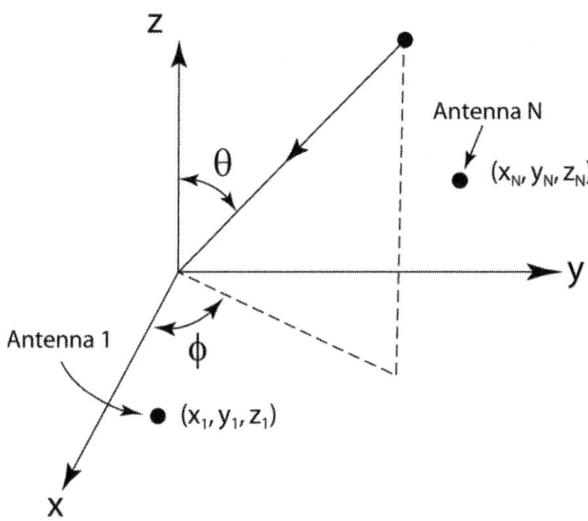

Figure 4.4 Location of N antennas.

direction of a GPS satellite of interest. We first need to determine the voltages that would be produced on all the taps in the architecture of Figure 4.2 if the array is illuminated by a jammer located in the direction (θ_j, Φ_j). The voltage this jammer would produce on an antenna located at the origin in Figure 4.4 is defined as

$$j_0(t) = j(t)\exp\left(i2\pi f_0 t\right) \tag{4.38}$$

where $j(t)$ is an envelope that varies slowly in comparison with f_0. In writing (4.38), it is assumed that the power spectral density is centered on the GPS carrier frequency, f_0. Therefore, if the antenna were located at (x_n, y_n, z_n), the voltage at its terminals is

$$j_n(t) = j\left(t - \frac{r_n}{c}\right)\exp\left[-i2\pi f_0\left(t - \frac{r_n}{c}\right)\right] \tag{4.39}$$

where r_n is the differential path length between the origin and the location of antenna n and is given by $r_n = x_n \sin\theta_j \cos\Phi_j + y_n \sin\theta_j \sin\Phi_j + z_n \cos\theta_j$. After downconversion (i.e., multiplication by $\exp(i2\pi f_0 t)$), we find that the jammer voltage on time tap q of antenna n is

$$j_{nq}(t) = j\left(t - (q-1)T - \frac{r_n}{c}\right)\exp\left(\frac{i2\pi f_0 r_n}{c}\right) \tag{4.40}$$

If we take the Fourier transform of the envelope $j(t)$, it is evident that in the frequency domain

$$J_{nq}(f) = J(f)\exp\left[i2\pi\left(f_0 + f\right)\frac{r_n}{c} + i2\pi(q-1)fT\right] \tag{4.41}$$

where $J(f)$ is the Fourier transform of the jammer envelope $j(t)$. In practice, all received signals are bandpass filtered after downconversion, so that (4.41) is multiplied by $H_B(f)$, where $H_B(f)$ is the transfer function of the bandlimiting brickwall filter. For the moment, we will ignore the effect of channel mismatch and multipath, but will include these effects later.

Likewise, a GPS satellite at (θ_s, Φ_s) radiating a signal that produces a voltage $s(t)\exp(-i2\pi f_0 t)$ on an antenna at the origin would produce a signal at time tap q of antenna n given by

$$s_{nq}(t) = s\left[t - (q-1)T - \frac{R_n}{c}\right]\exp\left(\frac{i2\pi f_0}{c}R_n\right) \tag{4.42}$$

with a corresponding Fourier transform

$$S_{nq}(f,\theta_s,\varphi_s) = S(f)\exp\left[i2\pi(f_0+f)\frac{R_n}{c}+i2\pi(q-1)fT\right] \tag{4.43}$$

where $R_n = x_n\sin\theta_s\cos\Phi_s + y_n\sin\theta_s\sin\Phi_s + z_n\cos\theta_s$. Again, we can readily include the effect of the bandlimiting filter by multiplying (4.43) by $H_B(f)$.

The total voltage present on time tap q of antenna n then is

$$v_{nq}(t) = j_{nq}(t) + s_{nq}(t) + \xi_{nq}(t) \tag{4.44}$$

where $\xi_{nq}(t)$ is the noise voltage at time tap q of antenna n. The output of the array (see Figure 4.2) is the weighted sum of all the antenna/tap voltages given by

$$u(t) = \sum_{n=1}^{N}\sum_{q=1}^{Q} w_{nq}v_{nq}(t) \tag{4.45}$$

If we define an $NQ \times 1$ weight vector

$$\mathbf{w}^T = \left[w_{11}\ w_{12}\ldots w_{1Q}\ \mathrm{w}_{21}\ldots\mathrm{w}_{2Q}\ \mathrm{w}_{31}\ldots\mathrm{w}_{NQ}\right] \tag{4.46}$$

and a total-voltage vector

$$\mathbf{v}^T = \left[v_{11}\ v_{12}\ldots v_{1Q}\ \mathrm{v}_{21}\ldots\mathrm{v}_{2Q}\ \mathrm{v}_{31}\ldots\mathrm{v}_{NQ}\right] \tag{4.47}$$

where the superscript T denotes a transpose, it is evident that (4.45) can be written in vector notation as

$$u = \mathbf{w}^T\mathbf{v} = \mathbf{v}^T\mathbf{w} \tag{4.48}$$

We now wish to find the weights that minimize the mean residual output power $\langle|u|^2\rangle$ while simultaneously preserving the signal from the GPS satellite. (This is equivalent to pointing a receive beam in the direction of the satellite.) The voltage at the output of the array that is received from the GPS satellite at (θ_s, Φ_s) is

$$u_s = \mathbf{w}^T\mathbf{s} \tag{4.49}$$

where

$$\mathbf{s}^T = \left[s_{11}\ s_{12}\ldots s_{1Q}\ \ s_{21}\ s_{22}\ldots s_{2Q}\ \mathrm{s}_{31}\ldots\mathrm{s}_{NQ}\right] \tag{4.50}$$

and expressions for s_{nq} for $n = 1$ to N and $q = 1$ to Q are given by (4.42).

Let us now minimize $\langle|u|^2\rangle$ subject to the constraint that the signal voltage u_s = 1 (i.e., signal voltage is preserved). If we use (4.48) to expand $\langle|u|^2\rangle$ as

$$\langle|u|^2\rangle = \mathbf{w}^H\langle\mathbf{v}^*\mathbf{v}^T\rangle\mathbf{w} \tag{4.51}$$

where the superscript H denotes a transpose conjugate, it is well known that this minimization can be accomplished by minimizing

$$\Gamma = \mathbf{w}^H C\mathbf{w} - \mu \mathbf{w}^H \mathbf{s}^* \tag{4.52}$$

where C is the covariance matrix defined as

$$C = \left\langle \mathbf{v}^* \mathbf{v}^T \right\rangle \tag{4.53}$$

μ is a Lagrange multiplier. For mathematical convenience, in (4.52) we have replaced the constraint equation $\mathbf{w}^T\mathbf{s} = 1$ by its complex conjugate $\mathbf{w}^H\mathbf{s} = 1$, which is clearly equivalent.

If we now differentiate (4.52) with respect to $\mathbf{w}^H$ and set the result equal to zero, we find

$$C\mathbf{w} - \mu \mathbf{s}^* = 0 \tag{4.54}$$

If we multiply (4.54) by C^{-1}, we obtain

$$\mathbf{w} = \mu C^{-1}\mathbf{s}^* \tag{4.55}$$

and if (4.55) is substituted into the constraint equation $\mathbf{w}^H\mathbf{s}^* = 1$, we find $\mu = (\mathbf{s}^T C^{-1}\mathbf{s}^*)^{-1}$. Consequently, the desired weight vector has been obtained. The processor that uses the weight vector calculated in (4.55) is sometimes called the constrained minimum-variance beamformer. It can be shown that the performance of this algorithm is relatively insensitive to small errors in the angular location of the steering direction (θ_S, Φ_S).

There is a subtlety when applying (4.55) to a GPS signal because each GPS satellite transmits a pseudorandom signal $s(t)$, so the question is "What do we do when $s(t)$ is a stochastic quantity?" In this case, we perform the optimization a little differently. We instead minimize the mean squared difference between the desired output of the array, $s(t)$, and the actual output $\mu(t) = \mathbf{w}^H\mathbf{v}$. That is, we minimize

$$\Gamma_1 = \left\langle \left| s - \mathbf{w}^T \mathbf{v} \right|^2 \right\rangle \tag{4.56}$$

where $s(t)$ is the scalar denoting the signal transmitted by the GPS satellite of interest (i.e., the satellite located at (θ_s, Φ_s)).

Let us rewrite Γ_1 as

$$\Gamma_1 = \left\langle ss^* \right\rangle - \left\langle s^* \mathbf{v}^T \right\rangle \mathbf{w} - \mathbf{w}^H \left\langle \mathbf{v}^* s \right\rangle + \mathbf{w}^H C\mathbf{w} \tag{4.57}$$

where $C = \left\langle \mathbf{v}^* \mathbf{v}^T \right\rangle$. If we differentiate Γ_1 with respect to $\mathbf{w}^H$, we obtain

$$C\mathbf{w} - \left\langle \mathbf{v}^* s \right\rangle = 0 \tag{4.58}$$

If both sides of (4.58) are multiplied by C^{-1}, we obtain

$$\mathbf{w} = C^{-1}\boldsymbol{\rho}^* \tag{4.59}$$

where

$$\boldsymbol{\rho}^* = \langle \mathbf{v}^* s \rangle \tag{4.60}$$

We can simplify (4.60) further by recalling from (4.44) that the vector **v** is

$$\mathbf{v} = \mathbf{j} + \mathbf{s} + \boldsymbol{\xi} \tag{4.61}$$

where $\mathbf{j}^T = [j_{11}\ j_{12} \cdots j_{NQ}]$, $\mathbf{s}^T = [s_{11}\ s_{12} \cdots s_{NQ}]$, and $\boldsymbol{\xi}^T = [\xi_{11}\ \xi_{12} \cdots \xi_{NQ}]$. Because the transmitted GPS sequence is statistically independent of both the jammer voltage and the noise, we have

$$\langle \mathbf{j}^* s \rangle = 0 \tag{4.62a}$$

$$\langle \boldsymbol{\xi}^* s \rangle = 0 \tag{4.62b}$$

so that (4.60) simplifies to

(4.63)

By using (4.42) for s_{nq}, it is possible to express the terms in (4.63) in terms of the covariance $\langle s(t) s^*(t+\tau) \rangle$ of the signal.

If we compare the solution in (4.59) with the result in (4.55), it is evident that both have the same structure; the weight vector is the inverse of the covariance matrix operating on a vector $\mathbf{s}^*$ or ρ, which is known as the steering vector.

Typically, the solutions given by (4.55) and (4.59) are complex to implement, because they require an approximate knowledge of the angular location of each GPS satellite in view and require a separate constraint and processing chain for each of these GPS satellites. A much less complex implementation results if we simply minimize the output power of the array, as we did for the simple two-element array in Section 4.2. That is, we recognize that the GPS signals are well below (typically 20 to 30 dB lower) the receiver noise floor, so there is no danger of canceling desired signals when output power is minimized unless the GPS satellites have an angular location that is close to the angular location of jammer. Thus, we can simply "take our chances" and hope that a sufficient number of GPS satellites are at angular locations outside of the pattern nulls, so that a position solution can be achieved. For this simple power minimization (also called minimum variance) algorithm, we minimize the output power of the array subject to the constraint that the weight on the center tap of antenna 1 is unity. That is, if we use an odd number Q time taps per antenna, then the weight w_{1P} is constrained to be unity, and all other weights are free, where $P = \frac{(Q+1)}{2}$. This constraint can be expressed mathematically as

$$\mathbf{w}^T \mathbf{h} = 1 \tag{4.64}$$

where $\mathbf{h}_T = [0\ 0 \ldots 1 \ldots 0 \ldots 0]$ and the unity value corresponds to tap $P = \dfrac{(Q+1)}{2}$ on antenna 1. We now minimize

$$\Gamma' = \mathbf{w}^H C \mathbf{w} - \mu_1 \mathbf{w}^H \mathbf{h}^* \tag{4.65}$$

where μ_1 is the Lagrange multiplier and we obtain the solution

$$\mathbf{w} = \mu_1 C^{-1} \mathbf{h}^* \tag{4.66}$$

where $\mu_1 = (\mathbf{h}^T C^{-1} \mathbf{h})^{-1}$. Observe that the optimum weight vector again has the generic form of the inverse covariance operating on a steering vector. There are a number of other optimization approaches used, but all lead to the same generic result $\mathbf{w} = KC^{-1}\mathbf{h}_e^*$ where K is a constant and $\mathbf{h}_e^*$ is the appropriate steering vector.

In order to show the differences in performance between constrained and unconstrained algorithms, let us consider a hexagonal, seven-element planar array with elements at positions (x_n, y_n, z_n) given by (0, 0, 0), $(\Delta, 0, 0)$, $\left(\Delta\cos\frac{\pi}{3},\ \Delta\sin\frac{\pi}{3},\ 0\right)$, $\left(\Delta\cos\frac{2\pi}{3}, \Delta\sin\frac{2\pi}{3}, 0\right)$, $(-\Delta,\ 0,\ 0)$, $\left(\Delta\cos\frac{4\pi}{3}, \Delta\sin\frac{4\pi}{3}, 0\right)$, $\left(\Delta\sin\frac{5\pi}{3}, \Delta\cos\frac{5\pi}{3}, 0\right)$, where Δ = one-half wavelength. Suppose further there are five strong (jammer power very much greater than thermal noise power) jammers at angles (θ_j, Φ_j), equal to (81°, –30°), (90°, –110°), (79°, 45°), (60°, 121°), (88°, 175°), and that each antenna employs only one time tap ($Q = 1$). Then, we can define the response of this array to a signal incoming from an arbitrary angle (θ_s, Φ_s), by using the magnitude squared of (4.49), which we rewrite as

$$P_s(\theta_s, \varphi_s) = \left|\mathbf{w}^T \mathbf{s}(\theta_s, \varphi_s)\right|^2, \tag{4.67}$$

where $\mathbf{s}^T = [s_1 \ldots s_7]$, $s_n = \exp\left(\dfrac{i2\pi f_0 R_n}{c}\right)$, $R_n = x_n \sin\theta_s \cos\Phi_s + y_n \sin\theta_s \sin\Phi_s + z_n \cos\theta_s$, and $n = 1, 2 \ldots 7$.

In Figures 4.5 and 4.6, we show this response pattern in polar form for the constrained (4.55) and unconstrained (4.66) algorithms, respectively. The center of each plot corresponds to $\theta_s = 0$ and the outer edge corresponds to $\theta_s = 90°$. The azimuth angle Φ increases counterclockwise from 0° to 360°. We note from Figure 4.5 that the constrained algorithm places a distinct null (shown shaded) on each of the jammers (denoted by an x), whereas from Figure 4.6, we see that the unconstrained algorithm places nulls that consume a larger solid angle. Reception from GPS satellites that lie in these shaded areas is denied. Using the more-complex constrained algorithm allows a larger portion of the upper hemisphere to be available to the GPS receiver.

4.3.3 Frequency-Angle Response

We now wish to obtain the response in the frequency domain to a signal from an arbitrary direction (θ, Φ). If we assume that only the signal is present and then take the Fourier transform of (4.45) and include the effect of the digital bandlimiting filter that is placed after downconversion in each channel, we obtain

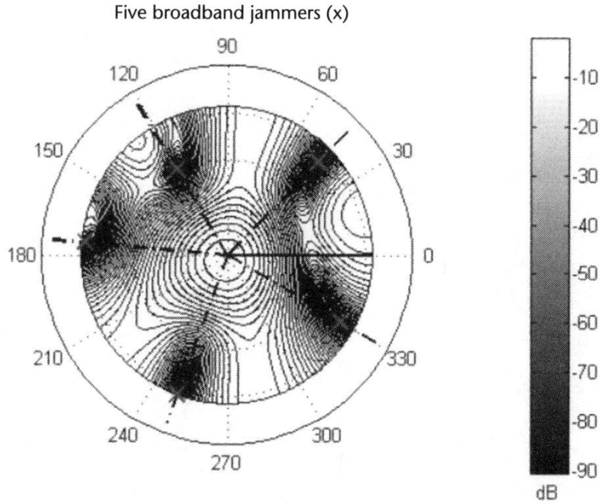

Figure 4.5 Adapted radiation pattern of adaptive arrays when five jammers are present and the constrained algorithm is used.

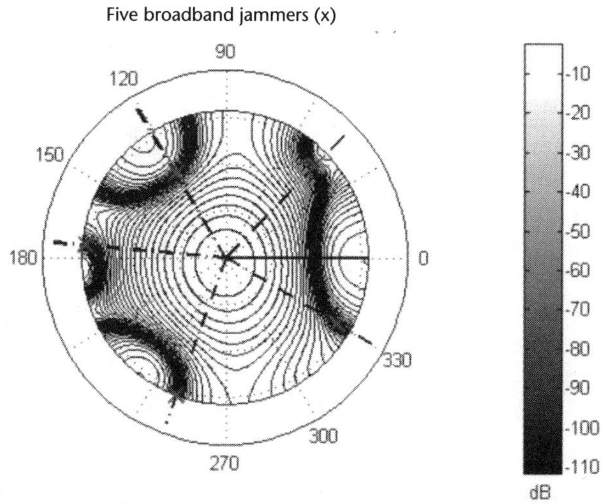

Figure 4.6 Adapted radiation pattern of adaptive arrays when five jammers are present and the unconstrained (power minimization) algorithm is used.

$$U(f,\theta,\varphi) = H_B(f)\sum_{n=1}^{N}\sum_{q=1}^{Q} w_{nq} S_{nq}(f,\theta,\varphi) \tag{4.68}$$

where $S_{nq}(f,\theta,\Phi)$ is given by (4.43) for the case when $\theta = \theta_s$, $\Phi = \Phi_s$. If (4.43) is used in (4.68), we then find

$$U(f,\theta,\varphi) = S(f)\ H(f,\theta,\varphi) \tag{4.69}$$

where $H(f, \theta, \Phi)$ is the frequency-angle response of the adaptive array given by

$$H(f,\theta,\varphi) = H_B(f)\sum_{n=1}^{N}\sum_{q=1}^{Q} \mathrm{w}_{nq}\exp\left[i2\pi\left(f_0 + f\right)\frac{R_n}{c} + i2\pi(q - QQ)fT\right] \tag{4.70}$$

$R_n = x_n\sin\theta\cos\Phi + y_n\sin\theta\sin\Phi + z_n\cos\theta$, $QQ = \dfrac{(Q+1)}{2}$ and we have shifted the time reference from the first tap to the center tap of each FIR filter. We caution that (4.70) is the response of an ideal array in which channel dispersion and multipath are negligible and mutual coupling is ignored. However, in practice, each channel will have a slightly different analog frequency response (see Figure 4.7).

We can account for these effects by generalizing (4.70) to

$$H(f,\theta,\varphi) = H_B(f)\sum_{n=1}^{N} \mathrm{H}_n(f,\theta,\varphi)\sum_{q=1}^{Q} \mathrm{w}_{nq}\exp\left[i2\pi\left(f_0 + f\right)\frac{R_n}{c} + i2\pi(q - QQ)fT\right] \tag{4.71}$$

where $H_n(f, \theta, \Phi)$ is the transfer function of antenna n and its associated analog hardware, including channel mismatch. Note that H_n depends on the angle of incidence of the incoming radiation.

It is useful to rewrite (4.71) in vector notation. If we define the vectors $\mathbf{a}^T = \left[a_1 \ldots a_n \ldots a_N\right]$, $\mathbf{b}^T = \left[b_1 \ldots b_q \ldots b_Q\right]$ and the weight matrix

$$W = \begin{bmatrix} w_{11} \cdots & w_{1Q} \\ w_{21} \cdots\cdots & \cdots \\ w_{N1} \cdots & w_{NQ} \end{bmatrix}$$

where $a_n = H_B(f)H_n(f,\theta,\varphi)\exp\left[\dfrac{i2\pi\left(f_0 + f\right)R_n}{c}\right]$ and $b_q = \exp[i2\pi(q - QQ)fT]$, then (4.71) can be rewritten more compactly as

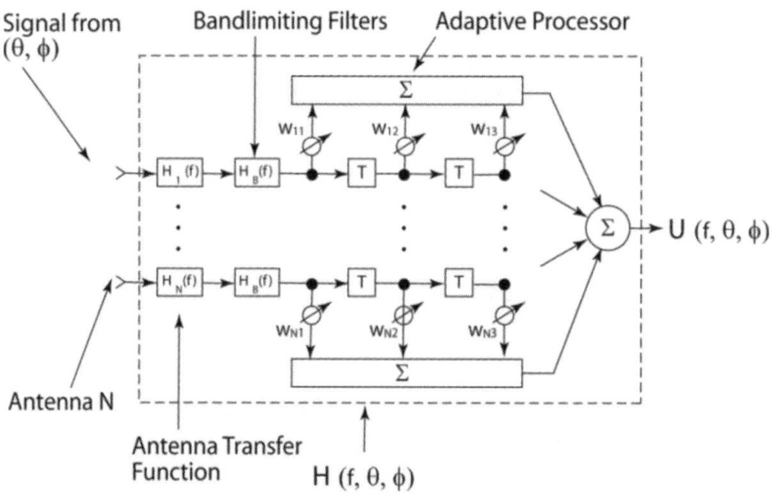

Figure 4.7 Overall transfer function of *N*-element, 3-tap adaptive processor.

$$H(f,\theta,\varphi) = \mathbf{a}^T W\mathbf{b} \tag{4.72}$$

where the superscript T denotes a transpose.

Furthermore, by using vector notation, it is straightforward to include mutual coupling. Let us assume that all antennas are well matched and that the mutual impedance between antennas is much less than the self-impedance. Then, if we define the antenna scattering matrix[5] in the frequency domain as

$$\Lambda(f) = \begin{bmatrix} \Lambda_{11}(f) & \dots & \Lambda_{1N}(f) \\ \cdot & & \\ \cdot & & \\ \cdot & & \\ \Lambda_{N1}(f) & \dots & \Lambda_{NN}(f) \end{bmatrix}$$

the transfer function with mutual coupling included is

$$H(f,\theta,\varphi) = \mathbf{a}^T (I + \Lambda)^T W\mathbf{b} \tag{4.73}$$

where I is the $N \times N$ identity matrix. In practice $\Lambda(f)$ will rarely be completely measured. For a well-designed antenna array, the following result is generally true:

$$|\Lambda_{nm}| \ll 1$$

for all n and m. For a very closely spaced (i.e., antenna elements spaced by a quarter wavelength or less) antenna array, these assumptions may no longer be valid.

4.4 Covariance Matrix Estimate

In (4.7), we indicated that the expectation of the various cross correlations that are required to calculate the covariance matrix, C, can be estimated using the temporal samples in each channel. Thus, the cross correlation between the voltage on time tap q of antenna n and the voltage on time tap p of antenna m is usually estimated as

$$\langle v_n(t-qT)v_m^*(t-pT)\rangle = \frac{1}{N_s}\sum_{k=0}^{N_s-1} v_n(t-qT-kT)v_m^*(t-pT-kT) \tag{4.74}$$

The number of samples, N_s, required to give a good (i.e., sufficient to result in jammer cancellation that is within 1 dB of the cancellation achieved using $N_s \to \infty$)

5. The antenna scattering matrix relates the reflected voltages to the incident voltages in an M-port network. For example, if (u_1, u_2) were the incident voltages on ports 1 and 2 and (r_1, r_2) were the reflected voltages on ports 1 and 2, then $\begin{bmatrix} r_1 \\ r_2 \end{bmatrix} = \begin{bmatrix} \Lambda_{11} & \Lambda_{12} \\ \Lambda_{21} & \Lambda_{22} \end{bmatrix} \begin{bmatrix} u_1 \\ u_2 \end{bmatrix}$

estimate of the expectation has been studied in detail and it was found that excellent jammer cancellation occurs if $N_S \geq 4NQ$, where N is the number of antennas and Q is the number of time taps per antenna.

The estimation required in (4.74) is usually achieved using matrix multiplication. Let us define the $N_s \times 1$ vector $e_n(t)$ as

$$\mathbf{e}_n(t) = \left[v_n(t)\ v_n(t-T) \ldots v_n\left(t-(N_s-1)T\right)\right] \tag{4.75}$$

where $v_n(t)$ is the voltage at time t on antenna n, after downconversion and bandlimiting. We now form the $NQ \times N_S$ matrix E, defined as

$$E = \begin{bmatrix} \mathbf{e}_1(t) \\ \mathbf{e}_1(t-T) \\ \mathbf{e}_1(t-(Q-1)T) \\ \mathbf{e}_2(t) \\ \cdot \\ \cdot \\ \cdot \\ \mathbf{e}_N(t-(Q-1)T) \end{bmatrix} \tag{4.76}$$

Then, it is readily shown that the required covariance matrix C can be generated as

$$\mathrm{C} = \frac{1}{N_s}\mathrm{E}^*E^T \tag{4.77}$$

where E^T is the transpose of E, given by

$$E^T = \left[\, \mathbf{e}_1(t)^T\ \mathbf{e}_1(t-T)^T \ldots\ \mathbf{e}_N(t-(Q-1)T)^T \,\right] \tag{4.78}$$

and $\mathbf{e}_1^T$ denotes the transpose of the vector $\mathbf{e}_1$, and so forth.

We now discuss some practical implementation issues in the estimation of the covariance. The formation of the matrix C in (4.77) requires a number of operations (and, hence, time to perform those operations), and then the inversion of the matrix C and the multiplication of this inverse by a steering vector (as in 4.55) requires even more operations, so that by the time the weights have been computed, the time samples present at the antenna terminals are very different from those used to calculate the weight vector. Thus, the weights are "stale," and if the interference is temporally nonstationary so that the voltage samples,, where T_L is the latency time to compute the weights, are significantly different from those at time t, the jammer cancellation will be imperfect. The simplest way to solve this difficulty is to buffer the data for an interval T_L, so that by the time the weights have been calculated, they can be applied back to those samples that were used to compute them. This solution may not always be able to be implemented perfectly, because in practice the buffer may not be sufficiently long to accommodate all of the samples we wish to store.

A second approach to deal with jammer nonstationarity and the possibility that the array may need to operate in a dynamic environment is to update the covariance matrix each time a new data sample arrives. Thus, suppose the covariance matrix estimate at time $t = kT$ is $C(k)$ and the data samples received at time $(k + 1)$T on antennas 1 to N are $v_1(k + 1)$, $v_N(k + 1)$, respectively. Then, if we form the vector

$$\mathbf{Y}^T(k+1) = \left[v_1(k+1)\;\; v_1(k)\ldots v_1(k-Q+2)\;\; v_2(k+1)\ldots v_N(k-Q+2)\right] \tag{4.79}$$

we can write the updated estimate of $C(k)$ as

$$C(k+1) = \beta_F C(k) + \mathbf{Y}^*(k+1)\mathbf{Y}^{\mathrm{T}}(k+1) \tag{4.80}$$

where $\beta_F \le 1$ is a "memory" parameter sometimes called the "forgetting" factor that allows us to weight the past data a little less heavily than the most recent data. Typically, β_F is chosen to be very close to unity.

There is one other point that should be made regarding the covariance matrix. In some situations, it is necessary to artificially add noise, so as to better condition the covariance. This leads to the addition of $\sigma_d^2 I$ to the covariance matrix C where I is an $NQ \times NQ$ diagonal matrix and σ_d^2 is the artificial noise known as diagonal loading. The reason that diagonal loading is sometimes required can be understood by recalling that the covariance matrix C is a square, Hermitian matrix that can be expanded in terms of its eigenvalues λ_k and eigenvectors v_k as

$$\mathrm{C} = \sum_{k=1}^{NQ} \lambda_{\mathrm{k}} \mathbf{v}_{\mathrm{k}}^* \mathbf{v}_{\mathrm{k}}^T \tag{4.81}$$

Consequently, its inverse C^{-1} can be expanded as

$$\mathrm{C}^{-1} = \sum_{k=1}^{NQ} \lambda_{\mathrm{k}}^{-1} \mathbf{v}_{\mathrm{k}}^* \mathbf{v}_{\mathrm{k}}^T \tag{4.82}$$

As can be seen from (4.82), the smallest eigenvalues may sometimes dominate the behavior of C^{-1}, causing it to become ill-conditioned. By adding $\sigma_d^2 I$ to C, a matrix C_1 is created with an inverse

$$\mathrm{C}_1^{-1} = \sum_{k=1}^{NQ} \left(\lambda_{\mathrm{k}} + \sigma_d^2\right)^{-1} \mathbf{v}_{\mathrm{k}}^* \mathbf{v}_{\mathrm{k}}^T \tag{4.83}$$

in which the smallest eigenvalues are deemphasized.

4.5 Dual-Polarized Antenna Array

Thus far, we have treated the antennas as point elements. However, real antenna elements have a gain that depends on both the angle and polarization of the incoming radiation. In this section, we will discuss how polarization can be used to our advantage in canceling jamming. In order to discuss this, let us return to the two-element array discussed in Section 4.1, except that now we consider elements that are small, square microstrip patches with each patch having two orthogonal feed ports, as shown in Figure 4.8. Let us assume that the linear dimension D of each of these patches is small in comparison with the signal wavelength λ, and temporarily restrict our attention to the patch located at the origin.

It can be shown that the vector voltage gain of the port directed along the y axis is

$$\mathbf{g}_y(\theta,\varphi) = g_\theta(\theta,\varphi)\hat{\theta} + g_\varphi(\theta,\varphi)\hat{\varphi} \tag{4.84}$$

where $\hat{\theta}$, $\hat{\varphi}$ are the unit vectors shown in Figure 4.8, and

$$g_\theta = G^{1/2}\text{sinc}(a_0)\cos(b_0)\sin\varphi \tag{4.85a}$$

$$g_\varphi = G^{1/2}\text{sinc}(a_0)\cos(b_0)\cos\theta\cos\varphi \tag{4.85b}$$

$$a_0 = \frac{\pi D}{\lambda}\sin\theta\cos\varphi \tag{4.85c}$$

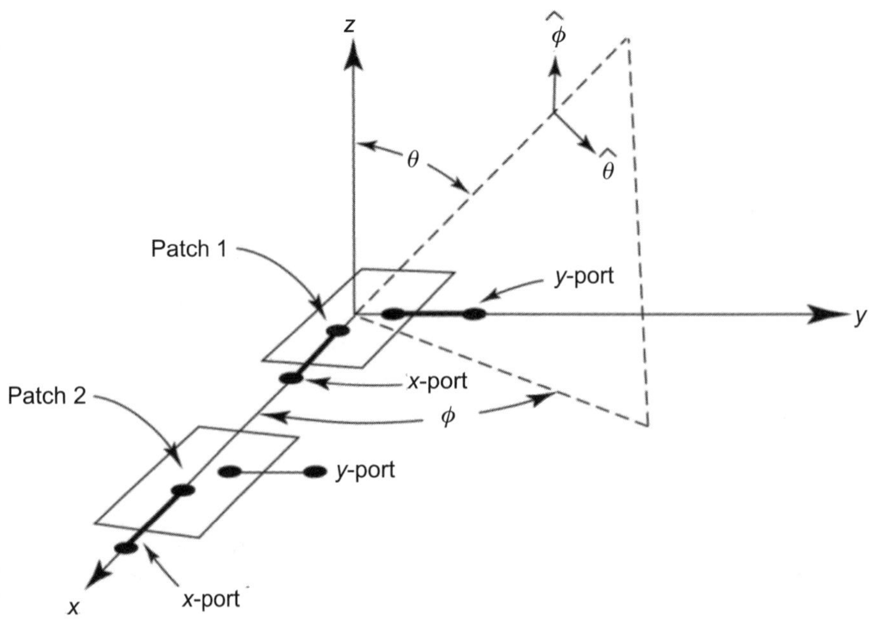

Figure 4.8 Two dual-polarized microstrip patches.

$$b_0 = \frac{\pi D}{\lambda}\sin\theta\sin\varphi \tag{4.85d}$$

D is the linear dimension of each square antenna patch and G is the total power gain. Thus, if a narrowband jammer were located in the direction (θ, Φ) with the polarization properties

$$\mathbf{J} = J_\theta\hat{\theta} + J_\varphi\hat{\varphi} \tag{4.86}$$

the voltage it would produce on the y-port is

$$u_y = \mathbf{J}\cdot\mathbf{g}_y = J_\theta g_\theta + J_\varphi g_\varphi \tag{4.87}$$

For example, a circularly-polarized, narrowband jammer would have $J_\theta = \frac{J}{\sqrt{2}}$, $J_\varphi = \frac{iJ}{\sqrt{2}}$, producing the voltage on the y-port

$$u_y = \frac{J}{\sqrt{2}}\left(g_\theta + ig_\varphi\right) \tag{4.88}$$

The voltage gain of the x-directed port in Figure 4.8 is

$$\mathbf{g}_x(\theta,\varphi) = h_\theta(\theta,\varphi)\hat{\theta} + h_\varphi(\theta,\varphi)\hat{\varphi} \tag{4.89}$$

where

$$h_\theta = G^{1/2}\mathrm{sinc}\left(b_0\right)\cos\left(a_0\right)\cos\varphi \tag{4.90a}$$

$$h_\varphi = -G^{1/2}\mathrm{sinc}\left(b_0\right)\cos\left(a_0\right)\cos\theta\sin\varphi \tag{4.90b}$$

so that a jammer with the polarization state given by (4.86) produces a voltage on the x-port given by

$$u_x = J_\theta h_\theta + J_\varphi h_\varphi \tag{4.91}$$

Now let us demonstrate that the two-element array in Figure 4.8 has the potential to cancel three narrowband jammers. Suppose there are three jammers at (θ_1, Φ_1), (θ_2, Φ_2), (θ_3, Φ_3), with polarization states $\left(J_\theta, J_\varphi\right)$, $\left(\tilde{J}_\theta, \tilde{J}_\varphi\right)$, and $\left(\hat{J}_\theta, \hat{J}_\varphi\right)$. Then the output of the y-port of the microstrip patch centered on the origin is

$$u_{1y} = g_\theta(1)J_\theta + g_\varphi(1)J_\varphi + g_\theta(2)\tilde{J}_\theta + g_\varphi(2)\tilde{J}_\varphi + g_\theta(3)\tilde{J}_\theta + g_\varphi(3)\tilde{J}_\varphi \tag{4.92}$$

where $g_\theta(1) = g_\theta(\theta_1, \Phi_1)$, $g_\Phi(1) = g_\Phi(\theta_1, \Phi_1)$, and so forth. Likewise, the voltage on the x-port of antenna 1

$$u_{1x} = h_\theta(1)J_\theta + h_\varphi(1)J_\varphi + h_\theta(2)\tilde{J}_\theta + h_\varphi(2)\tilde{J}_\varphi + h_\theta(3)\tilde{J}_\theta + h_\varphi(3)\tilde{J}_\varphi \tag{4.93}$$

If the center of antenna 2 is displaced from the center of antenna 1 by a distance d_x in the x-direction, then the voltages out of the x- and y-ports of antenna 2 are

$$u_{2y} = \left[g_\theta(1)J_\theta + g_\varphi(1)J_\varphi\right]e^{i\psi_1} + \left[g_\theta(2)\tilde{J}_\theta + g_\varphi(2)\tilde{J}_\varphi\right]e^{i\psi_2} + \left[g_\theta(3)\hat{J}_\theta + g_\varphi(3)\hat{J}_\varphi\right]e^{i\psi_3} \quad (4.94)$$

$$u_{2x} = \left[h_\theta(1)J_\theta + h_\varphi(1)J_\varphi\right]e^{i\psi_1} + \left[h_\theta(2)\tilde{J}_\theta + h_\varphi(2)\tilde{J}_\varphi\right]e^{i\psi_2} + \left[h_\theta(3)\hat{J}_\theta + h_\varphi(3)\hat{J}_\varphi\right]e^{i\psi_3} \quad (4.95)$$

where $\psi_n = k_0 d_x \sin\theta_n \cos\Phi_n$, $n = 1, 2, 3$, and $k_0 = \dfrac{2\pi f_0}{c}$.

We now wish to combine the four voltages u_{1x}, u_y, u_{2x}, u_{2y} such that (neglecting noise)

$$w_1 u_{1y} + w_2 u_{1x} + w_3 u_{2y} + w_4 u_{2x} = 0 \quad (4.96)$$

where w_1 is a constrained to be unity. Let us suppose the three jammers are circularly polarized so that $J_\theta = \dfrac{J}{\sqrt{2}}$, $J_\varphi = \dfrac{iJ}{\sqrt{2}}$, $\tilde{J}_\theta = \dfrac{\tilde{J}}{\sqrt{2}}$, $\tilde{J}_\varphi = \dfrac{i\tilde{J}}{\sqrt{2}}$, $\hat{J}_\theta = \dfrac{\hat{J}}{\sqrt{2}}$, and $\hat{J}_\varphi = \dfrac{i\hat{J}}{\sqrt{2}}$. Then if we define

$$\tilde{g}_n = \frac{\left[g_\theta(n) + ig_\varphi(n)\right]}{\sqrt{2}} \quad (4.97)$$

$$\tilde{h}_n = \frac{\left[h_\theta(n) + ih_\varphi(n)\right]}{\sqrt{2}} \quad (4.98)$$

and require that the coefficients of J, $\tilde{J}$, and $\hat{J}$ vanish independently, we find that the weights w_2, w_3, w_4 must satisfy

$$\begin{bmatrix} \tilde{h}_1 & \tilde{g}_1 e^{i\psi_1} & \tilde{h}_1 e^{i\psi_1} \\ \tilde{h}_2 & \tilde{g}_2 i\psi_2 & \tilde{h}_2 i\psi_2 \\ \tilde{h}_3 & \tilde{g}_3 i\psi_3 & \tilde{h}_3 i\psi_3 \end{bmatrix} \begin{bmatrix} w_2 \\ w_3 \\ w_4 \end{bmatrix} = -\begin{bmatrix} \tilde{g}_1 \\ \tilde{g}_2 \\ \tilde{g}_3 \end{bmatrix} \quad (4.99)$$

A solution for the unknown weights exists as long as the determinant of the 3 × 3 matrix in (4.99) does not vanish. In general, this determinant is nearly always nonzero. However, because g_Φ and h_Φ vanish for jammers on the horizon (i.e., at $\theta = \dfrac{\pi}{2}$), there is concern that the array cannot cancel three circularly-polarized jammers at $\theta = \dfrac{\pi}{2}$ because at $\theta = \dfrac{\pi}{2}$, the array receives only vertical polarization. Therefore, let us consider this case. Setting $\theta = \dfrac{\pi}{2}$ gives $g_\Phi = h_\Phi = 0$ so that (ignoring the $\sqrt{2}$)

$$\tilde{g}_n = G^{1/2}\text{sinc}(Q_0 \cos\varphi_n)\cos(Q_0 \sin\varphi_n)\sin\varphi_n \tag{4.100}$$

$$\tilde{h}_n = G^{1/2}\text{sinc}(Q_0 \sin\varphi_n)\cos(Q_0 \cos\varphi_n)\cos\varphi_n \tag{4.101}$$

where $Q_0 = \dfrac{\pi D}{\lambda}$ and $n = 1, 2, 3$. We now note that even though g_Φ and h_Φ vanish, the output $\tilde{g}_n$ of the y-port of each antenna is different than the output of the x-port unless $\sin\Phi_n = \cos\Phi_n$. Thus, just because the Φ-component of the voltage gain vanishes doesn't necessarily mean that degrees of freedom are lost. We will demonstrate later via simulation.

It is interesting to note that even a single dual-polarized element can cancel an interferer. For example, suppose there is a single interferer at (θ_1, Φ_1) with an arbitrary polarization state $J(\alpha_\theta\hat{\theta} + \alpha_\varphi\hat{\varphi})$, where $\alpha_\theta^2 + \alpha_\varphi^2 = 1$. Then, if we combine the outputs of the x and y ports of the element at the origin in Figure 4.8, we obtain

$$u = u_{1y} + wu_{1x} = \left[\left(g_\theta(1)\alpha_\theta + g_\varphi(1)\alpha_\varphi\right) + w\left(h_\theta(1)\alpha_\theta + h_\varphi(1)\alpha_\varphi\right)\right]J \tag{4.102}$$

Thus, if w is chosen as

$$w = -\frac{g_\theta(1)\alpha_\theta + g_\varphi(1)\alpha_\varphi}{h_\theta(1)\alpha_\theta + h_\varphi(1)\alpha_\varphi} \tag{4.103}$$

the jammer residue at the output vanishes. Furthermore, although there is a null on the jammer, the element can still receive signals from other angles (θ, Φ) and with different polarization states. Thus, a unit-amplitude signal with polarization state $\left(\beta_\theta\hat{\theta} + \beta_\varphi\hat{\varphi}\right)$, where $\beta_\theta^2 + \beta_\varphi^2 = 1$ located at (θ_2, Φ_2) produces an output

$$u(\theta,\varphi) = g_\theta(2)\beta_\theta + g_\varphi(2)\beta_\varphi + w\left(h_\theta(2)\beta_\theta + h_\varphi(2)\beta_\varphi\right) \tag{4.104}$$

which does not vanish unless $\theta_2 = \theta_1$, $\Phi_2 = \Phi_1$, $\beta_\theta = \alpha_\theta$, and $\beta_\Phi = \alpha_\Phi$. That is, even if the signal is coming from the same direction as the jammer, it is not completely cancelled unless it also has the same polarization state as the jammer.

The algorithms used for the space-time-polarization adaptive array are the same as those described in Section 4. 3.2, except that the weight vector and voltage vectors in (4.46) and (4.47) are replaced by the $2NQ \times 1$ vectors

$$\mathbf{w}^T = \left[w_{11}^x \; w_{11}^y \; w_{12}^x \; w_{12}^y \; \ldots \; w_{1Q}^x \; w_{1Q}^y \; \ldots \; w_{NQ}^x \; w_{NQ}^y\right] \tag{4.105}$$

$$\mathbf{v}^T = \left[v_{11}^x \; v_{11}^y \; v_{12}^x \; v_{12}^y \; \ldots \; v_{1Q}^x \; v_{1Q}^y \; \ldots \; v_{NQ}^x \; v_{NQ}^y\right] \tag{4.106}$$

and everything then proceeds as described in Section 4.3.2.

In Table 4.1, we show simulated results using the minimum variance algorithm with $Q = 1$ and 5 time taps for the two dual-polarized element array shown in Figure 4.8. We examine the case when strong (jammer-noise ratio = 40 dB per jammer per element) circularly polarized, broadband jammers are placed randomly in azimuth in an elevation band of 0° to 20° above the horizon (i.e., the $z = 0$ plane). Each value in Table 4.1 is the average of 200 jammer realizations. As expected, up to three broadband jammers can be cancelled, but four or more broadband jammers overwhelm the array. In general, an array of N dual-polarized patches can cancel up to $2N$-1 broadband jammers, even if those jammers are near the horizon where the patches receive only the vertical component of the interference.

4.6 Performance Measures

4.6.1 Some Traditional Measures

Now that we have discussed the various types of adaptive arrays and some of the algorithms used for jammer cancellation, it is desirable to specify some measures of adaptive array performance. The simplest is the signal-to-interference ratio after adaptation. From (4.51), it is evident that the expected value of the interference (jammer plus noise) power at the array output (for any jammer-cancellation algorithm) is

$$\left\langle |u|^2 \right\rangle = \mathbf{w}^H C \mathbf{w} \tag{4.107}$$

where C is defined in (4.53). Using (4.49), we see that the expected value of the signal power is

$$\left\langle |u_s|^2 \right\rangle = \mathbf{w}^H \left\langle \mathbf{s}^* \mathbf{s}^T \right\rangle \mathbf{w} \tag{4.108}$$

so that the signal-to-interference ratio at the array output is

$$\frac{S}{I} = \frac{\mathbf{w}^H \left\langle \mathbf{s}^* \mathbf{s}^T \right\rangle \mathbf{w}}{\mathbf{w}^H C \mathbf{w}} \tag{4.109}$$

Now recall from Section 4.3.2 that the weight vector has the form

Table 4.1 Jammer Cancellation Performance of Two Dual-Polarized Patches

Number of Broadband Jammers	*Increase in Noise Floor (dB) if Q = 1 Tap*	*Increase in Noise Floor (dB) if Q = 5 Taps*
0	0	0
1	0	0
2	0.2	0
3	1	0
4	24	23
5	28	28
6	31	30

$$\mathbf{w} = \mu C^{-1}\mathbf{d}^* \tag{4.110}$$

where μ is a constant and the steering vector **d** depends on the algorithm chosen. If (4.110) is used in (4.109), we obtain

$$\frac{S}{I} = \frac{\mathbf{d}^T C^{-1} \left\langle \mathbf{s}^* \mathbf{s}^T \right\rangle C^{-1} \mathbf{d}^*}{\mathbf{d}^T C^{-1} \mathbf{d}^*} \tag{4.111}$$

If the signal is deterministic and one chooses the algorithm where the output power is minimized subject to the constraint that a beam is pointed towards the incoming signal, we see from (4.55) that $\mathbf{d} = s$, so that (4.111) simplifies to

$$\frac{S}{I} = \mathbf{s}^T C^{-1} \mathbf{s}^* \tag{4.112}$$

However, for stochastic signals, such as GPS waveforms, we must use the general result in (4.111), which we rewrite as

$$\frac{S}{I} = \frac{\mathbf{d}^T C^{-1} C_{ss} C^{-1} \mathbf{d}^*}{\mathbf{d}^T C^{-1} \mathbf{d}^*} \tag{4.113}$$

where

$$C_{ss} = \left\langle \mathbf{s}^* \mathbf{s}^T \right\rangle \tag{4.114}$$

is the signal covariance matrix.

A second measure of adaptive algorithm effectiveness is the increase in the thermal noise floor produced by uncanceled jamming. If we define the noise covariance matrix as

$$C_{\xi\xi} = \left\langle \boldsymbol{\xi}^* \boldsymbol{\xi}^T \right\rangle \tag{4.115}$$

then the jammer-plus-noise-to-noise ratio after adaptation is

$$\frac{I}{N} = \frac{\mathbf{w}^H C \mathbf{w}}{\mathbf{w}^H C_{\xi\xi} \mathbf{w}} \tag{4.116}$$

This is the quantity used to calculate the results presented in Table 4.1.

Another useful measure of performance is called satellite availability. For this measure, imaginary GPS satellites are randomly placed in the upper hemisphere (minus a small elevation mask at the horizon) and Monte-Carloed to calculate the probability that the signal-to-interference ratio is above a specified value. The value specified is typically that below which the GPS receiver loses lock on a GPS satellite signal.

4.6.2 System Performance Measures

We will now derive a useful measure of the system performance of an adaptive GPS receiver. Let us refer to Figure 4.9, where we show the code-tracking[6] portion of a GPS receiver for the case when an adaptive space-time processor is inserted to cancel interference.

We show radiation arriving from an arbitrary direction (θ,Φ) in a spherical coordinate system. This incoming radiation is received by each of the N elements in the antenna array, downconverted to baseband and then bandlimited using a digital filter with transfer function $H_B(f)$, where f denotes the baseband frequency. In the GPS antijam system block diagram, we also include the transfer function $H_n(f, \theta, \Phi)$ for each antenna and its associated analog[7] hardware. The function H_n also includes any uncorrected mismatch from channel-to-channel. The voltages v_1 to v_N at the output of the bandlimiting filters are then fed into the adaptive processor, which was shown in expanded form in Figure 4.7. The output of the adaptive filter, consisting of desired GPS signals, residual jammer interference and noise is then cross-correlated with a stored, delayed replica of the signal $s(t)$ that was transmitted by the GPS satellite. In Figure 4.9, we show only the signal from one GPS satellite, but of course there are signals present from every GPS satellite in view. Likewise, in Figure 4.9, we show only the cross correlator for one signal, but there

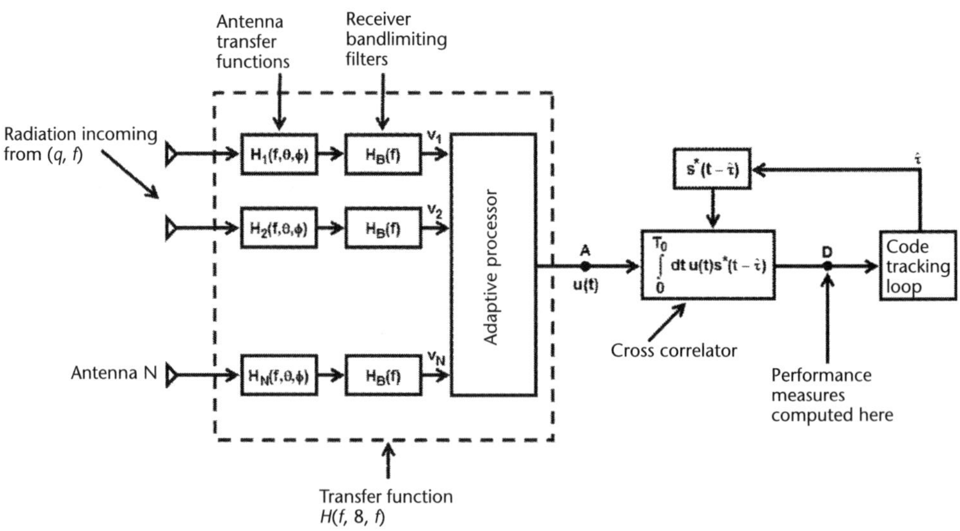

Figure 4.9 GPS antijam system block diagram (s^* = stored code, $\tau^\wedge$ = estimate of delay).

6. Every GPS satellite emits an unique code and the GPS receiver attempts to lock onto the code of each GPS satellite in view in order to estimate the delay between the time when that code was emitted and the time when it was received.
7. In practice, one tries to equalize the analog portion of each channel across the operating bandwidth so as to make H_n the same for all channels, but this cannot be achieved perfectly because the antenna itself has a response that depends on (θ, Φ).

are actually multiple cross correlators operating in parallel, each using appropriate stored replicas for other GPS satellites. For each satellite, the tracking loop adjusts the estimated delay so as to match the true delay for that satellite signal. For simplicity, we do not show the phase-locked loop that corrects for phase, but rather assume the carrier phase of the replica has been matched to the carrier phase of the received signal.

Suppose a signal incident from the direction (θ_s, Φ_s) produces a voltage (after downconversion) given by $A_0 s(t - \tau)\exp(i\psi)$ on the reference element of the antenna array, where $s(t)$ is a random, unit-amplitude sequence of chips, A_0 is the rms amplitude, τ is the delay (the quantity the GPS receiver must estimate), and ψ is the phase of the carrier signal, which is also estimated by high-precision GPS systems. The Fourier transform of this signal is by $A_0 S(f)\exp(-i2\pi f\tau + i\psi)\exp(i\psi)$, where $S(f)$ is the Fourier transform of $s(t)$. By using (4.71), it is evident that at the output of the adaptive filter (i.e., at point A in Figure 4.9), the Fourier transform of the total signal from all antennas is

$$U_s(f) = A_0 H(f,\theta_s,\varphi_s)S(f)\exp[-i2\pi f\tau + i\psi] \tag{4.117}$$

and the time-domain signal at point A is

$$u_s(t) = A_0 \int_{-\infty}^{\infty} df\, S(f)H(f,\theta_s,\varphi_s)\exp[i2\pi f(t-\tau) + i\psi] \tag{4.118}$$

Next, suppose a jammer located in the direction (θ_j, Φ_j) produces a voltage $j(t)$ on the reference element. Then, paralleling the analysis in the last paragraph, it is evident that the total jammer voltage at point A is

$$u_j(t) = \int_{-\infty}^{\infty} df\, J(f)H(f,\theta_j,\varphi_j)\exp(i2\pi ft) \tag{4.119}$$

where $J(f)$ is the Fourier transform of $j(t)$.

The system noise must be treated differently than the incoming radiation because it is not an external voltage. If $\xi_n(t)$ is the noise at the input to the bandlimiting filter in channel n and $\Gamma_n(f)$ is its Fourier transform, then the total thermal noise voltage at point A is

$$u_\xi(t) = \int_{-\infty}^{\infty} df\, U_\xi(f)\exp(i2\pi ft) \tag{4.120}$$

where

$$U_\xi(f) = H_B(f)\sum_{n=1}^{N}\sum_{q=1}^{Q} w_{nq}\Gamma_n(f)\exp[i2\pi f(q-QQ)T] \tag{4.121}$$

The total voltage at point A then is

$$u(t) = u_s(t) + u_j(t) + u_\xi(t) \tag{4.122}$$

The voltage $u(t)$ is multiplied in the receiver by a stored, delayed replica $s^*(t-\hat{\tau})$ of the transmitted signal, which produces the following signal (the residual jamming and thermal noise will be treated later) at the cross-correlator output

$$Z_s(\hat{\tau}) = \int_{t_n}^{t_n+T_0} dt\; u_s(t) s^*(t-\hat{\tau}) \tag{4.123}$$

where t_n is an arbitrary start time and T_0 is the coherent integration time. Next, substitute (4.118) in (4.123) and make the transformation $t' = t - \hat{\tau}$. This gives

$$Z_S(\hat{\tau}) = A_0 e^{i\psi} \int_{T_n}^{T_n+T_0} dt' S^*(t') \int_{-\infty}^{\infty} S(f) H(f,\theta_s,\phi_s) e^{i2\pi f(t'-\Delta\tau)} \tag{4.124}$$

where $\Delta\tau = \tau - \hat{\tau}$ and $T_n = t_n - \hat{\tau}$. We now write $S(f)$ in terms of its inverse Fourier transform, take the expectation of Z_S and use the fact that for a stationary random signal

$$\langle s(t) s^*(t') \rangle = \rho_{ss}(t-t') \tag{4.125}$$

where ρ_{SS} is the signal autocorrelation function and $\langle ... \rangle$ denotes an expectation. If this is done, and we recall that the signal power spectrum $\Phi_{SS}(f)$ is related to ρ_{SS} via

$$\Phi_{ss}(f) = \int_{-\infty}^{\infty} d\eta\; \rho_{ss}(\eta) \exp(-i2\pi f \eta) \tag{4.126}$$

we finally obtain for the complex cross-correlation function at point D in Figure 4.9

$$\langle Z_s(\hat{\tau}) \rangle = A_0 T_0 e^{i\psi} \int_{-\infty}^{\infty} df\; \Phi_{ss}(f) H(f,\theta_s,\varphi_s)\, e^{-i2\pi f \Delta\tau} \tag{4.127}$$

Because $s(t)$ has unit amplitude, it is evident that $\rho_{SS}(0) = 1$, so that $\Phi_{ss}(f)$ satisfies[8]

$$\int_{-\infty}^{\infty} \Phi_{ss}(f) df = 1 \tag{4.128}$$

The tracking loop estimates the delay as the value of $\hat{\tau}$ where $|\langle E(Z_s) \rangle|$ is a maximum. For ideal antennas and in the absence of the adaptive processor (i.e., $H = 1$), Φ_{SS} is a real function, so this maximum clearly occurs at $\Delta\tau = 0$, so that $\hat{\tau} = \tau$, and the delay estimate is correct. However, when an adaptive processor is used, there is no a priori guarantee that $H(f, \theta_S, \Phi_s)$ is a real function when multiple jammers are present. In fact, we expect that $H(f, \theta_S, \Phi_s)$ will be complex, so let us see what this implies. Suppose we write

8. In writing (4.128), we have implicitly assume that the transmission bandwidth, B_T, of the GPS satellite is sufficiently large that the integral from $-\frac{B_T}{2}$ to $\frac{B_T}{2}$ can be approximated by $-\infty$ to ∞.

$$H(f,\theta_s,\varphi_s) = \left|H(f,\theta_s,\varphi_s)\right| \exp\left[i\alpha(f,\theta_s,\varphi_s)\right] \tag{4.129}$$

and then expand the phase α in a Taylor series as $\alpha = \alpha_0 + \alpha_1 f + \alpha_2 f^2 + \ldots$. Then, if we ignore the quadratic and higher order terms in f, it can be seen that (4.127) becomes

$$\langle Z_s \rangle = A_0 T_0 e^{i(\psi+\alpha_0)} \int_{-\infty}^{\infty} df\ \Phi_{ss}(f) \left|H(f,\theta_s,\varphi_s)\right| e^{-i2\pi f\left(\Delta\tau - \frac{\alpha_1}{2\pi}\right)} \tag{4.130}$$

By examining (4.130), the following points are evident: (1) there is a carrier phase error introduced by the adaptive processor because the carrier phase is now $(\psi + \alpha_0)$ instead of the correct value, ψ, and (2) the maximum of $\left|\langle Z_s \rangle\right|$ now occurs at $\Delta\tau - \alpha_1/2\pi = 0$, or at $\hat{\tau} = \tau - \alpha_1/2\pi$, so that there is an error $\alpha_1/2\pi$ in the estimated delay.

In order to correctly calculate the error in the estimated delay and the carrier phase error, it is necessary to return to (4.127), take its absolute value and then find[9] the value of $\Delta\tau$ where $\left|\langle Z_s \rangle\right|$ is a maximum. Let us denote this value by $\Delta\tau_0$. Then the peak signal power at the cross-correlator output is given by

$$P_{\max} = (A_0 T_0)^2 \left| \int_{-\infty}^{\infty} df\ \Phi_{ss}(f) H(f,\theta_s,\varphi_s)\, e^{-i2\pi f \Delta\tau_0} \right|^2 \tag{4.131}$$

and the carrier-phase error then is

$$\Delta\psi = \tan^{-1}\left[\frac{\int_{-\infty}^{\infty} df\ \Phi_{ss}(f)\left|H(f,\theta_s,\varphi_s)\right| \sin(\alpha - 2\pi f \Delta\tau_0)}{\int_{-\infty}^{\infty} df\ \Phi_{ss}(f)\left|H(f,\theta_s,\varphi_s)\right| \cos(\alpha - 2\pi f \Delta\tau_0)} \right] \tag{4.132}$$

where $\alpha(f, \theta_s, \Phi_s)$ is defined in (4.129).

We next need to calculate the expectation of the residual jammer power at the cross-correlator output (point D in Figure 4.9). By paralleling the analysis used to compute $\langle Z_s \rangle$, we can show that

$$\left\langle \left|Z_j\right|^2 \right\rangle = T_0 \int_{-\infty}^{\infty} df\ \Phi_{jj}(f)\ \Phi_{ss}(f) \left|H(f,\theta_j,\varphi_j)\right|^2 \tag{4.133}$$

where $\Phi_{jj}(f)$ is the power spectral density of the jamming waveform, and has the property

$$\int_{-\infty}^{\infty} \Phi_{jj}(f)\, df = P_j \tag{4.134}$$

9. In practice, this needs to be done numerically, and there is no simple analytical result for δT_0.

where P_j is the jammer power. If M jammers are present at angles (θ_{jm}, Φ_{jm}) with power spectra Φ_{jjm}, then (4.133) is generalized to

$$\left\langle \left|Z_j\right|^2 \right\rangle = T_0 \int_{-\infty}^{\infty} df\ \Phi_{ss}(f) \sum_{m=1}^{M} \Phi_{jjm}(f) \left|H\left(f, \theta_{jm}, \varphi_{jm}\right)\right|^2 \tag{4.135}$$

Finally, the expected thermal noise power at the cross-correlator output is

$$\left\langle \left|Z_\xi\right|^2 \right\rangle = T_0 \int_{-\infty}^{\infty} df\ \Phi_{\xi\xi}(f)\ \Phi_{ss}(f) \left|H_\xi(f)\right|^2 \tag{4.136}$$

where $\Phi_{\xi\xi}(f)$ is the power spectrum of the thermal noise,

$$\left|H_\xi(f)\right|^2 = \left|H_B(f)\right|^2 \sum_{n=1}^{N} \left| \sum_{q=1}^{Q} w_{nq} \exp\left(-i2\pi\left(q - QQ\right) fT\right) \right|^2 \tag{4.137}$$

and we have assumed that the thermal noise in each channel is statistically independent of the thermal noise in all other channels.

We are now in a position to calculate several useful performance measures. The jammer-to-noise ratio at the cross-correlator output (point D in Figure 5.9) is

$$JNR = \frac{\int_{-\infty}^{\infty} df\ \Phi_{ss}(f) \sum_{m=1}^{M} \Phi_{jjm}(f) \left|H\left(f, \theta_{jm}, \varphi_{jm}\right)\right|^2}{\int_{-\infty}^{\infty} df\ \Phi_{ss}(f)\ \Phi_{\xi\xi}(f) \left|H_\xi(f)\right|^2} \tag{4.138}$$

For well-designed adaptive arrays, one usually finds $JNR < 1$, as long as $M < N$.

The signal-to-interference (we define interference as jammer residue plus thermal noise) ratio at point D in Figure 5.9 is

$$SIR = \frac{A_0^2 T_0 \left| \int_{-\infty}^{\infty} df\ \Phi_{ss}(f) H\left(f, \theta_s, \varphi_s\right) e^{-i2\pi f \Delta\tau_0} \right|^2}{\left(1 + JNR\right) \int_{-\infty}^{\infty} df\ \Phi_{ss}(f)\ \Phi_{\xi\xi}(f) \left|H_\xi(f)\right|^2} \tag{4.139}$$

where JNR is given by (4.138).

For the case when the thermal noise is white ($\Phi_{\xi\xi}$ = constant = N_0), it is traditional in the GPS community to express results in terms of the carrier-to-noise ratio, which is commonly denoted by C/N_0, where $C = A_0^2$. For white noise, we can rewrite (4.139) in terms of C/N_0 as

$$SIR = \left(\frac{C}{N_0}\right)_{eff} T_0 \tag{4.140}$$

where

$$\left(\frac{C}{N_0}\right)_{eff} = \frac{\frac{C}{N_0}\left|\int_{-\infty}^{\infty} df\ \Phi_{ss}(f)H(f,\theta_s,\varphi_s)e^{-i2\pi f\Delta\tau_0}\right|^2}{(1+JNR)\int_{-\infty}^{\infty} df\ \Phi_{ss}(f)\ \left|H_\xi(f)\right|^2} \tag{4.141}$$

Equations (4.127), (4.131), (4.132), and (4.138) through (4.141) are some measures of the GPS antijam system performance. In the limit when the adaptive processor is absent, and the channels are perfectly matched, so that $H(f,\theta,\varphi) = H_B(f)$, $\left|H_\xi(f)\right|^2 = \left|H_B(f)\right|^2$, (4.141) reduces to the result presented in (6.7) of Section 6.2.2.5 in the text by Kaplan and Hegarty cited in the bibliography.

4.6.3 Spatial Adaptive Processor

It is interesting to examine our results in the limit when the adaptive processor uses spatial degrees of freedom only, so there is only one time tap per antenna (i.e., $Q = 1$). In order to simplify our results, we also assume that the maximum linear dimension Δ of the antenna array is such that $\pi B\Delta/c << 1$, where B is the operating bandwidth. This allows us to approximate $\exp(i2\pi fR_n/c)$ by unity in (4.71). Finally, if we assume that there is very little variation of the (analog) antenna-transfer functions H_n with frequency, we can approximate $H_n(f, \theta, \Phi)$ by $H_n(0, \theta, \Phi)$, its value at the center of the band. If all of the assumptions are employed, $H(f, \theta, \Phi)$ in (4.71) reduces to

$$H(f,\theta,\phi) = H_B(f)H_A(\theta,\phi) \tag{4.142}$$

where

$$H_A(\theta,\varphi) = \sum_{n=1}^{N} H_n(0,\theta,\varphi)w_{n1}\exp\left[\frac{i2\pi f_0 R_n(\theta,\varphi)}{c}\right] \tag{4.143}$$

Likewise $\left|H_\xi\right|^2$ in (4.137) reduces to

$$\left|H_\xi(f)\right|^2 = \left|H_B(f)\right|^2 \sum_{n=1}^{N}\left|w_{n1}\right|^2 \tag{4.144}$$

If we now approximate $H_B(f)$ by a brickwall filter of bandwidth B, and substitute (4.142) into (4.131), we find that the peak power for $Q = 1$ is

$$P_{\max} = (A_0T_0)^2\left|H_A(\theta_s,\varphi_s)\right|^2\left|\int_{-B/2}^{B/2} df\ \Phi_{ss}(f)e^{-i2\pi f\Delta\tau_0}\right|^2 \tag{4.145}$$

Because $\Phi_{ss}(f)$ is a real, nonnegative definite function, it is readily seen that the right-hand side of (4.145) achieves its maximum when $\Delta\tau_0 = 0$. Furthermore, if (4.142) is substituted into (4.132), along with $\Delta\tau_0 = 0$, it is evident that

$$\Delta\psi = \tan^{-1}\left(\frac{\sin\alpha_A}{\cos\alpha_A}\right) = \alpha_A\left(\theta_s, \varphi_s\right) \tag{4.146}$$

where $\alpha_A(\theta_s, \Phi_s)$ is the argument of $H_A(\theta_s, \Phi_s)$. We observe that even though the peak of the cross correlation occurs at the correct delay $\hat{\tau} = \tau$, the adaptive processor may introduce a carrier-phase error, as given by (4.146).

Finally, if we use $\Delta\tau_0 = 0$ and substitute (4.142) and (4.144) into (4.141), we obtain for the signal-to-interference ratio when the noise is white

$$SIR = \left(\frac{C}{N_0}\right)_{eff} T_0 \tag{4.147}$$

where

$$\left(\frac{C}{N_0}\right)_{eff} = \frac{\left(\frac{C}{N_0}\right)\left|H_A\left(\theta_s, \varphi_s\right)\right|^2 \int_{-B/2}^{B/2} df\ \Phi_{ss}(f)}{(1 + JNR)\sum_{n=1}^{N} \left|w_{n1}\right|^2} \tag{4.148}$$

Typically, the system bandwidth B is sufficiently large that the integral over Φ_{ss} in the numerator is very close to unity (see (4.128)). Furthermore, in a well-designed adaptive processor, the JNR after adaptation is usually less than unity, so $(1 + JNR)$ can often be approximated by unity. Thus, the result for the SIR becomes quite simple.

4.7 Direction Finding

In the presence of strong (jammer-to-noise ratio >> 1) jamming, the information contained within the covariance matrix allows one to estimate the direction $(\hat{\theta}, \hat{\varphi})$ to each jammer, provided the number of antenna elements is greater than the number of jammers. Two of these estimators are

$$E_1\left(\hat{\theta}, \hat{\varphi}\right) = \frac{1}{\left|\boldsymbol{\rho}^T\left(\hat{\theta}, \hat{\varphi}\right) C^{-1}\mathbf{u}_0\right|^2} \tag{4.149}$$

and

$$E_2\left(\hat{\theta}, \hat{\varphi}\right) = \frac{1}{\boldsymbol{\rho}^T\left(\hat{\theta}, \hat{\varphi}\right) C^{-1}\boldsymbol{\rho}^*\left(\hat{\theta}, \hat{\varphi}\right)} \tag{4.150}$$

where for an N-element planar array in which each element has only one time tap ($Q = 1$), we have

$$\boldsymbol{\rho}^T(\hat{\theta},\hat{\varphi}) = [\rho_1 \dots \rho_N] \tag{4.151}$$

$$\rho_n = \exp\left[\frac{i2\pi f_0}{c}\sin\hat{\theta}\left(x_n\cos\hat{\varphi} + y_n\sin\hat{\varphi}\right)\right] \tag{4.152}$$

$$\mathbf{u}_0^T = [1 \quad 0 \quad 0 \dots 0] \tag{4.153}$$

and (x_n, y_n) is the position in the $z = 0$ plane of element n. The estimator E_1 is called the autoregressive (AR) estimator and E_2 is called the Capon estimator.

In order to see how these estimators work, we recall that the covariance matrix can be decomposed in terms of its eigenvectors v_n and eignevalues λ_n as

$$C = \sum_{n=1}^{N} \lambda_n \mathbf{v}_n^* \mathbf{v}_n^T \tag{4.154}$$

where for N_J narrowband jammers, the first N_J values of λ_n are very large and the last $(N - N_J)$ eigenvalues are much smaller and equal to the noise σ_ξ^2. The inverse covariance can be written as

$$C^{-1} = \sum_{n=1}^{N_J} \frac{\mathbf{v}_n^* \mathbf{v}_n^T}{\lambda_n} + \frac{1}{\sigma_\xi^2} \sum_{n=N_J+1}^{N} \mathbf{v}_n^* \mathbf{v}_n^T \tag{4.155}$$

where the term

$$\sum_{n=N_J+1}^{N} \mathbf{v}_n^* \mathbf{v}_n^T$$

represents the noise subspace. Consequently, if the trial steering vector $\rho^T(\hat{\theta},\hat{\varphi})$ lies completely in the noise subspace, it is evident that

$$\left(\boldsymbol{\rho}^T C^{-1}\mathbf{u}_0\right) \rightarrow \left(\boldsymbol{\rho}^T C^{-1}\mathbf{u}_0\right)_{\text{noise}} = \frac{1}{\sigma_\xi^2} \sum_{n=N_J+1}^{N} \left(\boldsymbol{\rho}^T \mathbf{v}_n^*\right)\left(\mathbf{v}_n^T \mathbf{u}_0\right) \tag{4.156}$$

whereas if ρ lies totally in the jammer subspace $\boldsymbol{\rho}^T C^{-1} \mathbf{u}_0$ is

$$\left(\boldsymbol{\rho}^T C^{-1}\mathbf{u}_0\right) \rightarrow \left(\boldsymbol{\rho}^T C^{-1}\mathbf{u}_0\right)_{\text{jam}} = \sum_{n=1}^{N_J} \frac{\left(\boldsymbol{\rho}^T \mathbf{v}_n^*\right)\left(\mathbf{v}_n^T \mathbf{u}_0\right)}{\lambda_n} \tag{4.157}$$

Because $\lambda_n >> \sigma_\xi^2$, it is seen that

$$\frac{1}{\left|\boldsymbol{\rho}^T C^{-1}\mathbf{u}_0\right|_{jam}} >> \frac{1}{\left|\boldsymbol{\rho}^T C^{-1}\mathbf{u}_0\right|_{noise}} \tag{4.158}$$

Thus, when $(\hat{\theta},\hat{\varphi})$ are such that $\boldsymbol{\rho}(\hat{\theta},\hat{\varphi})$ lies in the jammer subspace $E_1(\hat{\theta},\hat{\varphi})$ in (4.149) is very much larger than when $(\hat{\theta},\hat{\varphi})$ are such that $\boldsymbol{\rho}(\hat{\theta},\hat{\varphi})$ lies in the noise subspace. Therefore, as $(\hat{\theta},\hat{\varphi})$ are varied, $I_1(\hat{\theta},\hat{\varphi})$ will be very large when $(\hat{\theta},\hat{\varphi})$ corresponds to a jammer direction, but much smaller otherwise.

In studies, it is found that I_1 performs better than I_2 in estimating jammer direction. It is also found that $I_1(\hat{\theta},\hat{\varphi})$ can localize up to $N-1$ narrowband jammers, but the performance degrades significantly as the jammer bandwidth increases because the noise and jammer subspaces in (4.155) are no longer distinct. Additionally, if the number of antennas is too much greater than the number of jammers (even if the jammers are narrowband), spurious sources may appear. Thus, too many degrees of freedom can actually be a bad thing.

4.8 Array Phase Center

For high-precision GPS applications, such as landing airplanes using GPS, it is extremely important to accurately estimate position, sometimes to a small fraction of a wavelength. As a result, there is considerable interest in the location of the phase center of an antenna.

Therefore, we now develop a theory that allows us to calculate the phase center of an arbitrary array in the limit when mutual coupling is ignored.

Our first goal is to define what we mean by the phase center of an antenna array. Let us suppose we have an antenna array with its geometric center at the origin of the coordinate system shown in Figure 4.4. This array radiates an electric field that, for observers located at a distance R that is in the far zone (also called the Fraunhofer zone), can be written as

$$E(\theta,\varphi) = A(\theta,\varphi)\exp\left[ikR + i\psi(\theta,\varphi)\right] \tag{4.159}$$

where R is the distance from the origin O to the observer, $A(\theta,\Phi)$ is the amplitude of the electric field and is real, k is the wavenumber $= 2\pi f_0/c$, and $\psi(\theta, \Phi)$ is the phase. If $\psi(\theta, \Phi)$ is not either a constant or a zero, then the surface of constant phase does not correspond to a spherical wave emanating from the origin O. A phase center at the origin would lead to the spherical wave $\exp(ikR)$, having constant phase when R = constant, so that the geometric and phase centers of the array would coincide.

When $\psi(\theta, \Phi)$ is not zero, the surface of constant phase is distorted. This can be visualized by taking a two-dimensional slice to obtain a surface of constant phase such as shown in the plane $x = 0$ by the dashed curve in Figure 4.10. This curve differs from a circle centered on the origin. For this curve, the phase center at each angle θ can be obtained by fitting a circle (for the three-dimensional case we fit a spherical wave) that has the same curvature as the phase $\psi(\theta, \pi/2)$ in the $x = 0$ plane

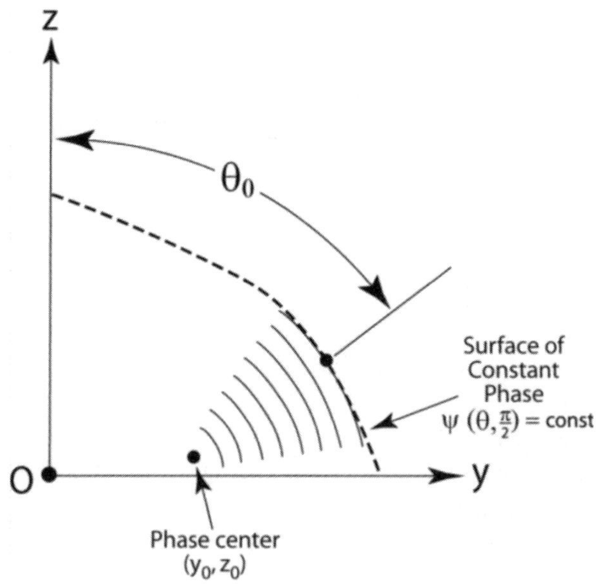

Figure 4.10 Illustration of a surface of constant phase and the offset of the phase center corresponding to the angle θ_0.

and then projecting back with a series of increasingly smaller circles to an apparent origin. This procedure is shown in Figure 4.10 for $\theta = \theta_0$, and the apparent phase center is denoted by (y_0, z_0). If we choose a different angle $\theta = \theta_1$ we would find a different phase center (y_1, z_1). In the three-dimensional case, we would find a spherical surface that has the same curvature as $\psi(\theta, \Phi)$ at (θ_0, Φ_0) and then project back using spheres of increasingly smaller radii until an apparent origin (x_0, y_0, z_0) is reached. We now develop a procedure to calculate (x_0, y_0, z_0).

In order to see how we determine the phase $\psi(\theta, \Phi)$ in (4.159), let us assume we have an array of N isotropic antennas, with the center of antenna n located at (x_n, y_n, z_n) in the coordinate system of Figure 4.4. If all of these elements are weighted and summed, we obtain the field (in the far zone)

$$E(\theta,\varphi) = K\sum_{n=1}^{N} w_n \exp(ikR + i\alpha_n) \tag{4.160}$$

where K is a real constant that depends on (θ, Φ), $\alpha_n = k(x_n \sin\theta \cos\Phi + y_n \sin\theta \sin\Phi + z_n \cos\theta)$ and w_n is the complex adaptive weight applied to element n. If we write $w_n = |w_n|\exp(i\beta_n)$, then we can write (4.160) in the same form as (4.159), where $\psi(\theta, \Phi)$ is given by

$$\psi(\theta,\varphi) = \tan^{-1}\left[\frac{\sum_{n=1}^{N} |w_n|\sin\gamma_n}{\sum_{n=1}^{N} |w_n|\cos\gamma_n}\right] \tag{4.161}$$

and $\Gamma_n = \alpha_n + \beta_n$.

Now in order to obtain the phase center for a specified angular direction (θ_0, Φ_0), we first postulate that the phase center is a point source at some unknown location (x_0, y_0, z_0). Then, in the far zone, this point source radiates an electric field given by

$$\hat{E}(\theta,\varphi) = A_0(\theta,\varphi)\exp\left[ikR + i\psi_0(\theta,\varphi)\right] \tag{4.162}$$

where A_0 is real and

$$\psi_0(\theta,\varphi) = k\left[x_0\sin\theta\cos\varphi + y_0\sin\theta\sin\varphi + z_0\cos\theta\right] \tag{4.163}$$

We now require that (x_0, y_0, z_0) be chosen so that the phase and curvature of the surface of constant phase, ψ_0 = constant, matches the phase and curvature of the surface ψ = constant (where ψ is given by (4.161)) at the angular location (θ_0, Φ_0). Equating (4.163) to (4.161) gives

$$\psi_0(\theta_0,\varphi_0) = \psi(\theta_0,\varphi_0) \tag{4.164}$$

and requiring that the derivatives with respect to θ and Φ are equal gives

$$\left(\frac{d\psi_0}{d\theta}\right)_{\theta_0,\varphi_0} = \left(\frac{d\psi}{d\theta}\right)_{\theta_0,\varphi_0} \tag{4.165}$$

$$\left(\frac{d\psi_0}{d\varphi}\right)_{\theta_0,\varphi_0} = \left(\frac{d\psi}{d\varphi}\right)_{\theta_0,\varphi_0} \tag{4.166}$$

If we substitute (4.163) into (4.164) through (4.166) and perform the differentiations, we obtain three equations in the three unknowns (x_0, y_0, z_0). This leads to the phase-center solution

$$\begin{bmatrix} x_o \\ y_o \\ z_o \end{bmatrix} = \frac{1}{k\sin\theta_o}\begin{bmatrix} S_0\cos\varphi_0 & Q_0\cos\varphi_0 & -\sin\varphi_0 \\ S_0\sin\varphi_0 & Q_0\sin\varphi_0 & \cos\varphi_0 \\ Q_0 & -S_0 & 0 \end{bmatrix}\begin{bmatrix} \psi(\theta_0,\varphi_0) \\ \left(\frac{d\psi}{d\theta}\right)_{\theta_0,\varphi_0} \\ \left(\frac{d\psi}{d\varphi}\right)_{\theta_0,\varphi_0} \end{bmatrix} \tag{4.167}$$

where $S_0 = \sin^2\theta_0$, $Q_0 = \sin\theta_0\cos\theta_0$, and $d\psi/d\theta$, $d\psi/d\Phi$ are obtained by direct differentiation of (4.161).

The phase center displacement (relative to the geometric center) in an adaptive array depends on the algorithm used to calculate the weights. In order to illustrate this, let us again consider the seven-element, planar array with five jammers present as described following (4.66). Let us assume that the unconstrained optimization (i.e., power minimization) algorithm is used, so that the weight vector is given by (4.66). In this case, it is found that as the angle θ_0 (measured from array boresight) is increased from $\theta_0 = 0$ along a constant azimuth Φ_0, there is no displacement of the phase center unless a pattern null is crossed. This is illustrated in Figure 4.11 for the case when $\Phi_0 = 15°$. Observe that the phase center is undisplaced from the geometric center (0, 0, 0) for $\theta_0 < 62.8°$, but suffers an abrupt displacement at $\theta_0 < 62.8°$. If we refer back to Figure 4.6, we see that if we set $\Phi_0 = 15°$ and we increase θ_0 from 0° (center of polar plot), we cross a null contour at $\theta_0 < 62.8°$. Therefore, when the power minimization algorithm is used, there could be a significant displacement of the phase center for some values of (θ_0, Φ_0).

The results for the constrained (beamforming) algorithm are different because the beam is always pointed to the GPS satellite of interest, so there is no null contour crossed except when the observation point is very close to an actual jammer location. For the constrained algorithm, the phase center is usually slightly displaced from the geometric array center, but the displacement is never very large (in wavelengths). For example, in Figure 4.12, we show the phase center displacement for the same jammer scenario as was used to compute Figure 4.11 (i.e., five strong jammers at (81°, –30°), (90°, –110°), (79°, 45°), (60°, 121°), and (88°, 175°)), but using the constrained jammer cancellation algorithm. Observe that the maximum phase-center displacement is approximately 0.011 wavelength (0.2 cm at the GPS L_1 frequency of 1575 GHz). This is good news because it means that a phase-center

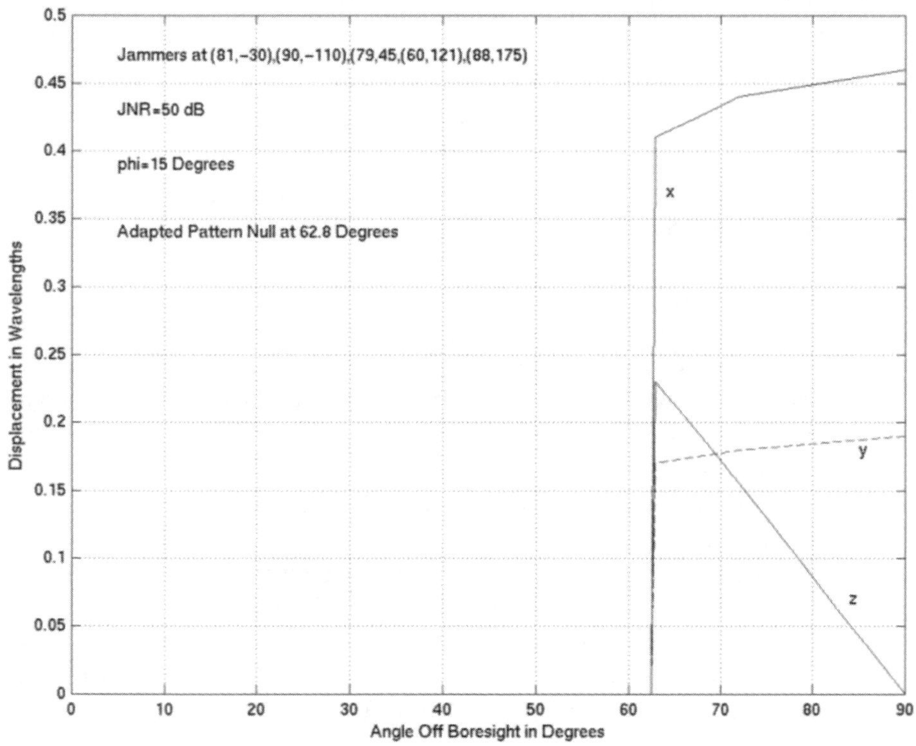

Figure 4.11 Phase center displacement for seven-element planar array using power min algorithm, five jammers.

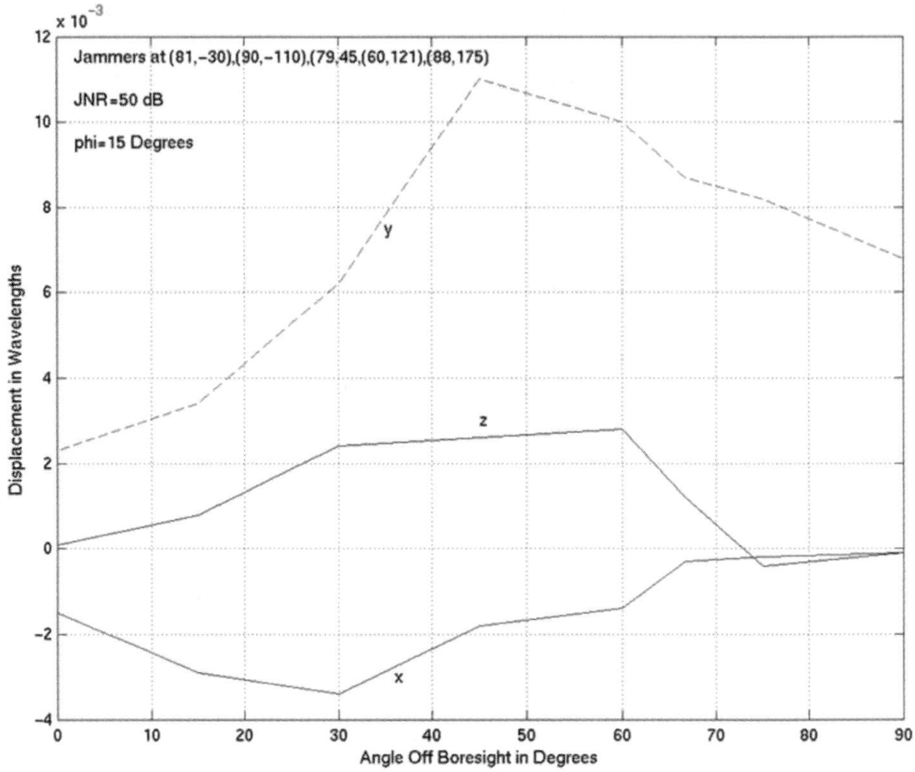

Figure 4.12 Phase center displacement for planar adaptive array using constrained beamforming algorithm, five jammers.

correction is unnecessary when a constrained (beamforming) jammer-cancellation algorithm is used to calculate the weight vector.

4.9 Typical Results

In this section we illustrate the frequency-angle response (FAR) for various jamming scenarios. The algorithm used to compute the adaptive weight vector is the power-minimization algorithm described by (4.66) and (4.70). The antenna is a hexagonal array with one element at the center and the remaining six on the vertices as described previously following (4.66).

To illustrate the benefits of using temporal taps, we compare the performance of a space-only processing (which corresponds to $Q = 1$) to a STAP architecture using seven time-taps ($Q = 7$). For simplicity, we use a jammer-to-noise ratio (JNR) of 50 dB per antenna element for each jamming source analyzed. The JNR is defined as the jammer power divided by the receiver thermal noise power. Also, a system bandwidth of 24 MHz is assumed. We also define narrowband (NB), partial-band (PB), and broadband (BB) jammers as interfering sources broadcasting white noise over bandwidths of 0.5, 5.0, and 24.0 MHz, respectively.

4.9.1 Effects of Temporal Taps

Consider an array receiving a single BB jammer located at ($\Phi = 140°$, $\theta = 45°$). The frequency-angle response of spatial-only adaptive processing (SAP) versus Φ and normalized frequency (denoted by 2f/B, where the system bandwidth is defined as -12 MHz < B < 12 MHz) in the elevation plane of the jammer ($\theta = 45°$) is depicted as a contour plot in Figure 4.13(a). This figure illustrates that the null is appropriately spatially aligned with the jammer at certain frequencies, but not over the entire bandwidth of the jammer. The noticeable improvement in the spatial alignment of the null for STAP is shown in Figure 4.13(b). The null is correctly positioned over virtually the entire operating bandwidth of the jammer. Note that the STAP null is more uniform (sharper and deeper) at the exact jammer angle, whereas the SAP filter has two local minima which are less accurately aligned. Similar results occur when examining performance for two jammers at the same azimuth with different elevations.

4.9.1.1 Summary

For BB jammers, increasing the number of temporal taps improves the spatial alignment, sharpness, and depth of the null across the entire jammer bandwidth.

4.9.2 Effects of Temporal Taps for Different Types of Jammers

Here we consider the benefits of STAP over SAP for different jammer types. The duality of the Fourier transform allows one to equivalently consider the space-time data vector, from which the filter is derived, as a space-frequency vector. Thus, more time-taps corresponds to more frequency samples, which logically allow the filter greater control of its shape in frequency.

Figure 4.14 illustrates the SAP and STAP FAR for a BB jammer ($\theta = 45°$, $\Phi = 70°$) in the same elevation plane as an NB or PB jammer ($\theta = 45°$, $\Phi = 250°$). The lack of time taps forces SAP to place a very broad null in frequency at the spatial location of a jammer regardless of its bandwidth. STAP is able to achieve much more precise nulling in frequency. The nulling of the NB and PB jammers are localized only to the operating bandwidths of the jammers. STAP results in a higher satellite

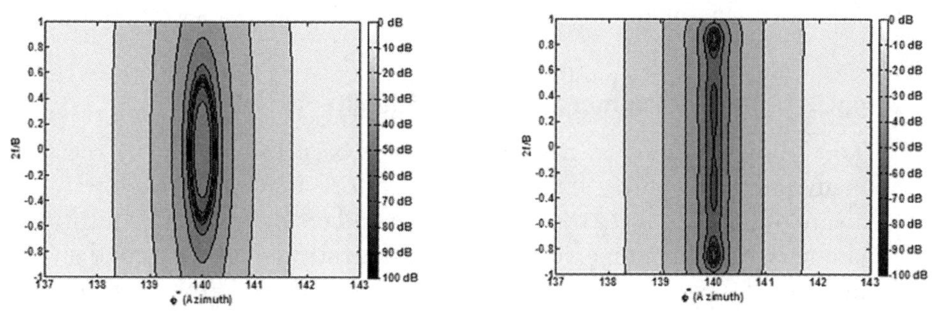

(a) (b)

Figure 4.13 FAR for BB jammer. (a) SAP, (b) STAP.

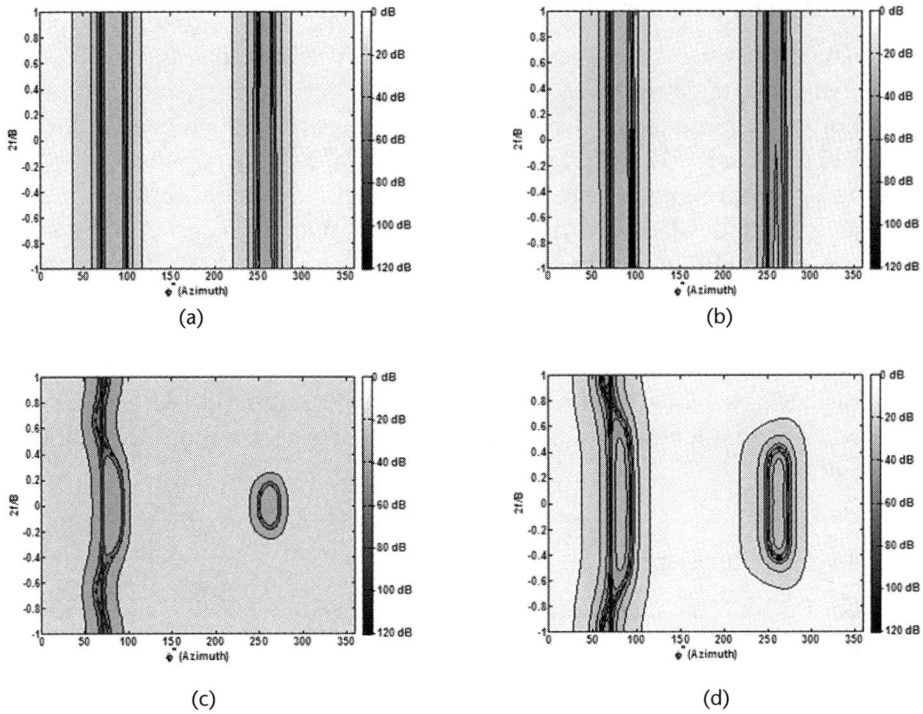

Figure 4.14 FAR for three types of jammers. (a) SAP: BB and NB, (b) SAP: BB and PB, (c) STAP: BB and NB, and (d) STAP: BB and PB.

availability than does SAP, increasing the overall probability that more satellites will be in view and that an accurate receiver location solution may be achieved.

4.9.2.1 Summary

SAP places a null that extends across the system bandwidth in the direction of any jammer. STAP promotes a higher satellite availability. As an example, a satellite signal coming from the direction of a narrowband jammer would not be received by a SAP system, since it forms a null across the entire system bandwidth. Conversely, a STAP system could receive this signal since the filter only nulls a small portion of the system bandwidth.

4.9.3 Effects of Jammer Spacing on Satellite Availability

The following question may have already been formulated by the reader: How many jammers can an algorithm null? Before answering this query, we must be aware of the spatial separation between jammers necessary to constitute "different jammers." If two jammers are located sufficiently close to one another, only a single null may be required to mitigate their effects. If they are significantly separated, however, two nulls may be necessary. Only STAP is examined via surface plots to illustrate this concept.

Notice in Figure 4.15 how the FAR changes for different jammer spacings in azimuth (all jammers in this example are located in the $\theta = 35°$ elevation plane and

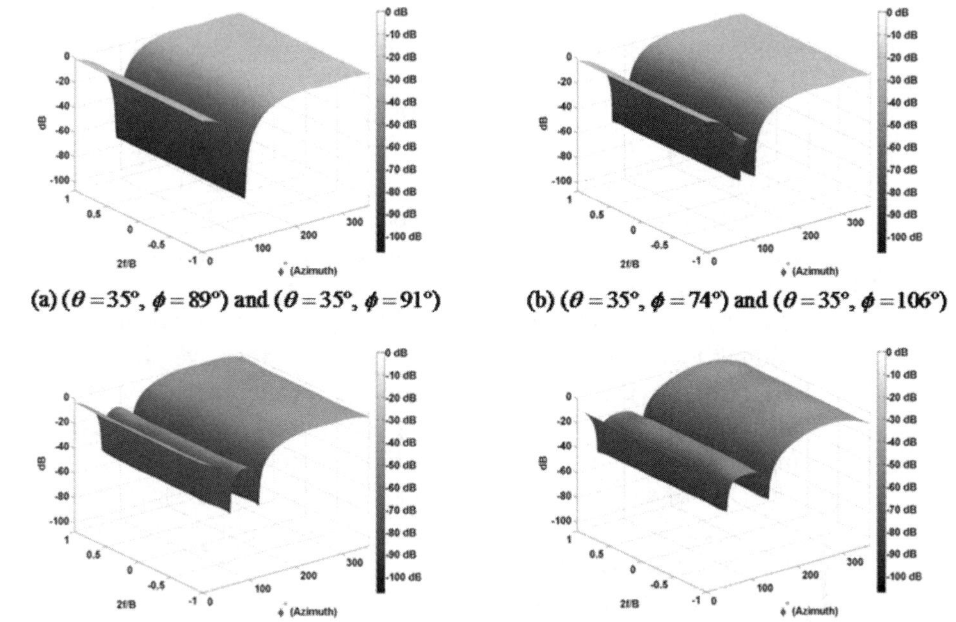

(a) ($\theta = 35°$, $\phi = 89°$) and ($\theta = 35°$, $\phi = 91°$)

(b) ($\theta = 35°$, $\phi = 74°$) and ($\theta = 35°$, $\phi = 106°$)

(c) ($\theta = 35°$, $\phi = 59°$) and ($\theta = 35°$, $\phi = 121°$)

(d) ($\theta = 35°$, $\phi = 44°$) and ($\theta = 35°$, $\phi = 136°$)

Figure 4.15 FAR for four different jammer spacing.

azimuthal locations are shown in the subplot captions). The upper-left subplot jammer azimuthal spacing is 2. The spacing subsequently increases by 30° for each subplot. For small spacing, only one null is required to mitigate the two jammers. As the spacing grows, two separate nulls start to form. The FAR, however, between the two jammers is less than optimal. As the jammer spacing increases, the FAR between the two jammers improves. Thus, for a particular antenna array configuration, there are corresponding jammer spacings that result in the highest and lowest overall satellite availability.

4.9.3.1 Summary

The number of jammers that an adaptive algorithm can null is a function of the jammer spacing. A single null can mitigate several jammers if they are nearly angularly colocated. When jammers are spaced sufficiently far apart, separate nulls are required to negate their effects. If jammers are located between these two extremes, the nulls may bleed into one another, causing a further reduction in satellite availability.

4.9.4 Maximum Number of Jammers to Null

As alluded to in the previous section, it is important to quantify the number of jammers that a particular system can nullify. This quantity explicitly depends on the jammer spacing. In this section, we will only investigate jammers spaced sufficiently far apart that the algorithm forms a distinct null for each jammer.

Here we only consider STAP and BB jammers. The theoretical limit for the number of randomly located BB jammers that the adaptive algorithm can mitigate is one less than the number of antenna elements. Thus, in our case, we examine the FAR for interference scenarios from four to seven jammers. Figure 4.16 illustrates the FAR for an increasing number of BB jammers via polar plots. The polar plots illustrate the FAR for all θ ($\theta = 0°$ at the center and $\theta = 90°$ at the outer edge) and Φ (depicted along the circumference) at the system center frequency. The exact jammer locations are indicated by an x for clarity. As expected, the null and jammer locations are well aligned until the number of jammers exceeds one less than the number of antenna elements. In particular, the jammers located at ($\theta = 85°$, $\Phi = 100°$) and ($\theta = 60°$, $\Phi = 200°$) are outside of any null.

Note that yet again the ability of an algorithm to mitigate jamming is a function of jammer location. Occasionally, the use of an adaptive filter results in reductions in the FAR that are not at the location of any jammers. These areas are known as sympathetic nulls and are an unavoidable result of the antenna array geometry. If the jammers are located in a sympathetic null, however, they are of course mitigated. Thus, if the jammer location is chosen in this manner, one may be led to believe (incorrectly) that the algorithm can always null more randomly located jammers than is theoretically possible. Such a case is depicted in Figure 4.17 where all of the jammers have the same elevation ($\theta = 65°$). The seventh jammer lies in a sympathetic null which can be seen for the six jammer case in Figure 4.17(d) at azimuth $\Phi = 300°$. Therefore, jammer locations should vary notably when testing system performance.

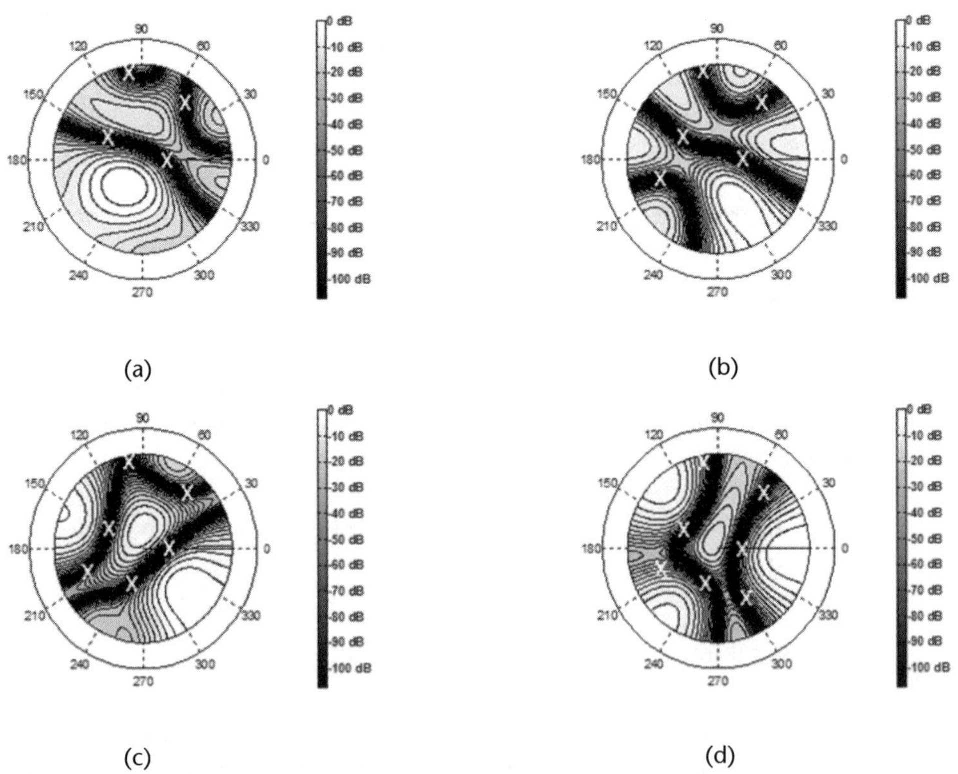

Figure 4.16 FAR for an increasing number of BB jammers. (a) four BB, (b) five BB, (c) six BB, and (d) seven BB.

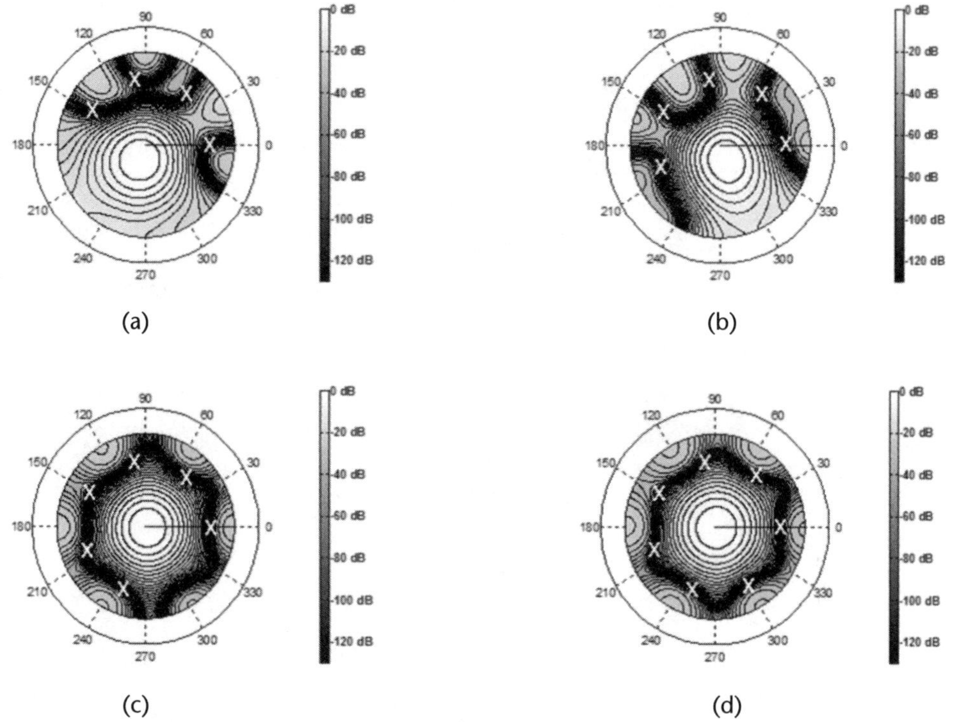

Figure 4.17 FAR for an increasing number of BB jammers.

4.9.4.1 Summary

The number of randomly located BB jammers that an adaptive algorithm can null is equal to one less than the number of antenna elements. As the number of jammers exceeds this limitation, the nulls become less spatially aligned with the jammer locations. Sympathetic nulls may permit the nulling of more jammers than this limit, depending on jammer location.

4.9.5 Frequency-Agile Jammers and Stale Weights

One of the most attractive aspects of STAP algorithms is that they are adaptive. Hence, as a jammer changes operating frequency or as the geometry between the array and the jammers change, the STAP filter continually adapts to mitigate the jamming. Unfortunately, a continual, rapid update of the filter comes at a high computational cost. Therefore, some systems calculate a filter and apply it not only to the data from which it was computed, but also to future times. This methodology assumes that the interference environment is not changing rapidly, and the system is said to have stale weights (see Section 4.4). In slowly changing environments, the technique tends to work well. Highly dynamic platforms as well as time or frequency agile jammers, however, tend to violate this assumption. For example, consider a swept-frequency jammer which sweeps over 24 MHz in 0.002 seconds. The adaptive processor should calculate the filter from data spanning at least 0.002 seconds. Thus, jamming at all frequencies will be properly accounted for and nulled. This case is depicted in Figure 4.18(a) and (c).

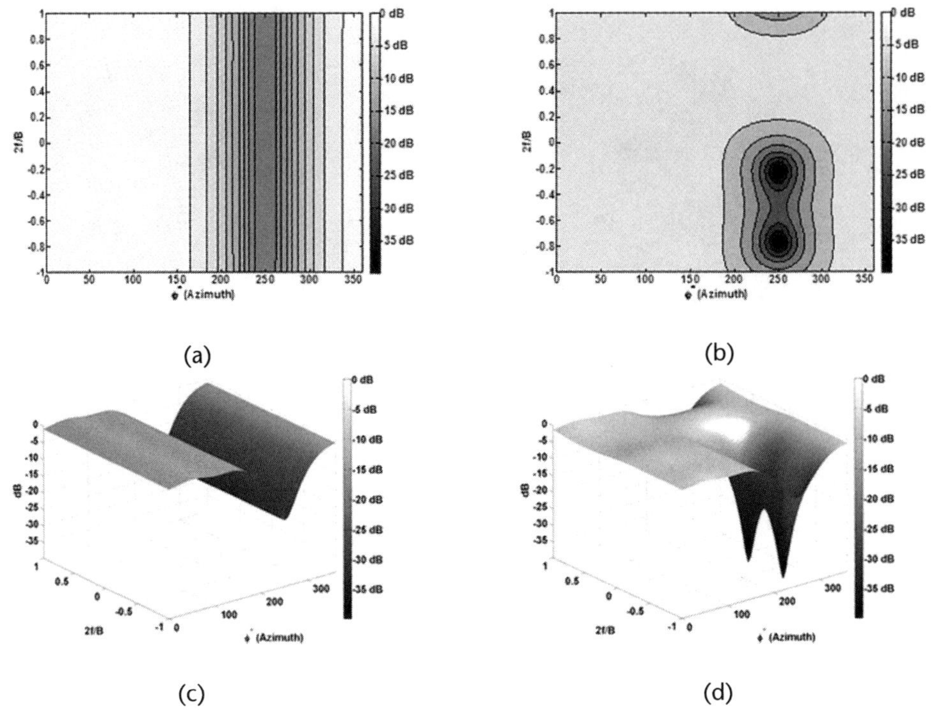

Figure 4.18 Stale vs. nonstale weights. (a) Full interval weight calculation: contour, (b) half-interval weight calculation: contour, (c) full interval weight calculation: surface, and (d) half-interval weight calculation: surface.

Consider a system that uses stale weights and assume that the filter is calculated from only the first 0.001 seconds of data. While the first half of the sweep will be appropriately nulled, the second half will pass through the system where there is no null ($0 \leq 2f/B \leq 1$). This scenario is shown in Figure 4.18(b) and (d).

Stale weight systems can suffer similar difficulties when handling pulsed or hopped-frequency jammers. In the pulsed case, a null may appear when the jammer is off, or a null will not be placed when the jammer is emitting. For the hopped-frequency jammer, the null may appear at the "old" frequency and instead of the current frequency.

4.9.5.1 Summary

Stale weights may be used in a stationary interference environment with limited degradation on performance to save on computational expense. Using stale weights against frequency or time agile jammers or in a highly dynamic environment, however, may drastically impact performance.

Bibliography

GPS Principles

Kaplan, E. (ed.), *Understanding GPS: Principles and Applications,* Norwood, MA: Artech House, 1996.

Kaplan, E., and C. Hegarty (eds.), *Understanding GPS: Principles and Applications, Second Edition,* Norwood, MA: Artech House, 2006.

Misra, P., and P. Enge, *Global Positioning System,* Lincoln, MA: Ganga-Jamuna Press, 2001.

Parkinson, B., and J. Spilker (eds.) *Global Positioning System: Theory and Applications,* Washington, DC: American Institute of Aeronautics and Astronautics, 1996.

Space-Time Adaptive Processing (General)

Applebaum, S., "Adaptive Arrays," *IEEE Transactions,* Vol. AP-24, 1976, pp. 585–598.

Brennan, L., and I. Reed, "Theory of Adaptive Radar," *IEEE Transactions on Aerospace and Electronic Systems*, Vol. AES-9, , 1973, pp. 237–252.

Brennan, L., J. Mallett, and I. Reed, "Adaptive Arrays in Airborne MTI Radar," *IEEE Transactions*, Vol. AP-24, 1976, pp. 607–615.

Carlson, B., "Covariance Matrix Estimation Errors and Diagonal Loading in Adaptive Arrays," *IEEE Transactions*, Vol. AES-24, 1988, p. 4.

Clarkson, P., "Optimal and Adaptive Signal Processing," CRC Press, 1993.

Compton, R., "On the Performance of a Polarization Sensitive Adaptive Array," *IEEE Transactions on Antennas and Propagation*, Vol. AP-29, 1981, pp. 718–727.

Compton, R., "Adaptive Antennas," Englewood Cliffs, NJ: Prentice Hall, 1988.

DiPietro, R., "Extended Factored Space-Time Processing for Airborne Radar Systems," Proceedings of the 26th Asilomar Conference, 1992, pp. 425–430.

Farina, A., "Antenna-Based Signal Processing Techniques for Radar Systems," Norwood, MA: Artech House, 1992.

Frost, O., "An Algorithm for Linearly-Constrained Adaptive Array Processing," *Proceedings of the IEEE*, Vol. 60, 1972, pp. 926–935.

Gabriel, W., "Adaptive Arrays—An Introduction," *Proceedings of the IEEE*, Vol. 64, 1976, p. 239.

Godara, L., *Smart Antennas,* Boca Raton: FL, CRC Press, 2004.

Griffiths, L., "A Simple Adaptive Algorithm for Real-Time Processing in Adaptive Arrays," *Proceedings of the IEEE*, Vol. 57, 1969, pp. 1696–1704.

Guerci, J., *Space-Time Adaptive Processing for Radar,* Norowood, MA: Artech House, 2003.

Gupta, I., and A. Ksienski, "Effect of Mutual Coupling on the Performance of Adaptive Arrays," *IEEE Transactions on Antennas and Propagation*, Vol. 31, 1983, pp. 785–791.

Hudson, J., *Adaptive Array Principles,* London: Peter Peregrinus, 1989.

Kaiser, T., et al. (eds.), *Smart Antennas: State of the Art,* Cairo, Egypt: Hindaivi Publishing, 2005.

Klemm, R., *Space-Time Adaptive Processing,* London: IEEE, 1999.

Klemm, R. (ed.), *Applications of Space-Time Adaptive Processing,* London: Institution of Electrical Engineers, 2004.

Monzingo, R., and T. Miller, *Introduction to Adaptive Arrays,* New York: Wiley, 1980.

Niocolau, E., and D. Zaharia, *Adaptive Arrays,* Elsevier, 1989.

Paulraj, A. J., and C. B. Papadias, "Space-Time Processing for Wireless Communications," *IEEE Signal Processing Magazine,* November 1997, pp. 49–83.

VanVeen, B., "Minimum Variance Beamforming," in *Adaptive Radar Detection and Estimation,* S. Haykin and A. Steinhardt (eds.), New York: Wiley, 1992.

Ward, J., "Space-Time Adaptive Processing for Airborne Radar," Technical Report 1015, MIT Lincoln Laboratory, December 1994.

Weiner, M. (ed.), *Adaptive Antennas and Receivers,* Boca Raton, FL: CRC Press, 2006.

Widrow, B., and S. Stearns, *Adaptive Signal Processing,* Prentice Hall, 1985.

Space-Time Adaptive Processing (GPS-Specific)

Fante, R, and J. Vaccaro, "Wideband Cancellation of Interference in a GPS Receive Array," *IEEE Transactions*, Vol. AES-36, April 2000, pp. 549–564.

Fante, R., and J. Vaccaro, "Ensuring GPS Availability in an Interference Environment," *Proceedings of the IEEE*, 2000 Precision Location and Navigation Symposium, 2000, pp. 37–40, (ISBN: 0-7803-5872-4, IEEE Catalog No. 00CH37062).

Ngai, E., and D. Blejer, "Mutual Coupling Analyses for Small GPS Adaptive Arrays," IEEE AP-S International Symposium, 2001, pp. 38–41.

Ngai, E., D. Blejer, T. Phuong, and J. Herd, "Anti-Jam Performance of Small GPS Polarimetric Arrays," IEEE AP-S International Symposium, 2002, pp. 128–131.

Antenna Phase Center

Best, S., "Distance-Measurement Error Associated with Antenna Phase-Center Displacement in Time-Reference Radio Positioning Systems," *IEEE Antennas and Propagation Magazine*, Vol. 46, 2004, pp. 13–22.

Cruz, J., B. Gimeno, E. Navarro, and V. Such, "Phase Center Position of a Microstrip Horn Radiating in an Infinite Parallel Plate Waveguide," *IEEE Transactions*, Vol. AP-42, August 1994, pp. 1185–1188.

James, J., et al., *Microstrip Antenna Theory and Design,* P. Peregrinus, Ltd., 1989, pp. 77–80.

Smith, D., D. Wilton, and J. Williams, *Compensating GPS Time Delay Due to Phase Center Variations,* University of Houston Report, 2003.

Wardelick, D., "Phase Center of Aperture Antennas," *IEEE Transactions*, Vol. AP-28, pp. 263–264, 1980.

CHAPTER 5

Ground Plane, Aircraft Fuselage, and Other Platform Effects on GPS Antennas

Basrur Rama Rao

5.1 Introduction

GPS antennas have certain unique characteristics that make them different from antennas typically used for point-to-point communications systems. They are RHCP, dual-band antennas with very broad radiation patterns. This enables them to simultaneously acquire at least four or more GPS satellites visible over the upper hemisphere extending down to an elevation angle as low as 5°. Reception of signals from low elevation satellites is critical for maintaining good PDOP. Coverage at even lower elevation angles may be necessary for ship and airborne systems to avoid losing signals from low elevation satellites due to vehicle dynamics. GPS modernization may also soon require these antennas to operate in three frequency bands L_1, L_2, and L_5 to meet the demands imposed by both civilian and military GPS systems. Ideally, these antennas are also required to provide adequate gain to allow acquisition of GPS satellites over this visible region, have very low backlobes to mitigate multipath and interference, and display good phase center stability and minimum group delay to allow accurate carrier phase and code processing. Due to their broad beamwidth, GPS antennas can cause significant illumination of the edges of the platform on which they are mounted; hence, their radiation patterns could be heavily influenced by the size, shape, and material of the platform. Due to the varied applications and growing demand for GPS antennas, the platforms on which the antennas are mounted can be in different sizes and shapes. They vary from planar metallic ground planes, either rectangular or circular in shape, generally only a few GPS wavelengths in cross section to very large, complex three-dimensional platforms such an aircraft, helicopter, or an automobile. The platform can also shrink to a very small size, much smaller than a wavelength when the antenna is integrated directly into the casing of a mobile phone or into other types of man-portable GPS systems. Ground planes serve different roles, depending on the type of GPS antenna. For many types of GPS antennas such as the microstrip patch, crossed slot, and monofilar helix, the ground plane becomes an integral part of the antenna structure. For some other antenna types, such as quadrafilar helix and log spiral, the ground plane is an auxiliary part of the antenna. These antenna types can perform without the ground plane, but need to use them to cut off the pattern below the horizon so as to suppress multipath appearing from low grazing angles down to nadir. In both these cases, the ground plane can have a significant influence on antenna performance, especially at lower elevation angles. In this chapter, we analyze the effects on the GPS antenna caused by the different platforms. We also consider techniques used for mitigating platform effects on antenna performance to improve the accuracy of GPS systems needed in many high-precision applications such as geodesy, geodynamics, and cadastral surveys [1]. Of particular concern is the variation in the phase center of the antenna as a function of the observation angle caused by edge diffraction from the platform.

The candidate GPS antenna selected for the evaluation of such platform effects is a microstrip patch antenna; the design and radiation characteristics of this type of antenna was described in Chapter 2. Due to its low profile, low cost, and small size, microstrip antennas are one of the most popular antennas used for many different GPS applications. The microstrip antenna is almost always used for airborne GPS systems on U.S. aircraft since it easily meets the ARINC 743A form factor requirement for antenna size imposed by the FAA. This requires that the cross-sectional dimensions of the GPS antenna be no longer than 4.7" × 2.9" and also be no more than 0.75" in height to minimize aerodynamic drag. The use of a ceramic substrate also allows the size of the GPS antenna to be reduced to very small dimensions of less than 25 mm in cross section, making it suitable for integration into a cell phone and for automobile and other civilian applications [2].

5.2 Microstrip Antenna on a Planar Ground Plane

5.2.1 Planar Ground Plane Effects

In this section, we will consider the effects of square and circular shaped planar ground planes on the microstrip antenna pattern. Microstrip antennas are mounted on finite-size ground planes to suppress multipath from ground reflections. Diffraction from the edges of the ground plane can influence the antenna performance in two ways. It can cause significant backlobes in the antenna pattern, making the GPS antenna susceptible to multipath from GPS signals reflected off the ground. The diffracted signals from the edge of the ground plane interfere with the direct radiated signals from the microstrip antenna above the ground plane causing a "scalloped" antenna pattern at higher elevation angles that can affect the gain and phase and enhance the frequency sensitivity of the antenna that results in increased group delay. These ground plane effects become important in high-precision GPS applications requiring submeter and even subcentimeter accuracy [1]. Different types of ground planes needed for mitigating ground plane effects on the antenna will be described and the results of relevant measurements will be shown to demonstrate their efficacy in suppressing such adverse effects.

The uniform geometric theory of diffraction (UTD) developed by Ohio State University will be used extensively in this analysis [3]. This theory correctly predicts the antenna pattern in upper hemispherical region above the ground plane as well as the backlobes occurring below the ground plane in the shadow region. The microstrip antenna will be represented by an equivalent parallel slot model [4] proposed by Huang, since it lends itself very well for UTD analysis.

The radiation from the antenna for this model occurs from the fringing electric fields between the edge of the slot and the ground plane. The length of each slot $a = \lambda_0/(2\sqrt{\varepsilon_r})$ where λ_0 is the wavelength of the resonant frequency of the antenna and ε_r is the relative dielectric constant of the substrate. Figure 5.1 shows the slot model representation for a RHCP square-shaped microstrip patch antenna of $L = W = a$, where W and L are the width and length of the microstrip patch. The height of the microstrip antenna above the ground plane is h and is determined by the thickness of the dielectric substrate. The microstrip patch antenna is mounted on a square ground plane with side dimension b.

To generate RHCP required for receiving GPS signals, the patch is excited by two orthogonally located coaxial probes 1 and 2 fed in phase quadrature by a 90° hybrid. The phase of probe 1 is 0° and probe 2 is -90°. The electric field for radiation from probe 1 that is resonant in the TM_{10} mode is polarized linearly and is parallel to the $+X$ axis. The electric field for radiation from probe 2 that is resonant in the TM_{01} mode is polarized linearly along the $+Y$ axis. The orthogonally polarized E field components generated by these two probes are of equal amplitude but 90° out of phase relative to each other. RHCP signals are generated when the inputs from these two probes are combined in a quadrature hybrid. The radiation behavior of this antenna occurs can be deduced by noting the equivalent magnetic current distribution along the periphery of the patch and fringing electric fields between the surface of the microstrip antenna and the ground plane along the four sides of the microstrip patch antenna. For resonance in the TM_{mn} mode [5], the equivalent magnetic current $\vec{M}$ flows along the four sides of the square microstrip patch antenna can be represented by $\vec{M} = 2\hat{n} \times \vec{E}$ where $\hat{n}$ is the normal to the magnetic wall of the cavity of the patch and $\vec{E} = \hat{z}|E|$ [5]. For the TM_{mn} mode, the modal function $_{mn}$ can be represented by the product of two cosine functions determined by the index m along the x axis and n along the y axis.

$$\varphi_{mn} \propto \cos(m\pi x/a)\cos(n\pi y/a) \tag{5.1}$$

For the TM_{10} mode generated by probe 1, the modal indices are $m = 1$ and $n = 0$. Hence, as shown in the top figure of Figure 5.1(a), magnetic current along the side of the patch at $x = 0$ and $x = W = a$ are oriented along the $+\hat{y}$ axis and is a constant. Similarly, for sides of the patch at $y = 0$ and $y = L = a$, the current varies as $\cos(\pi x/a)$ with oppositely directed currents along $\pm x$ axis. Hence, there is no radiation along the two sides of the patch antenna parallel to the X axis—the resonant direction for probe 1 excited in the TM_{10} mode. The radiation occurs from the two sides of the patch parallel to the y axis. Hence for probe 1, the microstrip antenna can be represented as two parallel slots, slot 1 and slot 2 oriented along the $+\hat{x}$ axis and separated by a low impedance microstrip transmission line whose width corresponds to the side dimension of the patch antenna. The width of each slot h is approximated by the thickness of the dielectric substrate used for the microstrip patch antenna, and the effective length l of each slot is equal to the physical length a plus the substrate thickness h. This approximation can be applied without

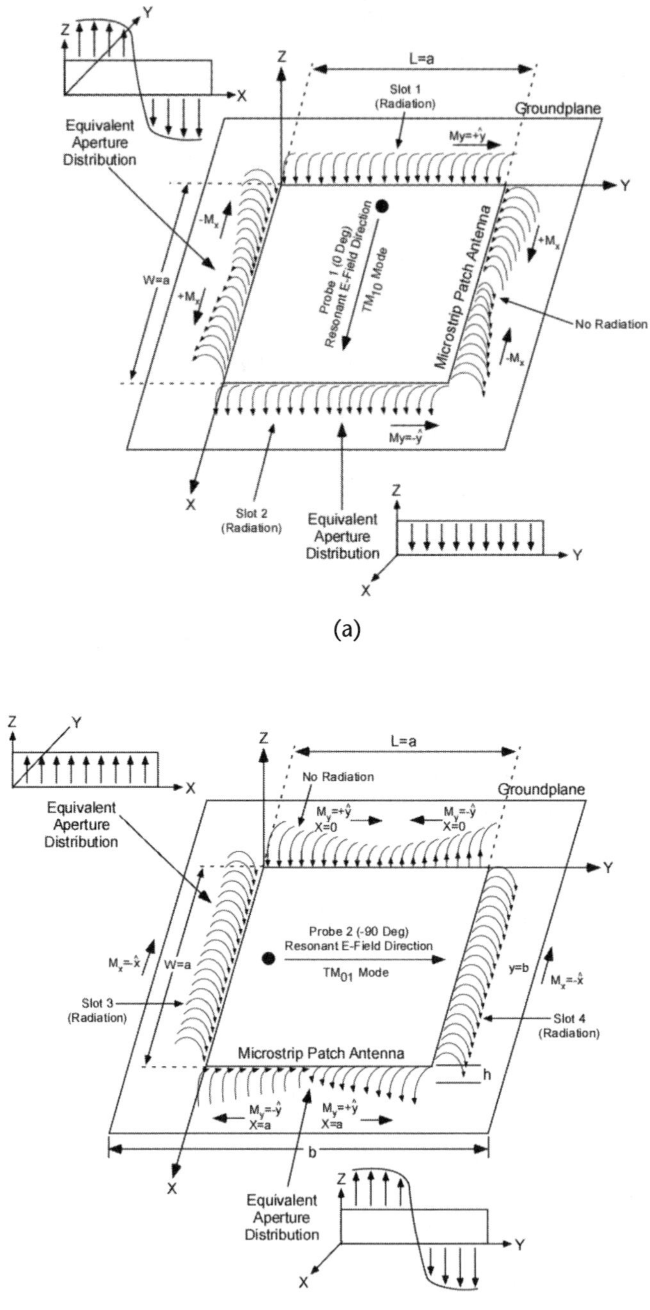

Figure 5.1 Slot model representation for RHCP microstrip patch antenna showing fringing electric field configuration.

modifications when the product of the substrate thickness in wavelength and the dielectric constant is much greater than 0.1; otherwise, the accuracy of this antenna model degrades [4]. For the corresponding TM_{01} mode generated by probe 2, the modal indices are $m = 0$ and $n = 1$, the directions of the currents in the two sides are reversed as shown in the lower Figure 5.1(b). Hence for probe 2, there is no radiation from the two sides of the patch parallel to the y axis, the resonant direction; the radiation occurs from the sides parallel to the x axis as shown in Figure 5.1(b). For probe 2, the radiation from the microstrip patch can be represented as coming from two slots, slots 3 and 4, that are parallel to the $+\hat{x}$ axis and separated by a distance a, the length of the microstrip patch. The radiation from the parallel pair of slots corresponding to each of the two probes can illuminate the edges of the ground plane and cause diffracted fields that can interfere with the direct radiation from the slots. The geometry of this problem is illustrated in Figure 5.2 for the case of probe 2.

The total radiation pattern at a far-field observation point P is a combination of four different rays—from the two parallel antenna slots A_1 and A_2 and the two edges of the ground plane E_1 and E_2. The contribution from the direct ray emanating from slot A_1 at the observation point P is E_{A1} and can be determined from the transmission line model [5].

$$E_{A1} = \hat{\mu}E_0\left\{\sin\left(kh*\cos\mu_{A1}\right)/\left(kh*\cos\mu_{A1}\right)\right\}\cdot\left(e^{-jkR_{A1}}/\sqrt{R_{A1}}\right) \tag{5.2}$$

where $k = (2\pi / \lambda_0)\sqrt{\varepsilon_r}$ is the wave number, ε_r is the relative dielectric constant of the substrate, and λ_0 is the wavelength in free space, and $\hat{\mu}$ is the orientation of the incident electric field. The amplitude dependence $1/\sqrt{R}$ indicates a cylindrical wavefront originating from the slot [6].

Similarly, the contribution from the direct ray emanating from slot A_2 is given by

$$E_{A2} = \hat{\mu}E_0\left\{\sin\left(kh*\cos\mu_{A2}\right)/\left(kh*\cos\mu_{A2}\right)\right\}\cdot\left(e^{-jkR_{A2}}/\sqrt{R_{A2}}\right) \tag{5.3}$$

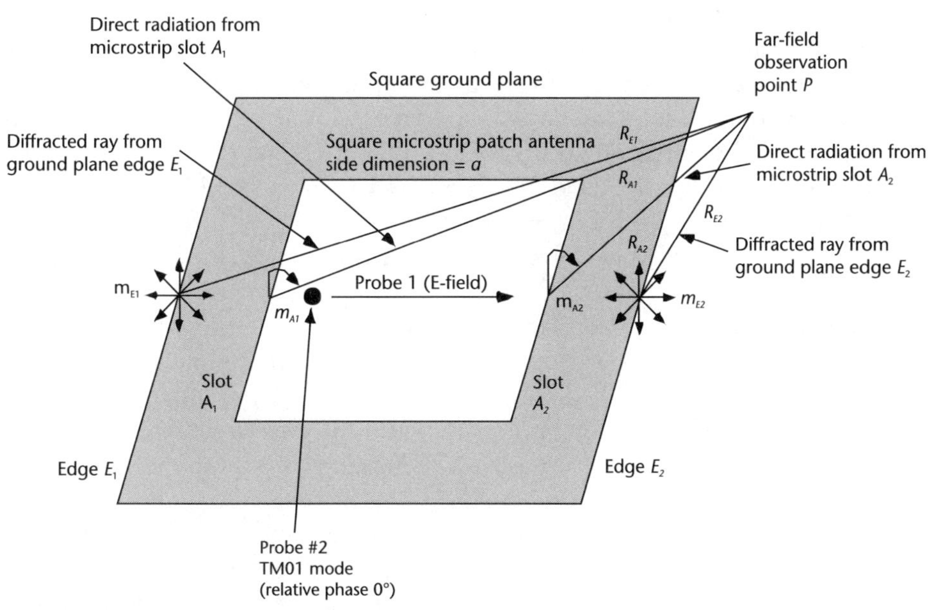

Figure 5.2 Effect of interference between direct radiation and ground plane edge diffraction on GPS microstrip antenna for Pprobe 1.

The diffracted field from the edge of the ground plane E_1 at the observation point P is given by

$$E_{E1}^{D} = \hat{\mu}\left\{\sin(kh * \cos\mu_{E1})/(kh * \cos\mu_{E1})\right\}\left(e_1^{-jkd}/\sqrt{d_1}\right)D\perp\left(e^{-jkR_{E1}}/\sqrt{R_{E1}}\right) \qquad (5.4)$$

The diffracted field from the edge of the ground plane E_2 at the observation point P is given by

$$E_{E2}^{D} = \hat{\mu}\left\{\sin(kh * \cos\mu_{E2})/(kh * \cos\mu_{E1})\right\}\left(e_2^{-jkd}/\sqrt{d_2}\right)D\perp\left(e^{-jkR_{E2}}/\sqrt{R_{E2}}\right) \qquad (5.5)$$

In (5.4) and (5.5), $D\perp$ represents the hard diffraction coefficient without accounting for the dielectric constant of the substrate [4, 6, 7]. The derivation of the diffraction coefficient is described in several excellent references [4, 6, 7]. The $1/\sqrt{R}$ amplitude dependence indicates that the diffracted fields have cylindrical wavefronts originating at the edges of the ground plane. There is also a small amount of diffraction from the edges contributed by the soft diffraction coefficient $D_{/\!/}$. This is a second order diffracted field and its contribution to the edge diffraction can be evaluated from the slope diffraction coefficient described [4, 6].

The total radiated field received at the observation point P is a combination of the direct rays from the two parallel slots of the microstrip antenna plus the contributions from the diffracted rays from the two opposite edges of the ground plane; the interference between these four rays can create a ripple in antenna gain pattern at higher elevation angles similar to that created by a four-element antenna array. The antenna backlobes of the patch antenna on the ground plane are caused almost entirely by the diffracted fields from the edges of the ground plane.

The interaction between the microstrip patch antenna and the finite square shaped ground plane was analyzed using the Numerical Electromagnetic Code–Basic Scattering Code (NEC-BSC) developed by the ElectroScience Laboratory at The Ohio State University [8]. This is a user-friendly computer code that can be used for the electromagnetic analysis of the radiation pattern of antennas in the presence of complex structures at high frequencies. The code can be used for analysis of microstrip antennas on ground planes that are larger than about two wavelengths in size. The GPS antenna used in these measurements and simulations was a square microstrip patch with dimensions 1.72" × 1.72" on a 0.1" thick dielectric substrate made from Roger's Duroid 6010 LM of dielectric constant 10.2. The patch was resonant at 1.5754 GHz. RHCP was obtained by using two coaxial probes fed in phase quadrature and placed at orthogonal locations on the patch. The location of each of the probes was optimized to yield an input impedance of 50 ohms. The patch antenna was placed at the center of a 26" × 26" square ground plane made from aluminum. The comparison between the measured antenna pattern and the calculated antenna pattern is shown in Figure 5.3 for both the vertical polarization and horizontal polarization.

The contribution to the diffracted signals from the edges of the ground plane for the vertical and horizontal polarizations comes from the hard and soft diffraction coefficients, respectively. The predominant interaction between the antenna and the ground plane that causes significant effects on the radiation pattern is from the vertical polarization. The results shown in Figure 5.3 indicate excellent agreement between the measured and calculated antenna patterns for the vertical polarization especially in predicting the scalloping of the pattern in the vicinity of zenith and also for the antenna backlobes occurring below the horizon. The results also show good agreement between the measured and calculated patterns for the

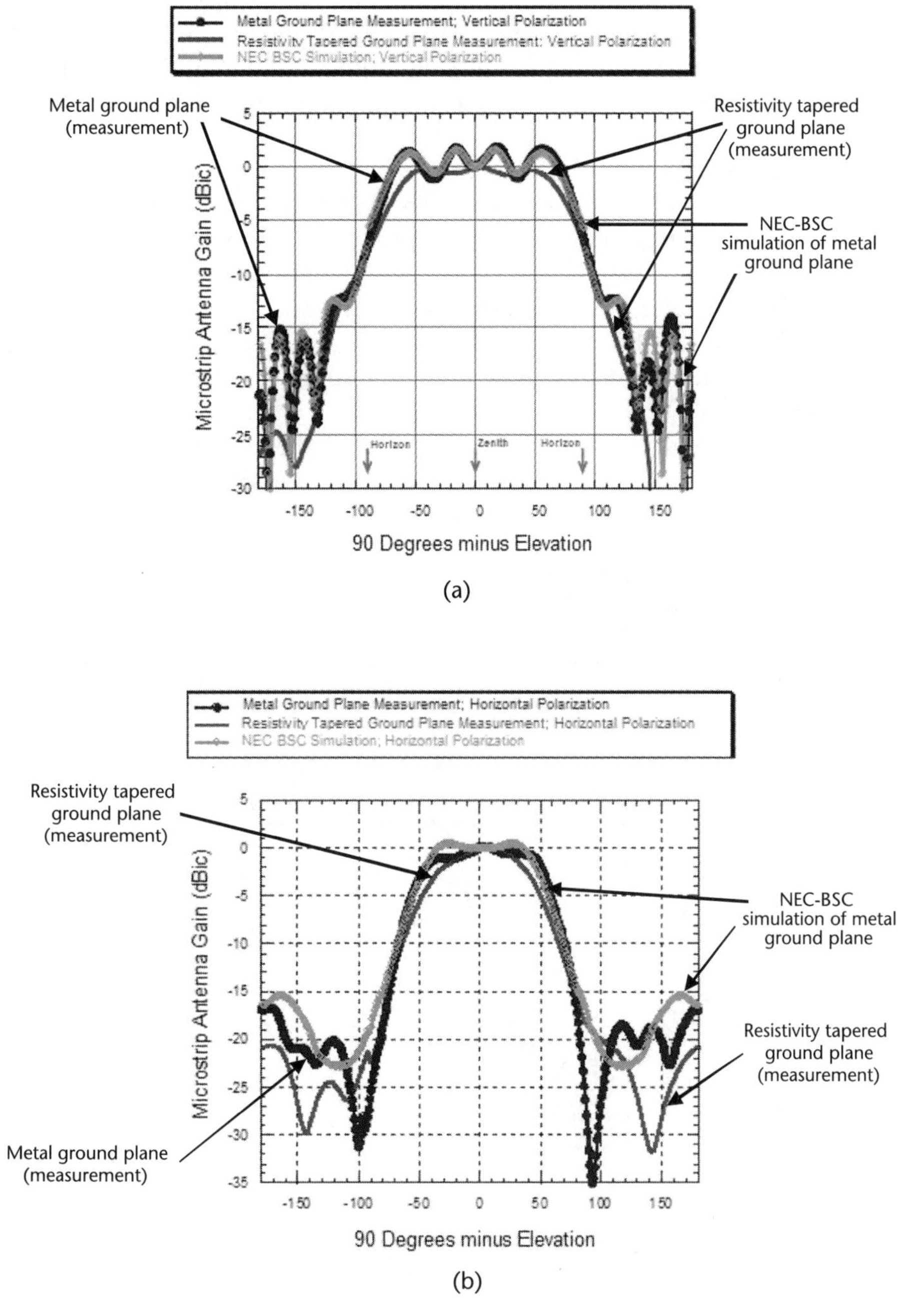

Figure 5.3 Ground plane edge diffraction effects on GPS microstrip antenna; comparison of measurement with NEC-BSC simulation; 1.5754 GHz; 26″ square ground plane. Vertical polarization = E_θ; horizontal polarization = E_ϕ.

horizontal polarization above the horizon. However, NEC BSC is unable to accurately predict the fine structure of the backlobes noticed below the horizon. Also shown in Figure 5.3 are measured results from an effective method for suppressing ground plane edge diffraction by the use of a specially designed resistivity tapered ground plane built by the MITRE Corporation [9]. The design of this ground plane is described in greater detail later in this chapter. In this type of ground plane, the currents travelling from the antenna towards the edges of the ground plane are effectively suppressed by increasing the surface resistivity of the surface of the ground plane in an exponential manner at the edges. From Figure 5.3, it is seen that the use of the resistivity tapered ground plane helps to get rid of the scalloping in the gain pattern at zenith and also eliminate the backlobes generated by diffraction from the edges of the ground plane.

Figure 5.4 shows the comparison between the measured and calculated patterns for RHCP. Notice again that there is excellent agreement between the two results. The pattern measured for the microstrip antenna placed on the resistivity tapered ground plane is also shown in this figure. The resistivity tapered ground plane is very effective in suppressing the antenna backlobes, thereby reducing the susceptibility of the GPS antenna to ground multipath effects. Scalloping in the antenna pattern at zenith caused by ground plane edge diffraction has also been eliminated.

The left-hand side of Figure 5.5 shows the calculated antenna patterns for vertical and horizontal polarizations when the microstrip antenna is placed on a

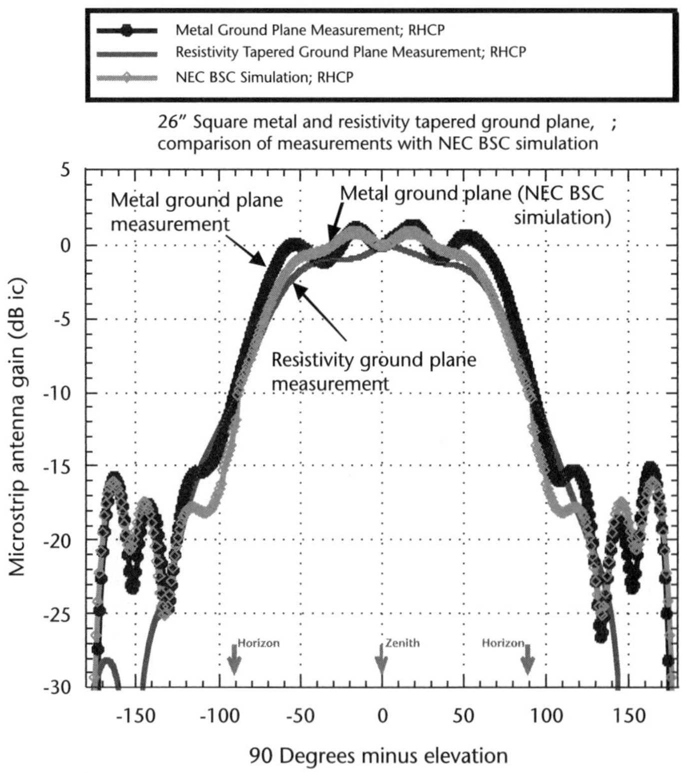

Figure 5.4 Ground plane edge diffraction effects on GPS microstrip antenna; comparison of measurement with NEC-BSC smulation; 1.5754 GHz; 26" square ground plane; RHCP.

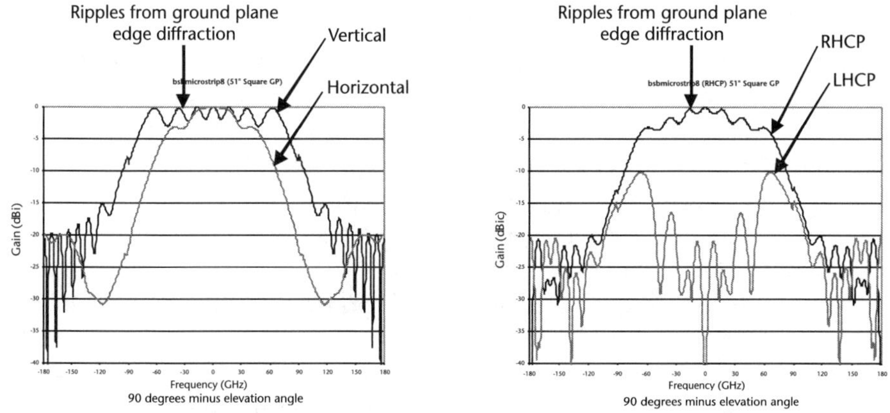

Figure 5.5 Measured radiation pattern of a GPS antenna on a 51″ square metal ground plane.

51"diameter square ground plane. On the right side of Figure 5.5 is the corresponding calculated pattern for RHCP. Because of the larger diameter ground plane, the interference ripples in the antenna pattern observed in the vicinity of zenith have increased in number but decreased in amplitude compared to the results shown in Figure 5.3 and 5.4 for a smaller size 26" square ground plane. This increase in the diffraction ripples in the antenna pattern is due to the larger separation distance between the edges of the radiating patch antenna and the two edges of the ground plane. By comparing Figures 5.3, 5.4, and 5.5, it can also be seen that increasing the size of the ground plane from 26" to 51" has resulted in suppressing the far-out antenna backlobes below the horizon by about 5 dB.

The results shown in this section were for a square-shaped conducting ground plane. The same type of analysis can also be used for estimating the effects of a circular-shape ground plane on the microstrip antenna. This requires replacing the linear edge diffraction coefficients in (5.2) through (5.5) by the corresponding diffraction coefficient for a circular edge [10].

The effects of a planar ground plane on the phase center of a GPS microstrip antenna has been investigated by Tranquilla and Colpitts [11] as a function of the observation angle for ground planes of different sizes. Large variations in the phase center especially at lower elevation angles have been measured as a result of ground plane edge diffraction effects. These antenna phase center variations are discussed in greater detail in Chapters 3 and 10.

5.3 Mitigation of Ground Plane Effects on Performance of GPS Antennas

Various types of modified ground planes have been designed to mitigate the edge diffraction effects discussed in the previous sections from affecting the performance of GPS antennas. These include choke-ring ground plane, EBG ground plane, rolled edge ground plane, and the resistivity tapered ground plane. A brief description of the design of these ground planes is given below and references are provided to obtain more detailed information.

5.3.1 Choke-Ring Ground Plane

The picture of a typical choke-ring ground plane made by Ashtech is shown in Figure 5.6. Choke rings are a well known method for shaping the radiation pattern to reduce multipath and stabilize the phase center. They consist of three to five narrow concentric metallic circular corrugations with the GPS antenna element located at the center [12]. The depth of the corrugation for the Ashtech ground plane is 63 mm, which is 0.25 wavelengths at 1.2156 GHz—the lowest frequency corresponding to the lower band edge frequency in the GPS L_2 band and 0.33 wavelengths at 1.587 GHz, the highest frequency corresponding to the upper band edge frequency in the GPS L_1 band. These circular corrugations operate approximately as quarter wavelength transmission lines in the L_2/L_1 bands, where the short circuit at the bottom of each groove is transformed into an open circuit at the top surface. They present high impedance to the electric field polarized perpendicular to the corrugations. They suppress surface wave propagation and prevent their diffraction from the edges of the ground plane, thereby providing a sharp pattern cutoff near the horizon and no evidence of main beam diffraction ripples near zenith. They also suppress LHCP levels over the entire upper hemisphere and promote a smooth main beam rolloff down to horizon.

Choke rings with the depth of the corrugation gradually changing towards the outer periphery have been built and other design modifications have also been made to improve frequency response and reduce size [13, 14]. A typical choke ring is about 15" in diameter, 2.5" tall, and generally weighs more than 10 lb. Figure 5.7 shows the measured radiation pattern of a NovAtel Model 503 L_1/L_2 GPS antenna on a choke-ring ground plane. The measured antenna pattern shown in Figure 5.7 indicates a sharp drop of in gain below the horizon and a peak cross-polarization level of –15 dB relative to RHCP over the entire upper hemisphere.

5.3.2 EBG Ground Plane

A disadvantage of the choke-ring ground plane for many applications is its size and weight. Compact and lightweight EBG ground planes whose thickness is much

Choke-ring ground plane

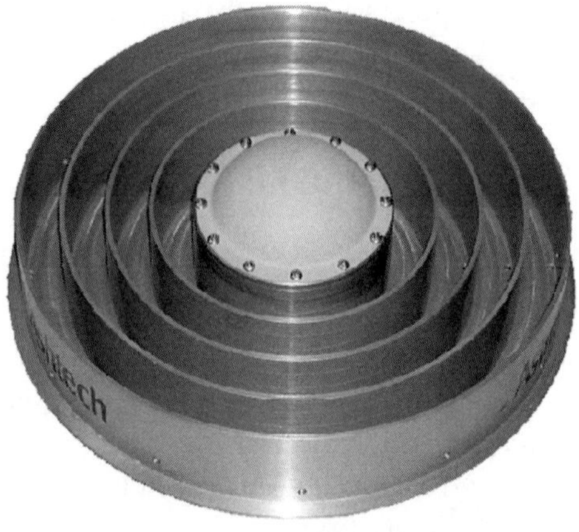

Figure 5.6 Ashtech choke-ring ground plane for GPS antennas to reduce edge diffraction effects on antenna pattern.

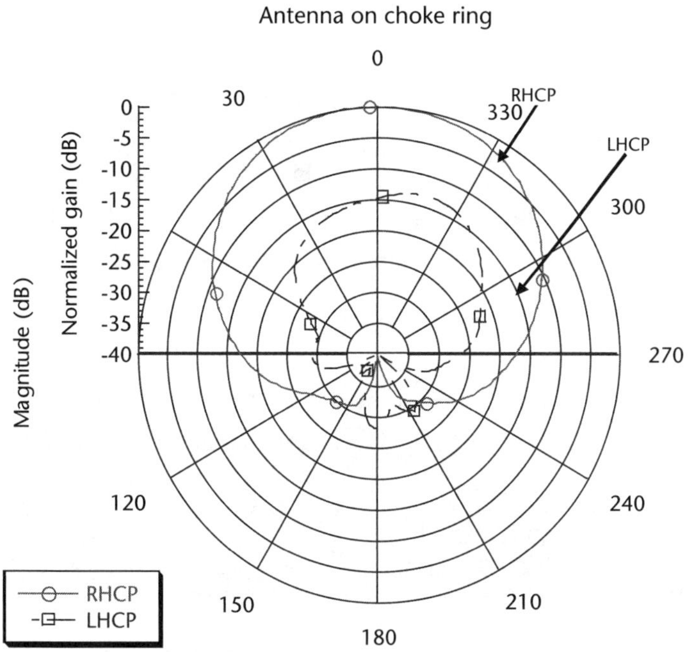

Figure 5.7 Measured radiation pattern of GPS antenna on a Novatel Model 503 choke ring ground plane.

less than a quarter wavelength have been built recently [15]. The surface of a EBG ground plane is a frequency selective surface and consists of a two-dimensional array of subwavelength metal discs with pins (or subcells) connecting them to the ground plane underneath. Their general appearance is that of an array of thumbtacks stuck to a ground plane. Each subcell in the structure can be considered as an LC resonant circuit. The reduction in height is achieved through capacitive loading. These EBG surfaces provide high impedance for both polarizations and for all propagation directions. They are sometimes also called artificial magnetic conductors, since the tangential magnetic field is zero at the surface rather than the electric field as in an electrical conductor. Shown in Figure 5.8 is a picture of the side and top views of the Model G 200a EBG ground plane for GPS applications built by the E-Tenna Corporation. The ground plane is 15" in diameter, weighs just 1.4 lb, and has a height of 0.59".

Figure 5.9 shows the comparison of the antenna patterns of a GPS antenna (EDO Corporation C 146 Model) measured on a 15" diameter ground plane (Figure 5.9(a)) and on the E-Tenna Model G 200a 15" diameter EBG ground plane (Figure 5.9(b)). Notice the very pronounced 3-dB deep diffraction ripples that occur when the antenna is mounted on the flat ground plane is absent in the pattern measured with the EBG ground plane. There is significant decrease in the LHCP cross-polarization levels in the lower hemisphere, although not as significant in the upper hemisphere. The sidelobe levels are also slightly lower, but the reduction in sidelobe level is not as great as observed in the choke-ring antenna shown in Figure 5.7. The E-Tenna Corporation has since gone out of business and no cost figures are available for this antenna to allow a comparison with the choke-ring ground plane.

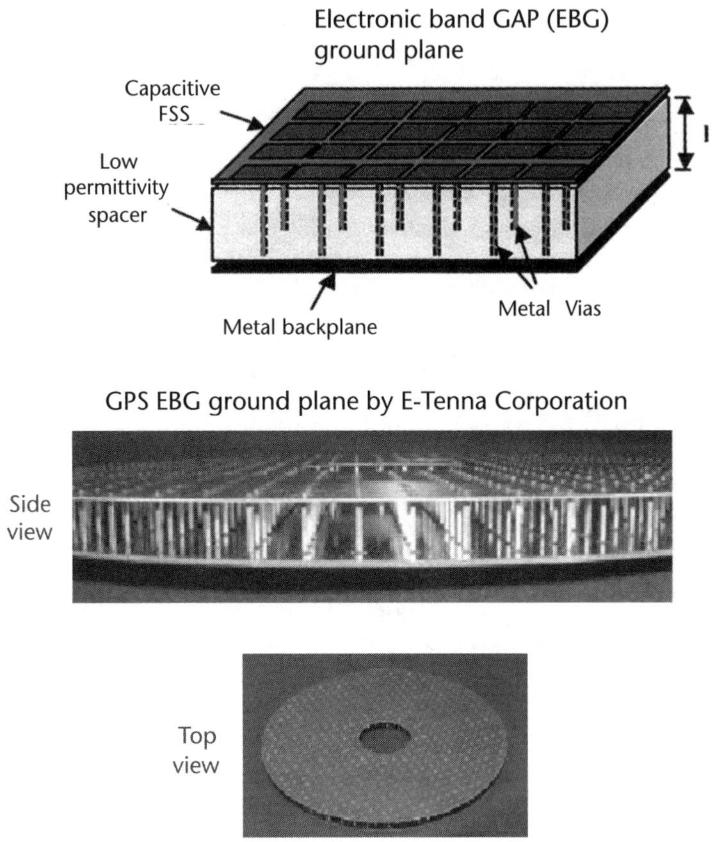

Figure 5.8 E-tenna Corporation electron band gap (EBG) ground plane for GPS antenna. (Derived from [15].)

5.3.3 Rolled Edge Ground Plane

The rolled edge ground plane built by The MITRE Corporation is shown in Figure 5.10. In this design, the signals diffracted from the sharp edges of the ground plane are prevented from affecting the performance of the GPS antenna by rolling the edge underneath the ground plane. This is a relatively large ground plane, but is relatively inexpensive and easier to build compared to either the choke-ring or EBG ground plane. The flat portion of the ground plane is 36" in diameter and the circumferential length of the curved section is 12".

Figure 5.11 shows the patterns of a GPS antenna operating at 1176 MHz (center frequency of the L_5 frequency band) measured on a 48" flat ground plane shown in the figure on the right side. The patterns of the same antenna measured on the 51" diameter rolled edge ground plane is shown on the left side of this figure. Diffraction ripples in the antenna pattern at higher elevation angles seen in the 48" ground plane (right) are minimized in the rolled edge ground plane (left). The rolled edge ground plane also provides lower LHCP cross polarization levels at the higher elevation angles. Unfortunately, the rolled edge ground plane does not provide much suppression of sidelobe levels in the lower hemisphere. This is because of creeping wave diffraction around the rolled edges of this type of ground plane. These backlobes can be suppressed by treating the rolled edge surface with radar

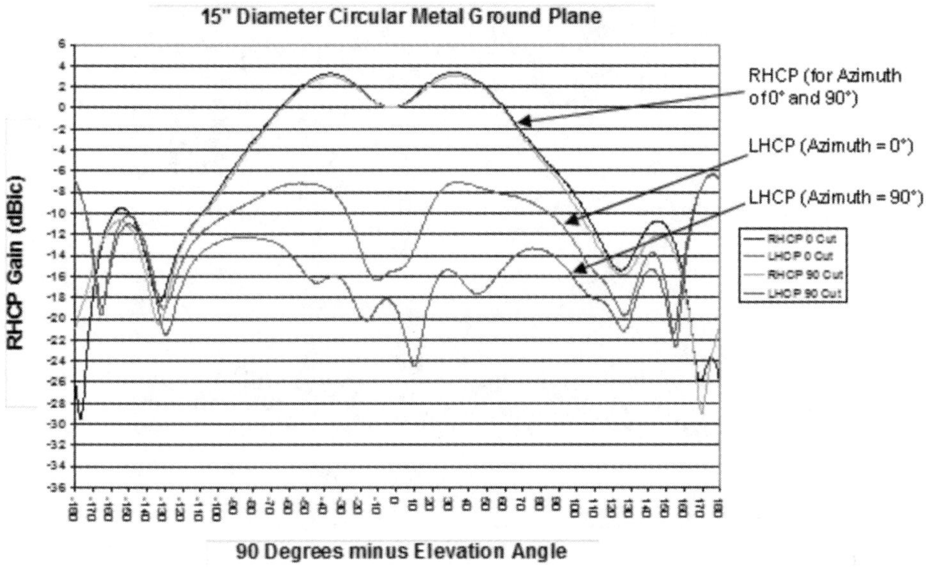

(a)

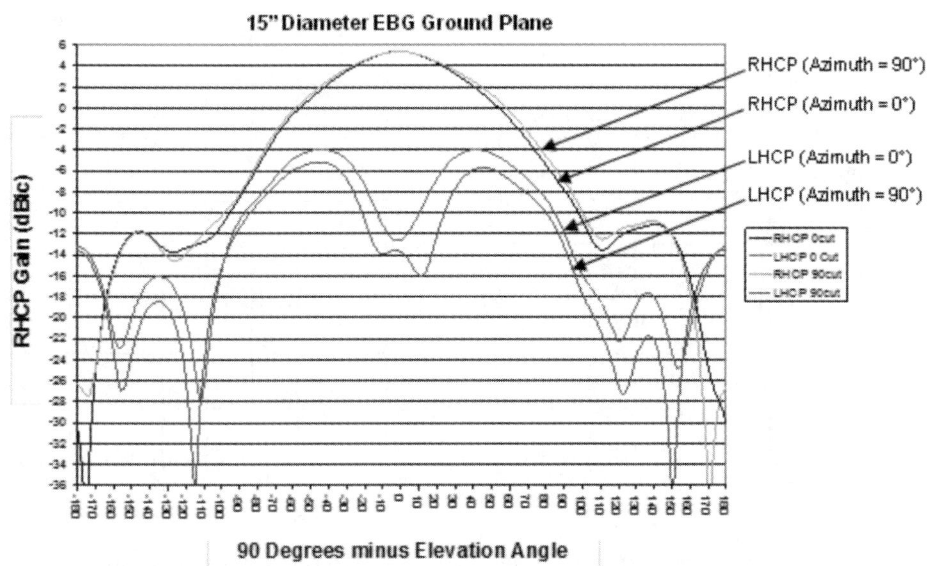

(b)

Figure 5.9 Comparison of elevation plane radiation patterns of a GPS microstrip antenna measured on a 15″ flat ground plane and on an EBG ground plane by E-tenna Corporation.

absorbing material (RAM), but could increase both weight and cost. The only application for this type of ground plane would be for an accurate measurement of the gain of a GPS antenna in an anechoic chamber where multipath effects are not of great concern.

Figure 5.10 MITRE designed rolled edge ground plane for GPS antennas.

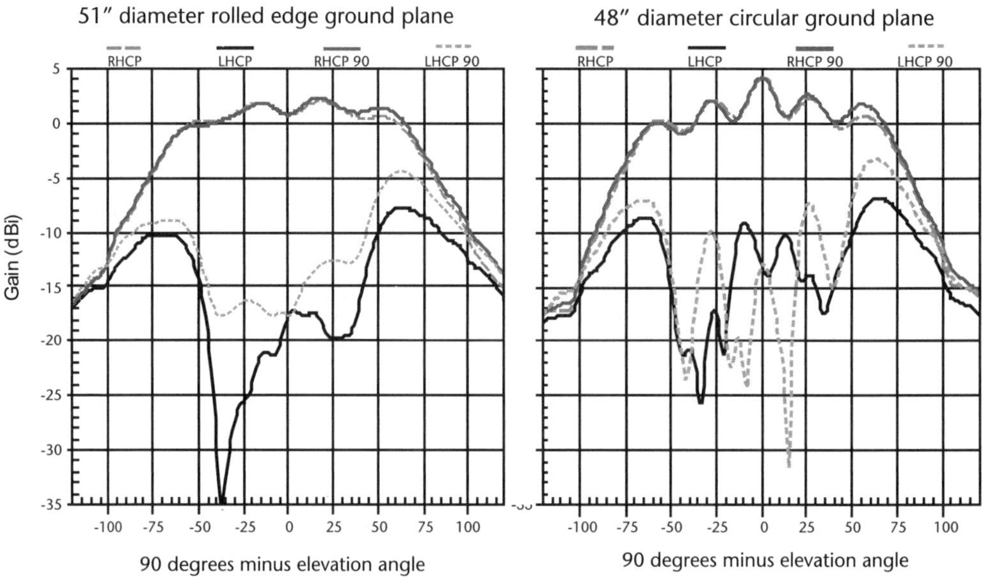

Figure 5.11 Comparison of radiation patterns of a GPS microstrip antenna measured on (a) 51" diameter rolled edge ground plane, and (b) 48" diameter circular ground plane.

5.3.4 Resistivity Tapered Ground Plane

Ground plane edge diffraction effects can be reduced by the use of resistivity tapered ground planes whose surface resistance gradually increases from the center to the outer edge of the ground plane [9, 16]. The advantage of this ground plane design is that it can be used over a very broad range of GNSS frequencies ranging from 1.150 to 1.6 GHz since it does not require concentric circular grooves that are tailored to certain frequency bands as in a choke-ring ground plane or frequency selective surfaces that can only operate in specific frequency regimes as in the EBG ground plane. To demonstrate this concept, antenna pattern measurements were taken of a RHCP GPS patch antenna placed at the center of a 26" square resistivity tapered ground plane. The GPS antenna used was the same as the one used for the measurements shown in Figures 5.3 and 5.4. The ground plane was formed by using resistivity tapered thin films of indium tin oxide (ITO) sputtered onto the surface of a thin sheet of Kapton that was bonded to the surface of a 26" plastic sheet. The surface resistivity of the ground plane increases from 0 (perfect conductor) at the center of the ground plane to about 2000 ohms/square at the outer edge of the ground plane. Figure 5.12 shows the resistivity profile of the 26" square ground plane. Pattern measurements were made of a GPS antenna mounted on both a perfectly conducting 26" square ground plane and on the resistivity tapered ground plane with the resistivity profile shown in Figure 5.12. The results of this comparison were shown earlier in Figures 5.3 and 5.4. These results show that the resistivity tapered ground plane is able to suppress both the diffraction ripples in the antenna pattern seen at higher elevation angles in the conventional metal ground plane measurements and also reduce the antenna backlobes of this antenna to below –27 dBic. This is a reduction in sidelobe level of about –12 dB when compared to that obtained with a conventional metal ground plane.

5.4 Radiation Patterns of Aircraft Mounted GPS Microstrip Antennas and Verification Through Scale Model Testing

In this section, we investigate the effects of the aircraft fuselage on the performance of a GPS antenna. Due to the interaction between the antenna and the complex and electrically large three-dimensional shape of the aircraft, the radiation pattern of a GPS antenna is expected to be significantly different from the radiation pattern of the same antenna when mounted on a flat ground plane that was the topic of the previous sections of this chapter. It is expected that the fuselage will affect both signal reception from the GPS satellites as well as its susceptibility to radio frequency interference from sources located either on the aircraft itself or on the ground. Of

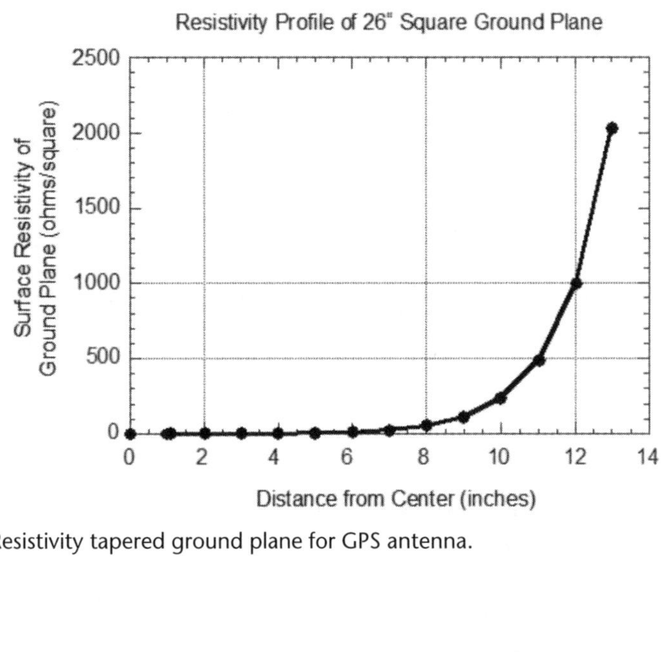

Figure 5.12 Resistivity tapered ground plane for GPS antenna.

particular interest in an airborne GPS system is the radiation of the antenna below the aircraft due to creeping waves or surface wave diffraction thatmakes the antenna susceptible to radio frequency interference from adjoining antennas. The GPS modernization program will soon require that GPS antennas mounted on top of aircraft receive the new L_5 signal (1164–1188 MHz) in addition to the legacy L_1 signal. Since it resides in the Aeronautical Radio Navigation Service Band (ARNS), the L_5 signal is particularly susceptible to in-band interference from non-GPS ARNS signals emitted by several U.S. commercial and military aeronautical navigation systems. Most prevalent are aircraft and ground-based pulsed DME and TACAN beacons (1025–1150 MHz), JTIDS/MIDS (969–1206 MHz), and ATC/ARNS airborne interrogators as well as harmonics from UHF and VHF transmitters [17]. Figure 5.13 shows the proximity of a GPS antenna located on top of a Boeing 737 aircraft and several ARNS antennas located directly below the aircraft that can be potential sources of interference.

The E5a and E5b frequency bands of the European Galileo Satellite Navigation System extend from 1164 to 1214 MHz and also fall within the ARNS band, making them subject to interference from the same ARNS sources that interfere with the U.S. GPS system. A critical parameter used for assessing the level of RFI on the airborne GPS and Galileo navigation systems from the various interference sources is the antenna gain in the direction of interference. RTCA has determined [17] that the most likely direction of interference for GNSS antennas occurs between –30° and –90° (nadir) from mostly ground-based interference. At these negative elevation angles well below the horizon, often defined as the "deep shadow region," the radiation pattern of the GPS antenna is heavily influenced by the actual antenna location on the aircraft. Diffraction, reflection, and shadowing from the aircraft appendages such as the wing, tail, vertical, and horizontal stabilizers, affect the installed antenna pattern. In addition, the reception of signals from the GNSS satellites in both the L_5 band as well as the L_1 band is affected by the change in the gain, phase, and polarization of the antenna caused by the fuselage. This could affect the accuracy of many high-precision airborne GPS systems such as those required for all weather aircraft precision and approach landing, aerial photogrammetry,

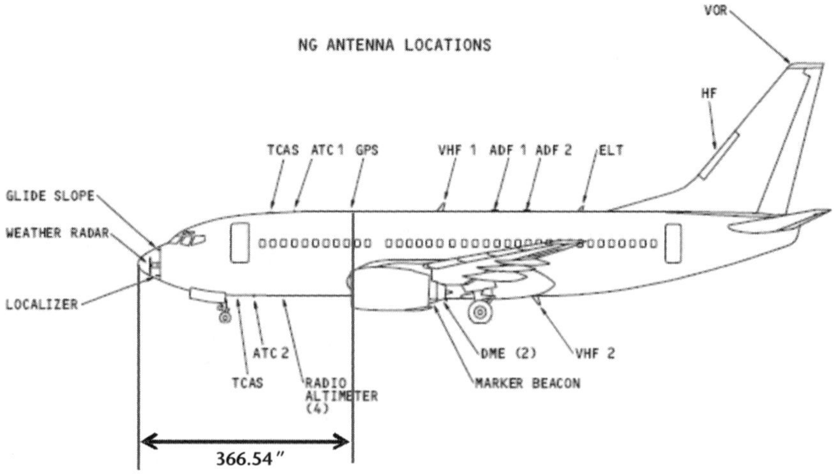

Figure 5.13 GPS microstrip patch antenna on Boeing 737 aircraft.

airborne precise gravimetry, and aerial surveying. Unfortunately, very little information is available on the measured radiation pattern of GNSS antennas mounted on full-scale aircraft. This is due to the difficulties and the expense of making such measurements on large-size aircraft and restrictions imposed by the FAA on the outdoor emission of GPS signals. To circumvent these problems, the radiation pattern of aircraft antennas is determined primarily through either scale model measurements or through computer simulations using electromagnetic codes. Scale modeling is a well-established technique for measuring the antenna pattern of aircraft in which the aircraft and antenna are scaled down in size and the measurement frequency is scaled up so as to maintain the ratio of size: wavelength constant. The scaling factor selected depends on the size of the aircraft and of the anechoic chamber used in these measurements. Scaling factors of up to 1/10 to 1/15 can be used for investigations in the GNSS frequency bands; higher scaling increases the scaled frequency into the K_U frequency band, making it difficult to build scaled GNSS antennas accurately because of their small size. In this chapter, we present the results of scale model measurements as well as electromagnetic simulations to investigate the volumetric radiation pattern of aircraft mounted microstrip GPS antennas operating in the L_1 and L_5 frequency bands on aircraft. Microstrip antennas are very popular for GPS airborne applications as mentioned earlier since they meet the ARINC 743 size requirements.

5.4.1 Scale Model Investigations on a GPS Antenna on a Beechcraft 1900C Aircraft

Scale model studies were conducted on a 1/10 scale model of an intermediate size commercial aircraft—the Beechcraft 1900C [18]. The results of these scale model investigations were compared with the results of electromagnetic simulations using Aircraft Code (Newair) developed by The Ohio State University. Good agreement between measurements and simulations confirms the accuracy of the Newair code for predicting the radiation patterns of GPS microstrip antennas on aircraft. The Newair code was then used to determine the radiation of a GPS antenna on a much larger aircraft, the Boeing 737, to determine susceptibility of the GPS antenna to interference from the ARNS sources located underneath the aircraft as shown in Figure 5.13. Figure 5.14(a) is that of the Beechcraft 1900C aircraft. The dimensions of this aircraft and the three locations of the GPS patch antennas on the aircraft selected for the antenna measurements are shown in Figure 5.14(b).

Along its widest dimension, the aircraft is 694" long (92.5 wavelengths at 1.5754 GHz) and 653.75" wide (87.2 wavelengths at 1.5754 GHz). The location of the aft antenna is well off to the side of the wings and will experience the smallest amount of body masking from the fuselage to interference from sources underneath the aircraft. The mid-antenna location is directly above where the wings meet the main body of the fuselage and will benefit from the maximum body masking provided by the aircraft fuselage. The forward location is just forward off the wings but in close proximity to the aircraft engines and is likely to suffer from signals scattered off the engines and the wings. Figure 5.15(a) shows a 1/10 scale model of the Beechcraft 1900 C aircraft installed in the near-field range of The MITRE Corporation for measurement of the roll plane antenna pattern. Figure 5.15(b) is closeup of the scaled aircraft showing the three antenna locations selected for

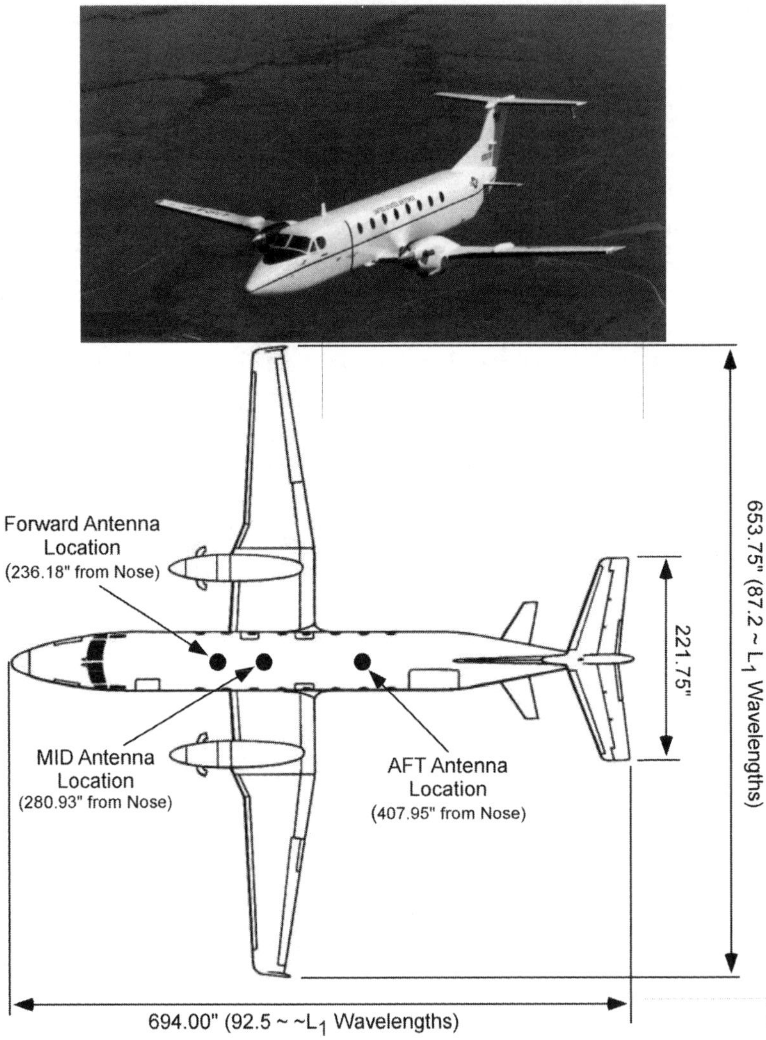

Figure 5.14 Scale model measurements and simulations on Beechcraft 1900C aircraft.

the antenna pattern evaluation. The scaled frequencies used in the measurements were 11.76 GHz and 15.75 GHz—ten times the frequency of the center frequency of the G PS L_5 and L_1 bands. Scaled microstrip patch antennas were built at these two frequencies on a 0.025" thick substrate made from Rogers TMM4 (dielectric constant of 4.5). The sizes of the patch antennas were 0.169" square at 15.75 GHz and 0.2264" square at 11.76 GHz.

The roll plane radiation pattern of the patch antennas on the scale model aircraft were measured using near-field cylindrical scanning techniques with a calibrated probe. In this technique, the near-field phase and amplitude of the radiated signal from the antenna is collected by the calibrated probe. The probe is robotically scanned at half-wavelength sampling intervals over a total length of 6 feet parallel to the rotational axis of the aircraft as the aircraft is rotated in 1° increment through a full 360° circle. The near-field data gathered by the calibrated probe is then transformed to the angular space of the far-field pattern of the antenna using Fourier transform techniques [19]. The principal plane antenna patterns in the roll,

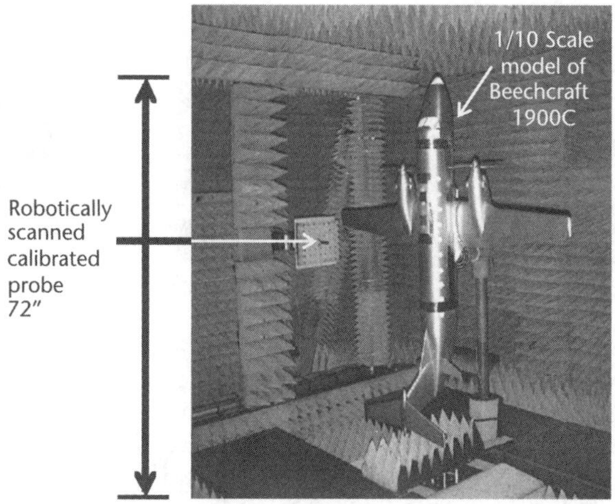

Picture of a 1/10 scale model of the Beechcraft 1900C aircraft installed in the MITRE near field antenna range for measurement of roll plane antenna patterns

(a)

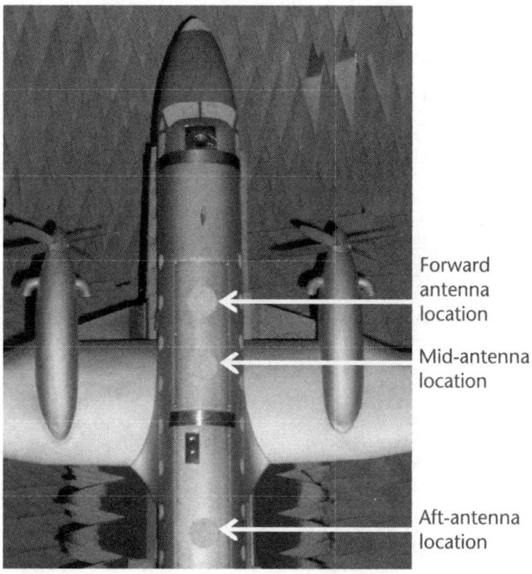

Picture showing the three selected GPS antenna locations on the scale model aircraft

(b)

Figure 5.15 (a) A 1/10 scale model of the Beechcraft 1900C aircraft installed in the MITRE near-field antenna range for measurement of roll plane antenna patterns, and (b) the three selected GPS antenna locations on the scale model aircraft.

pitch, and yaw planes have been measured. Figure 5.16 shows the orientation of the scale model aircraft relative to the calibrated probe for these pattern measurements. For the pitch and yaw plane measurements, the rotational axis is perpendicular to the rotational axis of the aircraft. The roll, pitch, and yaw plane axes are 90° relative to each other, as shown in Figure 5.16. This figure also shows the orientation of the E_θ and E_ϕ components of the far-field antenna pattern for the three measurements.

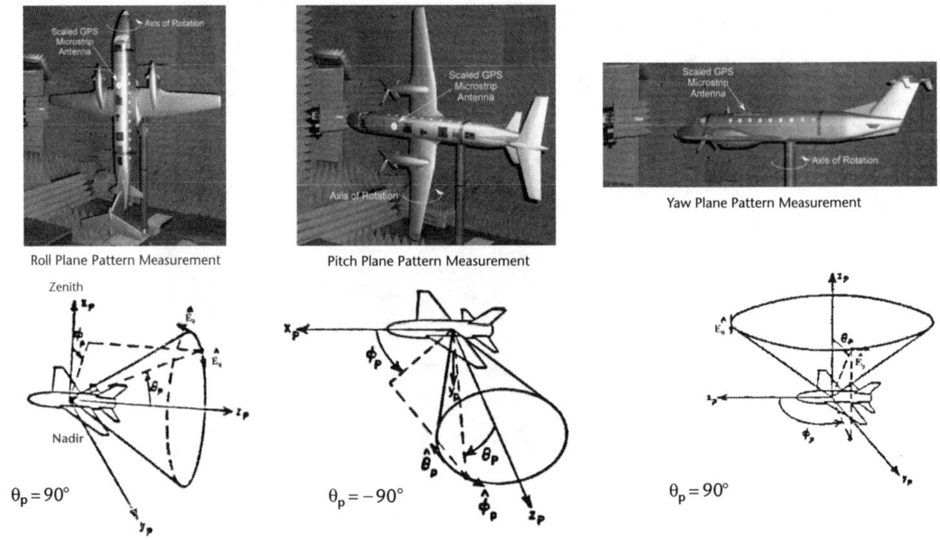

Figure 5.16 Roll, pitch, and yaw plane antenna pattern measurements using near-field cylindrical scanning on 1/10 scale Beechcraft 1900C aircraft.

The measured roll and pitch plane patterns for all three antenna locations at the scaled L_5 frequency (11.76 GHz) are shown in Figure 5.17. The roll plane pattern that is orthogonal to the longitudinal axis of the aircraft provides information on the maximum radiation from the antenna below the aircraft. The roll plane pattern for the aft antenna is particularly of interest since it has the greatest radiation below the aircraft and hence is most susceptible to RFI. There is a large difference in the contribution the radiation below the aircraft from the vertical and horizontal polarizations components for the aft antenna. The aft antenna location shown at the top left of the figure has little body masking from the wings. The amplitude of

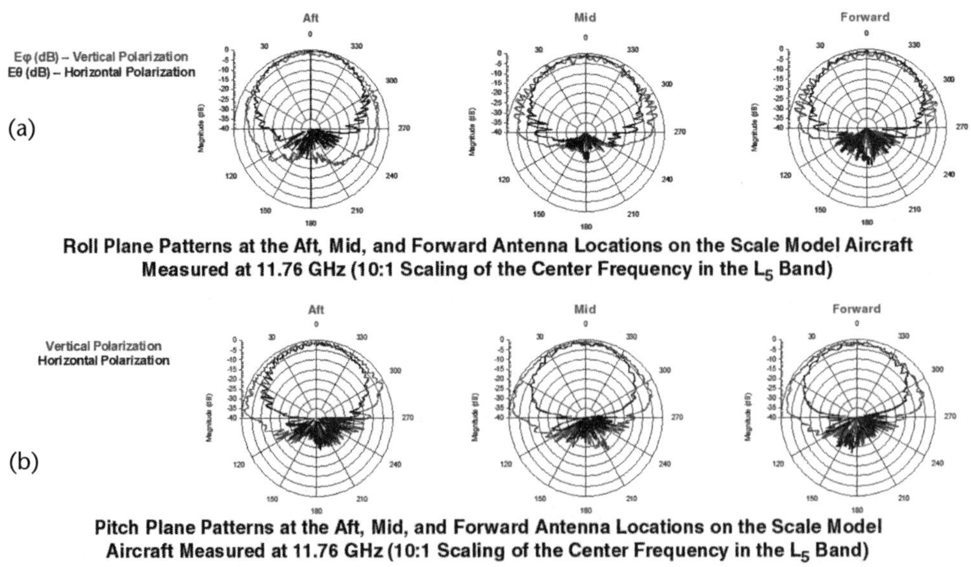

Figure 5.17 Scale model measurement of (a) roll and (b) pitch plane antenna patterns at 11.76 GHz (10:1 scaling of the center frequency in the L_5 band at the aft, mid, and forward antenna locations.

the horizontal polarization component that is parallel to the conducting fuselage of the aircraft at this aft antenna location is greatly attenuated below the horizon; the principal contribution to the radiation below the horizon is from the vertical polarization component, which is only –10 dB relative to the antenna gain at zenith. Significant radiation in the lower hemisphere, at negative elevation angles between –30° and –90° is primarily caused by "creeping waves" or by surface wave diffraction. These creeping waves propagate along a geodesic path around the cylindrical surface of the aircraft fuselage shedding energy as they propagate. They are exponentially attenuated as they propagate but can cause enough coupling to other ARNS located underneath the aircraft to cause an RFI problem in the L_5 band. At the mid-antenna location, the radiation below the aircraft is reduced even for the vertical polarization because the antenna is located directly above the wings benefiting from the large body masking provided by them to signals coming from the interference sources located under the aircraft. The antenna at the forward location is similar to that at mid-location. The antenna patterns of the antenna at the mid and forward locations also show significant multipath diffraction ripples at the lower elevations above the horizon. This is due to interference between the direct radiation and signals scattered of the wings and the horizontal and vertical stabilizers of the aircraft. The pitch plane patterns for the three antenna locations are shown in Figure 5.17(b). These patterns are significantly different from the roll plane patterns shown in the top half of this figure. The main body of the aircraft fuselage provides significant attenuation of radiation below the aircraft negative elevation angles for all three antenna locations. The beamwidth of the antenna patterns in the pitch plane is much broader and has a different shape than for the roll plane. These results indicate that phase and amplitude of the aircraft antenna pattern does not have symmetry in either elevation or azimuth; this could lead to differences in the antenna phase center as the look angle towards the satellite changes. This could be a source of potential errors in high-precision GPS systems that utilize differential carrier processing to achieve the highest accuracy in GPS observables and will be discussed in greater detail later in this chapter.

Figure 5.18 shows the measured roll and pitch plane patterns for the three antenna locations at 15.75 GHz (10:1 scaling of the center frequency of the GPS L_1 band).

5.4.2 Validation of the Aircraft Newair Code for Predicting Antenna Patterns of Fuselage-Mounted GPS Patch Antennas

The measured antenna patterns obtained from the scale model measurements for the aft and mid locations were used for validating the accuracy of the electromagnetic Newair aircraft code developed by The Ohio State University [20]. The Newair code, which uses UTD, calculates the total field of the antenna on the aircraft by a vector summation of the fields received by the primary source GPS antenna as well as the first- and second-order fields scattered off the fuselage including contributions from the creeping waves propagating around the surface of the main fuselage of the aircraft [21]. The electromagnetic model of the aircraft developed for the Newair simulations is shown in Figure 5.19. The main body of the aircraft is modeled as a composite ellipsoid with the fuselage chopped off at the back. The larger main horizontal stabilizer at the top of the tail section, the smaller horizontal

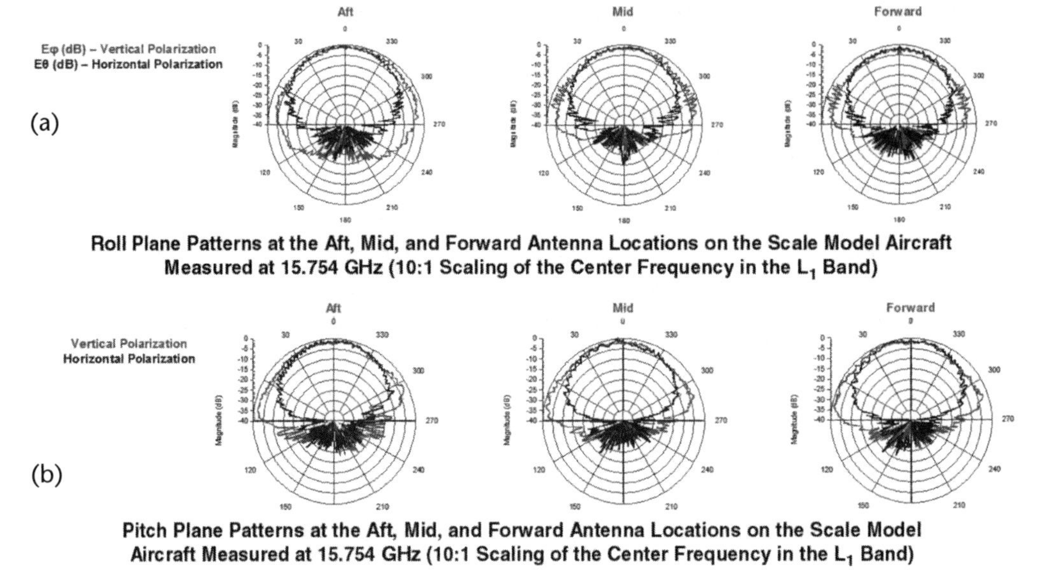

Figure 5.18 Scale model measurement of (a) roll and (b)pitch plane antenna patterns at 15.75 GHz (10:1 scaling of the center frequency in the L_5 band) at the aft, mid, and forward antenna locations.

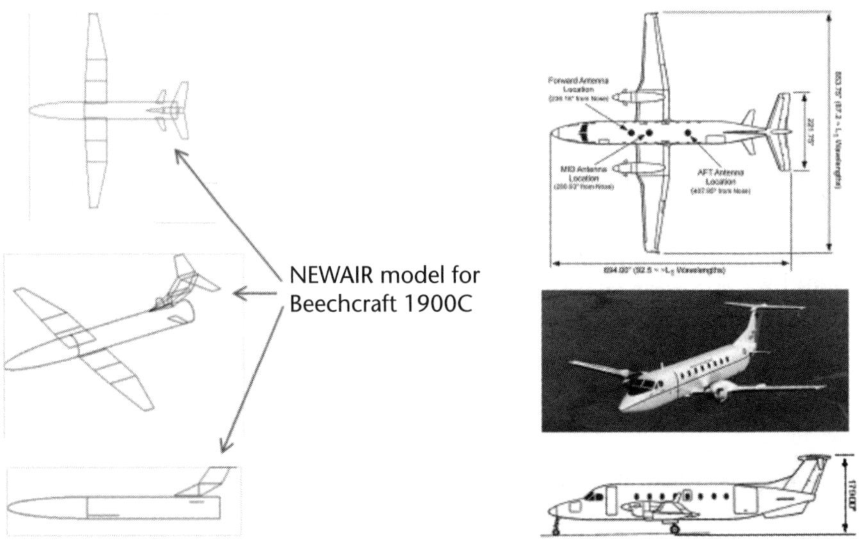

Figure 5.19 Newair Electromagnetic Model of Beechcraft 1900C for GPS antenna pattern calculations.

stabilizer below it, the vertical stabilizer and the right and left wings of this aircraft were modeled with a total of 21 plates of different dimensions and attached to the main body of the fuselage at their appropriate locations. The GPS microstrip antenna was modeled as four dielectrically loaded slot. Each orthogonal pair of slots was excited to be of equal magnitude, but 90° out of phase relative to each other to generate a RHCP signal. Referring to Figure 5.19, the pictures shown on the left side are the Newair electromagnetic models of the aircraft used in these simulations.

The corresponding pictures of the actual aircraft for these three viewing angles are shown on the right side of this figure.

Figure 5.20 shows the dominant rays that were used in the Newair code for computing the GPS airborne antenna. These are the direct radiation, the reflected, diffracted, and creeping wave contributions that are accounted for in these computations.

Figure 5.21 shows the comparison of the measured and computer roll plane patterns for the aft and mid antenna location at the center frequency of the GPS L_5 band. The comparison shown is for the E_ϕ (vertical polarization) and for E_θ (horizontal polarization).

The corresponding agreement between scale model measurements and simulations for the aft antenna location for the center frequency of the GPS L_1 band are shown in Figure 5.22. Once again there is excellent agreement between the scale model measurements and the Newair simulations.

Figures 5.23(a) and (b) show the agreement between the measured and calculated patterns for RHCP polarization for the pitch and roll plane patterns of the aft antenna location of the antenna on the aircraft. The Newair simulations are only shown in the upper hemisphere to evaluate the antenna-aircraft interaction in the upper hemisphere to determine impact on the signals received from the GPS satellite.

Figure 5.24 shows the comparison of the measured and computed phase angle of the antenna at 1.5754 GHz, the center frequency of the GPS L_1 frequency band. Notice that the antenna phase varies greatly as a function of elevation angle, unlike that of a GPS microstrip antenna mounted on a flat ground plane where the phase changes very little with elevation except at the lower elevation angles close to the ground plane. This result indicates that the phase enter of the antenna can vary

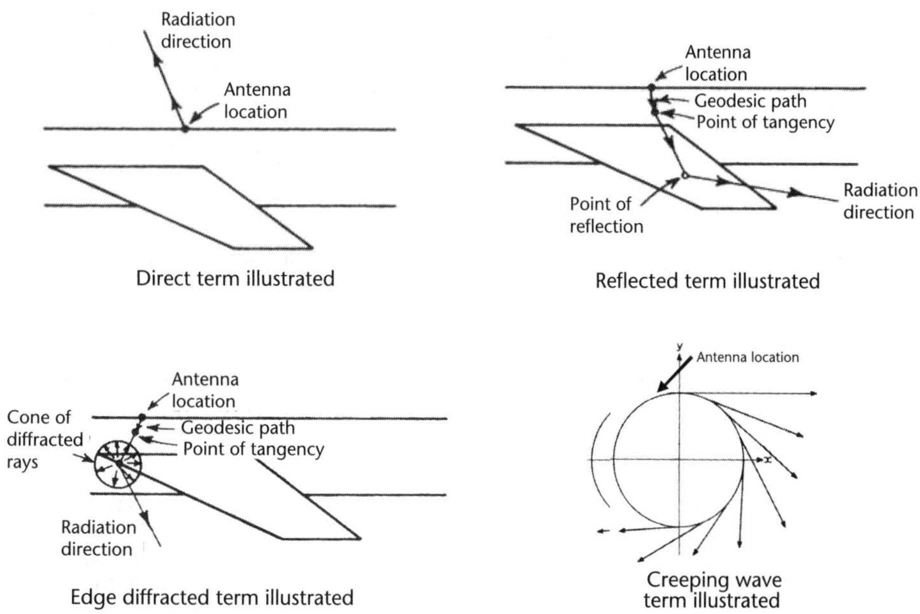

Figure 5.20 Dominant rays used in the OSU-Newair Code for computing airborne antenna patterns. (©1978 IEEE [18].)

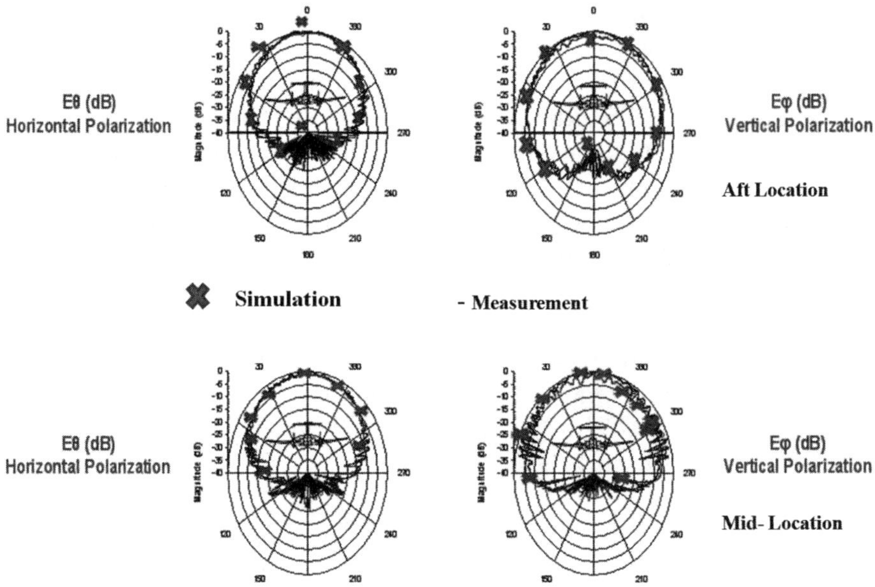

Figure 5.21 Comparison of the measured and computed roll plane patterns for the aft and mid locations at the L_5 center frequency

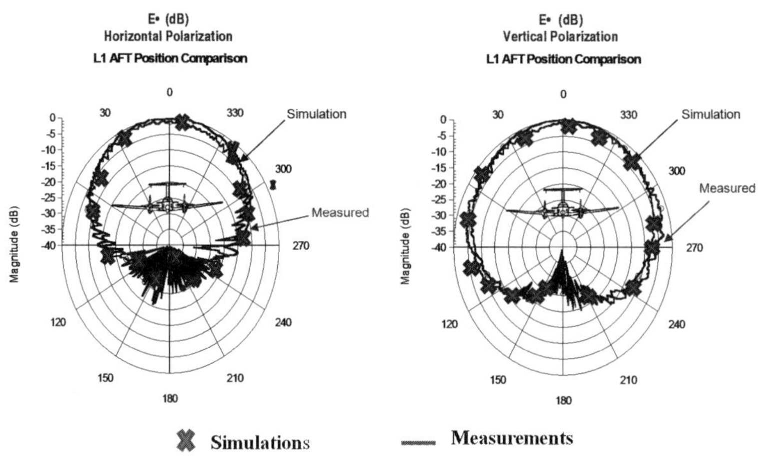

Figure 5.22 Comparison of measured and computed roll plane antenna patterns for the aft location at the L_1 center frequency.

greatly with elevation leading to position errors in differential carrier phase GPS systems. This is discussed in greater detail in Chapter 3.

5.4.3 Newair Analysis of GPS Antenna on a Boeing 737 Aircraft

The Newair code was used to conduct an analysis of a GPS antenna on top of the Boeing 737 aircraft in the L_5 band to determine its susceptibility to RFI by estimating the antenna gain at elevation angles between –30° and –90°. Figure 5.13 shows the location of the GPS antenna on the Boeing 737 aircraft. Notice there

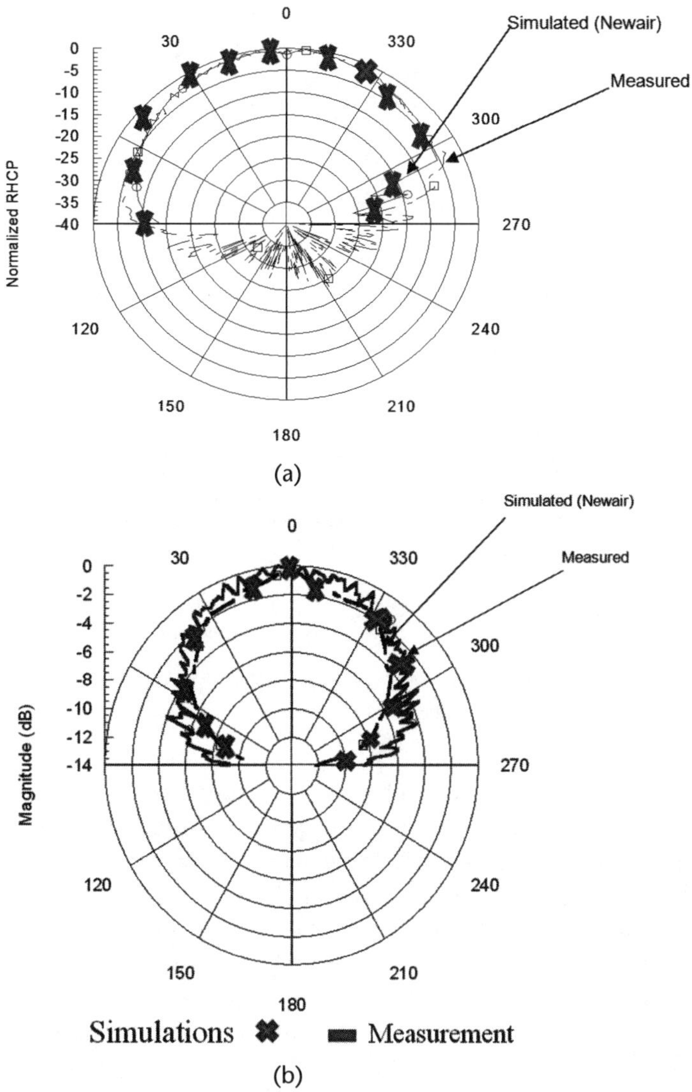

Figure 5.23 (a) Comparison of scale model measurements and computed pitch plane antenna pattern (upper hemisphere only), RHCP, L_1 band (1.5754 GHz), and (b) roll plane RHCP antenna patterns; comparison of scale model measurements with Newair simulations, L_1 band (1.575 GHz).

are several ARNS antennas located on the belly of this aircraft whose emissions could potentially interfere with the GPS antenna located on the top. The size of the Boeing 737 aircraft, shown in Figure 5.25(a), is too large for making scale model measurements.

A very large scaling factor would be needed to reduce the size of the model aircraft to a size small enough for such measurements. This would require building a scaled GPS antenna operating in the 30-GHz frequency range, which is difficult to do accurately. Hence, the only option available for analysis of the performance of GPS antennas on such large aircraft is through simulations using the Newair code whose accuracy was earlier confirmed through measurement on the Beechcraft 1900C aircraft. Figure 5.25(b) shows a simplified electromagnetic model of

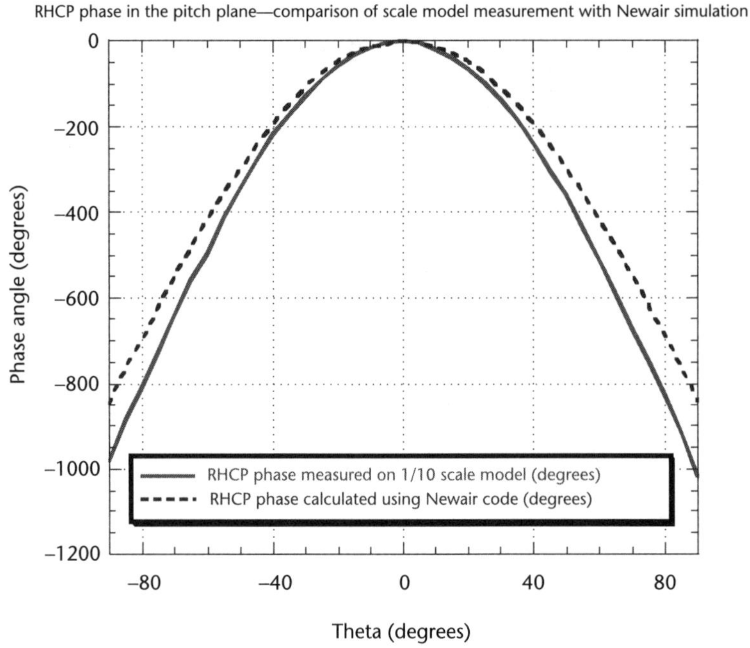

Figure 5.24 RHCP Phase in the pitch plane; comparison of scale model measurement with Newair simulation.

the Boeing 737. Figure 5.25(c) shows the simulated roll plane antenna pattern at 1575 and 1176 MHz, the center frequencies of the GPS L_1 and L_5 bands.

These patterns once again show a very slow rolloff in gain for the E_ϕ component below the horizon. This is attributed to the radiation from the creeping waves. This can result in coupling to the ARNS antennas below the Boeing 737 since they are also vertically polarized and couples strongly to the creeping waves. The E_θ component on the other hand, since it is horizontally polarized and tangential to the conducting surface of the aircraft fuselage, is rapidly attenuated below the horizon.

Based on the data obtained for the Beechcraft 1900C and the Boeing 737, the maximum radiation at the negative elevation angles between −30° and −90° is about 10 dB lower than the peak antenna radiation at zenith and the polarization of the radiation from the antenna in this region is vertical. This is the same polarization as the various interference sources in the L_5 band such as DME, TACAN, ATC, and JTIDS/MIDS, raising the potential for maximum RFI.

5.5 Radiation Pattern Analysis of a GPS Antenna on an Automobile

Several studies have been conducted to investigate the effects of the automobile frame on the radiation pattern of a GPS antenna [22–25]. Dai and coworkers [22] investigated the effects of antenna placement in an automotive environment on satellite availability, satellite lock, dilution of precision, and position fix. Data was gathered from measurements in an antenna range as well as road tests performed in an urban canyon in downtown Detroit. The roof antenna and the package tray

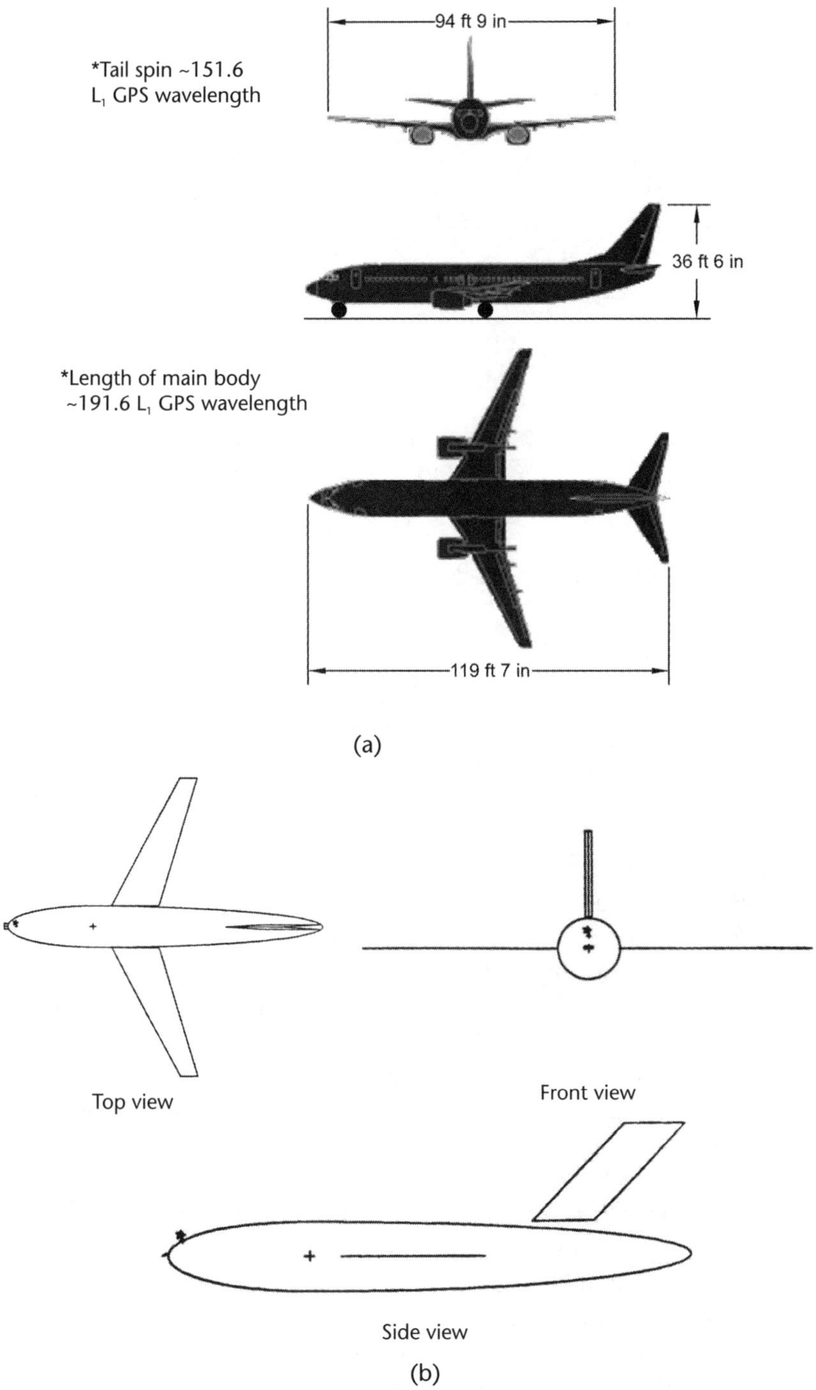

Figure 5.25 (a) Dimensions of, (b) Newair computation model for predicting GPS antenna pattern on, and (c) computed roll plane patterns of the GPS antenna on the Boeing 737 at 1.176 GHz (L_5 center band frequency) and 1.5754 GHz (L_1 center band frequency).

antenna in a Lincoln Town Car were tested with a Motorola Oncore 8 channel receiver. The type of antenna used in these tests was not mentioned in the paper. Not surprisingly the test results showed that the best location for a GPS antenna is either

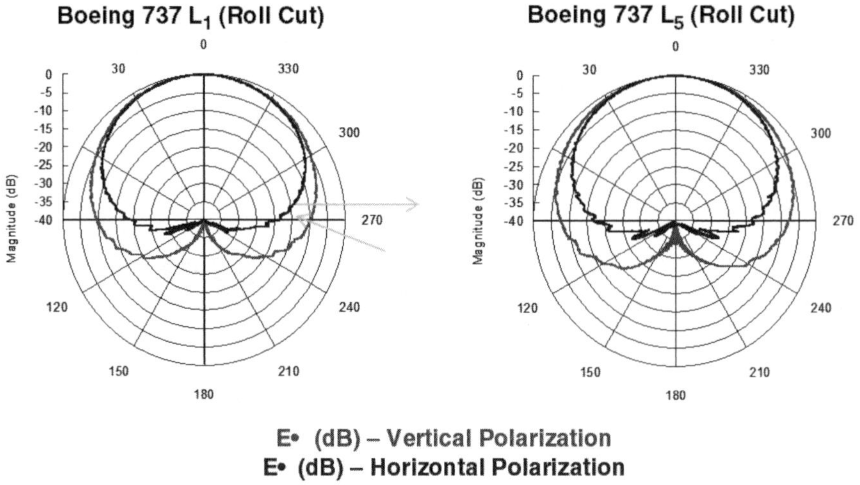

(c)

Figure 5.25 (continued)

on the roof of the car or on the trunk where the antenna has a full view of the sky. Placing the antenna inside the vehicle will degrade system performance due to the shielding effects of the body of the automobile. System-level testing is recommended when choosing and placing a GPS antenna inside the vehicle. The GPS antenna should be placed as close as possible to the receiver to minimize cable loss and RF interference from noise coupling into the coaxial cable. Windshield-mounted GPS antennas have a very small ground plane that can cause significant edge diffraction and coupling to adjacent metallic structures [22]. The UTD code was used in the analysis conducted by Natsuhara and coworkers [23] and the simulated results were compared against scale model measurements performed on a 1/6 scale model of the car at a frequency of 9.452 GHz with excellent agreement between the measurement and calculations for all directions. The effects of multipath reflections from the earth on the antenna pattern were also investigated. When the microstrip antenna is located near the edge of the automobile, narrow lobes of approximately ±6 dB appear in the gain pattern of the antenna due to reflections from the earth. The engine hood and the trunk can also affect the antenna pattern when the antenna is located near the edge. They can cause antenna ripples of about ±2 dB at elevation angles as high as 60°.

The conclusion from these studies show that the roof of an automobile is considered the ideal location for mounting a GPS antenna since it is the highest point on the automobile and can provide an unobstructed view of all visible GPS satellites [22–24]. It also has a nearly horizontal mounting surface; hence, it does not skew the elevation pattern of the antenna. To minimize edge diffractions effects on antenna performance of the type alluded to by Natsuhara [23], it is best to mount the antenna several inches away from the edges of the roof. If the vehicle has a sun roof, the antenna should be mounted, if possible, several inches away from the edges of both the sun roof and the roof of the automobile. Automobile manufacturers have developed their own proprietary set of "on-vehicle" tests for the GPS antenna when mounted on the vehicle to determine if it meets gain requirements [24]. These generally require a minimum passive antenna gain of 2 dBic in the zenith direction and –10 dBic for an elevation angle of 10°. An LNA is generally also integrated into the antenna to improve the noise figure and reduce the effect of cable loss [24, 25]. The LNA generally provides a gain between 15 to 35 dB and a low noise figure around 1 to 2 dB [24].

5.6 Body Interaction with a Handheld GPS Antenna

The performance of handheld GPS antennas, especially its radiation efficiency, resonant frequency, and antenna pattern can be strongly influenced by its interaction with the human body. The antenna in modern handsets is required to perform over multiple frequency bands and be able to receive GNSS as well as wireless communications signals. This is of particular importance in location-based services and in E-911 applications mandated by the FCC. This subject has commanded special attention in recent years and is discussed in much greater detail in Chapter 3.

References

[1] Seeber, G., *Satellite Geodesy, 2nd Edition,* New York: Walter de Gruyter, 2003.

[2] DeMartin, R., "A GPS Solution," *Microwave Product Digest,* June 1993, http://www.mpdisgest.com.

[3] Marhefka, R. J., and W. D. Burnside, "Antennas on Complex Platform," *Proceedings of the IEEE,* Vol. 80, No. 1, January 1992, pp. 204–208.

[4] Huang, J., "The Finite Ground Plane Effect on the Microstrip Antenna Radiation Pattern," *IEEE Transactions on Antennas and Propagation*, Vol. AP 31, No. 4, July 1983, pp. 649–653.

[5] Bahl, I. J., and P. Bhartia, *Microstrip Antennas,* Dedham, MA: Artech House, 1980, pp. 42–44.

[6] McNamara, D. A., C. W. I. Pistorius, and J. A. G. Malherbe, *Introduction to the Uniform Geometric Theory of Diffraction,* Norwood, MA: Artech House,1990, pp. 180.

[7] Stutzman, W. L., and G. A. Thiele, *Antenna Theory and Design,* New York: John Wiley & Sons, pp. 472–488.

[8] Marhefka, R. J., NEC-Basic Scattering Code, User's Manual (Version 3.2), Department of Electrical Engineering, The Ohio State University ElectroScience Laboratory, Final Report 718422-4, prepared under Contract N60530-85-C-0249 for the Naval Weapons Center.

[9] Rama Rao, B., and M. N. Solomon, et al., "Research on GPS Antennas at MITRE," *Proceedings IEEE Position Location and Navigation Symposium*, Palm Springs, CA, September 1998, pp. 634–661.

[10] Cockrell, C. R., and P. H. Pathak, "Diffraction Theory Techniques Applied to Aperture Antennas on Finite Circular and Square Ground Planes," *IEEE Transactions on Antennas and Propagation*, Vol. AP 22, May 1974, pp. 443–448.

[11] Tranquilla, J. M., and B. G. Colpitts, "The Microstrip Antenna with Finite Ground Plane for Use with Satellite Positioning Systems," *Proceedings Symposium on Antenna Technology and Applied Electromagnetics,* Winnipeg, Manitoba, Canada, August 1986.

[12] Tranquilla, J. M., J. P. Carr, and M. A. Hussain, "Analysis of a Choke Ring Ground Plane for Multipath Control in Global Positioning System (GPS) Applications," *IEEE Transactions on Antennas and Propagation*, Vol. 42, No. 7, July 1994, pp. 905–911.

[13] Kunysz, W., A Three Dimensional Choke Ring Ground Plane Antenna, Novatel Application Note.

[14] Ashjahee, J., J. Fillipov, and V. S. Tatarnikov, et al., Dual Frequency Choke Ring Ground Plane, U.S. Patent #6278407, issued August 21, 2001.

[15] McKinzie, W. E, R. Hurtado, and W. Klimczak, "Artificial Magnetic Conductor Technology Reduces Size and Weight for Precision GPS Antennas," *Proceedings Institute of Navigation National Technical Meeting*, January 28–30, 2002, San Diego, CA, pp. 1–12.

[16] Westfall, B. G, Antenna with R-Card Ground Plane, Trimble Navigation, U.S. Patent #5,694,136, issued December 2, 1997.

[17] SC-159, RTCA, Inc., "Assessment of Radio Frequency Interference Relevant to the GNSS L_5/E5A Frequency Band," RTCA/DO-292, July 29, 2004.

[18] Rama Rao, B, E. N. Rosario, and R. J. Davis, "Radiation Pattern Analysis of Aircraft Mounted GPS Antennas and Verification Through Scale Model Testing," *Proceedings 2006 IEEE PLANS Conference,* April 2006, San Diego, CA.

[19] Evans, G. E., *Antenna Measurement Techniques,* Boston, Artech House, Chapter 6.

[20] Burnside, W. D., J. J. Kim, and B. Grandchamp, et al., Airborne Antenna Radiation Pattern Code User's Manual, Department of Electrical Engineering, The Ohio State University ElectroScience Laboratory, Technical Report 716199-4, September 1985, prepared under Grant NSG 1498 for the National Aeronautics Space Administration.

[21] Yu, C. L., W. D. Burnside, and M. C. Gilreath, "Volumetric Pattern Analysis of Airborne Antennas," *IEEE Transactions on Antennas and Propagation*, Vol. AP-26, No. 9, September 1978, pp. 636–641.

[22] Dai, Y., T. Talty, and L. Lanctot, "GPS Antenna Selection and Placement for Optimum Automotive Performance", *IEEE Antennas and Propagation Society International Symposium, 2001 Digest,* Part 1, Vol. 1, Spring 2001, pp. 132–135.

[23] Natsuhara, K., M. Ando, and N. Goto, et al "Radiation Pattern Analysis of a GPS Microstrip Antenna Mounted on the Roof of a Car Model," *IEICE Transactions*, Vol. E-77B, No. 6, June 1994, pp. 823–830.

[24] Nagy, L. L., "Automobile Antennas," in *Antenna Engineering Handbook, Fourth Edition,* New York: McGraw Hill, 2007, pp. 30-19–39-21.

[25] M/A-COM, Inc. GPS Antenna Considerations for Automotive Applications, Application Note GPS01, www.globalspec.com/reference/7133/GPS01-GPS-Antenna-Considerations-for-Auto.

CHAPTER 6

Measurement of the Characteristics of GNSS Antennas

Basrur Rama Rao

6.1 Introduction

This chapter will describe techniques that are currently being used for testing of GNSS antennas to determine either their performance or to ensure that they comply with the requirements needed for a specific application. Techniques that will be discussed include those used for measuring the following performance parameters:

- Measurement of far-field radiation patterns and the amplitude and phase of the orthogonal polarization components in different types of indoor antenna test ranges. These include measurement of the front-to-back (FBR) and multipath ratio (MPR), which are metrics for the determination of the multipath mitigation capability of the antenna.
- The gain and the axial and cross-polarization ratios of the RHCP antenna.
- The measurement of three different versions that are used for defining the bandwidth of a GNSS antenna: the minimum gain bandwidth, the axial ratio band width, and the impedance/return loss bandwidth.
- The minimum- gain bandwidth is a better and more reliable gauge of antenna performance in GNSS antennas because wideband hybrids that are often integrated into the antennas can lead to misleading results.
- Phase center offset (PCO) from the antenna reference point (ARP) and phase center variation (PCV) as a function of the elevation, azimuth angles, and the frequency, which are critical for accurate geodetic measurements.
- Group delay verification (GDV) with frequency and with aspect angle (i.e., elevation and azimuth angle), which is important for evaluating antenna response to new signal such as binary offset carrier (BOC) waveforms used in modernized GPS and Galileo.

The testing of GNSS antennas presents some unique problems as described below:

- Due to restrictions imposed by the FAA to prevent interference with avionics systems that are dependent on GPS, outdoor testing of a GNSS antenna using a signal transmitted in the GPS frequency band is prohibited. These types of tests are only allowed at authorized U.S. Department of Defence (DOD) test facilities during specific preauthorized time periods. As a result of these restrictions, antenna radiation pattern measurements on GNSS antennas are conducted almost exclusively inside an enclosed anechoic chamber whose walls, ceiling, and floor are all lined with microwave absorbers to prevent outside radiation. These microwave absorbers in these anechoic chambers provide very low reflection levels in the GNSS band. These anechoic chambers also offer obvious additional advantages such as providing a capability for measurement under all weather conditions, security, and minimum RFI from outside sources. CW signals in the GNSS band can be generated by various signal generators or more accurate replicas of GNSS signals can also be generated inside the anechoic chamber by several commercial satellite signal simulators for more sophisticated testing. Alternatively, they can also be channelled into the anechoic chamber from signals received from an outside antenna to check on the response of the antenna and receiver under test if needed.
- However, on-site field calibration measurements of PCV and PCO of GNSS antennas can be performed outdoors of using relative calibration using two antennas (a reference and a test antenna) [24–26] and also robot-based absolute calibration techniques where the antenna under test is tilted and rotated using a fast moving automated robot[27, 28].

  In both of these measurement techniques the GNSS antenna under test is used only in the receive mode and there is no external transmission of GNSS signals.
- Antenna test ranges most frequently used for these pattern measurements include microwave anechoic chambers that are rectangular in shape, compact antenna test ranges (CATR), and a near-field/far-field (NF-FF) antenna test ranges. All three types of test ranges have been used for evaluating the performance of GNSS antennas as discussed later in this chapter; a brief description of each of these three types will be provided. For practical and cost considerations it is desirable to make these antenna test ranges as small as possible. The last two types, CATR and NF-FF ranges, allow accurate measurements on even very large size GPS antennas such as antenna arrays containing multiple elements, reflector antennas, or antennas mounted on large choke-ring type ground planes. By using spherical scanning techniques, NF-FF ranges also allow a very quick and accurate mapping of the complex electromagnetic fields of both the upper and lower hemisphere regions of a GNSS antenna. Far-field ranges are used primarily for small-size antennas in order to restrict their size and hence are restricted to measurements on smaller-size GNSS antennas.

- The radiation pattern of a GNSS antenna that is of interest from a measurement perspective covers almost the entire the upper and lower hemispheres surrounding the antenna. To comply with the minimum C/N_0 and also DOP requirements, a GNSS antenna is required to have a required minimum gain that extends over almost the entire upper hemisphere, from zenith down to a very low masking angle, generally only about 5 to 10 degrees above the horizon. The broadbeam width will make it possible for the antenna to acquire as many visible GNSS satellites as possible. At the same time, the antenna should also have a very sharp rolloff in gain below the horizon and also very low backlobes to minimize multipath from ground reflections as well as potential interference and jamming from outside radiation sources. The antenna performance in the lower hemisphere is important for evaluating the FBR or MPR.

  The performance of a GNSS antenna is therefore greatly affected by its entire surrounding environment as a result of all these requirements. The platform on which an antenna is located needs to be taken into consideration while measuring the radiation pattern as well as its surrounding environment. Some types of GNSS antennas most commonly used in GNSS, such as microstrip antennas, need ground planes for its very operation; others, such as spiral and drooping dipoles, generate bidirectional radiation patterns and also need ground planes, placed at a quarter of a wavelength beneath them, for reflecting the downward-directed radiation upwards. The wide beamwidth of GNSS antennas strongly illuminate the edges of these ground planes, creating diffraction effects that are deleterious to their radiation pattern and need to be suppressed. Special ground planes such as choke-ring, rolled edge, or resistive ground planes may become necessary for reducing these ground plane edge diffraction effects; these were discussed in Chapters 3 and 5. Although a basic GNSS antenna may be electrically quite small in size, the ground plane with which the antenna is being used could be electrically quite large in terms of wavelength and now becomes a part of the total antenna system. This could increase the minimum dimensions of the antenna range needed for conducting an accurate measurement of the far-field antenna pattern and gain.

  For GNSS antenna placed on larger platforms such as an aircraft for example, it may be necessary to do scale model testing since very few antenna ranges are large enough to accommodate even small-size aircraft for far-field measurements; some of these scale model measurements were described in Chapter 5.

  Handset GNSS antennas present a particularly challenging problem since their radiation pattern, bandwidth, gain, and efficiency are all affected by the presence of not only the receiver chassis but all other adjacent electronic components and also by the presence of the human operator, as explained in Chapter 4. Thus handset GNSS antennas will require special measurements techniques including the possible use of phantom models and replicas of the handset chassis and so forth to get a realistic measurement of its performance.

- Measuring the impedance bandwidth of a commercial antenna may also present a problem since many antennas come with an integrated wideband 90° hybrid built into the antenna package to make the antenna RHCP. This prevents an accurate measurement of its true impedance bandwidth via return loss measurements using a network analyzer since the signal reflected at the input terminals of the antenna is now diverted into the isolation port of the hybrid. The measured results for the return loss for such an antenna would indicate a deceptively low VSWR as a function of frequency, indicating a much wider bandwidth than its true bandwidth. To avoid such problems, the true bandwidth of commercial GNSS antennas with built-in hybrids are best measured by measuring the variation in gain versus the frequency (its gain bandwidth) rather than the return loss via a network analyzer.

- Active GNSS antennas present a special measurement challenge because of several active components integrated inside the antenna package itself whose contributions are impossible to separate from the passive antenna; for these, an effective measure of its performance is the G/T ratio described in Chapter 3. The group delay in active antennas also present a problem since the LNA, passband filters, and other devices built into the antenna package have their own individual frequency response characteristics of their own that can skew the group delay.

6.2 Radiation Pattern Measurements of GNSS Antennas

6.2.1 Near-Field and Far-Field Regions of an Antenna

The electromagnetic wave that is incident on the receiving antenna from a GNSS satellite is a plane wave; hence an antenna measurement range should provide far-field phase and amplitude conditions that can come close to replicating such a plane wave at the aperture of the GNSS antenna being tested. The desired goal is to have a phase variation of no greater than $\frac{\pi}{8}$ or 22.5 degrees, which corresponds to phase change over 1/16 of a wavelength, and an amplitude variation of less than 0.5 dB over the aperture area of the antenna under test. This is shown in Figure 6.1. The amplitude and phase variations across the aperture of the antenna being measured are a function of the D/R ratio where D is the largest dimensions of the GNSS antenna under test and R is the distance from the phase center of the source

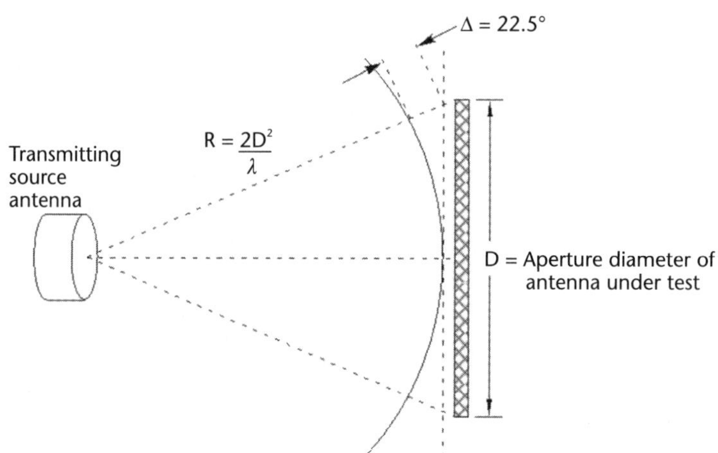

Figure 6.1 Maximum phase deviation error (22.5°) across aperture of antenna under test when illuminated by spherical wave from source antenna in the far field.

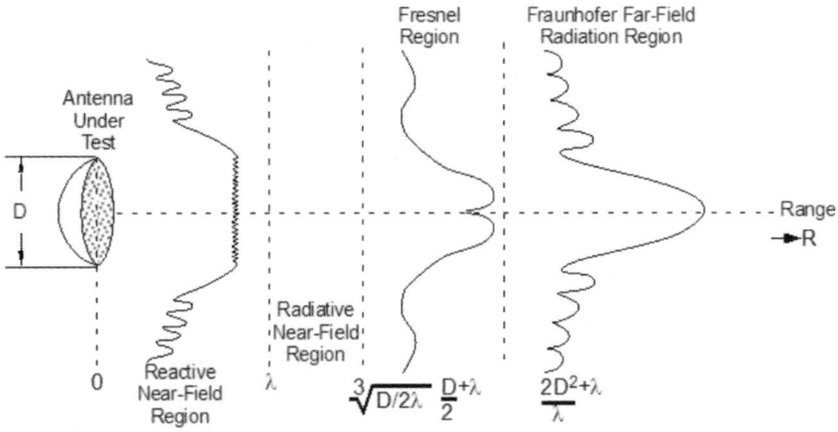

Figure 6.2 Field regions exterior to a radiating antenna.

(i.e., the transmitting antenna) to the GNSS receiving antenna. As mentioned earlier, the largest dimension in a GNSS antenna being tested could be determined by the diameter of the choke-ring or another type of ground plane on which the antenna is located since it is an essential part of the antenna. There are three principal regions to be considered depending on this (D/R) ratio and the wavelength while making these antenna measurements, as shown in Figure 6.2. The characteristics of the fields around the antenna are radically different in these three regions [1]. The far-field region, also known by its optical term as the Fraunhofer region, is the region of space in front of the antenna where the radial dependence of the electric and magnetic fields varies approximately as $\frac{e^{ikr}}{r}$, where $k = \frac{2\pi}{\lambda}$ is the wave-number and is the wavelength and r signifies the radial distance from the center of the antenna. The inner radius of the far field is a critical parameter for antenna pattern measurements since it used for deciding the minimum spacing between the transmit antenna and the antenna under test for conducting accurate far field measurements of the antenna pattern. This distance has been estimated by Yaghjian [1] to be at $R = \frac{2D^2}{\lambda} + \lambda$. The added term λ is to ensure that it covers the possibility of the maximum dimension D of the antenna being smaller than a wavelength—which is generally the case with most GNSS antennas. Note that the conventional definition for the far-field distance from the test antenna is generally given in most books as being just $R = \frac{2D^2}{\lambda}$; however, Yaghjian [1] has suggested that it is more accurate to add as an extra term to it for reasons mentioned above. The free-space region from the surface of the antenna to the inner boundary of the far-field region is referred to as the near field region of the antenna. It is further subdivided into two regions: the reactive near field and the radiating near field. The reactive near-field region is commonly defined as extending from the physical surface of the antenna to a distance of $R=\frac{\lambda}{2\pi}$, although experiences with near-field measurements indicate that a distance of a wavelength to be a more appropriate limit to defining this boundary [1]. In this reactive near-field region the fields, as its name indicates, are predominantly reactive; the electric field component does not propagate in a radial direction and its decay could be more rapid than given by 1/R. The region from to a distance $R = \frac{D^{\frac{1}{3}}}{2\lambda}\left(\frac{D}{2}\right) + \lambda$ is defined as the radiating near-field region. In this region the electric and magnetic fields tend to propagate predominantly in phase but do not exhibit the $\frac{e^{ikR}}{R}$ dependence until the far field is reached. The Fresnel region extends from $R = \frac{D^{\frac{1}{3}}}{2\lambda}\left(\frac{D}{2}\right) + \lambda$ to the far-field region defined by $R = \frac{2D^2}{\lambda} + \lambda$, where the Fraunhofer far- field region begins. The radiation pattern of the main beam of the antenna in this Fresnel region can be broad, as shown in Figure 6.2. The radiation pattern of the antenna is equal to the Fourier transform of the aperture distribution in this region with a phase error of more than 22.5°.

Accurate pattern measurements of the far-field patterns of GNSS antennas are measured by setting the distance of the antenna under test at a distance at least as small but generally much longer than the far-field distance given by $R = \frac{2D^2}{\lambda} + \lambda$. At this distance the radius of curvature of the spherical wave front emerging from the transmit antenna is so large that the maximum phase deviation over the aperture area of the GNSS antenna being tested is less than 22.5° and can be considered to be almost a planar wave as shown in Figure 6.1. In these far-field test ranges the distance between the transmitting antenna and the antenna under test that is needed for meeting the far-field distance criterion can become very large and expensive when measurements on large GNSS antennas, such as for example measurement on a multielement GNSS antenna array on a large ground plane is required. A CATR can create a plane-wave field at considerably shorter distance than set by the far-field criterion by generating a uniform plane wave from a large parabolic reflector fed by an offset fed feed horn or sometimes by using by dual reflector antenna system as described later. The dimensions of a conventional far-field antenna range for measuring the antenna pattern can also be reduced by making measurements in the radiating near-field region or the Fresnel region and then using fast Fourier transform (FFT) techniques for generating the corresponding far-field radiation patterns from these measured results. This is the underlying principle behind NF-FF antenna test ranges. A brief description of three different types of antenna test ranges that have been used by various organizations used for measuring the radiation pattern, gain, and axial ratio of GNSS antennas is given below. These include far-field test ranges, two different types of CATR, and the NF-FF antenna test range using planar, cylindrical, and spherical scanning.

6.2.2 Indoor Far-Field Antenna Test Ranges

The most popular antenna test range for testing GNSS antennas is a microwave anechoic chamber [2–8]. The size of rectangular anechoic chamber is chosen so to be large enough to simulate free-space conditions; this is ensured by placing the transmit antenna and the antenna under test (AUT) at opposite ends and at a sufficient distance apart to meet the far-field distance criterion. The antennas are placed near the side walls of the chamber where there the quiet zone is maximum and where the reflected energy from the sidewalls is minimized. Figure 6.3(a) is a schematic diagram of a microwave anechoic chamber and Figure 6.3(b) shows a picture of the

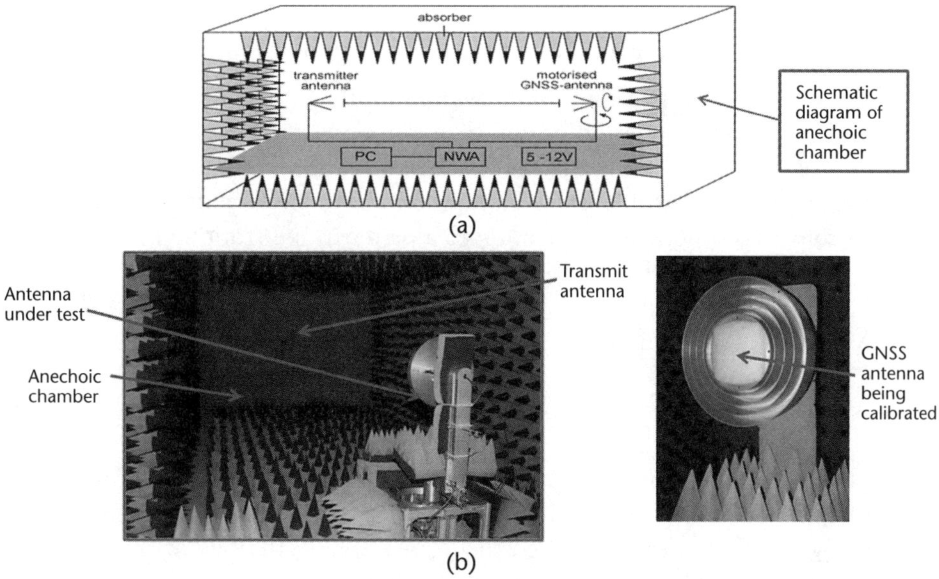

Figure 6.3 Microwave anechoic chamber used for GNSS antenna calibration at the University of Bonn, Germany. (Picture courtesy of Prof. Zeimetz and Prof. M. Becker.)

microwave anechoic chamber at the University of Bonn (IGG) in Germany, which has been used extensively for the calibration of GNSS antennas [2–5]. An even larger anechoic chamber, perhaps the largest anechoic chamber in Europe, located at the Technical Center for Information Technology and Electronics in Greding, Germany, has also been used by these same investigators for calibration of GNSS antennas. This anechoic chamber is 41m in length, 16m in width, and 14m in height and has microwave absorbers designed to go down to as low as 0.5 GHz. [5].

A major advantage of the anechoic chamber measurements versus other on-site relative as well as absolute robot-based antenna calibrations to be described later is that it is possible to conduct an absolute antenna calibration across the entire GNSS band from 1.164 to 1.606 GHZ by using synthetic signals without the need for having access to signals transmitted by a constellation of satellites of a fully deployed system. This is an important advantage while evaluating the wideband performance of these antennas at new GNSS frequencies of new systems that are yet to be deployed This has already been demonstrated by Becker and other investigators [2, 3].

A disadvantage of the far-field microwave anechoic chamber is that they can be quite large for meeting the far-field criterion and hence expensive especially when large-size GNSS antennas such as antenna arrays mounted on special types of ground planes need to be measured. It also may take considerable time to measure the entire 3-D pattern around the GNSS antenna in a microwave anechoic chamber.

Two excellent alternatives to far-field antenna test ranges that are considerably more compact than far-field ranges but also offer the same advantages for multi-frequency measurements are CATR and NF-FF ranges, which are described next. Both have been used for measurements of the characteristics of GNSS antennas.

6.2.3 CATR

The $\frac{2D^2}{\lambda}+\lambda$ far-field criterion for the separation distance between the transmitting antenna and the AUT is determined primarily by the maximum allowable phase and amplitude deviation required to simulate plane wave conditions across the aperture of the AUT. The CATR is able to reduce this separation distance between the antennas by using a large parabolic reflector fed by an offset-fed horn antenna to convert a spherical wavefront into a planar wave-front that is incident across the aperture of the AUT [9, pp. 982–995]. A schematic diagram of a single reflector CATR is shown in Figure 6.4. Figure 6.5 shows a picture of a CATR developed by MI Technologies, one of the leading companies in CATRs.

The phase center of the feed horn used in a CATR is located at the prime focus of a large parabolic reflector. The rays reflected off the surface of the reflector arrive at a planar surface that is transverse to the axis of the parabola with a uniform phase since they have all traveled an equal distance. They thus generate an incident plane wavefront at the AUT. The usable part of this test zone where uniform plane wave conditions exist is called the "quiet zone"; the size of this quiet zone could typically be 50% to 60% of the dimensions of the main reflector. The CATR provides the same type of advantages as the far-field ranges described above but can be much smaller in size and allow measurements on much larger antennas depending on the size of their quiet zone regions. CATRs have been used for the measurement of group delay variation of GPS antennas by Murphy [10] and by Van Grass and Bartone [11]; these techniques will be described later in this chapter.

Some of the drawbacks of a CATR are aperture blockage, direct radiation from the feed antenna of the reflector to the AUT, diffraction from the edges of the large reflector and the feed support, depolarization coupling between the two antennas, and reflections from the wall of the chamber. The magnitude of the imperfections of the phase and amplitude in this quiet zone is the primary figure of merit of the CATR. Probing of quiet zone region is often conducted to estimate phase and amplitude deviations from ideal plane wave conditions and to estimate its true extent. The probe can be a smaller antenna or open-ended waveguide that is scanned

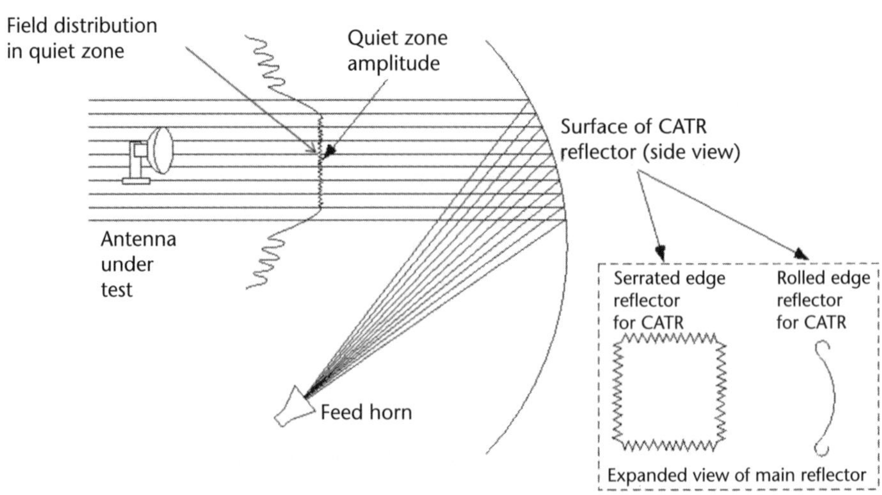

Figure 6.4 Schematic diagram of a single reflector CATR.

Figure 6.5 Compact antenna test range developed by MI Technologies.(Courtesy of J. Kendall of MI Technologies.)

both vertically and horizontally in the plane perpendicular to the fields propagating from the main reflector. The main causes of amplitude and phase ripples in the quiet zone are from diffraction effects from the edge of the parabolic reflector. To minimize these effects the edges of the reflector are serrated as shown in Figures 6.4 and in 6.5; sometimes the edge of the reflector can also be blended and rolled as shown in the side figure of Figure 6.4[12]. For most applications phase deviations in the quiet zone of less than 10° and peak-to-peak amplitude ripples of less than 1 dB and an amplitude taper of less than 1 dB are considered adequate. A large number of studies have also been conducted to better understand the causes of errors in the quiet zone of a CATR with a serrated reflector and their impact on the accuracy of antenna measurements. In addition to front-fed, single parabolic reflectors shown in Figure 6.4, dual reflector systems using Cassegrain and Gregorian style reflectors have also been used in CATR. A CATR using a Cassegrain type dual reflector system is shown in Figure 6.6.

6.2.4 NF-FF Antenna Test Ranges

The dimensions of the antenna test range can also be reduced by first sampling the field over a scan surface very close to the antenna, in the radiating near-field region, by a calibrated sampling probe. The phase and amplitude data that is collected from such sampling is then used to compute the far-field antenna pattern of the antenna entirely through analytical methods using Fourier transform techniques. The sampling intervals need to meet the Nyquist sampling requirements if grating lobes are to be avoided. The minimum scan density set by the Nyquist sampling theorem states that the state of the phase of any spatial frequency component must shift less than 180° between adjacent samples. The near-field to far-field transformations

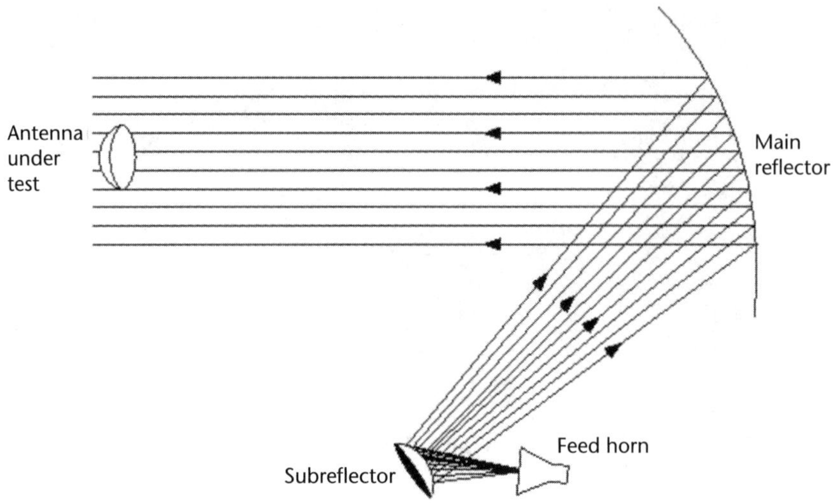

Figure 6.6 Schematic diagram of a dual reflector CATR.

used in these computations for generating the far-field pattern depend on the shape of the surface over which these sampling measurements are taken. Planar, cylindrical, and spherical surfaces are commonly used for scanning the fields around the AUT [13–17]; these scanning surfaces are illustrated in Figures 6.7, 6.8, and 6.9, respectively. Details on the theory for calculating far fields from the sampled near-field data is quite complex and beyond the scope of this book but they are well described in several recent books [13, 14, 17].

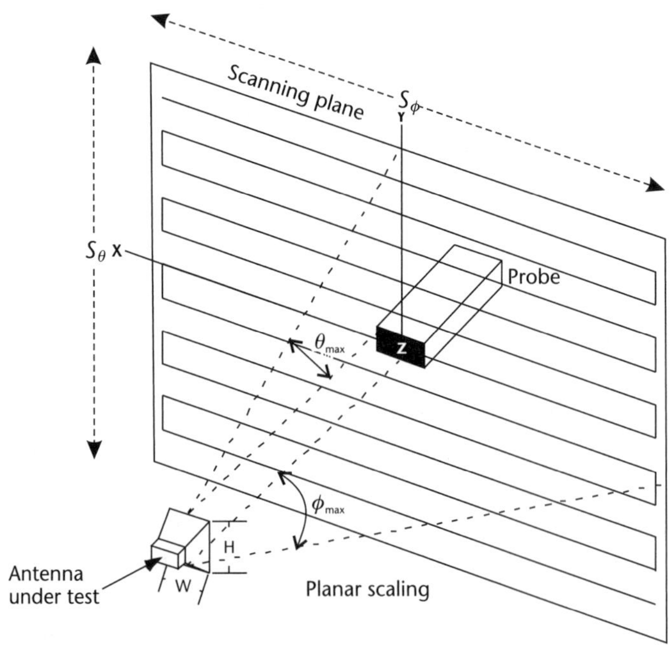

Figure 6.7 Near-field planar scanning for measuring the radiation pattern of an antenna.

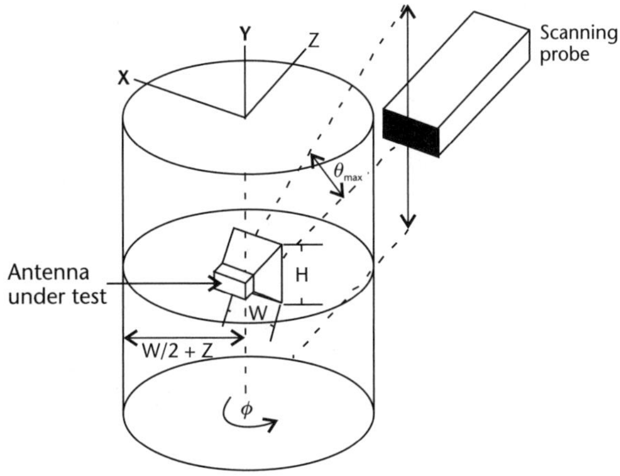

Figure 6.8 Cylindrical near-field scanning for determining the radiation pattern of an antenna.

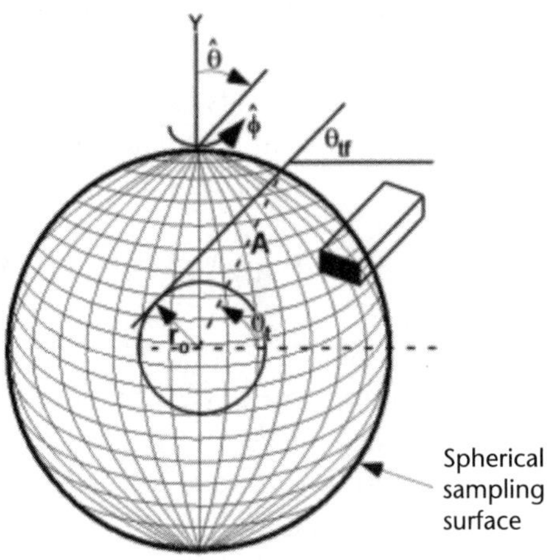

Figure 6.9 Spherical near-field scanning for determining the radiation pattern of an antenna.

6.2.4.1 Planar Near-Field Scanning

Planar scanning [13, 17] is suitable for measuring either the upper or lower hemispheric patterns of large-size GNSS planar arrays since it requires the least amount of computations and no movement of the AUTs is required during the collection of the sampling data. Using a large planar scanner, the sampling probe is scanned in a linear manner along the *X* and *Y* axes over a large area of the plane located in front of the AUT. The planar scanning process is shown in Figure 6.7. The AUT remains aligned to this scan plane and is not moved while the sampling data is being collected.

The theory assumes that the scan plane is infinite, which is impractical, requiring the scan plane to be truncated. The calculated patterns are only valid within certain maximum angular limits in orthogonal planes. The pattern of the antenna that is calculated beyond these angular limits is invalid. These maximum angular limits are given by:

$$\theta_{MAX} = \arctan\left(\frac{S_\theta - H}{2Z}\right) \text{ and } \varphi_{MAX} = \arctan\left(\frac{S_\phi - W}{2Z}\right)$$

as shown in Figure 6.7.

The total area to be scanned is then:

$$(S_\theta)*(S_\varphi) = [H + 2Z\tan(\theta_{MAX})]*[W + 2Z\tan(\varphi_{MAX})]$$

H and W are the height and width of the AUT and θ_{MAX} and ϕ_{MAX} are the elevation and azimuth angular limits of interest in the far field pattern and are assumed to be less than 90°. The limitation in the angular coverage of the test antenna is due to the truncation of the scanner's dimensions. Z is the distance between the sampling probe and the AUT and is typically between 1 to 10 wavelengths to avoid sampling of the reactive near-field energy of the antenna. Assuming that the spacing between sampling data points is a half-wavelength, the total number of scanned sampling points N_P at each frequency is given by [13]:

$$N_P = 4[H + 2Z\tan(\theta_{MAX})][W + 2Z\tan(\varphi_{MAX})]\left(\frac{1}{\lambda^2}\right)$$

Note that planar scanning can only provide pattern in the upper hemisphere down the low elevation masking angle. For measuring the pattern in the lower hemisphere the antenna would need to be turned by 180 degrees and the measurements would need to be repeated.

6.2.4.2 Cylindrical Near-Field Scanning

In the cylindrical NF-FF method the near-field of the AUT is sampled on the surface of a circular cylinder surrounding the test antenna as shown in Figure 6.8 [13, 15]. The sampling is performed by moving the sampling probe parallel to the cylinder axis parallel to the z axis while the antenna is being rotated in the azimuth plane. The sampling probe needs to be calibrated using cylindrical probe coefficients. The theory and computations are somewhat more complex than for the planar case and is based on representing the fields outside the cylinder as a superposition of cylindrical modes. If we assume a complete azimuth rotation with vertical motion of the probe over $H + 2Z\ \tan(\theta_{MAX})$ and if the radius of the cylindrical scanned surface measured from center of the AUT surface to the probe is $\frac{W}{2} + Z$, the number of sampled points per frequency is:

$$N_P = 4\pi(W + 2Z)[H + 2Z\tan(\theta_{MAX})]\left(\frac{1}{\lambda^2}\right)$$

Cylindrical computational time is similar to planar computational time except that approximately twice as many sampling points are involved, thus doubling the time per frequency. Cylindrical scanning would allow measurements of both upper and lower hemispheres of a GNSS antenna except for the radiation pattern close to the horizon (i.e., in the vicinity of zenith or nadir of the AUT, which is mounted vertically to face the sampling probe). This is not possible since radiation along the longitudinal axis of the scanned cylindrical surface will not be captured by the sampling probe due to truncation of sampling along the vertical z direction. Unfortunately, this missing section can be of great interest in certain GNSS applications for determining susceptibility of the antenna to multipath and interference.

6.2.4.3 Spherical Near-Field Scanning

In the spherical NF-FF method the near field around the AUT is sampled on a spherical surface surrounding the antenna as shown in Figure 6.9 [13–15]. The AUT is installed on a mount that is capable of providing spherical scanning. The sampling probe is normally at a fixed location while the AUT is scanned in one angular axis and stepped in an orthogonal angular axis. The two most common ways of performing such spherical scanning are (1) scans in ϕ with steps in θ, (2) scans in θ with steps in ϕ, or (3) continuous " ball of yarn" scanning with scanning in both θ and ϕ. This last method is less common. The differences between these three techniques are discussed in Hansen [14]. The antenna under test is mounted so that the radius r_t of the minimum sphere is as small as possible. In practical measurements the size of the minimum sphere that encloses the radiating RF currents of significance may not be obvious but guidelines for determining this are provided by Hansen [14]. The highest significant wave mode present in the test antenna is given by:

$$N = k_{r_i} + 10$$

The maximum permissible sampling increment in theta is then given by

$$\Delta\theta = \frac{2\pi}{2N+1} \cong \frac{\pi}{N} \quad (N >> 1)$$

In ϕ, the maximum sampling increment

$$\Delta\varphi \cong \frac{\pi}{M} \text{ with } M >> 1$$

where

$$M \cong k_{r_C} + 10$$

where r_C is the radius of the smallest cylinder parallel to the Z axis and enclosing the test antenna but is smaller the r_t. Further details can be found in Hansen [14, pp. 188–206].

The measurement time can be reduced if the sampling truncation surface is truncated judiciously. The truncation principle according to Hansen is that "the near field must be measured over at least that part of the surface which includes all the rays which radiate from any part of the test antenna to any of the desired far-field directions." This truncation principle indicates that the near field should be sampled up to $\theta = \theta_t$ if the far field is sought for $\theta \leq \theta$. The scan truncation region is given by [14, pp. 232–234]:

$$\theta_{tf} = \theta_t - \arcsin\left(\frac{r_0}{A}\right)$$

In the above equation, A is the radius of the measurement sphere and r_0 is the radius of the minimum sphere surrounding the AUT, as shown in Figure 6.9.

Spherical scanning is the most general and accurate of the three scanning techniques and also the most complex since it involves expressing the spectrum in terms of spherical waves. It is best suited for making measurements of low gain, wide-beam antennas typically used in GNSS systems. It is also the most accurate NF-FF method for measuring the gain and the polarization state of the circularly polarized antenna since both the phase and amplitude of E_θ and E_ϕ components of the field in almost the entire spherical volume around the antenna is sampled within the limits set by scan truncation. The total scanned surface is that of a sphere with a diameter of $2r_0$, as shown in Figure 6.9. This allows the CP gains of the antenna to be determined quickly through comparison with a standard gain antenna [14, pp. 206–214]; the software for these CP gain calculations is provided by several companies making NF-FF antenna test ranges. This asset is very useful while investigating the gain and axial ratio of RHCP GNSS antennas over the entire upper hemisphere down to a low masking angle in elevation. Figure 6.10 shows a picture of a GPS mounted on a large 51" diameter rolled edge ground plane that is being tested in a near-field antenna range at the MITRE Corporation using spherical scanning techniques. This near-field antenna range was purchased from Near-Field Systems, Incorporated, of Torrance, California.

Figure 6.10 Measurement of a GPS antenna on 48" diameter circular ground plane in a near-field antenna range using spherical scanning techniques.

The disadvantage of spherical scanning is that it is the slowest of the three NF-FF scanning techniques. The time taken by the planar scanner is the shortest followed by the cylindrical scan with the spherical taking the longest time of the three methods; it is about $\frac{\pi}{2}$ times slower than the cylindrical case to account for motion of the sampling probe along the curved vertical paths of the spherical scanning surface. It is estimated that spherical scanning computations takes longer than the corresponding cylindrical case by a factor of approximately $\frac{N_P^{\frac{1}{2}}}{\log_2(N_P)}$, which amounts to a factor of 8 for an array 100 by 100 sampling points.

Yaghjian [1] has made a comparison of computational times taken by the three methods; a Cyber 750 computer was used in these calculations.

The principal sources of errors in NF-FF scanning are multiple reflections between the AUTs and the sampling probe, nonlinearity in the receiving system, and scattering and absorption from the pedestal supporting the antenna being tested.

Spherical scanning techniques can also be used for measuring patterns of frequency-scaled GNSS antennas placed on scale models of aircraft as described in Chapter 5.

The types of electrical type errors that can occur while measuring antenna patterns in NF-FF antenna test ranges include the following: mutual coupling between the sampling probe and the AUT, nonlinearity in the receiver, multipath reflections generated by the scanner, scattering from the AUT mount and the walls of the anechoic chamber, and leakage from the transmitting and receiving systems. These errors can be reduced by microwave absorbers placed at the appropriate locations and by using proper shielding and cables. The mechanical errors can arise due to an imperfectly shaped scan surface, misalignment of the scan surface relative to the AUT, and errors in positioning of the sampling probe.

6.2.5 Radiation Pattern Cuts

Measurement of the radiation pattern of a circularly polarized GNSS antenna involves determining the spatial distribution of the amplitude and phase of two orthogonal polarization components of the radiation field $\vec{E}_\theta$ and $\vec{E}_\varphi$ along the $\hat{\theta}$ and $\hat{\phi}$ unit vector directions of a spherical coordinate system, as shown in Figure 6.11(a). This data can then be used for determining a number of critical parameters that characterize performance of the antenna—such as the RHCP and LHCP gain and the required minimum gain—beamwidth, the axial and cross-polarization ratios, the MPR and FBR and phase center variation of the antenna as a function of look angle, and so forth. If the phase is also measured as function of frequency within a specific GNSS band, it will permit a measurement of the group delay of the antenna. The far field radiated of the antenna is denoted in terms of the unit vectors for RHCP and LHCP.

$$\vec{E}(\theta,\varphi) = \left[E_R(\theta,\varphi,f)\right]\left[\hat{e}_r\right] + \left[E_L(\theta,\varphi,f)\right]\left[\hat{e}_L\right] \tag{6.1}$$

In the above equation the unit vector for RHCP—the principal polarization—is given by:

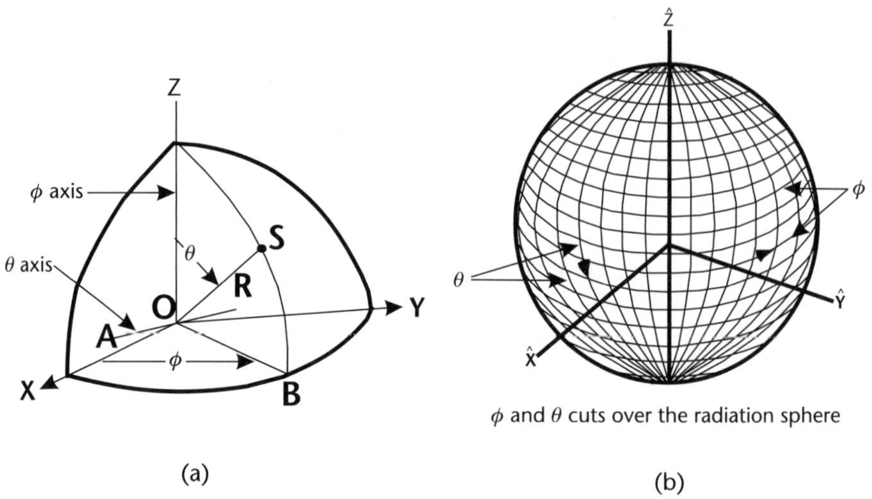

Figure 6.11 Spherical coordinate system and conical and great circle cuts for radiation pattern.

$$\hat{e}_r = \frac{e^{-j\varphi}}{\sqrt{2}}\left[\left(\hat{\theta}\right) - j\left(\hat{\varphi}\right)\right] \tag{6.2}$$

The unit vector for LHCP—the cross polarization—is given by:

$$\hat{e}_L = \frac{e^{+j\varphi}}{\sqrt{2}}\left[\left(\hat{\theta}\right) + j\left(\hat{\varphi}\right)\right] \tag{6.3}$$

Note that $\left|E_R(\theta,\phi,f)\right| = \overline{E}(\theta,\phi)\cdot(\hat{e}_r^*)$ and $(E_L(\theta,\phi,f)| = \overline{E}(\theta,\phi) * (\hat{e}_L^*)$

In the above equations:

θ, φ, and f are the zenith angle = (90–elevation angle), the azimuth angle, and the frequency, respectively.

In measuring the radiation pattern of the antenna over the radiation sphere, two standard types of pattern cuts are generally taken, as shown in Figure 6.11(b) [18]. The path formed by the locus of points for which the elevation angle θ is a specified constant and the azimuth angle ϕ is the variable is called a conical cut or as a ϕ cut. The patch formed by the locus of points for which the azimuth angle ϕ is a specified constant and θ is a variable is called a great circle cut or as a θ cut. The measurement of these radiation patterns allows the calculation of the directivity and the gains G_{RHCP} for RHCP and G_{LHCP} for LHCP allows the calculation of important performance factors of a GNSS antenna such as the required minimum gain beam width, the MPR given by:

$$MPR = \frac{G_{RHCP}(\theta)}{G_{RHCP}(180-\theta) + G_{LHCP}(180-\theta)} \tag{6.4}$$

The up/down ratio or FBR given by:

$$\text{Up/Down Ratio=}\ \frac{G_{RHCP}(\theta)}{G_{LHCP}(180-\theta)} \tag{6.5}$$

6.3 Measurement of Axial Ratio of the RHCP GNSS Antenna

The axial ratio R of a circularly polarized antenna is defined by

$$R = \frac{|E_R| + |E_L|}{|E_R| - |E_L|} = \frac{|\rho_c| + 1}{|\hat{\rho}_c| - 1} \tag{6.6}$$

where the circular polarization ratio $\hat{\rho}_c$ is defined by

$$\hat{\rho}_c = \frac{E_R}{E_L} = \rho_c e^{i\delta_c}$$

The tilt angle of the polarization ellipse $\tau = \frac{\delta_c}{2}$ [18, 19].

Other relevant terms used for characterizing the polarization of the antenna are CPR and the cross-polarization discrimination (XPD).

$$CPR = \frac{|E_L|^2}{|E_R|^2} = \left[\frac{|R| - 1}{|R| + 1}\right]^2 = \frac{1}{XPD} \tag{6.7}$$

There are two methods that are commonly for measuring the axial ratio of a circularly polarized antenna:

1. The dual linear amplitude-phase pattern method [19, pp. 81–91; 40] allows very precise measurement of the axial ratio and also the tilt angle of the polarization ellipse over the entire upper hemisphere of the GNSS antennas as a function of elevation and azimuth angles during the radiation pattern cuts. This involves measuring the amplitude and phase of each of the two orthogonally polarized components of the antenna along $\hat{\theta}$ and $\hat{\phi}$ directions and then comparing these against a second measurement performed with a purely linearly polarized standard gain horn (SGH) antenna with a linear axial ratio of typically around 40 dB or greater. Two separate measurements with the SGH antenna are necessary—the first with the SGH antenna oriented along the vertical and the second with the SGH antenna oriented along the horizontal direction. If the SGH antenna is rotated manually for these two measurements, great care is needed to ensure that the SGH is oriented precisely at the polarization null for the orthogonal polarization. These measurements can be conducted in either the far-field antenna range or CATR and also in a near-field antenna range especially using spherical scanning techniques. If these measurements are conducted

in a NF-FF antenna test range the orientation of the SGH is done automatically via a robotic controller. It is a standard procedure for measuring CP antennas and it is built into the software provided by the manufacturer of the NF-FF range. More details of these polarization evaluation techniques using spherical scanning are described in much greater detail by Hansen [14, Section 5.2.4.2]. If these measurements are performed in either a far-field range or a CATR range it greatly adds to complexity since multiple accurate measurement of phase and amplitude for both polarizations may need to be done manually.

2. The second method for measuring the axial ratio is known as the spinning linear method shown in Figure 6.12(a) and (b) [13, 18, 19]. This is much simpler than the above method and involves only measurement of the amplitude (power) and is also quicker since it is conducted with a rapidly rotating probe that is linearly polarized. In this method the GNSS AUT is rotated slowly through a desired angular sector as during a conventional antenna pattern measurement while a linearly polarized antenna, generally used as a transmitting antenna in such measurements, is spun relatively rapidly so that the axial ratio of the antenna being tested can be determined from the modulation rate of the resulting pattern data. The ratio of the amplitudes of adjacent maxima and minima represent the complex

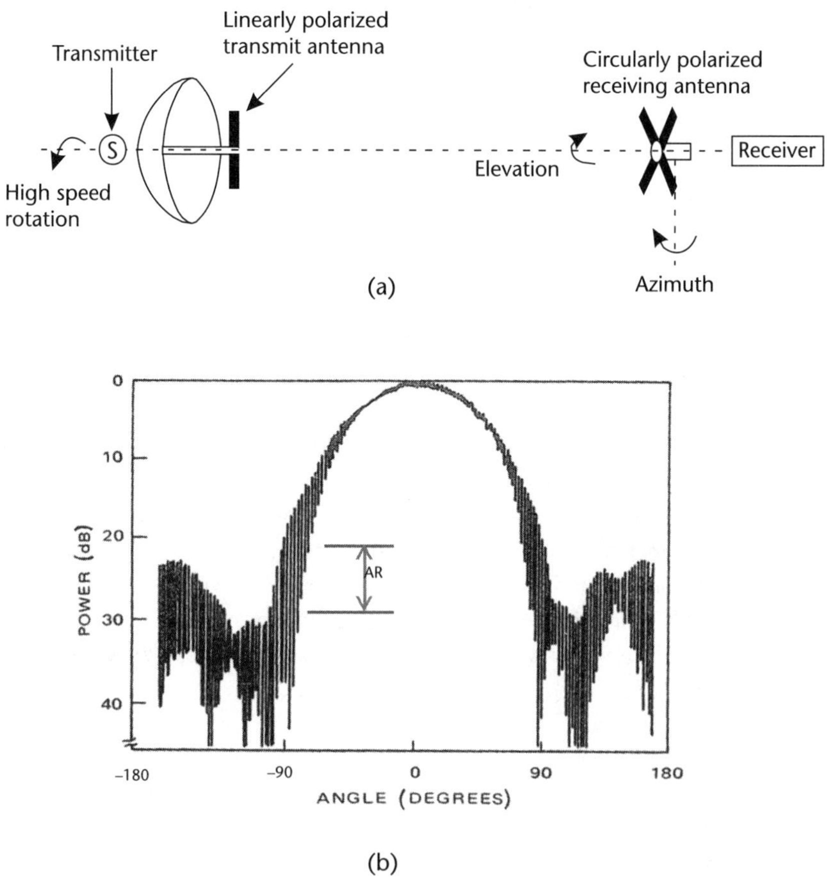

Figure 6.12 Measurement of axial ratio of an RHCP GNSS antenna using a spinning linear dipole.

polarization ratio ρ_c of the antenna under test if the fast-spinning transmit antenna is linearly polarized. The spin rate of this linearly polarized antenna should be adjusted so that the pattern of the GNSS antenna being tested does not change appreciably during one-half revolution of the linearly polarized spinning antenna while the test antenna is being rotated slowly. An example of the resulting modulation in the antenna pattern is shown in Figure 6.12(b). The axial ratio can be estimated by measuring the ratio of adjacent maximum and minimum the modulation envelope—or sometimes by taking the average of two maxima and the associate minima. The complex circular polarization ratio of the antenna $\hat{\rho}_c$ is given by [18, pp. 32–60]

$$\rho_c = \frac{\hat{j} - \hat{K}}{\hat{\rho}_{CS}\left(j - \hat{K}\right)}$$

where $\hat{K} = \dfrac{\hat{V}_{0^\circ}}{\hat{V}_{90^\circ}}$ = ratio of the response of the linear spinning probe when oriented at 0° to its response when oriented at 90°. $\hat{\rho}_{CS}$ is the complex polarization ratio for the linearly polarized sampling antenna. If the spinning probe is linearly polarized with then $\hat{\rho}_{CS} = 1$.

The tilt angle of the polarization ellipse using this method can be determined only if orientation of the spinning probe corresponding to the points on the pattern of the test antenna is known precisely, which is often difficult to measure and is generally not available.

In addition to these two methods, a third method called the three-antenna absolute method can also be used [18]. This method does not require the use of a polarization standard: however, there is restriction that no two antenna can be circularly polarized or nearly circularly polarized; for the third antenna the polarization can be arbitrary. Given these restrictions the third antenna needs to be the GNSS antenna since it is circularly polarized. This method needs a total of six measurements to determine the polarization of the three antennas. The response needs to be measured twice for each combination of antennas—the first with the receiving antenna oriented 0° and the second with the receiving antenna oriented 90°.

6.4 Gain Measurement of a GNSS Antenna

There are three techniques that are commonly used for measuring the gain of a RHCP GNSS antenna: (1) the two-antenna gain transfer method (also called the gain comparison or gain substitution method)[13, 18–20, 40], (2) the three-antenna absolute gain method (these two types of measurements can be conducted either in a far-field antenna range or a CATR [13, 18, 20]), and (3) the method for measuring CP gain conducted in an NF-FF antenna range using spherical scanning techniques [14]. All three techniques require separate measurements at two orthogonal polarizations to fully measure the gain of the GNSS antenna.

These first two types of far-field measurements are used to determine the total CP gain of a GNSS antenna since the total radiated power in the CP antenna is contained in the sum of any two orthogonal components. They are sometime also called the partial gain method and are relatively easy and quick to make since they require a comparison of the relative powers received by the AUT and a standard gain antenna for two orthogonal polarizations. However, for determining the RHCP and LHCP components of the antenna gain necessary for evaluation of GNSS performance of the antenna, additional information on either cross-polarization ratio or axial ratio of the antenna is needed [20]. This can only be determined only through pattern measurement and measurements of the amplitude and phase of the orthogonal vector components of the radiated field to calculate the directivity.

The third method using spherical NF-FF scanning is the most comprehensive of these three methods and provides all the information needed on the GNSS antenna: RHCP and LHCP gains, the cross polarization (or the CP axial ratio), and the radiation patterns for RHCP and LHCP over the entire spherical region from the phase and amplitude data that is collected. An example of the results obtained from this method will be shown later.

The two-antenna and three-antenna methods used for measuring the antenna gain that will be described next are both based on the transmission formula by Friis' for power transfer between two antennas. Consider two antennas A and B separated by a distance R, as shown in Figures 6.13. The ratio of the power transferred between these two antennas is given by

$$\frac{P_R}{P_T} = G_T G_R \left[\frac{\lambda}{4\pi R}\right]^2 \tag{6.8}$$

In the above equation,

P_T = power level at the input terminal of the transmit antenna
G_T = maximum gain of the linearly polarized transmit antenna
G_R = maximum gain of the linearly polarized receiving antenna
P_R = power level at the input terminals of the receiving antenna
R = separation distance between the two antennas

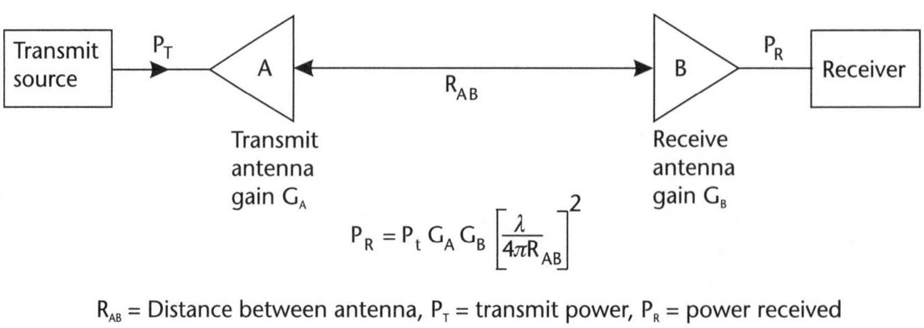

Figure 6.13 Power transferred between two antennas using Fris's formula. R_{AB} = distance between antennas, P_T = transmit power, P_R = power received.

λ = wavelength of the signal from the transmit antenna

In (6.8) it is assumed that both the transmitting and the two receiving antennas, AUT and SGH antennas, are all well matched to their respective power source and receiver. If this not the case, additional correction terms are needed to account for the impedance mismatch; this is discussed in [18].

6.4.1 Two-Antenna Gain Transfer Method

The two-antenna gain transfer method consists of measuring sequentially the power received by two antennas: the RHCP GNSS antenna (i.e., the AUT and a linearly polarized SGH antenna) for two orthogonal polarizations, as shown in Figure 6.14.

Good circularly polarized SGH antennas with very low axial ratios are not easily available; hence SGH antennas most often used in measuring the gain of circularly polarized GNSS antennas are linearly polarized pyramidal horn antennas. They offer very good linear axial ratios of around 40 dB or better; hence the relative error in the measured gain caused by the finite axial ratio of the SGH antenna is low—only between +0.086 and -0.109 dB. The use of a linearly polarized SGH antenna presents a problem while measuring the gain of a circularly polarized GNSS antenna requiring two successive measurements: with the SGH oriented for vertical polarization during the first set of measurements and for horizontal polarization during the second set of measurement. The transmitting antenna used in these measurements also needs to be linearly polarized so that its polarization can be changed from vertical to horizontal by means of a simple rotation in its

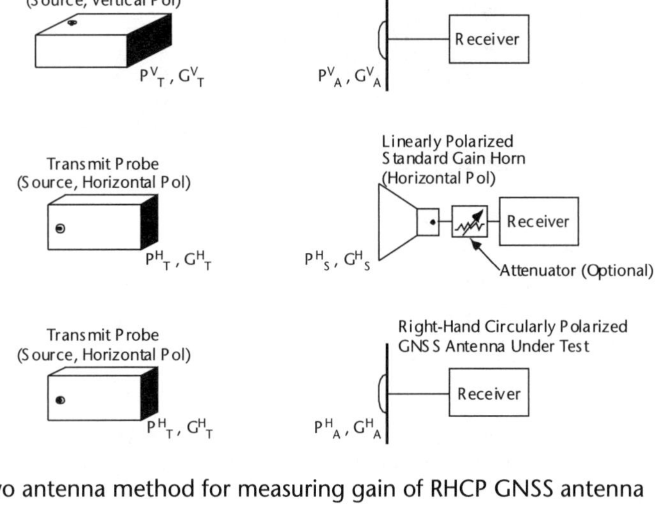

Figure 6.14 Two-antenna method for measuring gain of RHCP GNSS antenna.

orientation. A comparison of the relative powers that are received can then used for calculating the total circularly polarized gain of the antenna [21]. This is the total CP power radiated by the test antenna that is contained in the two linear orthogonal components. The corresponding RHCP and LHCP gain of the GNSS antenna can then be determined if the cross-polarization ratio or the axial ratio of the GNSS is determined from a prior measurement of the radiation pattern, as described in Section 6.3.

Figure 6.15(a) shows a picture of a typical L-band, linearly polarized, pyramidal horn antenna (model number NSI-RF-SG650) made by Near Field Systems Inc. of Torrance, CA, which can be used as an SGH antenna for measuring the gain of GNSS antenna. Its gain, shown in Figure 6.15(b), has been precisely calibrated by the manufacturer over a frequency range covered from 1.12 to 1.7 GHz and varies between 14.3 dBi to 16.8 dBi. The horn antenna is 21.93" wide, 16.25" high and is 21.8" in length and weighs 14 lbs.

The measurement is conducted in two steps: first with both the transmitting and the SGH oriented for vertical polarization as shown in Figure 6.14(a); the power received by the SGH antenna is then measured. The SGH antenna is then replaced by the GNSS AUT antenna being tested and the power received by the GNSS antenna is measured again.

From the measurements taken when the SGH and transmit antennas are both oriented for vertical polarization, we obtain the following:

$$\frac{P_A^V}{P_T} = \left(\frac{\lambda}{4\pi R}\right)^2 G_T^V G_A^V \tag{6.9}$$

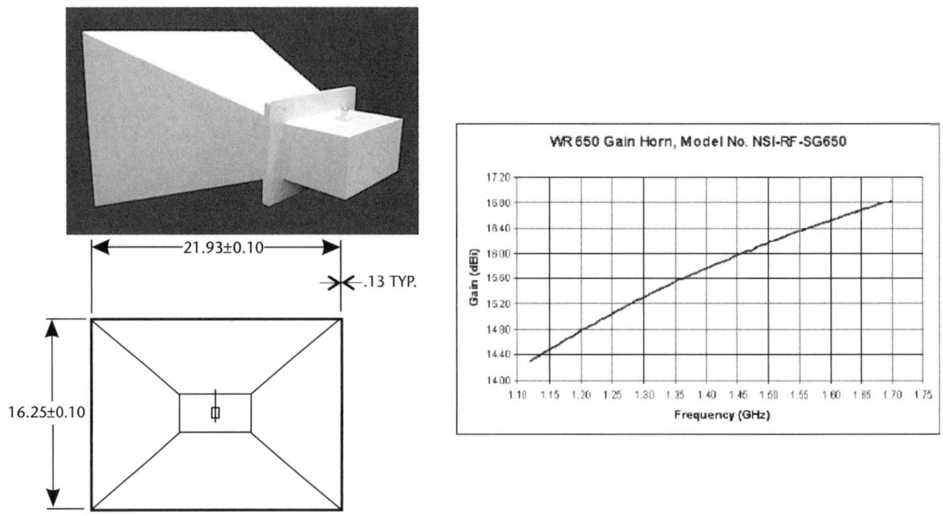

(a) (b)

Figure 6.15 L-band standard gain horn antenna for GNSS antenna calibration; product of Near Field Systems, Torrance, CA, model number WR 650.

$$\frac{P_S^V}{P_T} = \left(\frac{\lambda}{4\pi R}\right)^2 G_T^V G_S^V \tag{6.10}$$

From (6.9) and (6.10) we can obtain the gain of the AUT for vertical polarization [20]

$$G_A^V(dB) = P_A^V(dBm) - P_S^V(dBm) + G_S^V(dB) \tag{6.11}$$

and

$$G_A^V = 10^{\frac{G_A^V(dB)}{10}} \tag{6.12}$$

Similarly, from the measurement taken when the SGH and transmit antennas are rotated for horizontal polarization as shown in Figure 6.14(b) we can also obtain the following:

$$\frac{P_A^H}{P_T} = \left(\frac{\lambda}{4\pi R}\right)^2 G_T^H G_A^H \tag{6.13}$$

$$\frac{P_S^H}{P_T} = \left(\frac{\lambda}{4\pi R}\right)^2 G_T^H G_S^H \tag{6.14}$$

Using (6.13) and (6.14), we obtain

$$G_A^H(dB) = P_A^H(dBm) - P_S^H(dBm) + G_S^H(dB) \tag{6.15}$$

and

$$G_A^H = 10^{\frac{G_A^H(dB)}{10}} \tag{6.16}$$

From the linear sum of the these two partial gains for vertical and horizontal polarizations, respectively, we can now calculate the total gain of the circularly polarized GNSS AUT antenna

$$G_A^T = G_A^V + G_A^H \tag{6.17}$$

It can expressed in decibels as

$$G_A^T(dB) = 10\log_{10}\left(G_A^T\right) \tag{6.18}$$

These partial gain measurements represent the total radiated power that is contained in the two linear orthogonal components. Note that up until now there has been no measurement of the relative phase of the two orthogonal polarization components into which the plane wave is resolved. The specific gains that relevant for GNSS are the copolarized RHCP gain and the cross-polarized LHCP gain. These can be calculated from the total gain if the CPR or the axial ratio (R) is known; this, as described above, is determined independently through antenna pattern/directivity measurements. The copolarized RHCP gain is then determined by:

$$G_{RHCP} = \frac{G_T}{(1 + CPR)} \tag{6.19}$$

$$G_{RHCP}(dB) = 10\log_{10}(G_{RHCP}) \tag{6.20}$$

Similarly the cross-polarized LHCP gain is then given by

$$G_{LHCP} = G_{RHCP} \cdot CPR \tag{6.21}$$

when expressed in decibels it is:

$$G_{LHCP}(dB) = 10\log_{10}(G_{LHCP}) \tag{6.22}$$

A large number of terms appear in the above equations; they are defined below. In these terms the superscript indicates the polarization state with V being the vertical polarization and H being the horizontal polarization; the subscript identifies the type of antenna: A represents the AUT, S represents the SGH antenna, and T represents the transmit antenna used in these tests.

P_A^V and P_A^H = power received by the circularly polarized AUT when the transmit antenna is oriented for vertical and horizontal polarizations, respectively

P_S^V and P_S^H = power received by the SGH antenna when it along with the transmit antenna are oriented for vertical and horizontal polarizations, respectively

G_A^V and G_A^H = gain of AUT for vertical and horizontal polarizations, respectively

G_S^V and G_S^H = gain of the SGH antenna for vertical and horizontal polarizations, respectively

G_T^V and G_T^H = gain of the transmit antenna for vertical and horizontal polarizations, respectively

In performing these gain calculations it may be necessary to take into account the gains of any amplifiers or insertion losses introduced by coaxial cables, attenuators, and other devices that may be inserted into either the transmit or receive sections while performing these measurements. The correction terms needed to account for their presence can be determined by removing the antennas from the RF chain and physically joining the transmitting and receiving circuitry together to estimate this correction term that can then be appropriately included in the respective equations.

Note also that the dual-linear amplitude-phase method described earlier for measuring CP axial ratio already contains all the information in it that is needed for calculating the CP gains by conducting further analysis of the measured results.

The gain of SGH horn is typically between 10 to 20 dBi; the SGH shown in Figure 10.15(a) and (b) varies between 14 to 16 dBi in the GNSS frequency band. Hence its gain is much greater than that of a typical GNSS FRPA antenna that is much lower—typically no greater than 4 to 5 dBic since it is required to generate a wide beam to capture nearly all of the GNSS satellites within its view. A more precise method for determining the gain of the AUT can be obtained by introducing a precisely calibrated variable attenuator that is introduced between the SGH antenna and the receiver for the configuration shown in Figure 6.14(a); the attenuator is adjusted until the received signal levels from both the AUT and the SGH antennas become equal. This eliminates any concerns generated by a nonlinearity in the receiver.

The two disadvantages of the two antenna method are (1) the need for a standard gain horn antenna, which can be expensive, and (2) the risk of misalignment of the antennas during measurements due to the rotation needed for changing polarization from vertical to horizontal.

6.4.2 Three-Antennas Absolute Gain Measurement Method

The three-antennas method, illustrated in Figure 6.16, obviates the need for the SGH antenna; instead three antennas with different combination pairs are used

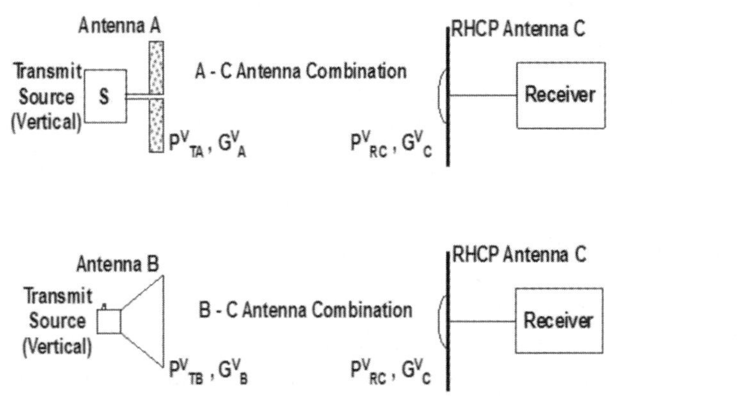

Figure 6.16 Three-antenna method for measuring gain of an RHCP GNSS antenna. Measured radiation pattern in the upper hemisphere of the center element in the L_1 band: (a) frequency = 1575 MHz, and (b) frequency = 1227 MHz.

in these measurements. Since one of three antennas is the GNSS AUT, which is RHCP, it requires the other two of these antennas to be linearly polarized to allow change in polarization from vertical to horizontal through a simple rotation in their orientation. Knowledge of the exact gain of the two linearly polarized antennas is not needed and can, in fact, be obtained from these measurements if necessary. The disadvantage is that this method will require a total of six different sets of measurements for different pairs of antenna combinations—a set of three for each orthogonal polarization.

In the discussion that follows the two linearly polarized antennas will be designated as antennas A and B and the circularly polarized GNSS AUT will be designated as antenna C. As in the previously described two-antennas method, the partial gain measurements for the two linear orthogonal components are first measured from which the total CP gain is determined. The RHCP and LHCP gains are determined from the total CP gain by using either the measured cross-polarization ratio or axial ratio of the antenna under test from separate far-field pattern measurements.

Once again the following equations from Friis's formula can be derived each set of antenna combinations for vertical polarization, as shown in Figure 6. 16:

Antenna Combination A-B

$$\left(G_A^V\right)_{dB} + \left(G_B^V\right)_{dB} = 20\log_{10}\left(\frac{4\pi R}{\lambda}\right) + 10\log_{10}\left(\frac{P_{RB}^V}{P_{TA}^V}\right) \tag{6.23}$$

Antenna Combination A-C

$$\left(G_A^V\right)_{dB} + \left(G_C^V\right)_{dB} = 20\log_{10}\left(\frac{4\pi R}{\lambda}\right) + 10\log_{10}\left(\frac{P_{RC}^V}{P_{TA}^V}\right) \tag{6.24}$$

Antenna Combination B-C

$$\left(G_B^V\right)_{dB} + \left(G_C^V\right)_{dB} = 20\log_{10}\left(\frac{4\pi R}{\lambda}\right) + 10\log_{10}\left(\frac{P_{RC}^V}{P_{TB}^V}\right) \tag{6.25}$$

Equations (6.23) through (6.25) represent three equations for the gains of the three unknown antennas: $\left(G_A^V\right)_{dB}$, $\left(G_B^V\right)_{dB}$, and $\left(G_C^V\right)_{dB}$ which can be determined knowing the other parameters appearing on the right-hand side of these equations: R, λ, and the three power ratios that appear in the above equations are measured. We are only interested in determining the value of $\left(G_C^V\right)_{dB}$ for the GNSS AUT from which we can determine $G_C^V = 10^{\frac{\left(G_C^V\right)_{dB}}{10}}$.

Similarly, by changing to horizontal polarization for the antennas used in all three of the above antenna combinations, we obtain for combinations A-B, A-C, and B-C, respectively, the following three equations:

$$\left(G_A^H\right)_{dB} + \left(G_B^H\right)_{dB} = 20\log_{10}\left(\frac{4\pi R}{\lambda}\right) + 10\log_{10}\left(\frac{P_{RB}^H}{P_{TA}^H}\right) \tag{6.26}$$

$$\left(G_A^H\right)_{dB} + \left(G_C^H\right)_{dB} = 20\log_{10}\left(\frac{4\pi R}{\lambda}\right) + 10\log_{10}\left(\frac{P_{RC}^H}{P_{TA}^H}\right) \tag{6.27}$$

$$\left(G_B^H\right)_{dB} + \left(G_C^H\right)_{dB} = 20\log_{10}\left(\frac{4\pi R}{\lambda}\right) + 10\log_{10}\left(\frac{P_{RC}^H}{P_{TB}^H}\right) \tag{6.28}$$

Again as before, (6.26) through (6.28) can be used to determine $G_C^H = 10^{\frac{\left(G_C^H\right)_{dB}}{10}}$.

From the linear sum of these two partial gains for vertical and horizontal polarizations, respectively, we can now calculate the total gain of antenna C—the circularly polarized GNSS AUT antenna.

$$G_C^T = G_C^V + G_C^H \tag{6.29}$$

It can expressed in decibels as

$$G_C^T(dB) = 10\log_{10}\left(G_C^T\right)$$

The copolarized RHCP gain of antenna C is then determined by

$$G_{RHCP} = \frac{G_C^T}{(1+CPR)} \tag{6.30}$$

$$G_{RHCP}(dB) = 10\log_{10}\left(G_{RHCP}\right) \tag{6.31}$$

Similarly, the cross-polarized LHCP gain of antenna C is then given by

$$G_{LHCP} = G_{RHCP} \cdot CPR \tag{6.32}$$

When expressed in decibels, it is:

$$G_{LHCP} = 10\log_{10}\left(G_{LHCP}\right) \tag{6.33}$$

The definitions for the various terms that appear in (6.23) through (6.25) are:

P_{RB}^V, P_{RC}^V = power received for vertical polarization by antennas B and C, respectively

P_{TA}^V, P_{TB}^V = power transmitted for vertical polarization by antenna A and B, respectively

G_A^V, G_B^V, G_C^V = gain for vertical polarization of the three antennas A, B, and C

The terms that correspond to the above terms that appear in (6.26) through (6.28), but for horizontal polarization, can be obtained from these above terms by simply replacing the V that appears in the superscript of each term by H.

6.4.3 Near-Field Spherical Scanning Techniques for Measuring Gain

The spherical near-field scanning techniques are perhaps the best and most comprehensive methods for measuring the 3-D RHCP and LHCP gain contours of the GNSS antenna [14]. Three separate steps are involved:

- Near-field scanning of the AUT;
- Near-field scanning of the SGH antenna;
- Comparison of the near-field measurements of the test antenna with the near-field measurements of the SGH antenna.

Data obtained from these three separate measurements are processed by the software provided by the manufacturer of these NF-FF antenna test ranges and yields all the important information needed on both the antenna gain in both RHCP and LHCP polarizations and the amplitude and phase of the radiation patterns. A detailed description of the theoretical details of this method are beyond the scope of this book and the reader can find these described [14, pp. 206–215] and also in the earlier chapters of this reference. The spherical scanning technique allows the gain of the GNSS antenna to be measured and mapped over the entire upper hemisphere. Figure 6.17 shows the RHCP gain contours of the radiated field of a GPS dual-band stacked circular microstrip patch antenna that has been

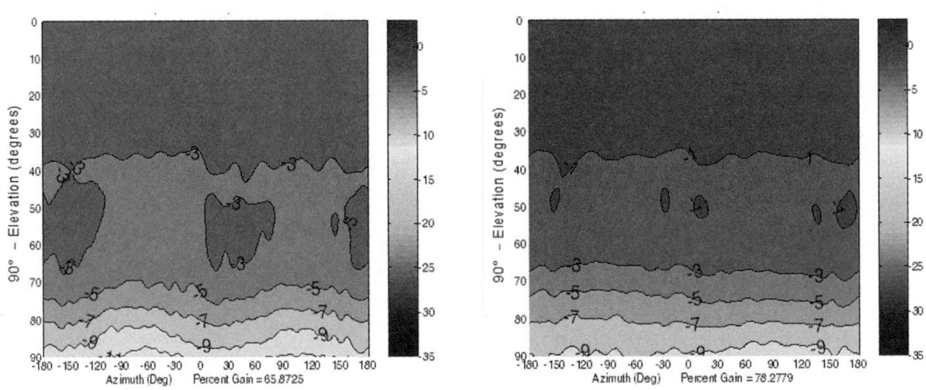

Measured radiation pattern in the upper hemisphere of the center element in the L_1 band (Frequency = 1575 MHz)

Measured radiation pattern in the upper hemisphere of the center element in the L_2 band (Frequency = 1227 MHz)

Figure 6.17 3-D radiation and gain pattern of a GPS dual-band stacked circular microstrip antenna measured in the near-field antenna range using spherical scanning techniques. (a) Measured radiation pattern in the upper hemisphere of the center element in (a) the L_1 band (frequency = 1575 MHZ), and (b) the L_2 band (frequency = 1227 MHz).

measured and then mapped for the entire upper hemisphere at two different GPS frequencies: 1.5754 GHz, the center frequency of the L_1 band, and 1.2276 GHz, the center frequency of the L_2 band.

These measurements were conducted at the MITRE Corporation in a NF-FF range built by Near Field Systems Inc. In the MATLAB graphical plot shown in this figure, the gain over the entire upper hemisphere from elevation angles from 0 to 90 degrees and +180 to −180 degrees in azimuth has been mapped; the top represents a single point—the zenith at azimuth angles varying from +180° to −180°. Note that this type of measurements provides a wealth of information allowing the calculation of PDOP and C/N_0 for determining satellite availability at any look angle over the upper hemisphere. It is far superior for the purposes of GNSS system analysis to an elevation plane pattern measured at single azimuth angle that is typically done with the standard far-field antenna ranges described earlier.

6.5 Measurement of the Bandwidth of GNSS Antennas

The bandwidth of a GNSS antenna is the frequency range over which a selected performance parameter of the antenna meets requirements that ensure satisfactory operation. The three performance parameters that are selected are (1) The minimum gain bandwidth, (2) the RHCP axial ratio bandwidth, and (3) the impedance or return loss bandwidth.

The minimum gain bandwidth is defined as the frequency range (defined by upper and lower frequency bounds) over which the antenna provides a gain that is equal to or better than the minimum threshold gain needed to acquire the GNSS satellites within a viewing region covered by the entire upper hemisphere down to minimum specified low-elevation masking angle above the horizon. There are two separate parameters in play here: (1) the beamwidth of the antenna determined by the low elevation cutoff angles, and (2) the RHCP gain of the antenna within this beamwidth sector. Since the gain of most GNSS antenna increases with elevation (or decreases from zenith), the antenna to be able to meet this requirement t should have a gain higher than the minimum gain at the lowest specified cut off elevation angles within this measured frequency band.

These gain requirements depend on the intended application and were discussed in Chapter 1. The minimum gain bandwidth is generally measured in any of the closed/anechoic chamber antenna test ranges described earlier. These include measurement of the amplitude and phase of the orthogonal polarization components, E_θ and E_ϕ at close frequency intervals throughout the GNSS band of interest from which the RHCP far-field pattern can be computed. It also requires measurement of the peak gain at boresight at all these frequency intervals using gain calibration techniques described earlier; the computed patterns can then be normalized relative to the peak gain value to boresight to determine the minimum gain frequency bandwidth. This requires a large number of measurements and calibrations but characterizes the performance of a GNSS antenna to its fullest and is the most meaningful in terms of determining GNSS performance.

The RHCP axial ratio bandwidth was defined in Section 6.2; it is the frequency range in which the antenna is able to keep the maximum LHCP cross-polarization level below its specified limit. A polarization axial ratio is important for GNSS antenna since it reduces the susceptibility to multipath. The LHCP cross-polarization level is a strong function of the elevation angle—it is generally lowest at zenith but increases rapidly for most GNSS antennas such as microstrip antennas since the antenna becomes linearly polarized (vertical polarization) at the horizon due to the presence of the ground plane or the conducting fuselage of an airplane or an automobile. The horizontal polarization parallel to the metal ground plane or fuselage is zero and the polarization of the antenna at horizon is almost vertical. The axial ratio is measured the same way as gain from the phase and amplitude of the two orthogonal polarization components at various frequency intervals from which the axial ratio can be computed.

The impedance bandwidth or the return loss bandwidth is the easiest of the three bandwidth parameters to measure since it can be measured using a network analyzer without requiring an antenna test range/anechoic chamber. The impedance bandwidth BW is determined by the maximum VSWR defined by the symbol S

$$BW = \frac{S-1}{Q_t\sqrt{S}}$$

Q_T is the total quality factor of the antenna and S is the input reflection coefficient that is defined in terms of the input reflection coefficient as

$$S = \frac{1+\Gamma}{1-\Gamma}$$

The input reflection coefficient is a measure of the reflected signal at the antenna feed point before it is attached to a quadrature hybrid or branch-line coupler that is typically used for generating RHCP. It is defined in terms of the input impedance Z_{in} of the antenna and the characteristic impedance Z_0 of the transmission line feeding the antenna

$$\Gamma = \frac{Z_{in} - Z_0}{Z_{in} - Z_0}$$

The impedance bandwidth is specified as the frequency range over which the voltage standing wave ratio S is less than 2. This corresponds to a loss of 0.454 dB in the power that is delivered from the antenna to the receiver or to a return loss of –9.5 dB or to 11% of power being reflected (i.e., only 89% of the signal power received by the antenna is transferred to the receiver). Generally the impedance bandwidth is much narrower than either the gain bandwidth or the axial ratio bandwidth.

Many GNSS antennas have a quadrature hybrid that is used for generating RHCP; this hybrid is embedded inside the antenna package and connected to the antenna and is not visible to the user. The impedance bandwidth of such antennas cannot be measured using a network analyzer since it is not possible to separate the effects of the hybrid from that of the antenna. The signal reflected back from the input port of the antenna due to an impedance mismatch is diverted under these conditions to the isolated port of the hybrid that is terminated in a matched 50-ohm load; hence this reflected power that is an indication of impedance mismatch between the antenna and the coaxial feed line is not detected by the network analyzer. This can result in a very low return loss being measured over a wide band of frequencies and could lead to the erroneous conclusion that the impedance bandwidth is much broader than its true bandwidth. Therefore, when a hybrid is included with the antenna package, the gain bandwidth is a safer and better metric of the true bandwidth of the antenna than one determined from return loss measurements alone.

6.6 Measurement of PCO and PCV of GNSS Receiver Antennas

The electromagnetic phase center of a GNSS antenna is the reference point at which all measurements derived from GNSS signal measurements are referred to. The phase of the antenna varies with the elevation and azimuth angle of the direction of arrival of the satellite signal and also with frequency. However it cannot be identified through a direct tape measurement and hence a geometrical point on the antenna is denoted as the ARP. The IGS had defined the ARP as the intersection of the vertical antenna axis of symmetry and the bottom of the antenna.

An antenna model that is particularly convenient for calibration of the absolute phase center of a GNSS antenna has been proposed by Gorres et al. [5, 29] and is shown in Figure 6.18.

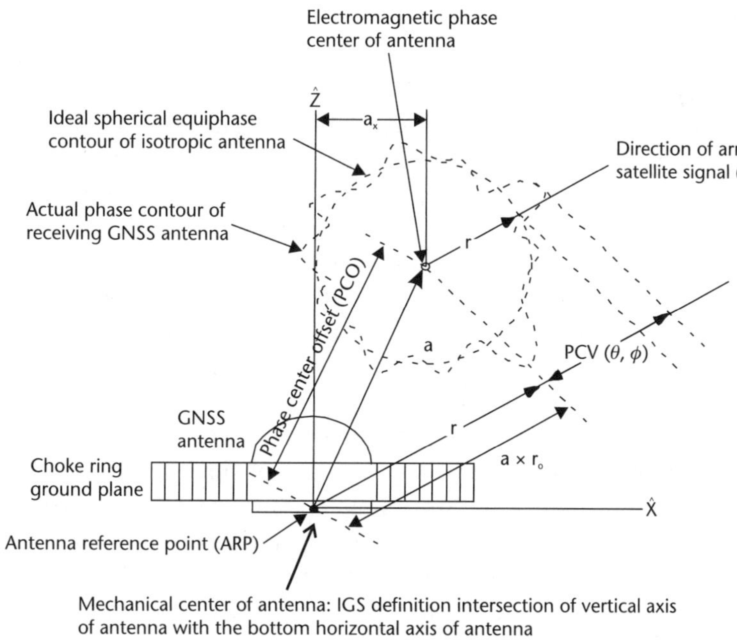

Figure 6.18 Calibration model for GNSS antenna depicting PCO and PCV. Mechanical center of antenna: IGS definition intersection of vertical axis of antenna with the bottom horizontal axis of the antenna.

The total antenna phase center correction for phase pseudorange measurements as depicted in this model consists of two parts as shown in this figure: (1) Δ_{PCO}, which is the projection of the vector $\vec{a}$ between the mean electrical phase center of the antenna and the ARP and the unit vector $\vec{r}_0$, the direction of arrival of the satellite signal at the receiver , and (2) Δ_{PCO} (θ, ϕ, f), which is phase variation of the antenna relative to the mean electromagnetic phase center as a function of the elevation θ and azimuth angle ϕ of the direction of arrival of the satellite signal and its frequency f. Note that in this model there are two separate sets of coordinate systems—the first whose origin is at the ARP, and a second system, in spherical coordinates, whose origin is located at the mean phase center of the antenna. This antenna model for the total phase correction from both PCO and is represented by the following equations:

$$\Delta r(\theta,\varphi) = \Delta_{PCO} + \Delta_{PCV} = \vec{a}\cdot\vec{r}_0 + d\tau(\theta,\varphi,f)\cdot\left(\frac{\lambda}{360^\circ}\right)$$

where the PCO is defined as

$$\vec{a} = \hat{x}(a_x) + \hat{y}(a_y) + \hat{z}(a_z)$$

In the above equation $\Delta r(\theta, \phi)$ = total correction for phase pseudorange in the direction of the satellite, θ, ϕ are the elevation angle and azimuth angle of the direction of arrival of the satellite signal, λ is the wavelength of the satellite signal, $\vec{r}_0$ is the unit vector between the satellite and the mean electromagnetic phase center of the antenna, $\vec{a}$ is the vector direction of the PCO between the ARP and the mean electromagnetic phase center, and $d\tau(\theta\ \phi)$ is the PCV in degrees which is a function of the elevation and azimuth angles of the satellite signal. Note that although a simple Cartesian x, y ,z coordinate system has been used in this model, it is more customary to use other LLSs such as the ENU coordinate system for antenna calibration [30, 31]; the north, east, up (NEU) system is shown in Figure 6.19.

The term PCV is not quite appropriate as has been pointed out by some investigators since it seems to suggest a varying center with spherical wavefronts. Instead, a more precise way of defining this term would be antenna phase variations (APVs) relative to the absolute mean phase center of the antenna. Unfortunately, the term PCV is more widely accepted despite its misleading description. In addition, the PCO and PCV also vary with wavelength and hence require calibration across the frequency bands of the four major international GNSS that vary from 1.1 to 1.65 GHz when multiband antennas are used for precision GNSS measurements.

An accurate calibration of both the PCO and PCV of a GNSS antenna is essential for high-precision applications so that positions determined from GNSS measurements can be related to physical antenna mounts. Since the phase center of the antenna changes with the azimuth angle ϕ and zenith angle θ (which is 90° minus the elevation angle) of the arriving satellite signal and also with its frequency, its calibration becomes especially important when different antennas and receivers are used in RTK differential measurements and in continuously operating permanent reference station networks [22, pp. 265–270]. In general the antennas at the reference station and for a given user will not be identical. Seeber has indicated [23] that

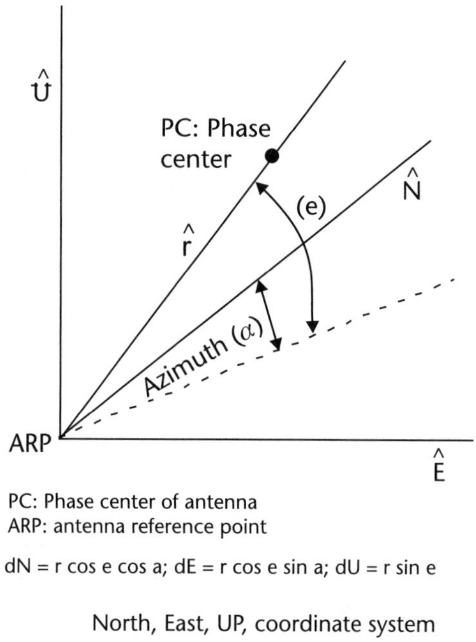

Figure 6.19 NEU coordinate system.

by neglecting PCV corrections one can introduce position errors of up to 3 cm and height errors of up to 5 cm. To correct for such errors the user needs to know the precise phase pattern of the reference antenna and also of the user antenna. Smaller errors also occur when two antennas that are used at opposite ends of a baseline are oriented at other than at their previously calibrated directions. Antenna calibration is also important when estimating the effects of the troposphere on GNSS measurements since both the PCV and delay from the troposphere depend on the elevation angle. PCO and PCV must be used together as a set since this will allow different sets to lead back to the same ARP. Generally the largest offset occurs in height with variations as large as 10 cm in the vertical plane and only about 3 cm in the horizontal plane; in the horizontal plane the variation in elevation can be larger than in azimuth.

There are basically two different ways to measure the PCO and PCV of antennas. The first is an absolute measurement conducted in an anechoic chamber that has been widely used and improved by many investigators since 1994 [2–8]; the second method relies on in situ field measurements using actual signals transmitted from the GNSS satellites. The field measurement techniques can be further divided into two groups: relative measurements with respect to a reference antenna first developed by Mader [24, 25] at the National Geodetic Survey (NGS) and later also several other investigators [26]; the second is an absolute measurement involving antenna rotations using a fast rotating and tilting robot developed by Geo++ of Garbsen, Germany, by Wubbena and Schmitz et al. [27, 28] in collaboration with University of Hanover (IfE). This method can eliminate errors due to far-field and near-field multipath that can be present in the results obtained from the first two measuring techniques. The designation of "absolute" in this context indicates that the PCV and PCO of the AUT are determined independently without recourse to a reference antenna.

These three techniques are described in greater detail below.

6.6.1 Microwave Anechoic Chamber Method of Antenna PCV and PCO Calibration

Absolute measurements are conducted inside an anechoic chamber using synthetic GNSS signals from either a signal simulator or a frequency source in lieu of actual signals transmitted by GNSS satellites. The phase center is determined by measuring the changes in the phase of the signal received by the AUT as it is being rotated and tilted to change its azimuth and elevation angles. This method was first investigated by Schupler dating back to 1994 [6, 7] and has since been investigated in greater detail and with more accuracy by several investigators in Germany, most notably by Gorres [5], Becker [2, 3], and Zeimetz and Kuhlmann [4]. The anechoic chamber method is particularly convenient for measuring the PCV and PCO across the entire GNSS band since it uses a synthetic GNSS signal from a signal simulator thus removing the need for actual signals from a satellite constellation. A picture of a typical anechoic chamber was shown in Figure 6.3(a) and (b).

The determination of the total phase center correction now involves two steps: the first step involves the estimation of the three components of the PCO with respect to the ARP. When using the ENU coordinates [30], this is often expressed as offset corrections in the vertical (up direction) and horizontal plane (north/east directions). The second is the phase center variation with respect to the elevation and azimuth angles of the arriving signal and its frequency relative to the mean electromagnetic phase center. These two parameters need to be used together as a set to uniquely define the antenna phase variation that can be traced back to the ARP. To determine the first parameter, the mean phase center location, a search is conducted to find a point for which the phase variations of the phase pattern of the antenna are minimized to determine a weighted average phase center position. The problem then becomes one of resolving the surface of a "best fit" sphere that most closely approximates the received phase and finding its center coordinates a_x, a_y, and a_z and also its radius, which is the PCO. This is determined in the anechoic chamber by measurement techniques first proposed by Schupler back in 1994 [6, 7]. A signal corresponds the satellite signal is transmitted within the anechoic chamber by a transmitting antenna that is fixed in space and phase of the signal received by the AUT is recorded by a network analyzer to determine the relative phase shift between the transmitted and received signals. A series of phase pattern measurements are recorded while this receiving antenna is rotated through zenith angles of –90° to +90° and also various azimuth angles for a fixed elevation using a remotely controlled positioner. Each selected position of the AUT corresponds to a different direction of the satellite. To rotate the test antenna as precisely as possible around the mean phase center for the actual measurements, the antenna is first shifted with respect to the center of rotation until the phase variations with elevation are minimal and as symmetrical as possible for zenith angles varying from -90° to +90° .The phase patterns at all these different position of the antenna relative to the rotation axis of the antenna are measured. The first parameter, the mean PCO with respect to the ARP, can be determined through least squares minimization.

$$\sum \Delta\Phi(\theta,\varphi)^2 = MIN$$

The phase pattern and PCV is obtained directly as the observed residuals from the above equation. As a second step in processing the collected data, the pattern can be modeled using harmonic functions on order to obtain an estimate of the measurement noise and smoothness of the phase patterns. The two-dimensional functional model for PCVs proposed by Gorres [5] is given by:

$$\Delta r(\theta) = \sum_k \left(a_k \cos(\theta) + b_k \sin(\theta)\right)$$

where the index $k = 0, \ldots, 3$, or 5.

The movement of the actual phase center from the estimated mean PCO is the phase center variation and indicates the relative smoothness or distortion of the phase front. The smoother or more spherical the wavefront, the smaller the PCV.

One advantage of the anechoic chamber measurement technique is that it allows calibration of the PCO and PCV of antennas across all GNSS frequencies ranging from 1.15 to 1.65 GHz as it does need an operating satellite constellation system that is necessary for field tests using either relative or robot positioner techniques; instead signals from a frequency source or a GNSS signal simulator can always be transmitted inside the anechoic chamber. This has been demonstrated recently by Becker [2, 3] and by Zeimetz and Kuhlmann [4]. Becker [2] and his colleagues at the TU Darmstadt and at the University of Bonn have measured the frequency-dependent PCV and PCO variations of several types of geodetic antennas from 1.15 to 1.65 GHz. A picture of the anechoic chamber used in their measurements along with a schematic diagram is shown in Figure 6.3(a) and (b). They tested several geodetic antennas including the Leica AR25, Trimble Zephyr Geodetic Model 2, and Leica AT504GG. They have also compared the results of their absolute anechoic chamber calibration against the two other techniques: relative calibration and absolute calibration using a robot. An example of the PCO variation measured for the Leica AR25 antenna is shown in Figure 6.20

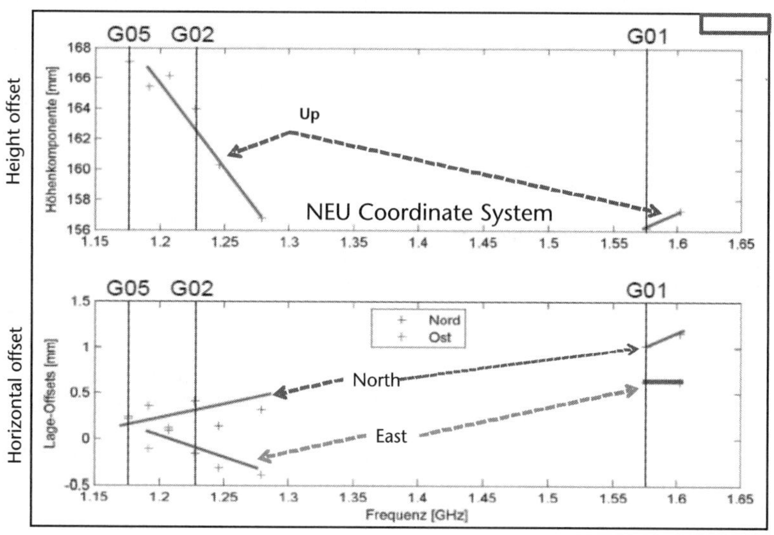

Figure 6.20 PCVs versus GNSS frequency for the Leica AR25 R3 geodetic antenna measured by Becker et al. (Courtesy of Dr. M. Becker et al. [2].

6.6.2 Relative Antenna Calibration by Comparison to a Reference Antenna

Relative field calibration procedure has been developed at the National Geodetic Survey (NGS) by Mader [24, 25] where the AUT is compared against the reference antenna. Similar techniques for relative antenna calibration have also been investigated byAkour, Santierre, and Geiger at the University of Laval [26] and at several universities in Germany such as Dresden (TUD) and Bonn (GIB) [5]. The first measurement technique developed by Mader [24, 25] will be described in some detail in this book. In this method both antennas are set up together separated by a very short baseline with accurately known positions and signals from the GPS satellites are then used to estimate the position of the phase center depending on the elevation and azimuth angles of the satellites. Figure 6.21(a) shows a diagram of the antenna layout for this type of calibration.

Since the relative measurements are performed using an interferometric technique, it only allows the differences in the phase center of the two antennas to be measured. The test range for conducting these relative antenna calibration consists of two stable 6“ diameter concrete piers rising about 1.8m above the ground. They are separated by a distance of 5m and are located in a flat field and lie along the N-S line; antenna mounting plates are firmly attached at the top of each. This

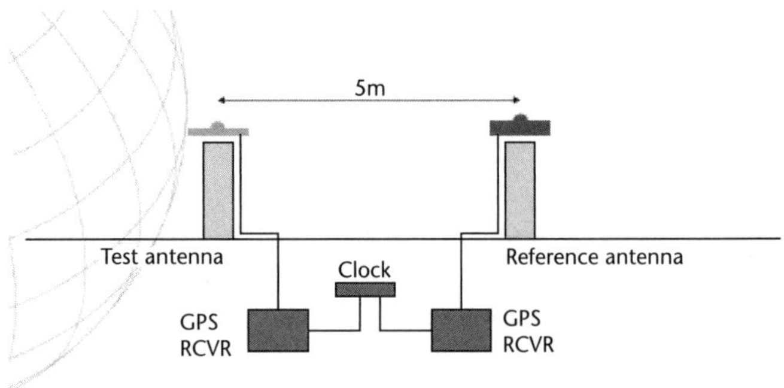

(a)

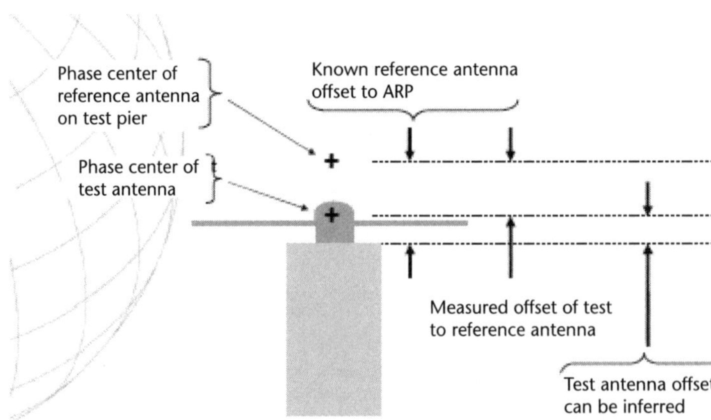

(b)

Figure 6.21 Relative GPS antenna calibration techniques developed by the National Geodetic Survey. (Courtesy of Dr. Gerald L. Mader, NGS [25].)

measurement site is located at NGS's Instrumentation and Methodologies Branch in Corbin, VA. Both the reference and test antennas are oriented towards the north during these calibrations. Note that because of this initial alignment in direction, the magnitude of the horizontal offset could change if one or both antennas were to be oriented other than in the preferred north direction used in this calibration. The reference antenna used in the measurements is a Dorne/Margolin choke-ring antenna of Type T (referred to as the JPL D/M-crT antenna in Mader's paper but in other papers this antenna is referred to as the AOAD/M_T antenna). The same reference antenna is used for calibrating all of the antennas being tested.

This reference antenna is mounted on the north pier and the AUT on the south pier. Both the reference and the antenna under test are connected to Ashtech Z 12 receivers and are set to track satellites down to a cutoff elevation of 10°. Both receivers use the same rubidium oscillator as external standard in all the tests.

The variation of the phase center as a function of elevation is determined separately for L_1 and L_2. Antennas identical to the reference antenna have been placed on the test pier in order to determine the location of this antenna's L_1 and L_2 phase centers. These positions determined from multiple measurements are then used as the a priori positions for the L_1 and L_2 phase centers for all antennas tested on this pier. The offsets from these a priori positions are then combined with the defined offsets for the reference antenna to find the L_1 and L_2 average phase centers. Figure 6.20(b) illustrates how these average phase center offsets are determined.

No azimuth dependence can be determined from these PCV solutions. The PCV is determined using L_1 and L_2 single differences rather than double differences to determine the relative PCV directly rather than from different satellites at different elevations. A standard elevation cutoff of 15° has been defined for the determination for the determination of the test antenna's average phase center locations. Since single differences are being used as the observable, the tropospheric and ionospheric errors cancel out; however, the clock differences between the two GPS receivers used by the test and reference antenna do not cancel out as they would with double differences when differences between the simultaneous observations of the two receivers and two satellites are used in the measurements. Therefore, a rubidium oscillator is used as an external frequency standard to remove most of the variation due to clock differences and time delays from the a priori single difference residuals. A least squares solution for a fourth-order polynomial is used for each measurement epoch to account for the clock offset and the elevation dependence. The procedure has been coded into a FORTRAN program called ANTCAL. Figure 6.22 shows a typical relative vertical antenna phase variation calibrations measured by NGS.

This technique is accurate as long as the elevation difference of a satellite as seen by both antennas is negligible; hence it is applicable for baselines lengths of several thousand kilometers before an appreciable change in elevation between the two antennas is noticed. Calibration is conducted in the GPS L_1 and L_2 frequency bands. This relative technique does not consider the azimuth component of PCV but only the elevation component. More details can be found in the original paper by Mader [24, 25]. A large number of GPS geodetic antennas have been calibrated by NGS using this relative method and information on these calibrations and is available from http://www.ngs.noaa.gov/ANTCAL/.

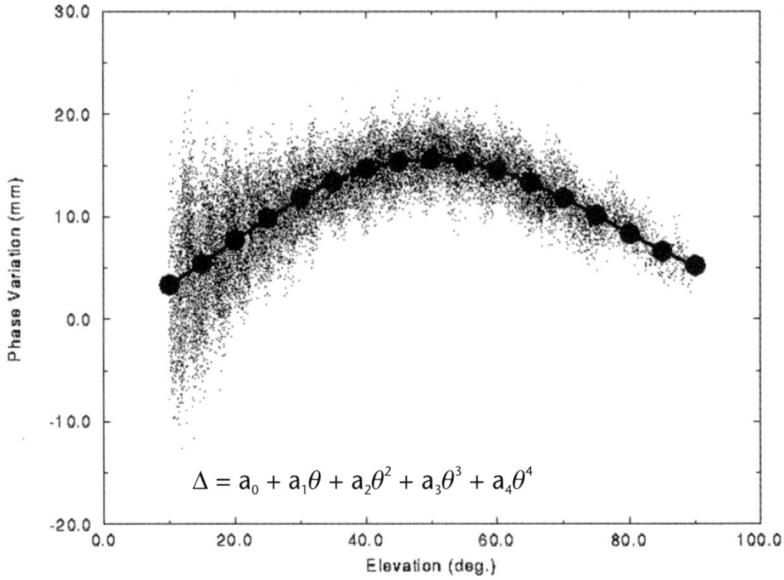

Figure 6.22 Vertical phase variation of GPS antenna determined by the relative antenna calibration method developed at NGS.

The advantage of this relative calibration method is that it allows for the calibration of a large number of antennas in a short time period. However, there are several drawbacks as well that have been pointed out by Alemu [31]:

- Corrections are dependent on a reference antenna whose phase center performance may need to be determined by one of the other two absolute calibration methods;
- Estimation of PCV at low elevations is not possible due to increase in noise level and the presence of strong multipath;
- Observations are never completely free from multipath, which has impact on the results of the calibration;
- The satellite constellation at the location of calibration might not evenly cover the entire upper hemisphere of the antenna;
- Relative calibrations are insufficient for baselines long enough for the curvature of the earth to introduce significant differences in elevation between the two antennas, which can be encountered in some global networks.

6.6.3 Absolute Robot-Based Antenna Calibration

An absolute robot-based field calibration of GNSS antennas has been developed by Geo^{++} of Garbsen, Germany, in cooperation with the Institut Fur Erdmessung Universitat Hannover [28]. Since this is an absolute measurement it does not need a reference antenna for calibration; more important, it is also able to eliminate errors from far-field and near-field multipath errors that can corrupt measurements obtained in the other two methods, although multipath can be small in anechoic chamber measurements. Precise calibration is achieved by using a fast-moving robot that tilts and rotates the AUT. Note that all the GNSS antennas being tested by this method are RHCP, which means that their phase varies as a function of the azimuth angle and increases by 2π radians for each 360-degree rotation in the azimuth angle [11]. However Geo++ applies phase windup corrections for every phase observation so that the impact of changes in the orientation of the antenna are properly accounted for. Hence even a slight azimuthal rotation of the antenna by the robot will results in an appreciable change in the phase. Far- field multipath effects have been limited further by applying a high elevation mask of 18° that is dynamically adopted to the tilted orientations. A picture of a typical robot used in this calibration is shown in Figure 6.23(a). The upper hemispheric sky coverage achieved by the rotating and tilting of the robot is shown in Figure 6.23(b).

These fast changes in antenna orientation are critical to this technique since it allows the elimination of multipath. Far-field multipath effects introduce slowly varying systematic errors with long periods. These errors can be eliminated by recognizing that these far-field multipath signals repeat at the site every mean sidereal day (i.e., every day the same systematics repeat themselves some minutes earlier with cross-correlation studies show a maximum of around 236 seconds). This fact is exploited in this technique to greatly reduce multipath errors. The undifferenced GPS observables of two consecutive days is subtracted— \keeping in mind the difference of 3 minutes 56 seconds between the mean solar and mean sidereal day. As a consequence all errors that repeat themselves after one sidereal day cancel out, in addition, unfortunately, to the required PCV information if the satellite information is received on both days with the same antenna orientation. This is avoided by tilting and rotating antenna on one of these two days. The PCV difference between the zero rotation position on the reference day and the rotated/tilted antenna

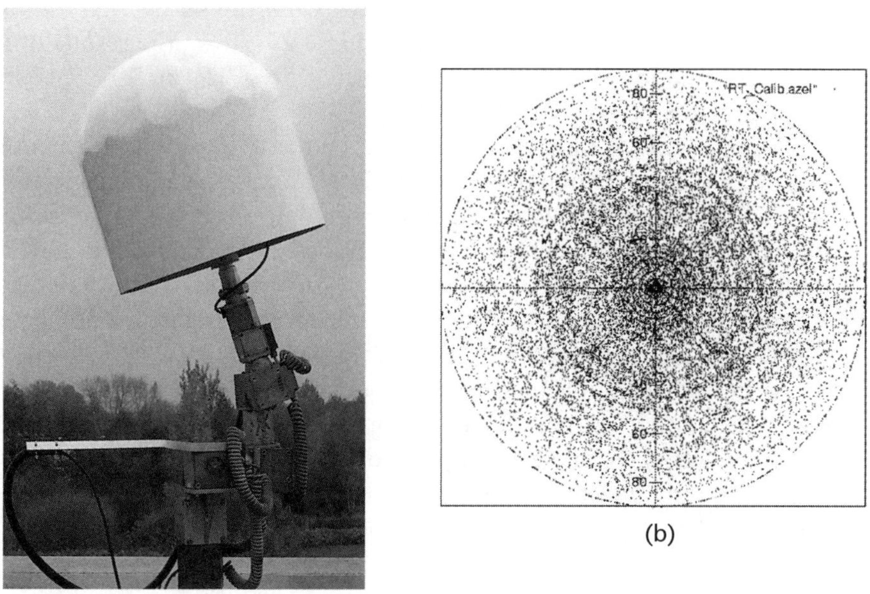

(a) (b)

Figure 6.23 (a) The robot developed by Geo+++ for automated absolute field calibration of PCV of GPS antennas. (b) Scatter plot coverage of the antenna hemisphere for calibration by the rotating and tilting robot (each point of this plot denotes a measurement at a different azimuth and elevation angle). (Courtesy of Dr. Martin Schmitz of Geo++ GmbH.)

orientation on the second day serves as the data that is used as an input for PCV determination using spherical harmonics. Rothacher [32] and Hoffman-Wellenhof [29, pp. 150–152] have proposed representing the PCV in the form of the following general spherical harmonic function:

$$\Delta\Phi(\theta,\varphi) = \sum_{n=0}^{\infty}\sum_{m=0}^{n}\left(A_{nm}\cos m\varphi + B_{nm}\sin m\varphi\right)P_{nm}(\cos\theta)$$

where A_{mn} and A_{mn} are coefficients to be determined and $P_{mn}(\cos\theta)$ are the Legendre functions where ϕ and θ are the azimuth and zenith angles, respectively, and $\theta = 90°$ minus elevation. If A_{mn} and B_{mn} are known then $\Delta(\theta, \phi)$ can be calculated for any arbitrary value of satellite arrival angle (θ, ϕ). If there are sufficient measurements of $\Delta\Phi(\theta, \phi)$ then the two spherical harmonic coefficients can be estimated from least squares minimization methods. Other errors due to the ionosphere, troposphere, and orbit biases cancel out by using a reference station that is in the near vicinity.

In addition to far-field multipath, near-field multipath errors can also be generated during these measurements, which are short-periodic systematic effects generated by diffraction and reflected signals from components in close vicinity to the antenna. These effects depend on the type of antennas and its design, the antenna mount and ground plane that is used, the presence or absence of a radome, and other environmental effects such as rain and snow. Recent in situ calibration techniques have also been developed by Wubbena et al. [33, 27] to eliminate these near-field multipath errors that involve difference calibrations requiring an optimal control of near-field effects and PCV. One key issue in these calibrations is the development of a near-field free station that is operated for the site analysis on a short baseline. This is achieved by using as an antenna mount for the near-field free station an optimized copy of the robot top and the setup used during individual absolute GNSS antenna calibration. Figure 6.24 shows the results of the absolute PCV calibrate ion measurements conducted on a TRPSCR.G3 geodetic antenna using an automated robot by Geo++.

More details of these techniques can be found in recent publications by Geo++ [27, 28, 33]. Information provided by Geo++ indicates that as of November 2008, a total of 154 different types of antennas including a total of 1,511 different individual antenna calibrations have been performed by the company using the absolute robot-based calibration techniques [27]. More information can be obtained from www.gnpcvdb.geopp.de; the PCV type mean computed from this type of calibration is available in their Geo++GNPCVDB database.

In November 2006 the International GNSS Service recommended the transition from relative to absolute calibration for antennas used in both receivers as well as in satellites [34]. For several applications and products, absolute corrections improved performance more than relative corrections because of separation of errors is enhanced and systematic errors can be removed by modeling.

6.6.4 Comparison of the Three Calibration Techniques for Determining PCO and PCV

One of the best comparisons of the relative accuracies of the three major antenna calibration methods described above was a study conducted in 2002 by several

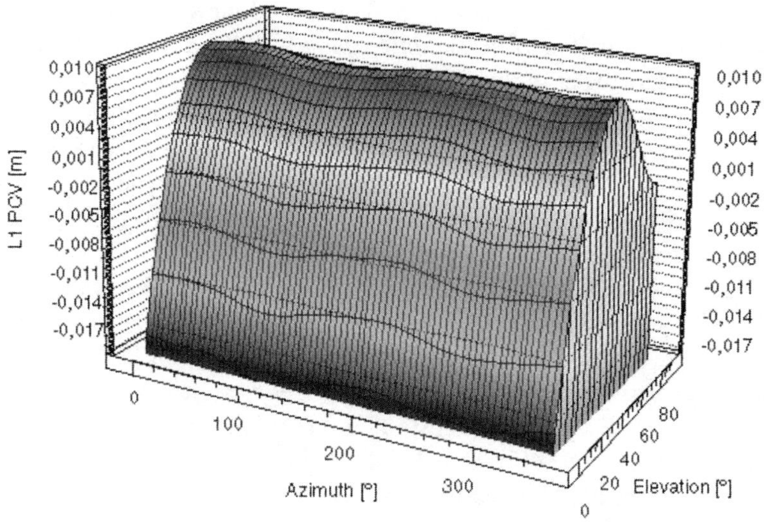

(a)

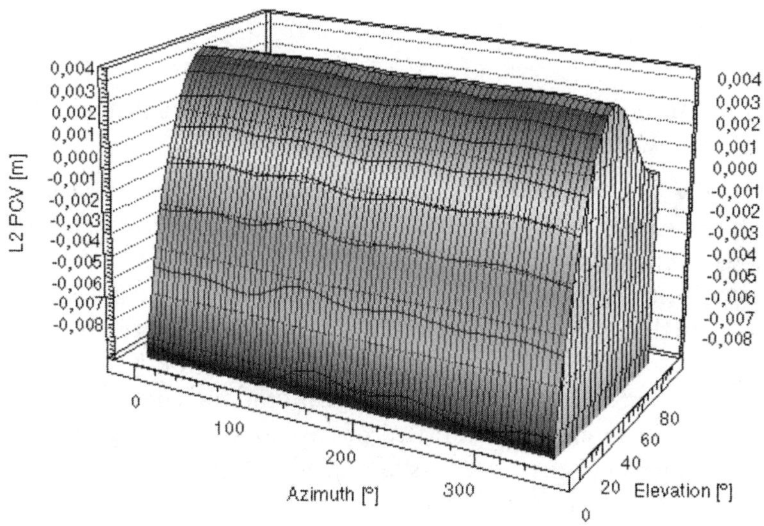

(b)

Figure 6.24 L1 and L2 PCV of the TPSCR.G3 antenna measured with the automated robot by Geo++. (Courtesy of Dr. M. Schmitz of Geo++ GmbH.)

institutions in Germany. This study is called the German Benchmark Test, and the results have been described in detail by Gorres et al. [5]. The antennas selected for this study were five different types of antennas made by three different companies. Three of these antennas were reference station antenna and two were rover antennas. These antennas were all calibrated at five different institutions in Germany listed in the paper by Gorres et al. [5]. Two of the institutions conducted absolute field measurement techniques with a robot developed by Geo++. Three other institutions used the relative mode calibration using a reference antenna. When the study was initially conducted in spring of 2002 there was no calibration testing done inside an anechoic chamber. This was done subsequently in 2008 by the University of Bonn. The official comparison of all the test results are compared in the paper by Gorres et al. [5]. The results of these studies showed that when the estimated parameters of identical antennas were measured using the two absolute calibration methods (i.e., the anechoic chamber and the Geo++ robot methods), they agreed to within 1 mm . for all elevations, whereas the relative field calibrations displayed variations of 2 mm at L_1 and 4 mm at $L_{2.}$ to the mean. The discrepancy appeared to be greater at the highest (near zenith) and lowest elevations (near the horizon)—which is to be expected because the smallest number of observations are at zenith and multipath and tropospheric effects are the strongest near the horizon. As expected, different antenna tested showed different PCV corrections. Additional details can be found in the paper by Gorres et al. [5].

A separate comparative study of relative and absolute antenna calibration method (using the robot) has been conducted by Mader [35]. This study has also indicated that absolute calibration using a robot is essential to correct for differences in elevation caused by the earth's curvature over a long baseline. The absolute measurements with the robot also indicated that it is able to correct for the effects introduced by the earth's curvature caused by tipping one of the antennas through a variety of angles, while the relative method produced increasing errors in height with increasing tipping angle.

6.6.5 Automated Absolute Field Calibration of the PCV and PCO of a Satellite Transmit Antenna Using a Robot

The PCV and PCO of the satellite transmit antenna can have an important impact on precise GPS, as explained in Chapter 1. In a cooperative effort between NGS in the United States and Boeing and Geo++of Garbsen, Germany, the calibration of the PCV of the GPS Block II/IIA satellite antenna was conducted in 2007 using the automated absolute field calibration technique using the robot that has been developed by Geo++. Figure 6.25(a) shows a picture of the GPS Block II/IIA antenna mounted on the automated robot for calibration. The array configuration used in this satellite antenna is comprised of two concentric rings of quadrifilar helix antennas. The inner circle consists of a quad of four elements and the outer ring is an octagonal array of eight elements. The desired antenna pattern is achieved by applying a 180° phase shift between the inner and outer concentric rings with a certain relative power ratio between the two rings. In the Block II/IIA satellite antenna, 90% of the of the L-band signal power is supplied to the inner four elements and only 10% to the outer elements; this is presumably to reduce the sidelobes of the array. The antenna array weighed 14.4 kg and is 1.34m in diameter. Modifications had to be made to the fast-moving automated robot to support the extra weight and size of this large satellite antenna that were well above smaller receiver antennas typically used in these calibration measurements. More details of the measurement procedure and layout can be found in Wuebbena and Schmitz [36].

The actual measurements of the antenna PCV were conducted for elevation angles of between 75° and 90° and conducted in the L_1, L_2, and L_0 bands. The L_0 frequency, in this case is an ionospheric-free linear combination of the carrier phases of L_1 and L_2 [37, 38], and is defined by the equation

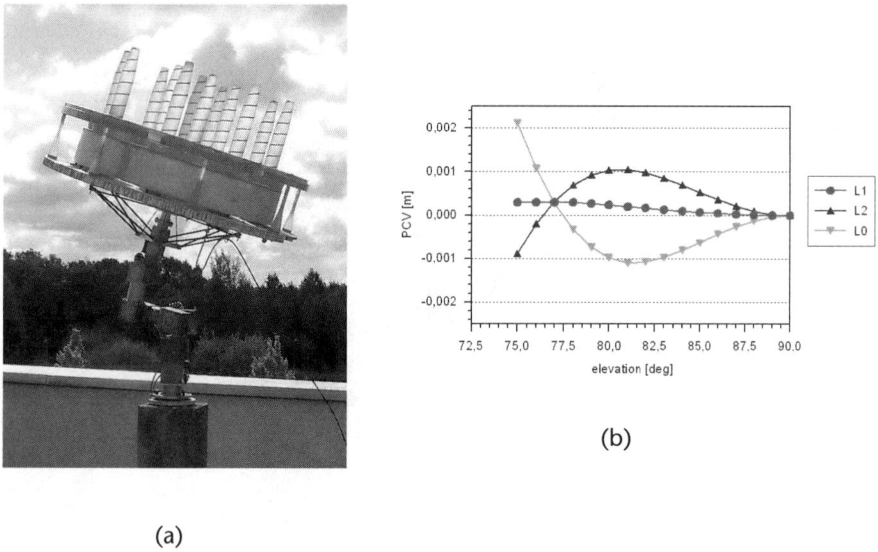

Figure 6.25 Block II/IIA satellite antenna (a) and PCV versus elevation at L_1, L_2, and L_0 measured using absolute robot based measurements (b). (Courtesy of Dr. M. Schmitz of Geo++ GmbH [27, 36].)

$$L_0 = 2.54728 * L_1 - 1.545728 * L_2$$

Only some sample results are given in this book and the interested reader should refer to the original paper [36] for more details. Figure 6.26 show the variations of the satellite antenna PCV at L_1 and L_2 as a function of azimuth and elevation angles. Significant azimuthal variation in the L_1 band was noticed, which is shown in Figure 6.26(a). The L_1 PCV ranged from −8 to +6 mm and also shows two significant maxima. The L_2 PCV shown in Figure 6.26(b) has even more maxima than at L_1 but the magnitudes are smaller with variation of between −4 to + 2 mm. The four maxima that are detected are believed to correspond to the four center elements of the antenna array. The offsets were also measured for both bands and are provided in the paper as are other details of many other PCV and PCO measurements of this satellite antenna.

6.7 Group Delay Variation in GNSS Antennas and Its Measurement

In (6.1) the total far-field radiation of a GNSS antenna was expressed in terms of the principal RHCP component and the cross-polarized, LHCP component.

The group delay of the GNSS antenna is defined in terms of the phase behavior of the RHCP component, which can be expressed in its complex form as:

$$\vec{E}_R = [\hat{e}_R] |E_R(\theta, \varphi, f)| \angle \Phi(\theta, \varphi, f)$$

where $\Phi(\theta, \phi, f)$ expressed in degrees represents the angular part of the phase pattern. The group delay is a measure of the time delay experienced by a narrowband signal packet [10, 11, 39] and is defined as

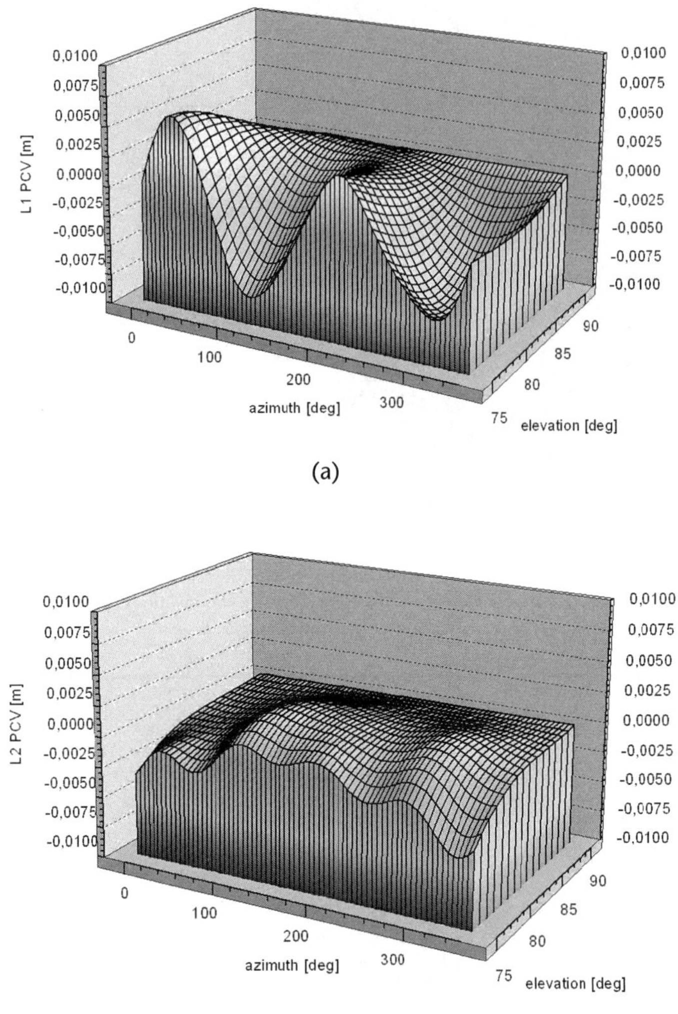

(b)

Figure 6.26 Block II/IIA satellite antenna PCV (m) in (a) L_1 and (b) L_2 (offset removed). (Courtesy of Dr. M. Schmitz of Geo++ GmbH.)

$$\delta\tau(\theta, \varphi, f) = \frac{1}{360}\frac{\partial\Phi(\theta,\varphi,f)}{\partial f} \quad \text{in seconds}$$

The group delay can also be expressed in terms of distance (meters) by multiplying it by the velocity of light *c*. If the phase of the antenna does not change with frequency its group delay is zero. But in reality the group delay does change with frequency and also with elevation and azimuth, which determines the aspect angle of the signal received by the antenna from the satellite.

There are two group delay parameters that are used in the evaluation of the performance of a GNSS antenna: the first deals with spectral variation of group delay (i.e., with frequency) within the operating bandwidth of a specific GNSS band, and the second deals with variation of group delay with the aspect angle (i.e., elevation and azimuth angles) of the arriving GNSS signal.

6.7.1 Requirements for Group Delay Variation with Frequency

The boresight differential group delay (BDGD) is defined individually for each of the GNSS frequency bands that were listed in Chapter 3.

$$\Delta T_B = \max\left\|\delta\tau\left(\theta_B,\varphi_B,f_i\right)-\delta\tau\left(\theta_B,\varphi_B,f_i\right)\right\|$$

where θ_B, ϕ_B denote the direction of boresight, f_i, and f_J are any two frequencies within the individual bandwidth of the GNSS bands on either side of the center frequency of the selected GNSS band. For example if the operating bandwidth of the GPS L_1 band and the center frequency is denoted by f_C, then f_i and f_J are any two frequencies such that:

$$f_{C-10.23MHz} \leq f_i, f_J \leq f_{C+10.23MHz}$$

For a GNSS antenna: $\theta_B = 0$; it can be averaged over several azimuth angles of θ at this fixed elevation angle if significant variation with an azimuth rotation is noticed.

The BDGD represents the combined effects of both the passive antenna as well as the built-in preamplifier and also includes other embedded filtering functions. The BDGD in any of the GNSS bands is expected to be less than 25 ns and applies equally to both passive as well as active antennas. For example, the new MOPS issued by RTCA requires that the group delay difference between any two frequencies within the operating bandwidth of ±10.23 MHz to be less than 25 ns [10, 39]. This new requirement is required to provide measurement consistency between satellite signals that may have different amplitude and phase characteristics. In the case of a simple passive antenna without any embedded additional filtering BDGD is expected to be less than 2 ns. The 25-ns specification is meant to be small relative to variations expected in the rest of the RF chain within a GNSS receiver where the maximum tolerable differential group delay variation for airborne equipment for example can be as large as 150 to 600 ns [39].

6.7.2 Group Delay Variation with Aspect Angle

The DGA versus angle specifies a requirement on the maximum variation in group delay as a function of the azimuth and elevation angle of the satellite signal arriving at the receiving GNSS antenna; it is specified as:

$$\Delta T_A = \max\left\|\delta\tau\left(\theta,\varphi_A,f_C\right)-\overline{\delta\tau\left(\theta=85^\circ,\varphi_A,f_C\right)}\right\|$$

where

f_C = center frequency of the GNSS band

ϕ_A = azimuth angle of the arriving satellite signal at the receiving antenna in the antenna reference frame

θ = 85° = Elevation Angle = 5° is the elevation angle of the arriving satellite signal at the antenna in the antenna reference frame. $\delta\tau(\theta = 85^\circ, \phi_A, f_c)$ or

$\delta\tau_{AVG}(\theta = 85°, \Phi_A, f_c)$ is the group delay averaged over all azimuth angle of the pattern at $\theta = 85° =$ Elevation Angle = 5°.

The requirements are that the DGA calculated for each GNSS band should not exceed the limit

$$\Delta T_A \leq 3 \text{ nanoseconds}$$

The requirement for maximum group delay set by the RTCA for the maximum DGA [10, 39]is :

$$\Delta T_A = 2.5 - 0.04625\ (\text{Elevation} - 5°) \text{ in nanoseconds for } 5° \leq \text{Elevation } 45°$$

and

$$\Delta T_A = 0.65 \text{ nanoseconds for Elevation} \geq 45°$$

6.7.3 Measurement Techniques for Calibrating Group Delay Variation with Frequency and Aspect Angle

The calibration of group delay variations can be conducted using different types of indoor antenna test ranges described earlier in this chapter; it is easier to measure variation of group delay with frequency since frequencies at small intervals that are needed for measuring the differential in phase can be easily generated using a stable frequency source. Group delay variation with aspect angle can be measured either in any of indoor test ranges or by using field measurements such as absolute robot-based outdoor field measurements. CATRs have been studied by Murphy [10] and also by Van Graas and Bartone [11] for calibrating the group delay of GPS antennas. The spherical NF-FF range has been used at the MITRE Corporation for similar type of measurements. The calibration method generally consists of measuring the amplitude and phase response of the antenna for the two orthogonal polarizations along the $\hat{\theta}$ and $\hat{\phi}$ at closely spaced frequency intervals in the GNSS band of interest. The complex RHCP pattern is then computed for each frequency measured. Since the derivative of the transfer function phase is to be measured and compared from the collected data, the frequency spacing that is selected should be small enough to give an accurate measure of the slope of the phase at each frequency. The linear slope between adjacent frequency pairs can then be determined through curve fitting or other interpolation methods to yield group delay.

Geo++ has also used field (i.e., outdoor) absolute robot-based antenna calibration techniques for measuring absolute GDVs versus elevation and azimuth angles of a large number of geodetic GNSS antennas [27].

The CATR calibration technique used by Murphy for measurements on two commercial GPS antenna in the GPS L_1 band will be described below. A picture of the CATR is shown in Figure 6.27(a).

The CATR shown in this figure has a quiet zone of 8' in the vertical plane and 12' in the horizontal plane [10]. At a frequency of 1.57542 GHz the axial ratio of

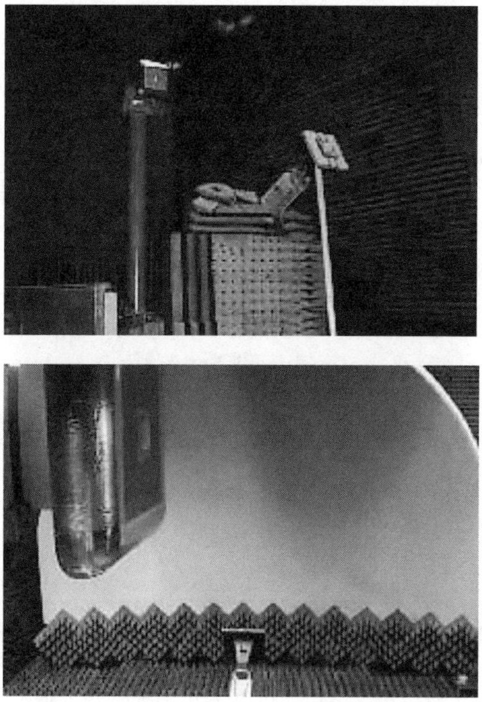

(a)

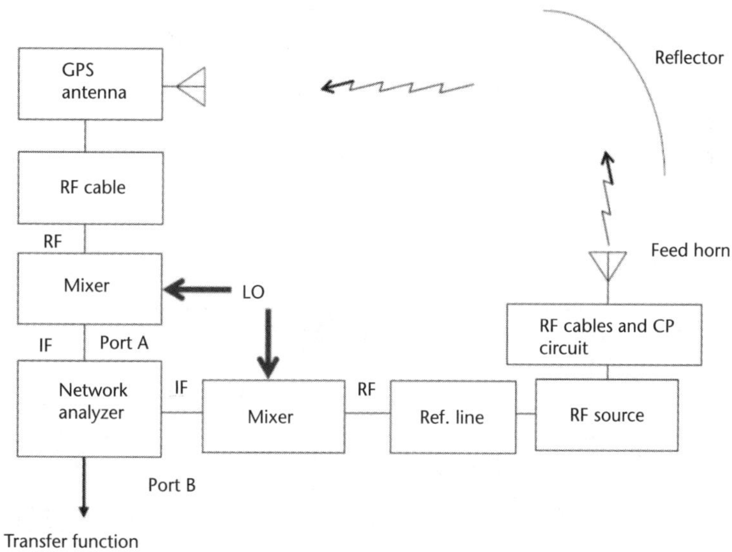

(b)

Figure 6.27 Group delay measurements of a GPS antenna in a CATR. (Courtesy of T. Murphy, Boeing Corporation [10].)

the illuminating field at the quiet zone was 0.09 dB; at the low/high edges of the frequency band the corresponding axial ratios were 0.35 and 0.30 dB, respectively. The phase center of the antenna was determined to within 0.2" of the rotation center of the positioner. As the antenna was scanned was being scanned to various elevation and azimuth angles the antenna center remained at the same position to within ±0.2".

A block diagram of the test setup used in these measurements is shown in Figure 6.27(b). The antenna being tested was mounted on a 2' square-shaped ground plane with rolled edges to minimize edge diffraction effects. The data was collected at every 0.5 MHz over a range of ±16 MHz in the GPS L_1 band. The elevation and

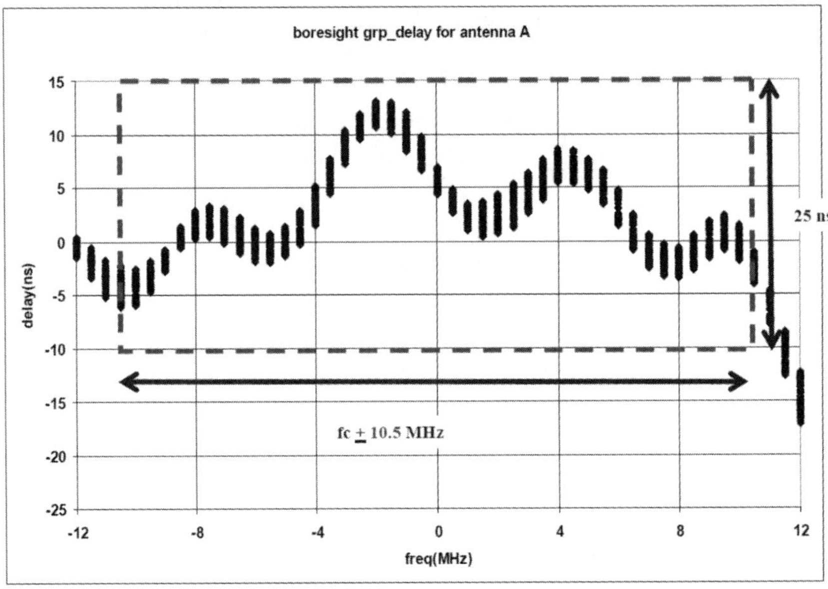

(a)

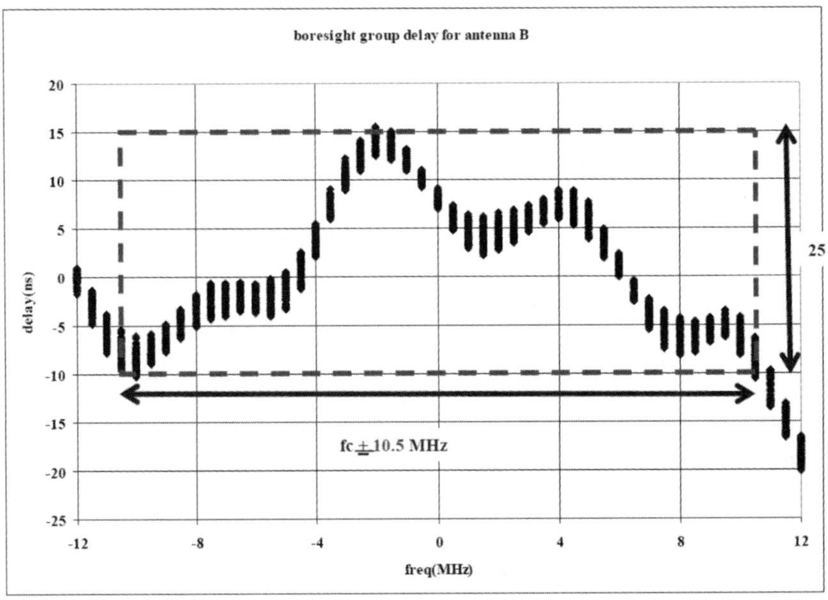

(b)

Figure 6.28 Measurement of group delay variation (GDV) with frequency for 2 commerical GPS antennas. (Courtesy of T. Murphy of Boeing Corp, [10].)

azimuth angles were sampled at every 5° from 0 to 90° in elevation and from 0 to 355 degrees in azimuth. Mixers are used to downconvert from RF to IF so that network analyzer measurements can be taken. A directional coupler is used to tap off a sample of the input feed signal so it can be compared with the signal received by the GPS antenna being tested and provides a reference signal for the transfer function. The net delay that is measured in the transfer function is due to the difference between the path length of from the RF source to port A on the network analyzer and that from the RF source to port B. The various cables and free-space lengths between the RF source driving the feed and the receive channel on the network analyzer introduced a delay of about 60 ns, which was removed to make the data reduction and display more convenient. The BDGD for two commercial GPS antennas were measured by Murphy [10] in the GPS L_1 band in a frequency range of ±10.23 MHz around a center frequency of 1.57542 GHz. These results, shown in Figure 6.28, indicate that both antennas tested were within the allowable specifications of 25 ns whose boundaries are indicated by the dashed lines.

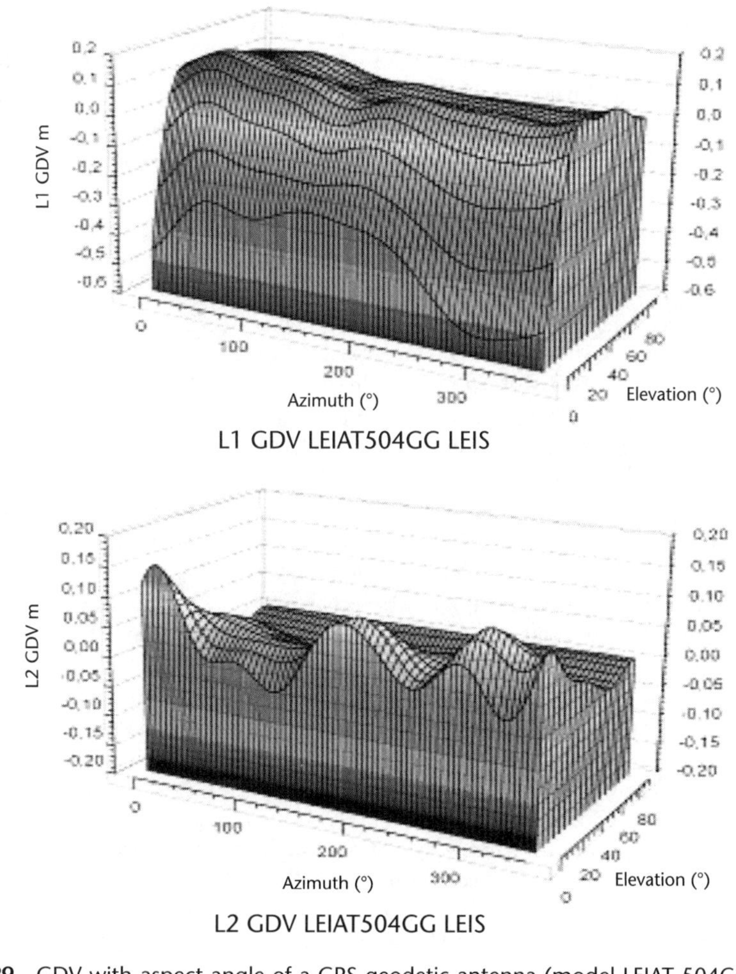

Figure 6.29 GDV with aspect angle of a GPS geodetic antenna (model LEIAT 504GG) measured using absolute robot-based calibration technique. (a) L1, and (b) L2. (Courtesy of Dr. Martin Schmitz of Geo++ GmbH.)

Results of similar measurements conducted in the GPS L_1 band for four typical GPS antennas—both active and passive—are also shown in Figure 3 of the paper by Van Dierendonck and Erlandson [39]; these results show that within a frequency range from 1565 to 1585 MHz the group delay variations were well with the 25-ns limit but varied significantly outside the limits of this frequency band. The BDGD of both passive and active commercial GPS antennas in the GPS L_1 band were also measured using the spherical near-field range by the MITRE Corporation. The measurements on both these antennas indicated results that were well within the maximum requirements of 25 ns.

Schmitz et al. of Geo++ [27] have measured the GDV as a function of the elevation angle of seven different geodetic antennas using the absolute robot-based calibration method developed at Geo++ described earlier. Measurements of GDV versus elevation angle being made on a LEIAT 504GG antenna using this absolute automated robot calibration method is shown in Figure 6.29.

These results show up to 1m variation in GDV can be seen at low elevation angles. These results are based on measurements made on a large number of antennas of each type, with samples varying from 3 to 14 for each type. Other measurements made by Geo++ on the variation of GDV versus aspect angle (both elevation and azimuth) for the LEIAT504GG LEIS antenna measured in the L_1 and L_2 bands are shown in Figure 6.29. Large GDVs are noticed at lower elevation angles, especially in the L_2 band. Other results of GDV with aspect angle can also be found in [27].

References

[1] Yaghjian, A. D., "An Overview of Near-Field Antenna Measurements," *IEEE Transactions on Antennas and Propagation,* Vol. AP-34, No. 1, January 1986, pp. 30–45.

[2] Becker, M., P. Zeimetz, and E. Schonemann, "Anechoic Chamber Calibrations of Phase Center Variations for New and Existing GNSS Signals and Potential Impacts on IGS Processing,"*IGS Workshop,* June 28–July 2, 2010, Newcastle Upon Tyne, England.

[3] Becker M., P. Zeimetz, and W. Schluter, et al., " Preparing for Galileo—New Results from Anechoic Chamber Absolute Antenna Calibration," abstract number G21A-0093, 2007 AGU Fall Meeting, December 11–14, 2007, San Francisco, CA.

[4] Zeimetz, P., and H. Kuhlmann, "On the Accuracy of Absolute GNSS Antenna Calibration and the Conception of a New Anechoic Chamber," Integrating Generations, FIG Working Week 2008, Stockholm, Sweden, June 14–19, 2008.

[5] Gorres, B., J. Campbell, and M. Becker, et al., "Absolute Calibration of GPS Antennas: Laboratory Results and Comparison with Field and Robot Techniques," *GNSS Solutions* Vol. 10, 2006, pp. 136–145, DOI 10.1007/s10291-005-0015-3.

[6] Schupler, B. R., R. L. Allhouse, and T. A. Clark, "Signal Characteristics of GPS User Antennas," *Navigation, Journal of the Institute of Navigation,* Vol. 41, No. 3, Fall 1994, pp. 277–295.

[7] Schupler, B. R., T. A. Clark, and R. L. Allhouse, "Characterization of GPS User Antennas: Reanalysis and New Results," from *GPS Trends in Precise Terrestrial, Airborne and Spacecraft Applications, Proceedings of the International Association of Geodesy Symposium,* No. 115, Springer-Verlag Publishers, 1996, pp. 328–332.

[8] Bartels, G.A., *GPS–Antenna Phase Center Measurements Performed in an Anechoic Chamber,* Report LR-791, Delft University of Technology, May 1995.

[9] Balanis, C. A., and C. R. Birtcher, "Antenna Measurements," in *Modern Antenna Handbook,* John Wiley & Sons, Inc., 2008, pp. 977–1033.

[10] Murphy, T., "GPS Antenna Group Delay Variations Induced Errors in a GNSS Based Precision Approach and Landing Systems," *Proceedings ION GNSS Conference 2007,* Vol. 3, September 2007, pp. 2974–2989.

[11] Van Graas, F., C. Bartone, and T. Arthur, "GPS Antenna Phase and Group Delay Corrections," *Proceedings ION NTM 2004,* January 26–28, 2004, San Diego, CA., pp. 399–408.

[12] Lee, T.-H., and W. D. Burnside,"Performance Trade-Off Between Serrated Edge and Blended Roll-Edge Compact Range Reflectors," *IEEE Transactions on Antennas and Propagation,* Vol. 44, No. 1, January 1996.

[13] Evans, G. E., *Antenna Measurement Techniques,* Norwood, MA: Artech House, 1990.

[14] Hansen, J. E. (ed.), *Spherical Near-Field Antenna Measurements,* London: Peter Peregrinus Ltd., 1998.

[15] Francis, M. H., and R. C. Whittman, " Near-Field Scanning Measurements: Theory and Practice," in *Modern Antenna Handbook,* C. A. Balanis (ed.), John Wiley & Sons, Inc., 2008, pp. 929–976.

[16] Slater, D., *Near Field Antenna Measurements*, Norwood, MA: Artech House, 1991.

[17] Gregson, G, J. McCormick, and C. Parini, *Principles of Planar-Near Field Measurements,* Stevenage, Herts, United Kingdom: The Institution of Engineering and Technology, 2007.

[18] Gillespie, E. S., "Measurement of Antenna Radiation Characteristics on Far-Field Ranges," in *Antenna Handbook—Theory, Applications and Design*, Y. T. Lo and S. W. Lee (eds.), New York: Van Nostrand Reinhold Company, Inc., 1988, pp. 32–39 to 32–63.

[19] Stutzman, W. L., *Polarization in Electromagnetic Systems,* Norwood, MA: Artech House, 1993.

[20] Brochu, C. J., G. A. Morin, and J. W. Moffat, *Gain Measurement of a Cavity-Backed Spiral Antenna from 4 to 18 GHz Using the Three Antenna Method,* Report Number 1337, Defence Research Establishment Ottawa, Canada, November 1998.

[21] Antenna Standards Committee of the IEEE Antenna and Propagations Group, *IEEE Standard Definitions of Terms For Antennas*, IEEE Std. 145–1973.

[22] Rizos, C., "GPS, GNSS and the Future," in *Manual of Geospatial Science and Technology, Second Edition,* J. Bossler (ed.), Boca Raton, FL: CRC Press, 2010, pp. 265–270.

[23] Seeber, G. "Real-Time Satellite Positioning on the Centimeter-Level in the 21st Century Using Permanent Reference Stations," paper given at the Nordic Geodetic Summer Scholl, Fevik, Norway, August 30, 2000, pp. 1–24.

[24] Mader, G., GPS Antenna Calibration at the National Geodetic Survey, National Geodetic Survey, NOS, NOAA, Silver Spring, MD, http://www.ngs.noaa.gov/ANTCAL/images/summary.pdf.

[25] Mader, G. L., and N. D. Weston, GPS Antenna Calibration at the National Geodetic Survey, slides presented at FIG Working Week 2008, Stockholm, Sweden, June 14–19, 2008.

[26] Boussaad, A., S. Rock, and G. Alain, "Calibrating Antenna Phase Centers—A Tale of Two Methods," *GPS World,* February 1, 2005.

[27] Schmitz, M., G. Wubbena, and M. Propp, "Absolute Robot-Based GNSS Antenna Calibration—Features and Findings," International Symposium on GNSS, Space-Based and Ground-Based Augmentation Systems and Applications, November 11–14, 2008, Berlin, Germany.

[28] Wubbena, G., and M. Schmitz, "Automated Absolute Field Calibration of GPS Antennas in Real-Time," *Proceedings ION GPS 2000,* September 19–22, 2000, Salt Lake City, UT.

[29] Hoffman-Wellenhof, B., H. Lichtenegger, and E. Wasle, *GNSS : Global Navigation Satellite Systems,* Wien, New York: Springer-Verlag, 2008, pp. 148–154.

[30] Misra, P., and P. Enge, *Global Positioning System: Signals, Measurements and Performance,* Lincoln , MA: Pratap Misra and Per Enge, Ganga-Jamuna Press, 2006, pp. 137–138.

[31] Alemu, G. T., Assessments on the Effects of Mixing Different Types of GPS Antennas and Receivers, master of science thesis in geodesy, No. 3106, TRIGAT–GIT- EX 08-009, Royal Institute of Technology (KTH), Stockholm, Sweden, September 2008.

[32] Rothacher, M., S. Schaer, and L. Mervart, et al., "Determination of Antenna Phase Center Variations Using GPS Data," *Proceedings of the IGS Workshop on Special Topics and New Directions,* Potsdam, Germany, May 15–18, Part 2, pp. 205–220.

[33] Wubbena, G., M. Schmitz, and N. Matzke, "On GNSS In-Situ Calibration of Near-Field Multipath," *International Symposium on GNSS, Space-Based Augmentation Systems and Applications,* November 29–30, 2010, Brussells, Belgium.

[34] Chatzinikos, M., A. Fotiou, and C. Pikiridas, "The effects of the Receiver and Satellite Antenna Phase Center Models on Local and Regional GPS Networks," *Proceedings International Symposium on Modern Technologies, Education and Professional Practice in Geodesy and Related Fields,* Sofia, Bulgaria, November 5–6, 2009.

[35] Mader, G. L., "A Comparison of Absolute and Relative GPS Antenna Calibrations," *GPS Solutions,* Vol. 4, No. 4, 2001, pp. 37–40.

[36] Wubbena, G., M. Schmitz, and G. Mader, et al., "GPS Block II/IIA Satellite Antenna Testing Using the Automated Absolute Field Calibration With Robot," *Proceedings ION GNSS 20th International Technical Meeting of the Satellite Division,* Fort Worth, TX, September 25–28, 2007.

[37] Schmitz, M., private communications to the author, May 2011.

[38] Seeber, G., *Satellite Geodesy, Second Edition,* Berlin: Walter de Gruyter GmbH & Co. KG, 2003, pp. 262–265.

[39] Van Dierendonck, A. J., and R. J. Erlandson, " RTCA Airborne GPS Antenna Testing and Analysis for a New Antenna Minimum Operational Performance Standards (MOPS)," *Proceedings ION NTM 2007,* Vol. 2, January 2007, pp. 692–701.

[40] Arai, H., *Measurement of Mobile Antenna Systems,* Norwood, MA: Artech House 2001, pp. 43–55.

CHAPTER 7

Antennas and Site Considerations for Precise Applications

Waldemar Kunysz

7.1 Introduction

This chapter describes how antennas can influence the performance of the overall ranging system. Antennas are not perfect devices and hence introduce various errors to overall system. Two types of errors affect the typical ranging system. The first type of error affects the received direct signal and the second type is associated with delayed in time multipath generated replicas of the original signal.

Spatial group delay variation and electrical phase-center displacement of the antenna are key parameters that effect reception of direct signal and hence accuracy of the overall system. Spatial group delay describes group delay variation of antenna phase pattern with changing azimuth and elevation angles. Ideally, that variation is constant for all angles in order to introduce the same amount of signal delay no matter what the direction of arrival of the ranging signal. Any variation in the group delay will be directly translated as a timing error and hence a ranging error. The signal processing unit of the receiver has no idea what the group delay values are; making it impossible for the receiver to resolve these variations. The known method of minimizing these variations is to design a wideband antenna with uniform radiation pattern in azimuth and elevation angle whose phase changes linearly with frequency. Satisfying such a stringent requirement will ensure that the electrical phase center does not move with changing frequency.

The electrical phase center location is an important property of a GPS antenna. Its location offset from a known mechanical center is often clearly marked and documented in most high-end precision antennas. There are specific measurement sites[1] devoted to calibration of high-end antennas that publish PCO results on their Web sites for general access.

Multipath is the dominant error source in ranging system solutions that are supposed to provide a high accuracy (mm range). GPS code-based ranging system

1. GEO++ website: http://anton.geopp.de/gnpcvdb/pcvdb/GNPCVDE; NOAA website: http://www.ngs.noaa.gov/ANTCAL.

provides limited ability for signal processing to resolve medium- and large-range multipath (several meters). It does not, however, provide sufficient bandwidth to resolve close-in (delayed by few meters) multipath. This type of multipath, in most cases, is generated near the antenna location. Careful antenna design and sitting consideration for antenna can reduce near-multipath effects significantly and hence improve overall system accuracy.

Four parameters allow antenna to "combat" multipath effect: "purity" of antenna polarization, cross-polarization level, FBR, and shape of amplitude pattern. The signal transmitted from the satellites is RHCP. Any reflection causes the GPS signal to change its polarization from right-hand to left-hand and vice versa. Antenna reception pattern should ideally be only RHCP, since most of multipath signal originates from ground reflections. To reduce reception of reflected multipath signals below antenna horizon (negative elevation angle), the so-called front-to-back and up-down ratios must be maximized. Front-to-back is the ratio of maximum values of amplitude pattern directivity at antenna boresight (elevation angle of 90°) to maximum directivity value of amplitude pattern at its opposite end (elevation angle of –90°). Note that the up value is associated with RHCP pattern, while the down value might be associated with RHCP or LHCP pattern (whichever has greater value at elevation angle of –90°). The up-down ratio is similar to the FBR concept, except the up value is measured at some positive elevation angle α while the down value is measured at its mirror image, which is the negative angle of α.

The down pattern of a typical antenna has usually stronger LHCP than RHCP amplitude, which does not help with rejection of LHCP-oriented multipath replicas of the original signal. Caution should be exercised when computing the up-down ratio or FBR. Normally, RHCP values will be used for the up and front values but in the case of down and back, the strongest value should be used (either RHCP or LHCP). This will be specific antenna-dependent.

The sitting consideration of an antenna is very important since it can affect its electromagnetic properties. For example, electromagnetic coupling to metallic mounting structure can alter antenna properties such as the back ratio or an additional antenna radome housing introduced during installation may introduce an additional group delay variation. This is discussed further in the following sections.

7.2 Antenna and Site Dependence

As mentioned in the introduction of this chapter, the sitting of an antenna plays an important role. A typical installation of a high-end precision GPS antenna is shown in Figure 7.1.

An antenna is usually mounted on a metallic or concrete monument or post and is often covered with an additional radome to protect it from the outside environment (rain, snow, birds, etc.). The antenna is elevated to height h with respect to its surroundings (ground, roof) in order to get an unobstructed view of the horizon. In addition, there is quite often an additional metal plate located underneath the antenna at distance d. There are few problems with such an antenna sitting arrangement as shown in Figure 7.1:

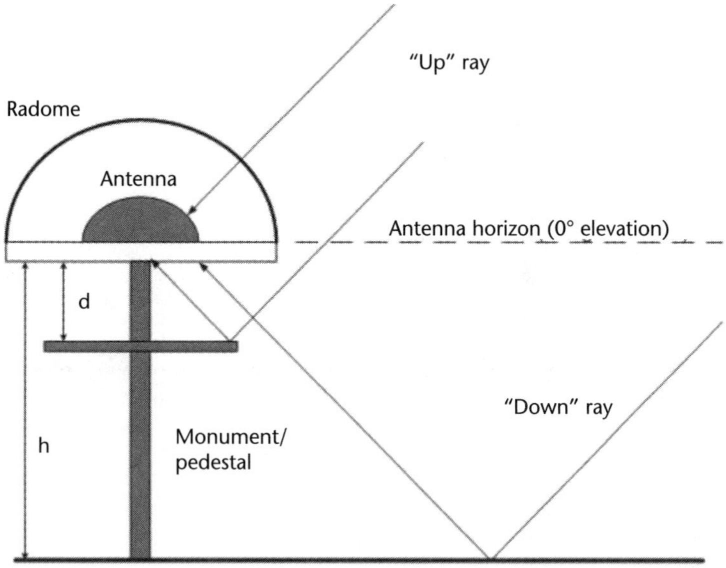

Figure 7.1 Typical antenna installation.

1. Increased susceptibility to multipath originating near the antenna. The multipath level increases linearly with height h.
2. Increased back and down directivity of antenna pattern due to a reflector placed below at distance d.
3. Altered antenna amplitude and phase pattern due to electromagnetic coupling between the antenna and the metal structure below.
4. Radome and size of ground plane can also significantly alter antenna amplitude and phase pattern.

7.2.1 Multipath Effects

All errors except multipath and noise can be reduced using techniques such as single-differencing, double-differencing, and DGPS corrections [1]. Multipath is the error caused by the reflected GPS signals entering the receiver front end and mixing with the direct signal [2]. Its effect will be more pronounced for static receivers close to large reflectors. It is specific to a receiver/antenna and depends on the surrounding environment. Hence, care has to be taken while installing GPS receivers for static applications,such as reference stations.

A typical single-ray multipath scenario is shown in Figure 7.2. The path length difference between the direct and reflected rays is $\Delta d = d_1 \text{-} d_2$.

$$\Delta d = d_1 - d_2 = d_1 - d_1 \cos 2\theta = d_1 (1 - \cos 2\theta) \tag{7.1}$$

$$d_3 = 2h \cdot \cos\theta \text{ and } d_3 = d_1 \cdot \sin 2\theta \Rightarrow d_1 = 2h \cdot \frac{\cos\theta}{\sin 2\theta} \tag{7.2}$$

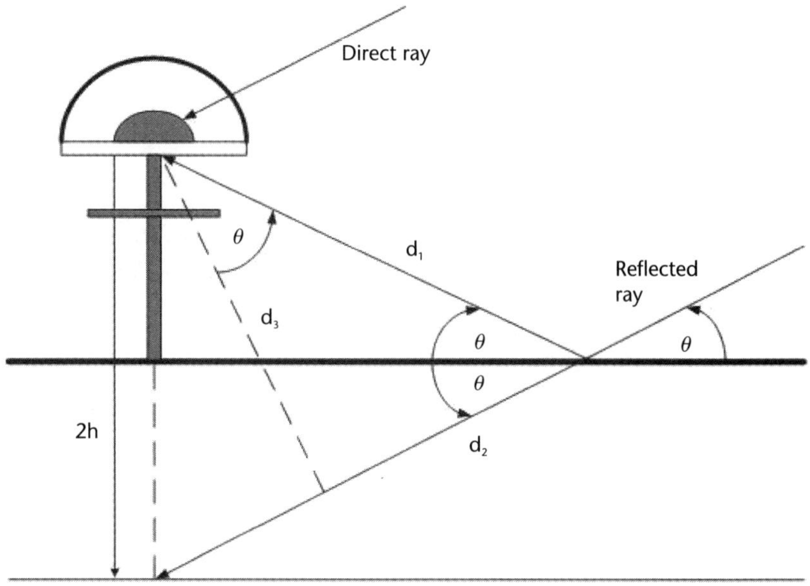

Figure 7.2 2-D model of typical antenna multipath scenario.

Substituting (7.2) into (7.1) yields:

$$\begin{aligned} \Delta d &= 2h \cdot \frac{\cos\theta}{\sin 2\theta}(1-\cos 2\theta) \\ \Delta d &= 2h \cdot \sin\theta \end{aligned} \tag{7.3}$$

We know that an electromagnetic wave exhibits a phase change of 2π for every one wavelength of traveled distance. We can convert equation (7.3) from meters to radians by the following translation:

$$\Delta\psi = 2h \cdot \sin\theta \cdot \frac{2\pi}{\lambda} = 4\pi hf \sin\theta / \lambda \tag{7.4}$$

Equation (7.4) gives us phase difference between direct and reflect signal. The received signal is the superposition of the direct and reflected signals. The phase of the received signal is perturbed by an amount given by [3]:

$$\sin\phi = \frac{\sin\psi}{\sqrt{G(\theta)^2 + 2G(\theta)\cos\psi + 1}} \tag{7.5}$$

where $G(\theta)$ is the up/down linear ratio of amplitude pattern value at positive angle θ to amplitude pattern value at negative angle θ. It is assumed that the amplitude pattern is relatively uniform in azimuth plane with little or no variation, which is the case for a high-precision geodetic antenna. Following Counselman [3] derivations and assuming that G>>1 and ϕ has small values, (7.5) can be simplified to

$$\phi = \frac{\sin \psi}{G} \tag{7.6}$$

Since ψ is function of frequency (f), we can now relate the group delay variation of antenna phase pattern to elevation angle θ and the up/down ratio G of the antenna amplitude pattern.

$$\begin{aligned} \frac{d\phi}{d\omega} &= \frac{d\phi}{2\pi df} = \frac{d(4\pi hf \sin\theta/c)}{df} \cdot \frac{1}{G(\theta)} \\ cd\tau &= \frac{2h\sin\theta}{G(\theta)} \Rightarrow \quad G(\theta) = \frac{2h\sin\theta}{cd\tau} \end{aligned} \tag{7.7}$$

Equation (7.7) can be now converted into logarithm form. Two sets of typical values for height (h) and group delay variation expressed in meters ($cd\tau$) are used to express (7.7) in logarithmic form:

$$cd\tau = 0.05m \text{ and } h = 1.5m \Rightarrow G(\theta) \geq 35.6dB + 20\log_{10}(\sin(\theta)) \tag{7.8}$$

and

$$cd\tau = 0.10m \text{ and } h = 10.0m \Rightarrow G(\theta) \geq 40dB + 20\log_{10}(\sin(\theta)) \tag{7.9}$$

Equations (7.8) and (7.9) are plotted in Figure 7.3. No planar GPS antenna can meet the plotted requirement. The only possible way to meet this requirement is to have a long or medium vertical array that will allow shaping the amplitude pattern that conforms to the up-down curve limit. More on this topic is discussed in Section 7.4. The third curve was computed using (7.7) with one substitution: $cd\tau$ is replaced with group delay limits (10) specified for WAAS network GPS antennas.

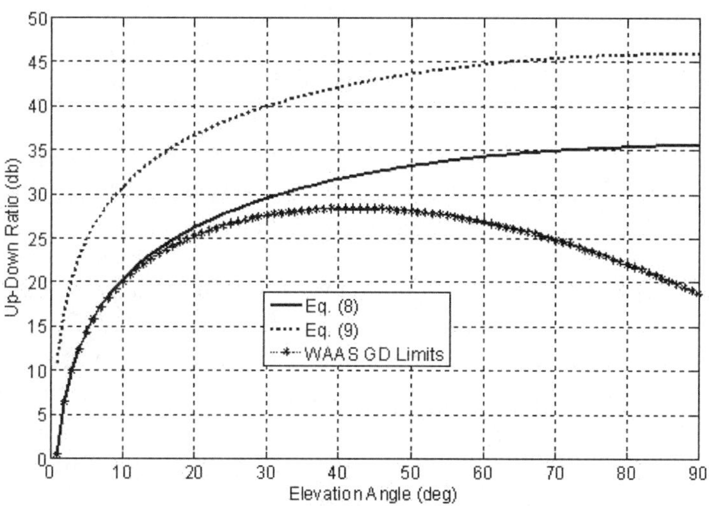

Figure 7.3 Up-down ratio limits as defined by (7.8) and (7.9).

$$Delay(m) = 0.045 + 0.3\, e^{-Elevation(\text{deg})/21} \tag{7.10}$$

where 5° ≤ Elevation ≤ 90°

The WAAS network that overlays North America requires antennas to conform to specific group delay requirements (referred to as WAAS GD limits) in order for the network to be able accurately model ionospheric corrections. The ionosphere introduces group delay variations to any signal piercing through Earth's atmosphere (including GPS). This variation is a function of frequency and causes GPS signal to be either advanced or delayed in time. Such variation causes timing and ranging errors. For high accuracy, ionospheric modeling it is very important for the antenna to have minimal spatial group delay variations. It is interesting to observe that WAAS group delay curve follows (7.8) limits for low elevation angle and then diverges (relaxes) at higher elevation angles to take advantage of the fact that group delay variations are smaller in this region. This may be easily explained by the fact that the signal-to-noise ratio (the C/N_0) of received GPS signal at higher elevation angles is much higher (in order of 10–15 dB) than at low elevation angles, and therefore the multipath causes smaller phase and code variations.

7.2.2 Monument Effects

Let us examine again the typical mounting installation shown in Figure 7.1. The monument can cause undesired effects to the antenna due to various phenomena such as:

- An electromagnetic coupling between metallic structures of the monument and antenna;
- Specular reflection from smooth surfaces in the first Fresnel zone;
- Diffuse reflections;
- Diffraction effects.

Antennas are devices that are very sensitive, by design nature, not just to the received signal but also to their surroundings. The behavior of the antenna can be easily altered if proper care is not taken during the design and commission of permanent antenna installation. An example of such unpredicted modification of an antenna pattern is shown in Figure 7.4.

In this case, the metallic plate is located a quarter-wavelength underneath the GPS antenna ground plane. It can be shown that RF rays in this configuration will exhibit multiple reflections between the plate and reflector (air travel –90° phase change, and each reflection –180° phase reversal) forming in-phase rays outside the plate. This will cause the antenna amplitude pattern to have higher directivity in the direction below the antenna—right towards the incoming multipath signal! This phenomenon is well known in antenna engineering to boost the gain of Yagi-Uda array antennas. In our case, we end up boosting undesired bottom gain of the antenna.

Specular reflection originates from smooth surfaces that are large in comparison to the first Fresnel zone. The area of the first Fresnel zone is approximately given by

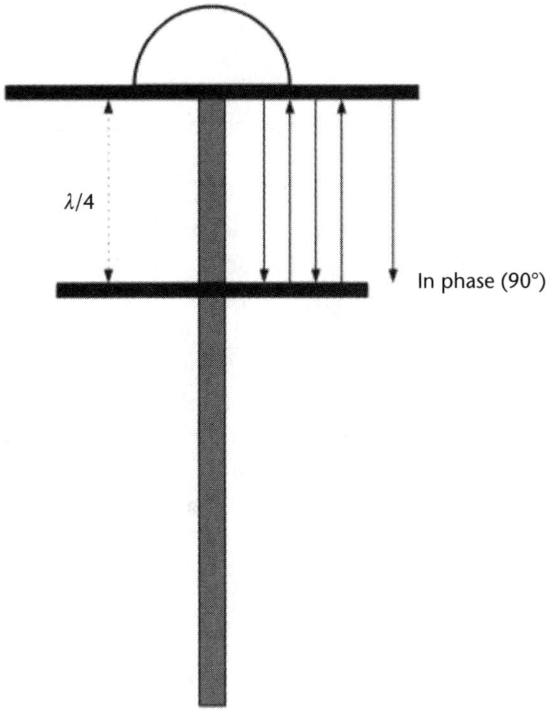

Figure 7.4 Electromagnetic effects of reinforced backward radiation.

$$A = \frac{\pi \cdot \lambda \cdot d}{\sin(\alpha)} \tag{7.11}$$

where

d = distance to the specular point on the reflecting surface
α = the grazing angle

Specular reflection causes the signal to undergo a polarization shift if the grazing angle exceeds the Brewster angle that is dependent on the reflective surface. For most metallic surfaces, the Brewster angle is less than 1 degree [4]. The area of the Fresnel zones varies between 1.5 to 35m. With a maximum grazing angle of about 1 degree, most reflections will be less than the Brewster angle. For these two reasons, it is unlikely that specular multipath is an issue in most typical installations.

Diffuse reflections occur when a reflective surface is not smooth. One can use Rayleigh criterion to distinguish smooth from rough surfaces. The difference in path length $\Delta\lambda$ may be written as function of the variation in the local surface height and the angle of incidence on the local surface area β

$$\Delta l = 2h\sin\beta \tag{7.12}$$

A microscopic smoothness criterion of $\Delta\lambda < \lambda/8$ yields a height difference limit of

$$\Delta h < \lambda/(16\sin\beta) \tag{7.13}$$

In most cases, the surface will be smooth when compared to GPS frequency wavelength, hence the diffuse reflections should not be a problem for a typical installation.

Diffraction occurs when an RF signal encounters a metallic edge and "bends" around it. Diffracted rays are produced when a ray strikes an edge, a vertex, or is incident tangentially to a curve surface such as sphere or cylinder. A portion of energy is trapped, resulting in a wave that propagates on along the surface and is known as a creeping wave. The effect of creeping waves is minimized if the ground plane edges are rolled down in a smooth fashion or the surface impedance of the ground plane is significantly increased near its edges. It can be demonstrated, either numerically or experimentally, that a circular smooth ground plane that has a radius of 1.2m will cause a deformation of an antenna gain pattern. This deformation tends to cause a randomization of signals approaching the antenna at a shallow angle.

7.2.3 Ground Plane Effects

A GPS antenna ground plane can help to reduce the effects of multipath from reflections off objects beneath the antenna. On land, this may be from the ground or the roof of a building, or reflections from the airplane or ship's various metal structures. The ground plane does not prevent multipath from objects located above the antenna horizon, such as from the side of a building or tall tower; fortunately these multipath signals are usually generated quite far (tens of meters or more) and hence are easily removed by the receiver signal processing. Therefore, we are interested to mitigate close-in multipath generated near antenna that cannot be removed by software (limitation caused by finite bandwidth of GPS signal).

The electrical size of the ground plane largely determines the interference pattern of the edge diffracted/scattering field and the field radiated from the antenna. The metal ground plane will radiate for example surface waves vertically if scattered by bends, discontinuities (i.e., edge or a corner of the antenna ground plane), or surface texture. Diffraction and scattering from the edges of finite ground planes have a significant impact on the far-field radiation from any antenna. These edge-generated fields have contribution from the following sources: leaky waves, surface waves, and space waves.

Leaky waves are substrate modes that radiate as they propagate along their traveled path and hence contribute to the overall far-field pattern. Because they radiate energy as they propagate, their contribution to the edge scattered field decrease as the ground plane size increases. Lossy substrates will decrease their edge scattered field contribution, but at the cost of energy transmitted along the "leakage" path. Again, proper edge shaping and or termination will help to reduce effect of leaky waves at the ground plane edges.

In many cases, the contribution from surface waves are the dominate component of the edge diffracted/scattered fields. Surface waves are modes in the substrate that do not radiate as they propagate, but at the ground plane edges they scatter and contribute to the far field. Thicker and/or higher dielectric constant substrates trap more of the energy into these modes. The negative effects are more pronounced for these geometries. Many antenna radiation structures suffer performance degradation due to surface wave excitation. These effects can be minimized if the ground plane edges are rolled down in a smooth fashion or the surface impedance of the ground plane is significantly increased near its edges. Another method for mitigating surface waves for etched structures is to surround the antenna structure with a 1-D or 2-D (PBG) [5, 6] embedded in the substrate with the same height as the substrate.

The space waves that diffract from the edges are the fields radiated by the antenna itself directly into free space that are incident upon the ground plane edge. Although this contribution to the diffracted field can never be eliminated, space waves decay faster with distance from the antenna than the surface waves and thus, their contribution is proportionally less for increased ground-plane size.

A flat metal sheet is used in many antennas as a reflector or ground plane [7]. The presence of a ground plane redirects one-half of the radiation into the opposite direction, improving the antenna gain and reducing the backlobe radiation. The flat metal ground "good" electrical properties are also its main weakness. If multipath signals reflected from the ground strike the underside of the ground plane at certain angles, the ground plane electrical properties can actually conduct multipath to the antenna through diffraction or reflection mechanism. In addition, if the antenna is an electric type and is placed too close to its conductive ground plane surface, the image currents will cancel the currents in the antenna, resulting in poor radiation efficiency. The remedy for this is to use magnetic antenna instead (i.e., the Pinwheel™ antenna from NovAtel).

The concept of undesired modes and waves on metals is not new. The behavior of surface waves on a general impedance surface is described in several books [8–10]. A corrugated metal slab [11, 12], commonly known as a choke-ring ground plane, or a metal sheet covered with small bumps are examples of various geometries used to suppress surface waves on metal structures. The choke-ring design is very effective for single-frequency applications. In case of dual-frequency operation, a compromise must be made. One interesting approach is a 3-D choke ring introduced by NovAtel, where the inner wall surface for each ring is a quarter-wavelength at higher frequency (L1) and the outer wall surface is a quarter-wavelength at lower frequency (L2).

Periodic dielectric, metallic, or metal-dielectric structures that prevent the propagation of electromagnetic waves are known as photonic crystals or bandgap structures [13, 14]. Antennas that implement planar high-impedance ground planes (i.e., the Zephyr antenna from Trimble), have been demonstrated to have smoother radiation pattern and with less energy radiated in the backward direction.

Choke-ring ground planes are a circular-shaped ground plane with quarter-wavelength slots that are shorted at the bottom and open at the top. This translates to a very high impedance ground plane that does not support image currents generated within the ground plane that normally would interfere with the currents generated within the patch antenna itself. This feature translates to very low sidelobes underneath the antenna horizon and very smooth amplitude and phase patterns generated by the antenna. In addition, very good axial ratio values (less than 3 dB) above 10° elevation angle are achieved. A typical field distribution around a choke-ring ground plane is shown in Figure 7.5 and for flat ground plane in

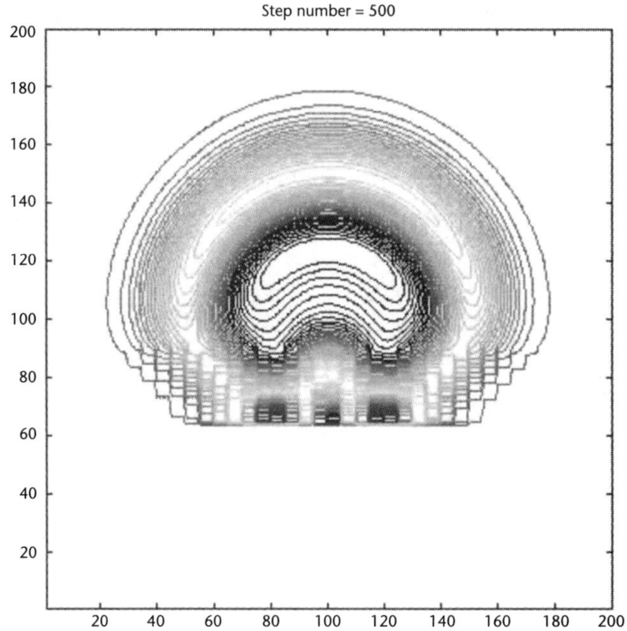

Figure 7.5 FDTD simulation of near-field (cross section) distribution around the choke ring ground plane.

Figure 7.6. Note the lack of contour lines below choke-ring ground plane indicates a very weak scattered field below its structure, which is an indication of good front/back or up/down ratios.

The concentric nature of field distribution above choke ring ground plane is an indication of vary stable phase center. The scattered fields plot shown in Figures 7.5

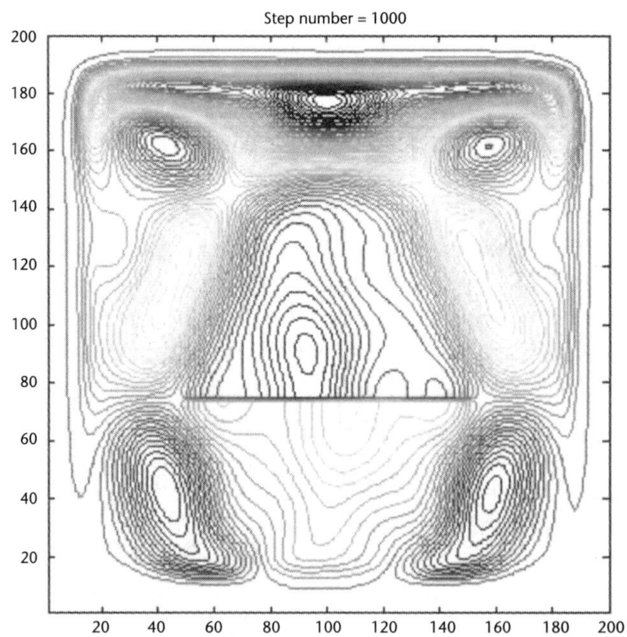

Figure 7.6 Field distribution around a flat ground plane.

and 7.6 were computed using the finite-difference-time-domain (FDTD) method with Gaussian pulse excited at the vertical distance of one-tenth of wavelength above the ground plane structure.

The large size of the ground plane translates to sharper amplitude roll-off (from zenith to the horizon) and increased main beam directivity of the antenna. Typical roll-off is in order of 10 to 12 dBi from zenith (90° elevation angle) to horizon (0° elevation angle) as compared to 3 to 5 dB roll-off of the patch antenna itself. Typical increase in the main beam directivity is in order of 1 to 2 dBi as compared to a stand-alone patch antenna without the choke-ring ground plane. This performance enhancement comes at the cost of the size and weight of the additional ground plane. Typical diameter size of the choke ring is in the order of 14" to 16" and a weight of 10 to20 lb.

7.2.4 Radome Effects

The radome serves mainly as environmental protection of the antenna element. It has to withstand harsh environment conditions (wind, hail, extreme temperatures, sun rays loading, petrochemical resistance, etc.) often for extended period of time (10–15 years for reference stations). It must prevent accumulation of precipitation and adherence of ice and water on antenna element itself. The radome surface must be therefore hydrophobic. It also must prevent living creatures such as birds and small animals from sitting on or occupying it. Its RF performance is quite often neglected and radome can either cause too much attenuation of incident RF signal or introduce a lenses effect by significantly changing the path that the RF signal takes from its surface to the antenna element. This causes GDVs in both azimuth and elevation angle and hence large PCV. Radome-induced errors of 15 mm or more have been reported in the literature [15]. The radome might also cause signal depolarization and degradation of antenna key parameters such as axial ratio.

Two design approaches are recommended for radome design. The first method is to have a very thin structure that is hemispherical in shape. Hemispherical shape will ensure that any signal distortion introduced by the radome is uniform for all angles and hence acts like a bias that is easily removed in the receiver software. The second method is to design a shaped radome that is filled with low-loss material. The dielectric constant can be altered as a function of radius, therefore yielding a desired outline shape of the radome. A similar approach is used to design Luneburg lenses for antennas.

7.3 Measurements and Ionospheric Modeling

Three different types of positioning information can be extracted from a GPS satellite signal: a pseudorange measurement, a carrier phase measurement, and Doppler.

A pseudorange is a range measurement between the GPS satellite and the user. This range measurement has inherent errors, which is different from the true range value [16]. The pseudorange is a measure of the time delay required to align the GPS signal received from the satellite with the local GPS signal generated by the receiver. This time delay is converted into a distance measurement using the speed of light. The receiver clock and satellite clock are not synchronized, which introduces an error in the range. Hence the measured range is different from the true range and is called a pseudorange [17]. The pseudorange is instantly available to compute position information and is given by (7.14) [18].

$$r(t) = \rho(t) + d_{orb} + cdT(t) + d_{trop} + d_{iono}(t) + \varepsilon_p \tag{7.14}$$

where

$r(t)$ is the pseudorange measurement at time t(m)
$\rho(t)$ is the true distance between satellite and receiver (m)
d_{orb} is the orbital error (m)
c is the speed of light (m/s)
$dT(t)$ is the system (receiver+satellite) clock error (s)
$d_{trop}(t)$ is the tropospheric error (m),
$d_{iono}(t)$ is the ionospheric error (m), and
ε_p is the code multipath and measurement noise (m).

A carrier phase measurement is a range measurement computed using the GPS carrier signal information. The total number of the carrier cycles from the GPS satellites to the user are measured and converted into a range measurement using the carrier wavelength [16]. The receiver cannot determine the number of integer cycles before the signal is acquired. This is referred to as the integer cycle ambiguity. This ambiguity must be resolved before the carrier phase measurement can be used for position computation. It can be represented by (7.15) [18].

$$\tau(t) = -\lambda\theta(t) = \rho(t) + d_{orb} + cdT(t) + d_{trop}(t) - d_{iono}(t) + \lambda N + \varepsilon_\tau \tag{7.15}$$

where

$\tau(t)$ is the carrier phase measurement at time t (m)
$\theta(t)$ is the carrier phase measurement (cycles)
λ is the carrier wavelength (m/cycle)
N is the integer carrier phase ambiguity (cycles)
ε_τ is the carrier multipath and measurement noise (m)

The carrier phase measurement with the ambiguity resolved to the correct integer provides a very accurate range measurement and is used to provide millimeter-level position accuracies.

Doppler is a measure of the instantaneous rate of the GPS carrier phase and is the instantaneous Doppler frequency shift of the incoming carrier. The Doppler shift results from the relative motion between the receiver and the satellite. The major role of the Doppler measurement in the navigation process is to compute a velocity estimate in order to be able to track the satellite and speed up its acquisition time.

Atmospheric errors (i.e., troposheric and ionospheric errors) introduce range and range-rate errors. When the GPS signal travels through the troposphere its path will bend slightly due to the refractivity of the troposphere [16]. The change of the refractivity from free space to the troposphere causes the speed of the GPS signal to slow down, which results in a delay of the GPS signal. Various models have been developed to reduce the tropospheric error by at least 90%. It is not a major cause of signal error anymore, so we will move on to the next subject—ionospheric errors.

The ionosphere is the layer of the atmosphere that extends from 60 to over 1,000 km of height above Earth's surface. It is an important source of range and range-rate errors for GPS users requiring high-accuracy measurements. The ionospheric variation is generally large compared to the troposphere and is more difficult to model. Ionospheric error can be eliminated using dual frequency measurements from GPS. Ionospheric error can be further reduced using better ionospheric estimation models and correction services provided by large networks like WAAS.

7.3.1 Phase Center Measurements

7.3.1.1 Rotation Test Using Satellite Observables

There are various ways to measure the mean location of electrical phase center of the antenna. The phase center will not be in exact one location due to the inherent small variation of phase pattern in azimuth and elevation angle. One method is to measure the phase pattern in an anechoic chamber and then compute its mean location from various orthogonal phase pattern cuts. Another way is to measure the mean location of phase center using a live GPS signal. However, certain steps must be taken in order for the GPS measurements to be repeatable. In addition, one must ensure that multipath level is low when doing a live GPS measurement campaign. One method to reduce the multipath effect was introduced by Geo++, where the AUT is mounted to a robotic arm that spins the antenna in a way to minimize the multipath reflections. Another method is to mount the antenna very low to the ground and rotate the antenna at discrete steps of 45 degrees every 24 hours (actually 23h 56 m–sidereal[2] day. Smaller data sets might be used (every 8 hours) at the price of repeatability. The data collected from each session must be then individually processed using the double-difference model. This approach yields a set of nine results for the 0, 45, 90, 135, 180, 225, 270, 315, and 360 (back to zero) rotations. If the electrical phase center is not aligned with the geometrical phase center, a circle with eight discrete points will be traced out. The radius of this circle is equivalent to the magnitude of PCO from the rotational center of the antenna. The mathematical details about this method are given next.

Rotating the antenna by a certain angle α from day to day produces a GPS signal information that includes PCV. Applying the method of double difference adjusted for sidereal time difference between consecutive days will give a relatively accurate estimate of PCV offset from the geometrical center of the antenna [19].

2. Sidereal day: Derived from mean solar day, which is 3 min 56 sec shorter than a standard Earth day.

The GPS phase observation (c_phase) equation is given by:

$$\begin{aligned} c\_phase^j_{\varphi_i} &= a^j_i x_i + cdT^j_i - \lambda N^j_i - d^j_{ION_i} \\ &+ d^j_{TROP_i} + d^j_{MP_i} + d^j_{PCV_i} + \varepsilon_\varphi \end{aligned} \tag{7.16}$$

where subscript i and j stand for receivers and satellites, ION is the error term due to ionosphere, $TROP$ is the error term due to troposphere, MP is the error term due to multipath, PCV is the error term due to antenna phase center variation, and ε_ϕ is thenoise of the carrier phase.

Creating a mean sidereal day time difference will yield and eliminate multipath PCV. To obtain antenna PCV, one antenna must be rotated by α deg from one day to the other to produce information that includes PCV. The linearized equation of the rotated antenna is given by:

$$\begin{aligned} \delta^{SID} c\_phase^j_{\varphi_i} &= c\delta^{SID} dT^j_i - \lambda\delta^{SID} N^j_i - \delta^{SID} d^j_{ION_i} \\ &+ \delta^{SID} d^j_{TROP_i} + d^{0\ \ j}_{PCV\,i} - d^{\alpha\ \ j}_{PCV\,i} + \delta^{SID}\varepsilon_\varphi \end{aligned} \tag{7.17}$$

The terms $d^0{}_{PCV}$ and $d^\alpha{}_{PCV}$ represent the PCV for the unrotated and rotated antennas. The double difference will contain PCV of only one antenna ($\nabla\Delta d_{PCV}$):

$$\nabla\Delta\delta^{SID} l^{j,l}_{\varphi_{i,k}} = -\lambda\nabla\Delta\delta^{SID} N^{j,l}_{i,k} + \underline{\nabla\Delta d^{0,\alpha\ j,l}_{PCV\ \ i,k}} + \nabla\Delta\delta^{SID}\varepsilon_\varphi \tag{7.18}$$

7.3.2 Tracking Performance Measurements

A simple method for measuring and comparing the tracking performance of any GPS receiver is to take a large (at least 24-hour) data sample and compare the total number of possible observations to the total number of actual observations contained in the file. The observed to expected ratio can be than computed, as the ratio of the total number of observations actually made to the total number of possible observations during that period. A perfect GPS receiver/antenna combination would always achieve a 100% observed-to-expected ratio, but in practice, the following occur: cycle slips, low-elevation obstructions, and losses of lock. These cause small periods of missing data in the recorded file. Thus, a simple percentage expresses the observed-to-expected ratio. Results close to 100% indicate a better overall ability to acquire and continuously track satellites.

At low elevation angles, the GPS signals travel very obliquely through the earth's atmosphere. Refraction effects reduce the signal power significantly, increasing the noise and making the signals much harder to track. At very low elevation angles (typically below 5–10 degrees) the number of lost data generally increase sharply. Therefore, it is common practice to estimate the statistics for high-elevation (10 to 90 degrees) and low-elevation tracking as separate data sets (0–10 degrees).

7.4 Perfect Antenna Pattern and Method of Realization

GPS antenna requirements differ in various applications. For precise surveying applications, ideally the antenna should receive only signals above the horizon and reject all signals below the horizon plane of the antenna, have a known and stable phase center that is colocated with the geometrical centre of the antenna, and have perfect circular polarization characteristics to maximize the reception of the incoming right-hand polarized (RHP) signal. For multiple frequency operation (E1, E5, L1, and L2), the antenna should also have a common phase centre for all frequencies and ideally the same radiation pattern and axial ratio characteristics. A perfect phase center antenna of an imaginary ideal antenna would have to have the same amplitude and phase pattern in every direction, whether it is zenith or horizon of the antenna (see Figure 7.7).

The antenna pattern will closely follow the phase pattern. Having same gain at horizon as in zenith would cause the antenna to be severely susceptible to multipath. In order to deal with the multipath a compromise must be made: shape the amplitude radiation pattern to reduce the reception of multipath signal generated at low elevation angles (see Figure 7.8).

This on the other hand will introduce phase center movement in the vertical plane that will be more significant for lower elevation angles (where change in phase and amplitude is largest).

7.4.1 Multipath-Limiting Antennas

We will now develop technical background to determine what ideal multipath limiting antenna would look like. The ideal pattern shown in Figure 7.8 can be realized

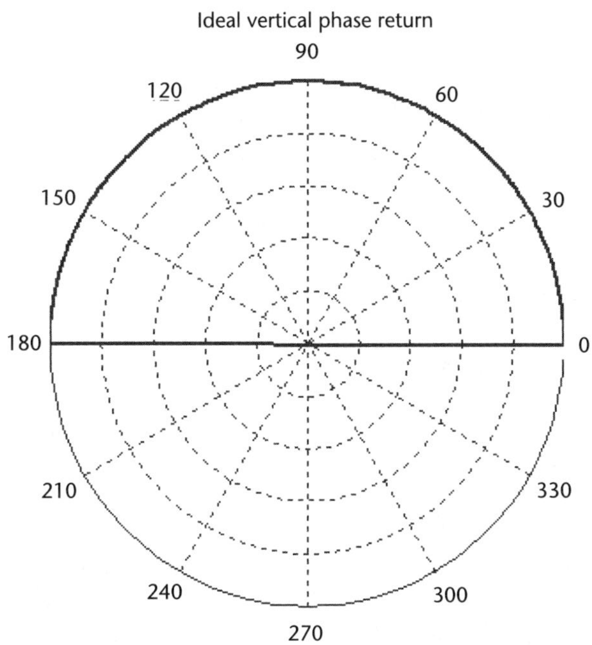

Figure 7.7 Ideal antenna radiation phase pattern.

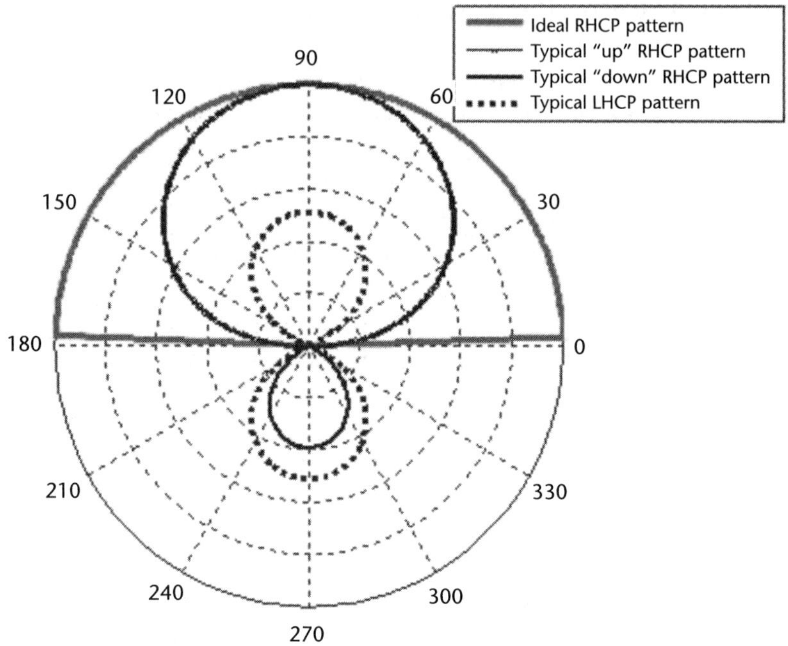

Figure 7.8 Ideal and typical antenna vertical amplitude radiation pattern.

using a certain current distribution that can be computed using Fourier transform. Other methods such as *z*-transform [3] can be used as well to determine the current distribution.

First we will transform the ideal pattern shown in Figure 7.8 using the following transformation:

$$w = \cos(\theta) \tag{7.19}$$

where θ is elevation angle.

Then $f(w)$ is a space factor that forms the pattern only above the horizon and then it truncates on the antenna horizon ($w = 0$ and $w = 1$).

$$f(w) : \Phi(w) = \Phi(w-1) \tag{7.20}$$

The current is determined from Figure 7.9 by carrying out Fourier transform integral (7.21):

$$Ideal(s) := \int_0^1 f(w) \cdot e^{-2j\pi sw}\, dw \tag{7.21}$$

The computed amplitude and phase of the current is shown in Figures 7.10 and 7.11. The computed current is a continuous function, therefore in practical realization it must be sampled at discrete points in order to build an antenna array.

One can perform various optimizations to minimize various design aspects such number of elements of total length of the array in order to yield amplitude pattern that is relatively close to the ideal. The computed ideal current is a function of antenna length. The spacing between L2 antenna elements must be different

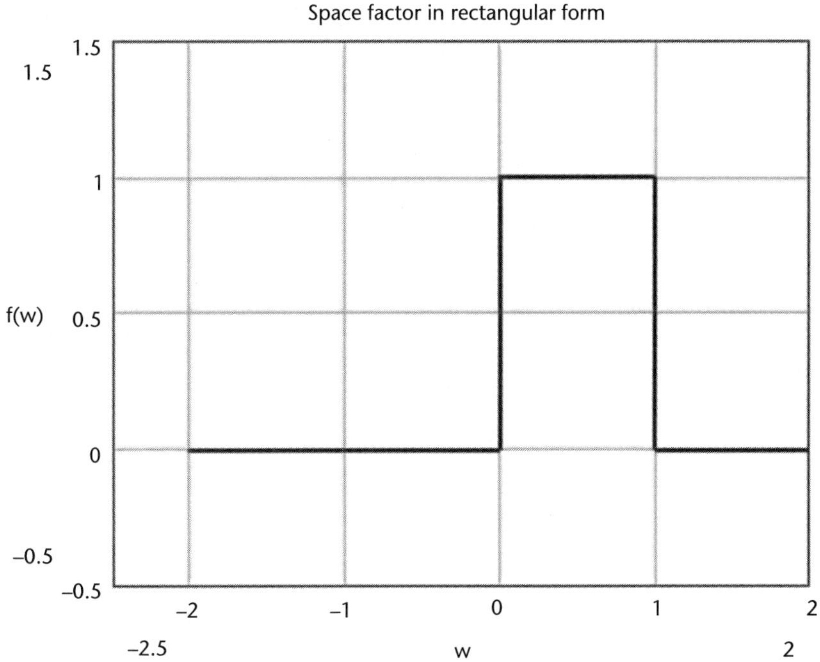

Figure 7.9 Ideal pattern of Figure 7.8 in rectangular form using transformation (7.19).

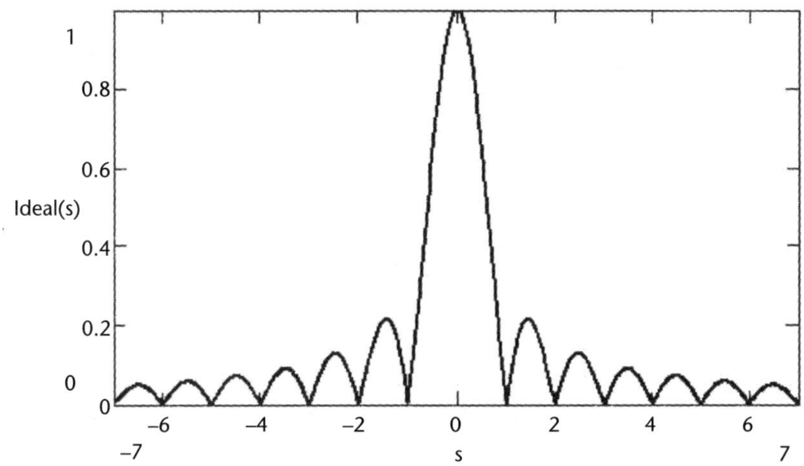

Figure 7.10 Antenna current amplitude of the ideal pattern of Figure 7.8.

than L1 antenna element (assuming the same current weights) in order to keep the same pattern shape at both frequencies. Such an approach was taken by Counselman [3] when he designed his three- and five-element vertical arrays. See Table 7.1 for design details. Table 7.2 shows a different approach to determine antenna pattern while keeping the same distance between L1 and L2 elements in order to simplify the design implementation.

The computed antenna amplitude pattern from excitations listed in Tables 7.1 and 7.2 are shown in Figures 7.12 and 7.13.

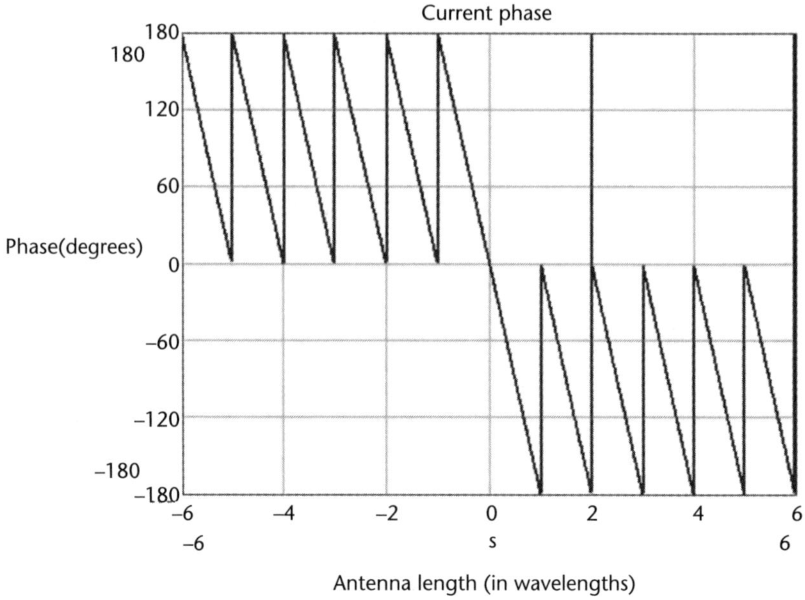

Figure 7.11 Antenna current phase of the ideal pattern of Figure 7.8.

Table 7.1 Counselman Design Parameters of Five-Element Vertical Array* 7.12.

L1 Element Distance (λ)	*L2 Element Distance (λ)*	*Element Amplitude*	*Element Phase*
1.000	1.000	−17.7 dB	−90°
0.333	0.333	−4.3 dB	−90°
N/A	0.07	−5.2 dB	0°
0.000	N/A	0.0 dB	0°
N/A	−0.07	−5.2 dB	0°
−0.333	−0.333	−4.3 dB	90°
−1.000	−1.000	−17.7 dB	90°

*See Figure 7.12.

Three vertical array designs have been used in the past with two used in actual field installation. The first design is the LAAS L1 system referred to integrated multipath-limiting antenna (IMLA) designed by the Avionics Group of Ohio State University. Initially it started as a frequency design that, while it effectively removed the multipath, had the inherent weakeness of not being able to track high elevation angle satellites. This is a single-frequency design that effectively removes multipath but had the inherent weakness of not being able to track high elevation angle satellites. The final (second array design) solution was to integrate two antennas in order to be able to remove multipath generated at medium- and low-elevation angles and track high-elevation angles where multipath is not dominant factor.

Table 7.2 Design Parameters of 15-Element Vertical Array

L1 Element Distance	*L2 Element Distance*	*Element Amplitude*	*Element Phase*
6.396	4.986	–34.3 dB	–105°
5.412	4.218	–31.7 dB	–120°
4.428	3.452	–26.4 dB	–125°
3.444	2.684	–20.3 dB	–102°
2.460	1.918	–17.7 dB	–90°
1.476	1.151	–13.9 dB	–90°
0.492	0.384	–3.9 dB	–90°
0.000	0.000	0.0 dB	0°
–0.492	–0.384	–3.9 dB	90°
–1.476	–1.151	–13.9 dB	90°
–2.460	–1.918	–17.7 dB	90°
–3.444	–2.684	–20.3 dB	102°
–4.428	–3.452	–26.4 dB	125°
–5.412	–4.218	–31.7 dB	120°
–6.396	–4.986	–34.3 dB	105°

*See Figure 7.13.

This approach requires two receivers for each antenna and system level process for satellite handoff operation between two antennas (see Figure 7.16).

7.5 PCV

In practice, the electrical phase centers of any GPS antenna move. However, there is a twofold difference between a precise, high-performance antenna (such as the choke ring) and a less accurate, low-performance antenna. In high-performance antennas, the variation of the phase center in the horizontal plane is limited to values of less than 1 to 2 mm regardless of the direction from which the signal is received or the rotation of the antenna. In low-performance antennas, the characteristics of an antenna are not consistent over the horizontal surface of the antenna, so the horizontal phase center is much less stable. Thus, the low-performance antenna experiences horizontal phase center shifts that vary depending on the direction to the satellite and the direction in which the antenna is pointing. Furthermore, two antennas of the same type at each end of a baseline may be manufactured to slightly different tolerances, and be pointing in different directions, creating inconsistent relative horizontal phase center variations. This unpredictable behavior does not tend to cancel during double-difference processing (See section 7.2.1). Its random nature also means that it cannot be effectively modeled either, leading to additional errors in the baseline solution.

All antennas, including the choke ring, have significant vertical PCVs as a function of satellite elevation angle in an absolute sense. However, in high-performance antennas, these variations are highly consistent, even across different antennas of the same type. This consistency means that two high-performance antennas at each end of a baseline will have almost identical PCVs. In double-difference processing,

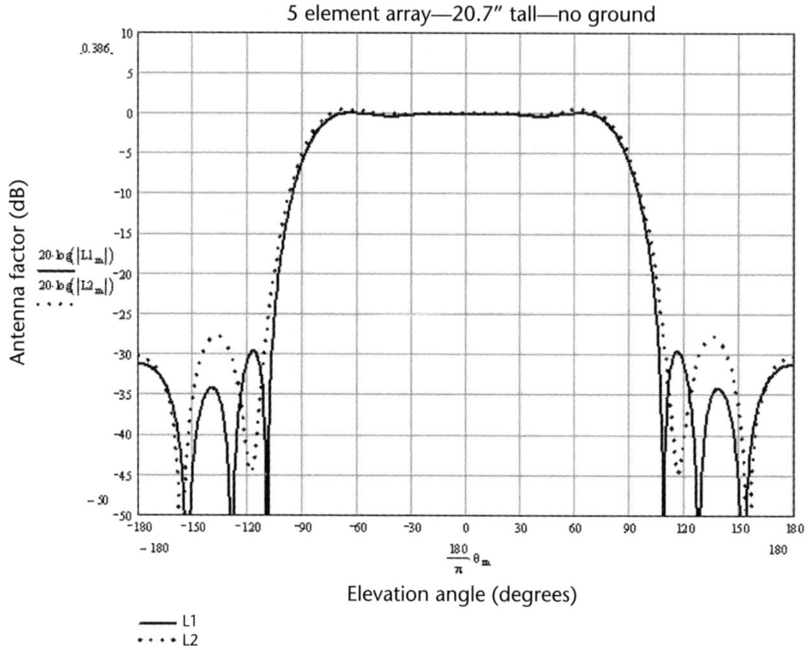

Figure 7.12 Counselman 5-element array antenna amplitude pattern.

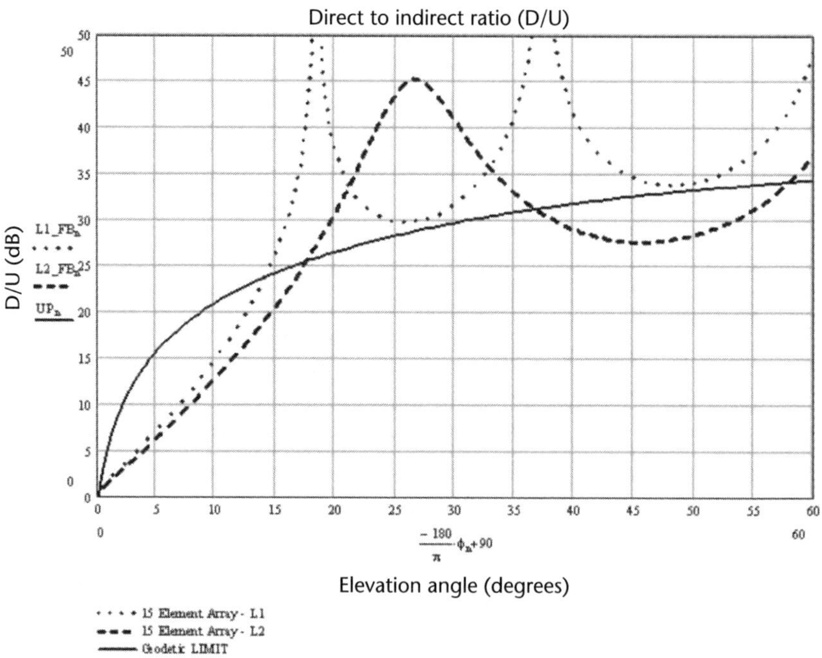

Figure 7.13 Up/down ratio of Counselman antenna.

this highly correlated error cancels out, effectively removing the error caused by the instantaneous offset. As baselines become significantly longer than 50 km, the apparent satellite elevation and azimuth gradually become different for the antennas at each end of the line; that is, the apparent elevation angle decorrelates with

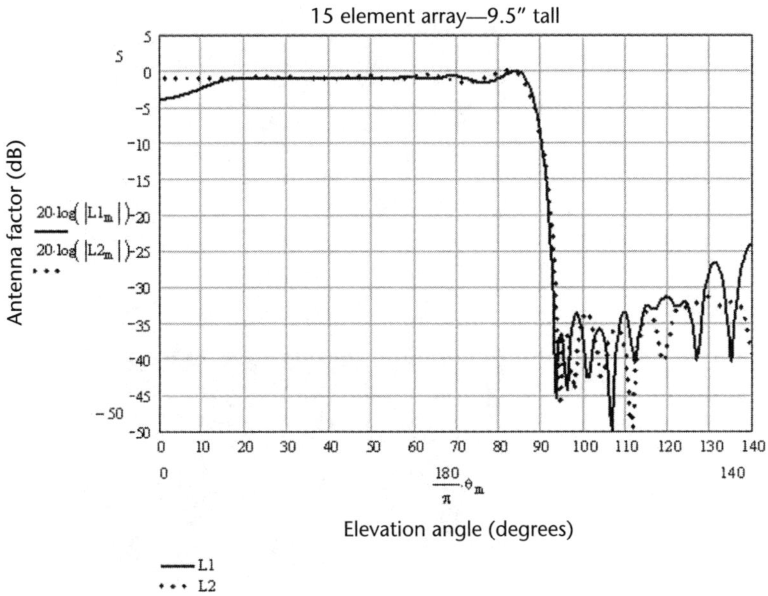

Figure 7.14 L1/L2 computed pattern for 15 element array of total length of 2.4m.

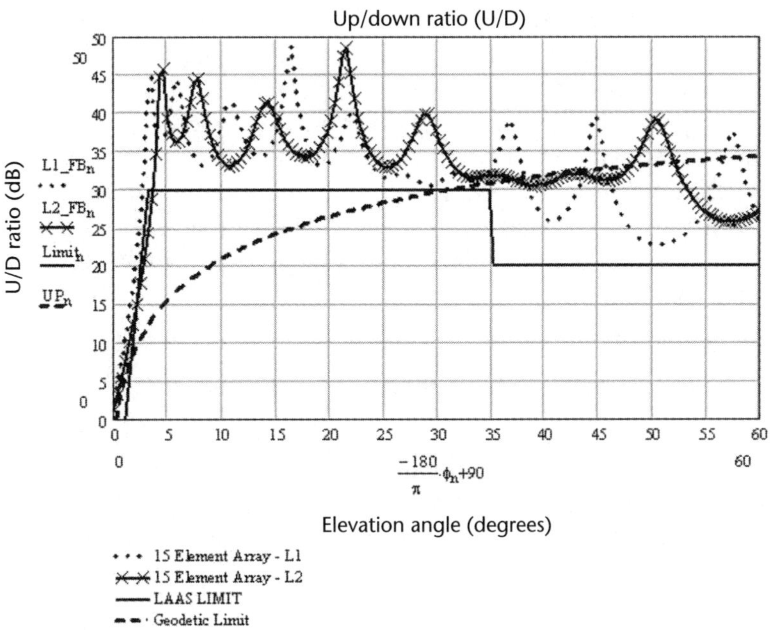

Figure 7.15 Up/down ratio of 15-element array based on the ideal antenna current distribution.

range. In high-performance antennas, this effect can be accurately predicted using a simple correction table for the antenna phase center, and then removed during processing. Although conventional antennas can also have correction tables, because the variations in different antennas of the same type are inconsistent, as mentioned earlier, the offset cannot be predicted accurately and leads to errors in the solution. If an antenna phase center moves very predictably as a function of satellite

Figure 7.16 IMLA veritical array antenna. (*Source:* dB Systems, Inc, 2006. Reprinted with permission.)

elevation, regardless of the azimuth of the satellite, and if this consistent behavior is held within very fine tolerances across all antennas of its type (even if thousands of them are produced), then the antenna has very good phase center repeatability.

Microstrip patch antennas with stable phase centers must be fed in at least two points, preferably in four points (for all four edges of rectangular/square patch). Since the antenna must be circularly polarized, a 90° phase gradient must be established between the feeding points. The feed network to establish a 90° phase gradient for a two point feed system is relatively simple (90° hybrid will do); however, as the number of feeds is increased the feed network becomes more complex and lossy.

7.5.1 Modeling and Measurement Methods of PCV

The PCV has two parts, the horizontal variation and the vertical variation. The PCV of an antenna is usually derived from averaged PCOs. The phase center is usually determined in two different ways: an absolute offset and a relative offset.

Absolute PCO and variation is derived from GPS observations using a high-precision robot arm. The robot arm tilts and spins antenna in a given azimuth and elevation angle during GPS data collection. See the Geo++ website for details. The average phase center location is a weighted average of all individual phase centers for each of the measurements.

The relative PCO and variation involves determination of the phase center location of an antenna relative to a known reference antenna using GPS observations gathered on very short, in addition to accurately known, baselines. The problem with that approach is that the reference antenna is not perfect and hence, errors are introduced.

7.6 Group Delay Variation Effects

Group delay variations between two given signals received from two satellites will be a function of phase difference variation of the antenna radiation pattern versus frequency change at the incidence azimuth and elevation angle. Group delay is defined as the derivative of the transfer phase response $\phi(\omega)$ versus frequency ω.

$$Group\_delay = \frac{d\varphi(\omega)}{d\omega} \tag{7.22}$$

Group delay versus frequency is of the most interest in microwave components, while in the case of an antenna it is useful to consider the variation of group delay versus elevation and azimuth angles. Thus, in GPS applications, maximum accuracy requires that the receiving antenna have a uniform group delay response for all angles of incidence in most of the upper hemisphere, otherwise signals from satellites at different elevation/azimuth angles will experience timing errors [3]. The group delay is usually expressed in terms of meters by multiplying its value by speed of light.

There are several methods available to measure antenna group delay. An antenna designer will usually rely on anechoic chamber measurements. Data is collected for many frequencies, allowing to compute any parameter variation with changing frequency. The method is well established; however close attention must be paid when aligning antenna on the turntable. The question is, what should the frequency separation be for group delay measurements? The author used a 16-MHz bandwidth (±8 MHz from center L1 frequency) since 90% of a spread spectrum signal is contained in such a bandwidth. A good correlation with GPS-type measurements of group delay was usually achieved.

The GPS signal can be used to estimate group delay introduced by the antenna from code-minus-carrier variation calculations. The amplitude and phase of the received RF carrier signal are used as quality measures of the GPS signal accuracy. All modern GPS receivers use carrier phase measurements to achieve subcentimeter accuracies, which is not the case with amplitude of the signal. Traditionally, the amplitude of the signal had found little use in GPS signal processing. Recently, more research has been carried out on using amplitude information in multipath mitigation techniques using SNR information [34, 40]. Aloi and Graas investigated the amplitude and polarization characteristics of the direct, multipath, and composite GPS signal upon reflection from various ground planes [34]. They found the multipath error can be reduced by adjusting amplitude and phase of the received signal responses from an antenna provides two orthogonal linearly polorized signals.

7.7 Methods of Improvement for Antenna Performance

Antenna performance can be controlled during the design stage, manufacturing, and installation (commission). Various improvements are summarized below.

7.7.1 Design Stage Improvement for Antenna Performance

Due to increasing band allocation for other satellite-based ranging systems (i.e., Galileo, GLONASS), it is desired to design wideband antenna elements that span frequency range from 1.15 to 1.65 GHz.

GDVs between different points in the radiation pattern of the antenna must be kept to minimum in order to minimize the timing and positioning errors. It is not possible to provide an ideal phase characteristic for practical antennas, but certain antenna designs have much better characteristics than others.

Phase center stability and polarization "purity" is best achieved by introducing orthogonal phasing network into the antenna element.

Complex ground plane structures help attenuating and removing unwanted radiation modes and undesired currents. This prevents from degradation of antenna radiation pattern.

An antenna radome should be ideally hemispherical in shape or flat. The nose-cone or spike-type approach may introduce undesired GDVs and phase center movement with elevation angles.

7.7.2 Manufacturing Stage Improvement for Antenna Performance

Thorough electrical test on every antenna shipped can ensure unit-to-unit repeatability. Antenna element tuning should be avoided for high-precision antennas.

7.7.3 Field Installation Stage Improvement for Antenna Performance

An antenna should not be elevated against its surrounding but preferably mounted flat. This minimizes any multipath coming from below the antenna. The sky view should be unobstructed and free of any large natural and manmade structures that can become large reflectors. A meshed metallic skirt should be employed below the antenna horizon. This arrangement serves two purposes: first, to extend antenna ground plane and redirect unwanted currents towards the ground plane, and second, to make the antenna surroundings look alike for any installation.

References

[1] Tiemeyer, B. M. E. Cannon, and G. Lachapelle, et al., "Satellite Navigation for High Precision Aircraft Navigation with Emphasis on Atmospheric Effects," *Position and Navigation Symposium*, 1994, pp. 394–401.

[2] Braash, M. S., "Isolation of GPS Multipath and Receiver Tracking Errors," *Proc. of the National Technical Meeting, ION*, 1994, pp. 511–521.

[3] Counselman, C. C. "Multipath-Rejecting GPS Antennas," *Proc. of the IEEE,* Vol. 87, No. 1, January 1999, pp. 86–9.

[4] Kelly, P. K., M. Piket-May, and I. Rumsey et al., "Microstrip Patch Antenna Performance on a Photonic Bandgap Substrate," *Proc. 1998 USN/URSI Nat. Radio Sci. Meeting Digest,* Atlanta, GA, June 1998, p. 5.

[5] Yablonovitch, E. ,"Inhibited Spontaneous Emission in Solid-State Physics and Electronics," *Phys. Rev. Lett.,* Vol. 58, 1987, p. 2059.

[6] Maradudin, A. A., and A. R. McGurn, "Photonic Band Structure Of A Truncated, Two-Dimensional, Periodic Dielectric Medium," *J. Opt. Soc. Amer. B,* Vol. 10, 1993, p. 307.

[7] Balanis, C., *Antenna Theory, Analysis, and Design, 2nd Edition,* New York: Wiley, 1997.

[8] Ramo, S., J. Whinnery, and T. Van Duzer, *Fields and Waves in Communication Electronics, 2nd Edition,* New York: Wiley, 1984.

[9] Collin, R., *Field Theory of Guided Waves, 2nd Edition,* New York: IEEE Press, 1991.

[10] Sievenpiper, D., and E. Yablonovitch, Circuit and Method for Eliminating Surface Currents on Metals, U.S. Patent 60/079963, issued March 30, 1998.

[11] Elliot, R., "On the Theory of Corrugated Plane Surfaces," *IRE Trans. Antennas Propagat.,* Vol. 2, April 1954, pp. 71–84.

[12] Lee, S., and W. Jones, "Surface Waves on Two-Dimensional Corrugated Surfaces," *Radio Sci.,* Vol. 6, No. l, 1971, pp. 811–818.

[13] Radisic, V., Y. Quian, and R. Coccioli, et al., "A Novel 2-D Photonic Band-Gap Structure for Microstrip Lines," *IEEE Microwave Guided Wave Lett.,* Vol. 8, February 1998, pp. 69–71.

[14] Radisic, V., Y. Quian, and R. Coccioli, et al., "Microstrip Patch Antenna Using Novel Photonic Band-Gap Structures," *IEEE AP-S Int. Symp.,* Atlanta, GA, June 21–26, 1998.

[15] Baker, T. F., and S. Zerbini, "Practical Experience and Consideration: Project Post and Planned Session Summary," *IEEE Trans. on Workshop on Methods for Monitoring Sea Level,* Vol. 50, March 1997, pp. 87–91.

[16] Kaplan, E. D. (ed.), *Understanding GPS: Principles and Applications.* Norwood, MA: Artech House, 1996.

[17] Parkinson, B. W., and J. J. Spilker (eds.), *Global Positioning System: Theory and Practice. Volumes I and II.* Washington, DC: American Institute of Aeronautics and Astronautics, 1996.

[18] Wells, D. (ed.), *Guide to GPS Positioning*, Fredericton, New Brunswick: Canadian GPS Associates, 1989.

[19] Banyai, L. "Investigation of GPS Antenna Mean Phase Centre Offsets Using a Full Roving Observation Strategy," *Journal of Geodesy,* Vol. 79, 2005, pp. 222–230.

[20] Kunysz, W., "A Novel GPS Survey Antenna," *Technical Session of ION,* Anaheim, CA, January 2000.

[21] Park, K. D., P. Elósegui, J. L. Davis, P. O. J. Jarlemark, B. E. Corey, et al., "Development of an Antenna and Multipath Calibration System for Global Positioning System Sites," *Radio Science,* Vol. 39, RS5002, 2004, pp. 1–13.

[22] Hebib, S., H. Aubert, O. Pascal, N. J. G. Fonseca, L. Ries, et al., "Multiband Pyramidal Antenna Loaded with a Cutoff Open-Ended Waveguide," *IEEE Transactions on Antennas and Propagation,* Vol. 57, No. 1, 2009, pp. 266–270.

[23] Hebib, S., H. Aubert, O. Pascal, N. J. G. Fonseca, L. Ries, et al., " Pyramidal Multiband Antennas for GPS/Galileo/MircoSat Application," *IEEE AP-S Int. Symp. Proceedings,* Honolulu, HI, 2007, pp. 2041–2044.

[24] Hebib, S., H. Aubert, O. Pascal, N. J. G. Fonseca, L. Rieset al., " Trap-Loaded Pyramidal Tri-Band Antenna for Satellite Applications," *IEEE AP-S Int. Symp. Proceedings,* San Diego, CA, 2008.

[25] Basilio, L. I., J. T. Williams, D. R. Jackson, and M. A. Khayat, "A Comparative Study of a New GPS Reduced-Surface-Wave Antenna," *IEEE Antennas and Wireless Propagation Letters,* Vol. 4, 2005, pp. 233–236.

[26] Basilio, L. I., J. T. Williams, D. R. Jackson, and R. L. Chen, "A new planar dual-band GPS antenna designed for reduced susceptibility to low-angle multipath," *IEEE Transactions on Antennas and Propagation,* Vol. 55, No. 8, 2009, pp. 2358–2366.

[27] Lopez, A. R., "GPS Ground Station Antenna for Local Area Augmentation System, LAAS," *ION NTM Proceedings,* Anaheim, CA, 2000.

[28] Lopez, A. R., "LAAS/GBAS Ground Reference Antenna with Enhanced Mitigation of Ground Multipath," *ION NTM Proceedings,* San Diego, CA, 2008.

[29] Kunysz, W., "A Three-Dimensional Choke Ring Ground Plane Antenna," *ION NTM Proceedings,* Long Beach, CA, 2003.

[30] Scire-Scappuzzo, F., and S.N. Makarov, "A Low-Multipath Wideband GPS Antenna With Cutoff or Non-Cutoff Corrugated Ground Plane," *IEEE Transactions on Antennas and Propagation,* Vol. 57, No. 1, 2009, pp. 33–46.

[31] Granger, R., and S. Simpson, "An Analysis of Multipath Mitigation Techniques Suitable for Geodetic Antenna," http:// www.roke.co.uk/.../Multipath-Mitigation-Techniques-for-Geodetic-Antennas.2008.pdf.

[32] Filippov, V., D. Tatrnikov, J. Ashjaee and I. Sutiagin, "The First Dual-Depth Dual-Frequency Choke Ring," *Javad Positioning Systems,* 1998.

[33] McKinzie III, W. E., R. B. Hurtado, B. K. Klimczak, and J. D. Dutton, "Mitigation of Multipath Through the Use of an Artificial Magnetic Conductor for Precision GPS Surveying Antennas," *IEEE Antennas and Propagation Soc. Intl. Symp.,* Vol. 4, 2002.

[34] Aloi, D. N., and F. Van Graas, "Ground-Multipath Mitigation via Polarization Steering of GPS Signal," *IEEE Trans. on Aerospace and Electronic Systems,* Vol. 40, No. 2, 2004, pp. 536–552.

[35] Aloi, D. N., and M. S. Sharawi, "C/No estimation In A GPS Software Receiver in the Presence of RF Interference Mitigation via Null Steering for the Multipath Limiting Antenna," *IEEE GLOBECOM proceedings,* 2006.

[36] Zhuang, W., and J. M. Tranquilla, "Effects of Multipath and Antenna on GPS Observables," *IEE Proc.-Radar, Sonar Navig.,* Vol. 142, No. 5, 1995, pp. 267–275.

[37] Lopez, A. R., "Calibration of LAAS Reference Antennas," ION GPS Proceedings, Salt Lake City, UT, 2001.

[38] Lopez, A. R., "LAAS Reference Antennas—Key Siting Considerations," *ION GNSS Proceedings,* Long Beach, CA, 2004.

[39] Lopez, A. R., "LAAS Reference Antennas—Circular Polarization Mitigates Multipath Effects," *ION AM Proceedings,* Albuquerque, NM, 2003.

[40] Comp, C. J., and P. Axelrad, "Adaptive SNR-Based Carrier Phase Multipath Mitigation Technique," *IEEE Trans. Aerosp. and Electron. Syst.,* Vol. 34, No. 1, 1998, pp. 264–276.

About the Authors

Dr. Rama Rao is a principal engineer at the MITRE Corporation, where his work is primarily related to antenna technology for GPS navigation and military satellite communications systems. He received his Ph.D. in applied physics from Harvard University, where he also served as an assistant professor. Prior to joining MITRE, he held technical staff positions at the Sperry Corporate Research Center and the MIT Lincoln Laboratory. Dr. Rao holds ten US patents; three of these patents are related to GPS antennas. He is the author of chapters 1, 2, 3, 5 and 6 of this book.

Dr. Keith F. McDonald received his B.S., M.S., and Ph.D. in electrical engineering from Lehigh University. Since then, he has been employed by the MITRE Corporation in Bedford, MA. Presently, he is the project lead for various efforts in the area of GPS user equipment. He also is the leader of the Navigation Group in MITRE's Signal Processing Department. His research interests include the application of signal processing theory to the design and analysis of navigation, communication, and radar systems. He is coauthor of Chapter 4 of this book.

Dr. Ronald L. Fante (1936–2006) was a nationally recognized expert in signal analysis and estimation, modern theories of electromagnetics, and adaptive antennas, with significant applications to radar, navigation, communications, and electronic counter-countermeasures. During his distinguished career, he made numerous fundamental contributions to science and engineering, particularly in the areas of electromagnetics and signal processing.

A prolific technical analyst and engineer, Dr. Fante published over 180 papers in refereed journals on the topics of antenna theory, electromagnetic theory, electromagnetic scattering, microwave breakdown, plasma physics, propagation imaging in random media, radar detection reentry effects on antennas, and wave propagation. In addition to Dr. Fante's numerous journal articles, his book, Signal Analysis and Estimation: An Introduction, was published in 1988.

Dr. Fante received his B.Sc. in electrical engineering from the University of Pennsylvania, his M.Sc. in electrical engineering from the Massachusetts Institute of Technology, and his Ph.D. in electrical engineering from Princeton University. He taught at Tufts University and Northeastern University, and he was an adjunct professor at the University of Massachusetts for seven years. In his career, he worked at Avco Corporation, the Air Force Cambridge Research Laboratory, and MITRE.

Fante was the recipient of a number of professional awards, including the Marcus O'Day Prize, the Atwater Kent Prize, and the IEEE Third Millennium Medal.

He was elected to Fellow of both the Institute of Electrical and Electronics Engineers (IEEE) and the Optical Society of America. As IEEE Fellow, he was recognized for "contributions to the understanding of electromagnetic wave propagation in turbulent media." For three years, he served as editor-in-chief of the IEEE Transactions on Antennas and Propagation. He was also an IEEE Distinguished Lecturer and Chairman of the Awards and Fellows Committee of the IEEE.

Dr. Fante is the author of Chapter 4 of this book, which he completed in 2006 with assistance from Dr. Keith F. McDonald. This chapter is being published for the first time in this book

Dr. Waldemar Kunysz is an RF/antenna engineer with over twenty years experience in research and new product development. His research interests include UWB, CDMA, GPS, phased array antennas, patch and slot antenna technologies, and RF circuit designs up to 10 GHz. He has published and presented several technical papers and holds eight US patents. He is a senior member of both IEEE and APEGGA. Dr. Kunysz is the author of Chapter 7 of this book.

Index

D

G

H

I

J

L

M

N

Q

R

The GNSS Technology and Applications Series

Elliott Kaplan and Christopher Hegarty, Series Editors

A-GPS: Assisted GPS, GNSS, and SBAS, Frank van Diggelen

Applied Satellite Navigation Using GPS, GALILEO, and Augmentation Systems, Ramjee Prasad and Marina Ruggieri

Digital Terrain Modeling: Acquisition, Manipulation, and Applications, Naser El-Sheimy, Caterina Valeo, and Ayman Habib

Geographical Information Systems Demystified, Stephen R. Galati

GNSS Applications and Methods, Scott Gleason and Demoz Gebre-Egziabher

GNSS Markets and Applications, Len Jacobson

GNSS Receivers for Weak Signals, Nesreen I. Ziedan

GNSS for Vehicle Control, David M. Bevly and Stewart Cobb

GPS/GNSS Antennas, B. Rama Rao, W. Kunysz, R. Fante, and K. McDonald

Introduction to GPS: The Global Positioning System, Second Edition, Ahmed El-Rabbany

MEMS-Based Integrated Navigation, Priyanka Aggarwal, Zainab Syed, Aboelmagd Noureldin, and Naser El-Sheimy

Navigation Signal Processing for GNSS Software Receivers, Thomas Pany

Principles of GNSS, Inertial, and Multisensor Integrated Navigation Systems, Paul D. Groves

Server-Side GPS and Assisted-GPS in Java™, Neil Harper

Spread Spectrum Systems for GNSS and Wireless Communications, Jack K. Holmes

Understanding GPS: Principles and Applications, Second Edition, Elliott Kaplan and Christopher Hegarty, editors

Ubiquitous Positioning, Robin Mannings

Wireless Positioning Technologies and Applications, Alan Bensky

For further information on these and other Artech House titles, including previously considered out-of-print books now available through our In-Print- Forever® (IPF®) program, contact:

Artech House Publishers
685 Canton Street
Norwood, MA 02062
Phone: 781-769-9750
Fax: 781-769-6334
e-mail: artech@artechhouse.com

Artech House Books
16 Sussex Street
London SW1V 4RW UK
Phone: +44 (0)20 7596 8750
Fax: +44 (0)20 7630 0166
e-mail: artech-uk@artechhouse.com

Find us on the World Wide Web at: www.artechhouse.com